Lecture Notes
in Control and Information Sciences   243

Editor: M. Thoma

Springer-Verlag London Ltd.

S.G. Tzafestas and G. Schmidt (Eds)

Progress in System and Robot Analysis and Control Design

Springer

Editors

S.G. Tzafestas, PhD
Department of Electrical and Computer Engineering, National Technical University of Athens, Zographou 15773, Athens, Greece

G. Schmidt, PhD
Department of Electrical Engineering and Information Technology, Technical University of Munich, D-80290 Munich, Germany

ISBN 978-1-85233-123-8

British Library Cataloguing in Publication Data
Progress in system and robot analysis and control design. -
  (Lecture notes in control and information sciences)
  1.system analysis 2.Control theory 3. Robotics
  I.Tzafestas, Spyros G., 1939- II.Schmidt, Günther, 1935-
  003.5'4
  ISBN 978-1-85233-123-8
Library of Congress Cataloging-in-Publication Data
European Robotics, Intelligent Systems, and Control Conference (1998 :
  Athens, Greece)
    Progress in system and robot analysis and control design / S.G.
  Tzafestas and G. Schmidt (eds.).
        p.   cm. -- (Lecture notes in control and information sciences
  ; 243)
    Includes bibliographical references.
    ISBN 978-1-85233-123-8   ISBN 978-1-84628-535-6 (eBook)
    DOI 10.1007/978-1-84628-535-6
    1. Robotics--Congresses. 2. Intelligent control systems-
  -Congresses. I. Tzafestas, S. G.., 1939- . II. Schmidt, Günther,
  1935- . III. Title. IV. Series.
  TJ210.3.E88 1998   98-46939
  629.8'92--dc21   CIP

Typesetting: Camera ready by contributors

69/3830-543210 Printed on acid-free paper

Preface

This book contains a selection of papers presented at the *"Third European Robotics Intelligent Systems and Control Conference "* (EURISCON '98) held in Athens, Greece (June 22-25, 1998). It is devoted to the analysis, design, optimization and control of technological systems and robots, and presents important results that reflect in a balanced way the research currently being conducted in the field.

The book is divided into six parts. Part I deals with system analysis, stability and identification problems. Part II is concerned with control design problems for systems with or without uncertainty. Part III is dedicated to the control system design via the Quantitative Feedback Theory (QFT) methodology. Part IV provides a number of important studies on driving simulators including state-of-art and practical issues. Part V involves a set of contributions on the analysis, design and control of robots (industrial manipulators, telemanipulators, flexible-joint robots). Finally, Part VI is devoted to mobile robots and legged robots. The particular topics covered by these parts are as follows:

PART I contains seven contributions that present new results concerning:
- a class of weakly positive singular (WPS) linear continuous-time systems (WPS properties and transformations)
- the frequency response analysis of continuous processes controlled by digital (microprocessor-based) controllers
- the singular value characteristics of state feedback discrete LQ regulators
- the parameter estimation of non-minimum phase FIR systems based on fourth-order cumulants of the output process
- the system identification via a frequency-domain technique based on a parametric model
- the detection of sensor faults via the innovations sequence approach
- the analysis of multitime-scale systems using a 2D system formulation.

PART II contains twelve contributions respectively dealing with established and new aspects of:
- control system design with significant nonlinearities and structured parametric uncertainty
- control system design through global optimization
- optimal model-following control of linear systems
- optimal control of time-varying systems
- optimization of the response of bang-bang control systems
- adaptive linear quadratic multirate tanker autopilot design
- adaptive linear quadratic optimal tracker design based on multirate output sampling
- control of an automaton using uncertain information
- knowledge based control via learning and logic-algebraic methods

- motion control of varying inertia systems
- model based H_2/H_∞ control design for active suspension systems
- Matlab-based graphical interface for control systems analysis and design.

PART III presents new results on the QFT control design methodology, namely:
- new criteria for Smith predictor design, in the QFT control framework, for systems not precisely known
- a new approach to nonlinear QFT robust control design based on local linearization about closed-loop acceptable outputs
- new algorithms for decentralized control structure design by which the system diagonal dominance is maximized
- a practical application of the QFT design method for pressure control of a gas compressor system.

PART IV contains six contributions dealing with state-of-the-art issues and implementation of driving simulators, namely:
- road safety issues discussed with the aid of five case studies concerning the improvement of road safety standards
- experimental evidence based on the driving simulator and the test track for selecting the driving activity that should be studied and the driving simulator to be used
- the evolution of the use of driving simulators in traffic and road safety studies
- the reproduction of realistic multimodal traffic in virtual urban environments on the basis of different models and tools
- state-of-art analysis of traffic simulation techniques oriented towards driving simulation, so as to achieve a good degree of control and a good degree of naturalism
- overview of the technical and software challenges in high fidelity driving simulators.

PART V covers several advanced aspects and problems of robotic systems, namely:
- the trends in remotely controlled robots with reference to the virtual reality contribution
- the robust stabilizing control design and implementation of an underactuated non-holonomic robot
- the synthesis of hand distributed kinesthetic feedback control on the human hand interacting with virtual environments
- the design issues of a modular and reconfigurable (fault tolerant) robot through the use of suitable software tools
- the design criteria for three test-bed mechanisms that help understand better the human involvement in powered orthotic devices
- the inclusion of a human arm model, along with the measurement of its neural input, and a predictor to enhance the robustness of a teleoperator scheme

- the design issues of an H_∞ robotic controller based on derived frequency response upper bounds for the unstructured additive uncertainties
- the optimization of the response of a flexible robotic arm carrying a variable concentrated mass at its end.

Finally *PART VI* contains nine contributions on mobile and walking robots that present the following:
- a novel platform for the development and testing of mobile robot units with reference to health care tasks
- a novel platform for integrating perception and localization components on mobile robots
- a method for structuring mobile robot environments via the use of intelligent (networked) building management systems
- a fluid mechanics method for computing sets of robust solutions of the robot path planning problem in uneven terrains
- a motion planning algorithm for a drift-free nonholonomic mobile robot where the steering is exact if the system is feedback nilpotentizable
- a technique for the navigation of an autonomous robot that can perform go-to (x,y) tasks and avoid obstacles not known beforehand
- the design of an omnidirectional mobile 3 DOF manipulator controller which can control the robot orientation irrespectively of the external force direction acting on the fingertip
- a new controllability benchmark of unstable nonlinear nonholonomic systems, applied to an extension-cableless robotic unicycle, via a new physical measure, namely the minimum entropy production of thermodynamic stability
- the introduction of additional support elements along with electromagnets that lead to force redistribution and allow to increase the minimal reserve of climbing robot stability towards sliding.

Taken together the forty-six contributions of this volume give a good picture of the recent progress in system and robot analysis and control, including both theoretical and practical issues.

The editors are deeply indebted to all colleagues who have contributed to the success of EURISCON '98, and especially to the authors of this volume for their high-level contributions and their last-minute work in revising and refining their manuscripts.

It is believed that the book will be an important addition to the current literature on robotics and control, inspiring young and senior researchers towards new achievements and new applications in modern life.

Spyros G. Tzafestas   December 1998   *Günther Schmidt*
Athens, Greece   *Munich , Germany*

Contents

PART II: *CONTROL SYSTEM DESIGN*

12. Response Optimization of a Discrete-Time Bang-Bang Optimal Control Problem
A.G. Petridis, G.N. Charalampopoulos and *A.E. Kanarachos*

13. Adaptive LQ Optimal Autopilots for Tankers Based on Two-Point Multirate Controllers
P.N. Paraskevopoulos, K.G. Arvanitis and *A.A. Vernardos*

14. Design of Adaptive LQ Optimal Trackers Based on Multirate Sampling of the Plant Output
K.G. Arvanitis and *G. Kalogeropoulos*

15. Control of an Automaton Using Uncertain Information
G. Tsirigotis and *M. Naranjo*

PART III: *QUANTITATIVE FEEDBACK THEORY (QFT) CONTROL SYSTEM DESIGN*

20. Smith Predictor for Uncertain Systems in the QFT Framework
M. Garcia-Sanz and *J.G. Guillen*

21. Nonlinear QFT Based on Local Linearization
A. Baños,O. Yaniv and *F.J. Montoya*

22. Frequency Domain Control Structure Design Tools
E. Kontogiannis, N. Munro and *S.T. Impram*

23. Quantitative Pressure Controller Design for a Gas Recovery System
E. Boje

PART IV: *DRIVING SIMULATORS*

PART V: *INDUSTRIAL ROBOT ANALYSIS, DESIGN AND CONTROL*

PART VI: *MOBILE AND WALKING ROBOTS*

Contributors

H. Abbassi
University of Annaba, Electronics Inst.,
BP12, Annaba, Algeria

K. Abderrahim
Ecole Natl. d' Ingenieurs de Gabès, 6029
Gabès, Tunisie

R. Ben Abdennour
Ecole Natl. d' Ingenieurs de Gabès, 6029
Gabès, Tunisie

L.Aguilar
Tijuana Tech. Inst., Calzada Tech. s/n,
Tijuana, B.C., Mexico

T. Akinfiev
Mech. Eng. Res. Inst., Russian Acad.
of Sci. 101 830 Moscow, Russia

M. Armada
Inst. Autom. Ind. (IAI), Carretera
de Campo Real km 0.200, La Poveda,
Madrid 28500, Spain

K. Arvanitis
C.S. Div., ECE Dept., NTUA, 15773
Zografou, Athens, Greece

A. Baños
Informatics & Systems Dept., Murcia
Univ., 3001, Murcia, Spain

S. Bayarri
INTRAS, Valencia Univ., 46010,
Valencia, Spain

A. Benmounah
University of Annaba, Electronics Inst.,
BP12, Annaba, Algeria

U. Berger
Univ. of Appl. Sci. Lüneburg,
Vollgershall 1, 21399, Lüneburg,
Germany

E. Blana
Transp. Planning & Eng. Dept., NTUA,
15773 Athens, Greece

E. Boje
Electr. Eng. Dept., Univ. of Natal,
Durban 4041, South Africa

J. Bokor
Comp. & Autom. Res. Inst., Hungarian
Acad. of Sciences, Budapest, Hungary

G. Bolmsjo
Production & Mater. Eng. Dept., Lund
Univ., PO Box 118, Lund, Sweden

C. Botan
Control & Ind. Info. Dept., "Gh. Asachi"
Iasi Tech. Univ., 6600 Iasi, Romania

Z. Bubnicki
Control & Syst. Eng. Inst.,
Wroclaw Tech. Univ.,
50-370 Wroclaw, Poland

M. Buss
LSR, Tech. Univ. of Munich,
80290 Munich, Germany

G. Charalampopoulos
Mech. Design & Control Div.,
Mech.Eng.Dept., NTUA, Box 64078,
Athens 15710, Greece

Ph. Coiffet
Paris Robotics Lab. (LRP), Univ. P&M
Curie, Velizy, France

R. De Keyser
Control Eng. Dept., Gent Univ., B-9052
Gent, Belgium

D. Famularo
DEIS, Calabria University, 87 030
Rende (Cs), Italy

M. Fernández
ARTEC-Robotics Inst., Valencia Univ.,
46010 Valencia, Spain

J. Fréchaux
Lab. de Psych. de la Conduite, INRETS,
BP34, 94114 Arcueil, France

M. Garcia-Sanz
Autom. & Comp. Dept, Navarra Public
Univ. (UPNA), 31006 Pamplona, Spain

O. Buckmann
Bremen Inst. of Ind. Tech. - BIBA,
Hochschulring 20, 28395 Bremen,
Germany

F. Caliskan
Electr. Eng. Dept., Istanbul Tech. Univ.,
Ayazaga, Istanbul, Turkey

F. Charrier
Production & Mater. Eng. Dept., Lund
Univ., PO Box 118, Lund, Sweden

I. Coma
ARTEC-Robotics Inst., Valencia Univ.,
46010 Valencia, Spain

S. Donikian
IRISA, Campus de Beaulieu, F-35042
Rennes, France

G. Favier
Lab.I3S,UNSA/CNRS, 2000 Route
des Lucioles, Sophia Antipolis, 06410
Biot, France

G. Foster
DS Group, Cybernetics Dept., Reading
Univ., Reading RG6 6AY, UK

K. Galkowski
Robotics & S/W Eng.Inst.,
TU Zielona Gora, Podgorna St.50,
65-246 Zielona Gora, Poland

P. Gáspar
Comp. & Autom. Res. Inst., Hungarian
Acad. Of Sciences, Budapest, Hungary

K. Geramanis
Mech. Design & Control Div.,
Mech.Eng.Dept., NTUA, Box 64078,
Athens 15710, Greece

L. Gonzalez
Tijuana Tech. Inst., Calzada Tech. s/n,
Tijuana, B.C., Mexico

A. Gramacki
CS & Electronics Inst.,TU Zielona Gora,
Podgorna St.50, 65-246 Zielona Gora,
Poland

J. Gramacki
CS & Electronics Inst.,TU Zielona Gora,
Podgorna St.50, 65-246 Zielona Gora,
Poland

J. Guillen
Autom. & Comp. Dept, Navarra Public
Univ. (UPNA), 31006 Pamplona, Spain

C. Hajiyev
Aeronaut. Eng., Istanbul Tech. Univ.,
Ayazaga, Istanbul, Turkey

R. Hanus
Service d' Automatique, ULB, CP165,
Brussels 1050, Belgium

W. Harwin
Human-Robot Interface Lab.,Cybernetics
Dept., Reading Univ., Reading
RG6 6AY, UK

P. Hedenborn
Production & Mater. Eng. Dept., Lund
Univ., PO Box 118, Lund, Sweden

C. H. Houpis
AFIT & FCT Air Force Research Lab.,
Wright Patterson, AFB
OH 45433, U.S.A.

S. Impram
Control Systems Centre, UMIST,
PO Box 88, Manchester M60 1QD, UK

K. Izumi
Mech. Eng. Dept., Saga Univ.,
Saga 840-8502, Japan

T. Kaczorek
Inst. of Control & Ind.
Electron.,Warsaw Tech. Univ., 00 662
Warsaw, Poland

T. Kaipio
Eng. School, Wolverhampton Univ.,
Wolverhampton WV1 1SB, UK

G. Kalogeropoulos
Maths Dept., Athens Univ.,
Athens 15784, Greece

A. Kanarachos
Mech. Design & Control Div.,
Mech.Eng.Dept., NTUA,
Box 64078, Athens 15710, Greece

A. Kheddar
Paris Robotics Lab. (LRP),
Univ. P&M Curie, Velizy, France

E. Kontogiannis
Control Systems Centre, UMIST,
PO Box 88, Manchester M60 1QD, UK

D. Kostis
IRAL, ECE Dept., Natl. Tech. Univ.
Athens, 15773 Athens, Greece

T. Koussiouris
ECE Dept., Electrosci. Div., NTUA,
Athens 15773, Greece

N. Koussoulas
LAR Lab., ECE Dept., Patras Univ.,
26500 Patras, Greece

M. Krömker
BIBA-Bremen Inst. of Ind. Tech.,
Hochshulring 20, 28395 Bremen,
Germany

M. Ksouri
Ecole Natl. d' Ingenieurs de Tunis,
BP 47, 1002 Tunis, Tunisie

Y. Kunitake
Mech. Eng. Dept., Saga Univ.,
Saga 840-8502, Japan

N. Leighton
Eng. School, Wolverhampton Univ.,
Wolverhampton WV1 1SB, UK

A. Liegeois
LIRMM, Univ. Montpellier II,
161 rue Ada, 34392 Montpellier, France

U. Lorentzon
Production & Mater. Eng. Dept., Lund
Univ., PO Box 118, Lund, Sweden

C. Louste
LIRMM, Univ. Montpellier II,
161 rue Ada, 34392 Montpellier, France

G. Malaterre
Lab. de Psych. de la Conduite, INRETS,
BP34, 94114 Arcueil, France

G. Mareczek
LSR, Tech. Univ. of Munich,
80290 Munich, Germany

G. Martin
ARTEC-Robotics Inst., Valencia Univ.,
46010 Valencia, Spain

F. Matia
UPM-DISAM, Jose Gutierrez Abascal 2,
E-28006 Madrid, Spain

F. Montoya
Informatics & Systems Dept.,
Murcia Univ., 3001 Murcia, Spain

G. Moreau
IRISA, Campus de Beaulieu,
F-35042 Rennes, France

C. Morgan
Eng. School, Wolverhampton Univ.,
Wolverhampton WV1 1SB, UK

F. Msahli
Ecole Natl. d' Ingenieurs de Gabès,
6029 Gabès, Tunisie

N. Munro
Control Systems Centre, UMIST,
PO Box 88, Manchester M60 1QD, UK

M. Naranjo
LASME, Univ.Blaise Pascal de
Clermont Ferrand, 63177 Aubitre,
France

H. Nasri
Production & Mater. Eng. Dept., Lund
Univ., PO Box 118, Lund, Sweden

F. O'Hart
DS Group,Cybernetics Dept., Reading
Univ., Reading RG6 6AY, UK
(Also: CS Dept., Robotics Group,
Trinity College, Dublin 2, Ireland)

T. Ohkura
Mech. & Control Eng. Dept.,
Electro-Communications Univ.,
Chofu,Tokyo 182-8585 Japan

M. Olsson
Production & Mater. Eng. Dept., Lund
Univ., PO Box 118, Lund, Sweden

A. Onea
Control & Ind. Info. Dept., "Gh. Asachi"
Iasi Tech. Univ., 6600 Iasi, Romania

Y. Papelis
Natl. Adv. Driving Simul. Center, Iowa
Univ., Iowa City, Iowa 52242 USA

P. Paraskevopoulos
C.S. Div., ECE Dept., NTUA, 15773
Athens, Greece

C. Peignot
UPM-DISAM, Jose Gutierrez Abascal 2,
E-28006 Madrid, Spain

A. Petridis
Mech. Design & Control Div.,
Mech. Eng. Dept., NTUA, Box 64078,
Athens 15710, Greece

D. Pollock
INTRAS, Valencia Univ.,
46010 Valencia, Spain

M. Prieto
Inst. Autom. Ind. (IAI), Carretera
de Campo Real km 0.200, La Poveda,
Madrid 28500, Spain

P. Prokopiou
IRAL, ECE Dept., NTUA, 15773
Athens, Greece
(Also: Human-Robot Interface
Lab.,Cybernetics Dept., Reading Univ.,
Reading RG6 6AY, UK)

E. Puente
UPM-DISAM, Jose Gutierrez Abascal 2,
E-28006 Madrid, Spain

P. Pugliese
DEIS, Calabria University, 87 030
Rende (Cs), Italy

K. Sato
Syst. & Control Eng. Dept., Saga Univ.,
Saga 840-8502, Japan

G. Schmidt
LSR, Tech. Univ. of Munich
80290 Munich, Germany

L. Smelov
Eng. School, Wolverhampton Univ.,
Wolverhampton WV1 1SB, UK

G. Thomas
IRISA, Campus de Beaulieu,
F-35042 Rennes, France

C. S. Tzafestas
Paris Robotics Lab. (LRP),
Univ. P&M Curie, Velizy, France
(Currently: IRAL, NTUA, Greece)

S.V. Ulyanov
R&D Div., Yamaha Motor Co. Ltd,
Iwata, Shizuoka 438, Japan

M. Uquillas
Inst. Autom. Ind. (IAI), Carretera
de Campo Real km 0.200, La Poveda,
Madrid 28500, Spain

E. S. Vicente
INTRAS, Valencia, Univ., 46010,
Valencia, Spain

K. Yamafuji
Mech. & Control Eng. Dept.,
Electro-Communications Univ., Chofu,
Tokyo 182-8585 Japan

Y. Sergeyev
Nizhni Novgorod Univ., pr. Gagarina
23, Nizhni Novgorod, Russia

P. Skiadas
LAR Lab, ECE Dept., Patras Univ.,
26500 Patras, Greece

I. Süssemilch
Human-Robot Interface Lab.,
Cybernetics Dept., Reading Univ.,
Reading RG6 6AY, UK

G. Tsirigotis
ET Dept., Tech.Educ.Inst.of Kavala,
St. Loukas, 65404 Kavala, Greece

S. G. Tzafestas
IRAL, ECE Dept., NTUA, 15773
Athens, Greece

V.S. Ulyanov
Mech. & Control Eng. Dept.,
Electro-Communications Univ., Chofu,
Tokyo 182-8585 Japan

A. Vernardos
C.S. Div., ECE Dept., NTUA, 15773
Athens, Greece

J. Yamé
Service d' Automatique, ULB, CP165,
Brussels, 1050 Belgium

O. Yaniv
Electr. Eng. & Systems Dept.,
Tel Aviv Univ., Tel Aviv, Israel

F. Wawak
UPM-DISAM, Jose Gutierrez Abascal 2,
E-28006 Madrid, Spain

K. Watanabe
Systems & Control Eng. Dept.,
Saga Univ., Saga 840-8502, Japan

S. Zhang
Electronic Eng. Dept.,
Harbin Eng. Univ., Harbin 15001, PRC

PART I

SYSTEM ANALYSIS, IDENTIFICATION AND STABILITY

1

Weakly Positive Continuous-Time Linear Systems

T. Kaczorek

1   Introduction

The analysis of singular (descriptor) discrete- time and continuous-time system has been considered in many papers and books [1,2,5,10-14]. Properties of the fundamental matrix of singular discrete-time linear system have been established and its solution has been derived in [14,4,7]. The reachability and controllability of singular systems have been considered in many papers [1,4]. Necessary and sufficient conditions for reachability and controllability of standard positive linear systems have been established in [3,15].

Positive singular discrete-time linear systems have been analysed in [4,7] The relationship between positive systems and electrical circuits has been considered in [6]. In this paper a new class of weakly positive singular systems will be introduced. Necessary and sufficient conditions will be established under which a weakly positive singular continuous-time system can be transformed by the strict equivalence to a positive system. It will be shown that linear electrical circuits consisting of resistances, inductances (capacitances) and source voltages are examples of positive singular continuous-time linear systems.

2   Preliminaries

Let $R^{n \times m}$ be the set of $n \times m$ real matrices and $R^n := R^{n \times 1}$. Consider the singular continuous-time linear system

$$E\dot{x} = Ax + Bu, \ x(0) = x_O \tag{1a}$$

$$y = Cx + Du \tag{1b}$$

where $x \in R^n$ is the semistate vector, $u \in R^m$ is the input vector, $y \in R^p$ is the output vector and $E \in R^{n \times n}$, $A \in R^{n \times n}, B \in R^{n \times m}, C \in R^{p \times n}$, $D \in R^{p \times m}$ with E possibly singular.

Definition 1. The system (1) is called standard if and only if $E = I_n$ (the identity matrix)

Definition 2. The system (1) is called regular if and only if

$$\det[Es - A] \neq 0 \text{ for some } s \in \mathbf{C} \text{ (the field of complex numbers)} \tag{2}$$

Let R_+^n be the set of n-dimensional real vectors with nonnegative components.

Definition 3. The system (1) is called positive if and only if for all $x_o \in R_+^n$ and $u(t) = u \in R_+^m$, $t \geq 0$ we have $x(t) = x \in R_+^n$, $t \geq 0$ and $y(t) = y \in R_+^p$, $t \geq 0$.

Definition 4. A matrix $A \in R^{n \times n}$ is called the Metzler matrix if all its off-diagonal entries are nonnegative.

It is easy to show [6] that $e^{At} \in R_+^{n \times n}$ if and only if A is a Metzler matrix. It is well-known [6] that the standard system (1) (with $E = I_n$) is positive if and only if A is a Metzler matrix and $B \in R_+^{n \times m}, C \in R_+^{p \times n}, D \in R_+^{p \times m}$. Taking into account this fact the following definition of weakly positive continuous-time linear systems is introduced.

Definition 5. The system (1) is called weakly positive if and only if A is a Metzler matrix and $E \in R_+^{n \times n}$, $B \in R_+^{n \times m}, C \in R_+^{p \times n}, D \in R_+^{p \times m}$.
If the system (1) is regular then [14]

$$[Es - A]^{-1} = \sum_{i=-\mu}^{\infty} \Phi_i s^{-(i+1)} \tag{3}$$

where $\mu = rank\ E - \deg(\det[Es - A]) + 1$ is the index of nilpotence and Φ_i is the fundamental matrix defined by

$$E\Phi_i - A\Phi_{i-1} = \Phi_i E - \Phi_{i-1} A = \begin{cases} I_n & for \ i=0 \\ 0 & for \ i \neq 0 \end{cases} \tag{4}$$

and
$$\Phi_i = 0 \quad for \ i < -\mu$$

Using (4) the following relations can be proved [14]

$$\Phi_0 A\Phi_i = \begin{cases} \Phi_{i+1} & for \ i \geq 0 \\ 0 & for \ i < 0 \end{cases} \tag{5a}$$

and
$$-\Phi_{-1} E\Phi_i = \begin{cases} 0 & for \ i \geq 0 \\ \Phi_{i-1} & for \ i < 0 \end{cases} \tag{5b}$$

Different methods of computation of Φ_i may be found in [14,4]

It is also well-known [5] that if (2) holds then there exist nonsingular matrices $P, Q \in R^{n \times n}$ such that

$$P[Es - A]Q = \begin{bmatrix} I_{n_1} s - A_1, & 0 \\ 0, & Ns - I_{n_2} \end{bmatrix} \tag{6}$$

where $n_1 := \deg(\det[Es - A])$, $n_2 := n - n_1$, $A_1 \in R^{n_1 \times n_1}$, $N \in R^{n_2 \times n_2}$ is a nilpotent matrix with its index μ

Theorem 1. If (2) holds then the equation (1a) and the equation

$$\dot{x} = \Phi_0 Ax + \Phi_0 Bu + \sum_{j=1}^{\mu} \Phi_{-j} \left(Bu^{(j)} + Ex_0 \delta^{(j)} \right), x(0) = x_0 \tag{7}$$

have the same solution

$$x = e^{\Phi_0 At} \Phi_0 Ex_0 + \int_0^t e^{\Phi_0 A(t-\tau)} \Phi_0 Bu(\tau) d\tau +$$

$$\sum_{j=1}^{\mu} \Phi_{-j} \left(Bu^{(j-1)} + Ex_0 \delta^{(j-1)} \right) \tag{8}$$

Proof. First we shall show that (8) is the solution of (1a). Application of the Laplace - transform (L) to (1a) yield

$$X(s) = [Es - A]^{-1} \left(Ex_0 + BU(s) \right) \tag{9}$$

Substituting of (3) into (9) and using the convolution theorem we obtain (8) since by (5a)

$$\Phi_0 A\Phi_i = \Phi_{i+1} \text{ for } i \geq 0 \text{ and } L\left[e^{\Phi_0 At}\Phi_0\right] = \sum_{i=0}^{\infty} \Phi_i s^{-(i+1)}$$

Using (8) and the relations (5) it is easy to check that

$$\Phi_0 Ax + \Phi_0 Bu + \sum_{j=1}^{\mu} \Phi_{-j}\left(Bu^{(j)} + Ex_0\delta^{(j)}\right) =$$

$$= \Phi_0 A\left(e^{\Phi_0 At} + \Phi_0 Ex_0 + \int_0^t e^{\Phi_0 A(t-\tau)}\Phi_0 Bu(\tau)d\tau + \right.$$

$$\left. + \sum_{j=1}^{\mu} \Phi_{-j}\left(Bu^{(j-1)} + Ex_0\delta^{(j-1)}\right)\right)$$

$$+ \Phi_0 Bu + \sum_{j=1}^{\mu} \Phi_{-j}\left(Bu^{(j)} + Ex_0\delta^{(j)}\right) =$$

$$= \Phi_0 Ae^{\Phi_0 At}\Phi_0 Ex_0 + \Phi_0 A\int_0^t e^{\Phi_0 A(t-\tau)}\Phi_0 Bu(\tau)d\tau +$$

$$\Phi_0 Bu + \sum_{j=1}^{\mu} \Phi_{-j}\left(Bu^{(j)} + Ex_0\delta^{(j)}\right) = \dot{x}$$

Then the solution (8) of (1a) satisfies also the equation (7). □
Remark 1. Using (5b) we may write (7) in the equivalent form

$$\dot{x} = \Phi_0 Ax + \Phi_0 Bu + \sum_{j=0}^{\mu-1}(-\Phi_{-1}E)^j \Phi_{-1}\left(Bu^{(j+1)} + Ex_0\delta^{(j+1)}\right) \quad (7')$$

3 Reduction of weakly positive linear systems

It is well-known [5] that there exist nonsingular matrices $\overline{Q}, \overline{P} \in R^{n \times n}$ such that

$$\overline{Q} E \overline{P} = \begin{bmatrix} I_r & 0 \\ 0 & 0 \end{bmatrix} \ (r < n) \ . \tag{10}$$

where $r = rank\ E$

Premultiplying (1a) by $\overline{Q}$ and introducing a new state vector $\bar{x} = \begin{bmatrix} x_1 \\ x_2 \end{bmatrix} = \overline{P}^{-1} x$ we

obtain

$$\begin{bmatrix} I_r & 0 \\ 0 & 0 \end{bmatrix} \begin{bmatrix} \dot{x}_1 \\ \dot{x}_2 \end{bmatrix} = \begin{bmatrix} A_1 & A_2 \\ A_3 & -A_4 \end{bmatrix} \begin{bmatrix} x_1 \\ x_2 \end{bmatrix} + \begin{bmatrix} B_1 \\ B_2 \end{bmatrix} u \tag{11}$$

where

$$\overline{Q} A \overline{P} = \begin{bmatrix} A_1 & A_2 \\ A_3 & -A_4 \end{bmatrix}, \ \overline{Q} B = \begin{bmatrix} B_1 \\ B_2 \end{bmatrix}, \tag{12}$$

$x_1 \in R^r, x_2 \in R^{n-r}, A_1 \in R^{r \times r}, A_4 \in R^{(n-r) \times (n-r)}, B_1 \in R^{r \times m}, B_2 \in R^{(n-r) \times m}$
From (11) we have

$$\dot{x}_1 = A_1 x_1 + A_2 x_2 + B_1 u \tag{13}$$

$$A_4 x_2 = A_3 x_1 + B_2 u_2 \tag{14}$$

If $\det A_4 \neq 0$ then from (1) we obtain

$$x_2 = A_4^{-1} A_3 x_1 + A_4^{-1} B_2 u_2 \tag{15}$$

Substitution of (15) into (13) yields

$$\dot{x}_1 = \overline{A}_1 x_1 + \overline{B}_1 u \tag{16}$$

where

$$\overline{A}_1 := A_1 + A_2 A_4^{-1} A_3, \ \overline{B}_1 := B_1 + A_2 A_4^{-1} B_2 \tag{17}$$

Definition 6. A matrix $\overline{P} \in R^{n \times n}$ is called a generalised positive permutation matrix if in each row and in each column it has only one positive entry and the remaining entries equal zero.

Lemma 1. [7] Let $A \in R^{n \times n}$ be nonsingular. The matrix $A^{-1} \in R_+^{n \times n}$ if and only if A is a generalized positive permutation matrix.

Theorem 2. The weakly positive system (1) can be reduced by the strict equivalence transformation (10), (12) to the positive standard system (16), (15) if and only if

1) $\overline{P}$ is a generalised positive permutation matrix

2) $\det A_4 \neq 0$

3) $\overline{A}_1$ is a Metzler matrix

4) $\quad \overline{B}_1 \in R_+^{r \times m}, \quad A_4^{-1} A_3 \in R_+^{(n-r) \times r}, \quad A_4^{-1} B_2 \in R_+^{(n-r) \times m}, \quad C \in R_+^{p \times m},$

$D \in R_+^{p \times m}$

Proof. Sufficiency. If 3) holds and $\overline{B}_1 \in R_+^{r \times m}$ then the solution of (16)

$$x_1 = e^{\overline{A}_1 t} x_{10} + \int_0^t e^{\overline{A}_1 (t-\tau)} \overline{B}_1 u(\tau) d(\tau) \in R_+^r \tag{18}$$

since $e^{\overline{A}_1 t} \in R_+^{r \times r}$ and $\begin{bmatrix} x_{10} \\ x_{20} \end{bmatrix} = \overline{P}^{-1} x_0$.

From (15) we have $x_2 \in R_+^{n \times r}$ since by 4) $A_4^{-1} A_3 \in R_+^{(n-r) \times r}$ and

$A_4^{-1} B_2 \in R_+^{(n-r) \times m}$. Hence $x = \overline{P} \overline{x} \in R_+^n$ and $y \in R_+^p$ since $C \in R_+^{p \times n}$ and

$D \in R_+^{p \times m}$.

Necessity. Let $x \in R_+^n$ and $y \in R_+^p$ for any $x_o \in R_+^n$ and $u \in R_+^m$ for $t \geq 0$. Note

that $x_1 \in R_+^r$ and $x_2 \in R_+^{n-r}$ only if $\overline{P}$ is a generalised positive permutation

matrix. From (16) and (15) it follows that $\overline{A}_1$ is a Metzler matrix, $\overline{B}_1 \in R_+^{r \times m}$ and

$A_4^{-1} A_3 \in R_+^{(n-r) \times r}, \quad A_4^{-1} B_2 \in R^{(n-r) \times m}$. Finally $y \in R_+^p$ implies $C \in R_+^{p \times n}$

and $D \in R_+^{p \times m}$. $\quad \square$

Theorem 3. If the conditions of Theorem 2 are satisfied then

1) the model is regular, i.e. (2) holds,

2) the nilpotence index $\mu = 1$ and $N = 0$

$$\Phi_{-1} B = \overline{P} \begin{bmatrix} 0 \\ A_4^{-1} B_2 \end{bmatrix} \in R_+^{n \times m} \text{ and } \Phi_{-1} E = 0 \tag{19}$$

$$\Phi_0 B = \overline{P} \begin{bmatrix} \overline{B}_1 \\ A_4^{-1} A_3 \overline{B}_1 \end{bmatrix} \in R_+^{n \times m}, \Phi_0 E = \overline{P} \begin{bmatrix} I_r & 0 \\ A_4^{-1} A_3 & 0 \end{bmatrix} \overline{P}^{-1} \in R_+^{n \times n},$$

$$\Phi_0 A = \overline{P} \begin{bmatrix} \overline{A}_1 & 0 \\ A_4^{-1} A_3 \overline{A}_1 & 0 \end{bmatrix} \overline{P}^{-1} \in R_+^{n \times n} \tag{20}$$

Proof. Taking into account that

$$\begin{bmatrix} I_r s - A_1, & -A_2 \\ -A_3, & A_4 \end{bmatrix} \begin{bmatrix} I_r & 0 \\ A_4^{-1} A_3 & I_{n-r} \end{bmatrix} = \begin{bmatrix} I_r s - A_1 - A_2 A_4^{-1} A_3, & -A_2 \\ 0, & A_4 \end{bmatrix}$$

and using (10) and (12) we obtain

$$\det[Es - A] = \det \left\{ Q^{-1} \begin{bmatrix} I_r s - A_1, & -A_2 \\ -A_3, & A_4 \end{bmatrix} P^{-1} \right\} = c \det \begin{bmatrix} I_r s - A_1 - A_2 A_4^{-1} A_3, & -A_2 \\ 0, & A_4 \end{bmatrix}$$

where $c = \det Q^{-1} \det P^{-1}$.

Note that $rank\,E = r$ and $\deg(\det[Es - A]) = r$. Then we have $\mu = rank\,E - \deg(\det[Es - A]) + 1 = 1$ and $N = 0$.

From (3), (10) and (12) we have

$$\Phi_{-1} = \lim_{s \to \infty} [Es - A]^{-1} = \overline{P} \left(\lim_{s \to \infty} \begin{bmatrix} I_r s - A_1, & -A_2 \\ -A_3, & A_4 \end{bmatrix}^{-1} \right) \overline{Q} \tag{21}$$

Note that

$$\lim_{s \to \infty} \begin{bmatrix} I_r s - A_1, & -A_2 \\ -A_3, & A_4 \end{bmatrix}^{-1} = \lim_{s \to \infty} \begin{bmatrix} [I_r s - A_1]^{-1} \left(I_r s + A_2 H^{-1} A_3 [I_r s - A_1]^{-1} \right) & [I_r s - A_1]^{-1} A_2 H^{-1} \\ H^{-1} A_3 [I_r s - A_1]^{-1} & H^{-1} \end{bmatrix} \tag{22}$$

where

$$H = A_4 - A_3 [I_r s - A_1]^{-1} A_2$$

Taking into account that

$$\lim_{s \to \infty} [I_r s - A_1]^{-1} = 0$$

and

$$\lim_{s \to \infty} H^{-1} = A_4^{-1}$$

from (21), (22) and (12) we obtain

$$\Phi_{-1} B = \overline{P} \begin{bmatrix} 0 & 0 \\ 0 & A_4^{-1} \end{bmatrix} \overline{Q} B = \overline{P} \begin{bmatrix} 0 & 0 \\ 0 & A_4^{-1} \end{bmatrix} \begin{bmatrix} B_1 \\ B_2 \end{bmatrix} = \overline{P} \begin{bmatrix} 0 \\ A_4^{-1} B_2 \end{bmatrix}$$

and

$$\Phi_{-1}E = \overline{P}\begin{bmatrix} 0 & 0 \\ 0 & A_4^{-1} \end{bmatrix}\overline{Q}E = \overline{P}\begin{bmatrix} 0 & 0 \\ 0 & A_4^{-1} \end{bmatrix}\begin{bmatrix} I_r & 0 \\ 0 & 0 \end{bmatrix}\overline{P}^{-1} = 0$$

In [4] it was shown that $\qquad \Phi_0 = \lim_{s\to\infty}\left\{\left(s[Es - A]^{-1}E\right)^\mu s[Es - A]^{-1}\right\}$

In this case for $\mu = 1$ using (10) and (12) we obtain

$$\Phi_0 B = \lim_{s\to\infty}\left\{s[Es - A]^{-1}Es[Es - A]^{-1}\right\}B =$$

$$\lim_{s\to\infty}\left\{\overline{P}s\begin{bmatrix} I_r s - A_1, & -A_2 \\ -A_3, & A_4 \end{bmatrix}^{-1}\overline{Q}E\overline{P}s\begin{bmatrix} I_r s - A_1, & -A_2 \\ -A_3, & A_4 \end{bmatrix}^{-1}\overline{Q}B\right\} =$$

$$\text{(23)}$$

$$= \overline{P}\left\{\lim_{s\to\infty}s^2\begin{bmatrix} I_r s - A_1, & -A_2 \\ -A_3, & A_4 \end{bmatrix}^{-1}\begin{bmatrix} I_r & 0 \\ 0 & 0 \end{bmatrix}\begin{bmatrix} I_r s - A_1, & -A_2 \\ -A_3, & A_4 \end{bmatrix}^{-1}\begin{bmatrix} B_1 \\ B_2 \end{bmatrix}\right\}$$

Note that

$$\lim_{s\to\infty}s\begin{bmatrix} I_r s - A_1, & -A_2 \\ -A_3, & A_4 \end{bmatrix}^{-1} = \lim_{s\to\infty}\begin{bmatrix} I_r - A_1 s^{-1}, & -A_2 s^{-1} \\ -A_3 s^{-1}, & A_4 s^{-1} \end{bmatrix}^{-1} =$$

$$\lim_{s\to\infty} = \begin{bmatrix} [I_r - A_1 s^{-1}]^{-1}\left(I_r + A_2 H_1^{-1}A_3[I_r - A_1 s^{-1}]^{-1}s^{-2}\right) \\ H_1^{-1}A_3[I_r - A_1 s^{-1}]^{-1}s^{-1} \end{bmatrix}$$

$$\begin{bmatrix} [I_r - A_1 s^{-1}]^{-1}A_2 H_1^{-1}s^{-1} \\ H_1^{-1} \end{bmatrix} = \begin{bmatrix} I_r, & A_2 A_4^{-1} \\ A_4^{-1}A_3, & * \end{bmatrix} \qquad \text{(24)}$$

since

$$\lim_{s\to\infty}[I_r - A_1 s^{-1}]^{-1} = I_r$$

$$\lim_{s\to\infty}s^{-1}H_1^{-1} = \lim_{s\to\infty}s^{-1}\left[A_4 s^{-1} - A_3[I_r - A_1 s^{-1}]^{-1}A_2 s^{-2}\right]^{-1} = A_4^{-1}$$

* denotes an entry not important in this considerations.

Using (24) from (23) we obtain

$$\Phi_0 B = \overline{P}\begin{bmatrix} I_r, & A_2 A_4^{-1} \\ A_4^{-1}A_3, & A_4^{-1}A_3 A_2 A_4^{-1} \end{bmatrix}\begin{bmatrix} B_1 \\ B_2 \end{bmatrix} =$$

$$\overline{P}\begin{bmatrix} B_1 + A_2 A_4^{-1}B_2 \\ A_4^{-1}A_3\left(B_1 + A_2 A_4^{-1}B_2\right) \end{bmatrix} = \overline{P}\begin{bmatrix} \overline{B}_1 \\ A_4^{-1}A_3\overline{B}_1 \end{bmatrix}$$

Similarily

$$\Phi_0 E = \overline{P}\begin{bmatrix} I_r, & A_2 A_4^{-1} \\ A_4^{-1} A_3, & A_4^{-1} A_3 A_2 A_4^{-1} \end{bmatrix}\overline{Q}E =$$

$$\overline{P}\begin{bmatrix} I_r, & A_2 A_4^{-1} \\ A_4^{-1} A_3, & A_4^{-1} A_3 A_2 A_4^{-1} \end{bmatrix}\begin{bmatrix} I_r & 0 \\ 0 & 0 \end{bmatrix}\overline{P}^{-1} = \overline{P}\begin{bmatrix} I_r, & 0 \\ A_4^{-1} A_3 & 0 \end{bmatrix}\overline{P}^{-1}$$

and

$$\Phi_0 A = \overline{P}\begin{bmatrix} I_r, & A_2 A_4^{-1} \\ A_4^{-1} A_3, & A_4^{-1} A_3 A_2 A_4^{-1} \end{bmatrix}\overline{Q}A =$$

$$\overline{P}\begin{bmatrix} I_r, & A_2 A_4^{-1} \\ A_4^{-1} A_3, & A_4^{-1} A_3 A_2 A_4^{-1} \end{bmatrix}\begin{bmatrix} A_1 & A_2 \\ A_3 & -A_4 \end{bmatrix}\overline{P}^{-1} = \overline{P}\begin{bmatrix} \overline{A}_1 & 0 \\ A_4^{-1} A_3 \overline{A}_1 & 0 \end{bmatrix}\overline{P}^{-1} \quad \Box$$

Collorary 1. If the assumptions of Theorem 3 are satisfied then the equations (1a) and

$$\dot{x} = \Phi_0 A + \Phi_0 Bu + \Phi_{-1}\left(B\dot{u} + Ex_0 \delta^{(1)}\right) \tag{25}$$

have the same solution

$$x = e^{\Phi_0 At}\Phi_0 Ex_0 + \int_0^t e^{\Phi_0 A(t-\tau)}\Phi_0 Bu(\tau)d\tau + \Phi_{-1}\left(Bu + Ex_0 \delta\right)$$

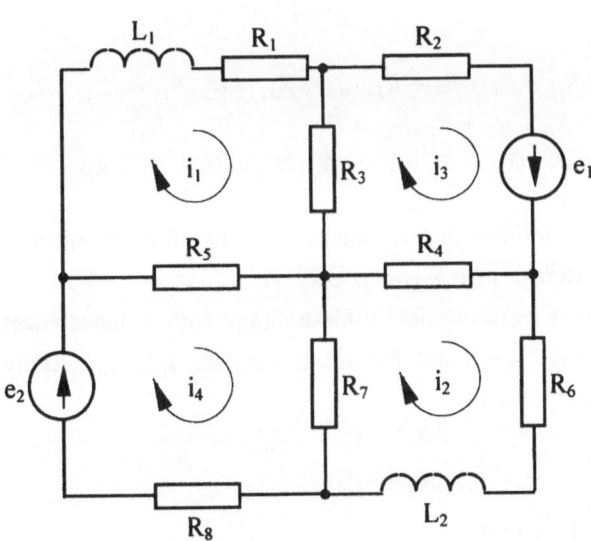

Figure 1: A four-mesh circuit

4   Linear electrical circuits

Consider the four-mesh circuit shown in Fig. with given resistances $R_1, R_2, R_3, R_4, R_5, R_6, R_7, R_8$, inductances L_1, L_2 and source voltages e_1, e_2. The mesh currents are denoted by i_1, i_2, i_3 and i_4, respectively.

Applying the mesh method to the circuit we obtain the equations

$$L_1 \frac{di_1}{dt} = -(R_1 + R_3 + R_5)i_1 + R_3 i_3 + R_5 i_4,$$

$$L_2 \frac{di_2}{dt} = -(R_4 + R_6 + R_7)i_2 + R_4 i_3 + R_7 i_4 \tag{26}$$

$$0 = R_3 i_1 + R_4 i_2 - (R_2 + R_3 + R_4)i_3 + e_1, \quad 0 = R_5 i_1 + R_7 i_2 - (R_5 + R_7 + R_8)i_4 + e_2$$

Assuming the mesh currents as the state variables $x_1 = i_1, x_2 = i_2, x_3 = i_3, x_4 = i_4$ we may write the equations (26) in the form (1a) with

$$
E = \begin{bmatrix} 1 & 0 & 0 & 0 \\ 0 & 1 & 0 & 0 \\ 0 & 0 & 0 & 0 \\ 0 & 0 & 0 & 0 \end{bmatrix}, \quad
x = \begin{bmatrix} i_1 \\ i_2 \\ i_3 \\ i_4 \end{bmatrix}, \quad
A = \begin{bmatrix} -\frac{R_{11}}{L_1}, & 0, & \frac{R_{13}}{L_1}, & \frac{R_{14}}{L_1} \\ 0, & -\frac{R_{22}}{L_2}, & \frac{R_{23}}{L_2}, & \frac{R_{24}}{L_2} \\ R_{31}, & R_{32}, & -R_{33}, & 0 \\ R_{41}, & R_{42}, & 0, & -R_{44} \end{bmatrix}, \quad
B = \begin{bmatrix} 0 & 0 \\ 0 & 0 \\ 1 & 0 \\ 0 & 1 \end{bmatrix},
$$

$$u = \begin{bmatrix} e_1 \\ e_2 \end{bmatrix} \tag{27}$$

where

$$R_{11} = R_1 + R_3 + R_5, R_{13} = R_{31} = R_3, R_{14} = R_{41} = R_5,$$

$$R_{22} = R_4 + R_6 + R_7, R_{23} = R_{32} = R_4, R_{24} = R_{42} = R_7 \tag{28}$$

$$R_{33} = R_2 + R_3 + R_4, R_{44} = R_5 + R_7 + R_8$$

From (28) it follows that the matrix E has the canonical form (10) with $I_r = I_2$, A is a Metzler matrix and $B \in R_+^{4 \times 2}$.

In general case let us consider n-mesh circuit with r inductances $L_1, \dots, L_r$, m source voltages $e_1, \dots, e_m$ and the mesh currents $i_1, \dots, i_n$. Applying the mesh method we obtain (1a) with

$$E = \begin{bmatrix} I_r & 0 \\ 0 & 0 \end{bmatrix} \in R_+^{n \times n}, \quad A = \begin{bmatrix} A_1 & A_2 \\ A_3 & -A_4 \end{bmatrix} \in R^{n \times n}, \quad B \in R_+^{n \times m}, \quad x = [i_1, i_2, \dots, i_n]^T,$$

$$u = [e_1, e_2, \dots, e_m]^T \tag{29a}$$

$$A_1 := \begin{bmatrix} -\dfrac{R_{11}}{l_1}, & \dfrac{R_{12}}{l_1}, & \cdots, & \dfrac{R_{1r}}{l_1} \\ \dfrac{R_{21}}{l_2}, & -\dfrac{R_{22}}{l_2}, & \cdots, & \dfrac{R_{2r}}{l_2} \\ \vdots & & & \vdots \\ \dfrac{R_{r1}}{l_r}, & \dfrac{R_{r2}}{l_r}, & \cdots, & -\dfrac{R_{rr}}{l_r} \end{bmatrix}, \quad A_2 := \begin{bmatrix} \dfrac{R_{1,r+1}}{l_1} & \cdots, & \dfrac{R_{1n}}{l_1} \\ \dfrac{R_{2,r+1}}{l_2} & \cdots, & \dfrac{R_{2n}}{l_2} \\ \vdots & & \vdots \\ \dfrac{R_{r,r+1}}{l_r}, & \cdots, & \dfrac{R_{rn}}{l_r} \end{bmatrix}$$

(29b)

$$A_3 := \begin{bmatrix} R_{r+1,1}, & \cdots, & R_{r+1,r} \\ R_{r+2,1}, & \cdots, & R_{r+2,r} \\ \vdots & & \vdots \\ R_{n1}, & \cdots, & R_{nr} \end{bmatrix}, \quad A_4 := \begin{bmatrix} R_{r+1,r+1}, & -R_{r+1,r+2}, & \cdots, & -R_{r+1,n} \\ -R_{r+2,r+1}, & R_{r+2,r+2}, & \cdots, & -R_{r+2,n} \\ \vdots & \vdots & & \vdots \\ -R_{n,r+1}, & -R_{n,r+2}, & \cdots, & R_{nn} \end{bmatrix}$$

where $\qquad R_{ij} = R_{ji} \begin{cases} > 0 & for \quad i = j \\ \geq 0 & for \quad i \neq j \end{cases} \quad i,j = 1,\ldots,n$

and T denotes the transposition.

With slight modifications the above considerations can be extended for electrical circuits consisting of resistances, capacitances and source voltages.

Firstly we shall show the following

Lemma 2. The inverse matrix R_n^{-1} of the mesh resistance matrix

$$R_n = \begin{bmatrix} R_{11} - R_{12} & \cdots & -R_{1n} \\ -R_{21} & R_{22} & \cdots & -R_{2n} \\ \vdots & & & \vdots \\ -R_{n1} - R_{n2}, & \cdots & R_{nn} \end{bmatrix}, \quad R_{ij} = R_{ji} \begin{cases} > 0 & for \quad i = j \\ \geq 0 & for \quad i \neq j \end{cases}, \quad R_{ii} \geq \sum_{\substack{j=1 \\ j \neq i}}^{n} R_{ij},$$

$$i,j = 1,\ldots,n \qquad (30)$$

has nonnegative entries, $R_n^{-1} \in R_+^{n \times n}$.

Proof. The proof will be accomplished by induction.

For $n = 1$ the hypothesis is true since $R_1 = R_{11} > 0$ and $R_1^{-1} = R_{11}^{-1} > 0$.

Assuming that the hypothesis is true for k we shall show that it is also true for $k+1$.

Let

$$R_{k+1} = \begin{bmatrix} R_k & u_k \\ v_k & R_{k+1,k+1} \end{bmatrix} \qquad (31)$$

where

$$v_k = \begin{bmatrix} -R_{k+1,1}, -R_{k+1,2}, \ldots, -R_{k+1,k} \end{bmatrix}, \quad u_k = \begin{bmatrix} -R_{1,k+1}, -R_{2,k+1}, \ldots, -R_{k,k+1} \end{bmatrix}^T$$

The inverse matrix R_{k+1}^{-1} of (31) is given by

$$
R_{k+1}^{-1} = \begin{bmatrix} R_k^{-1} + \dfrac{R_k^{-1} u_k v_k R_k^{-1}}{\overline{R}_k}, & -\dfrac{R_k^{-1} u_k}{\overline{R}_k} \\[3mm] -\dfrac{v_k R_k^{-1}}{\overline{R}_k}, & \dfrac{1}{\overline{R}_k} \end{bmatrix}
\tag{32}
$$

where $\overline{R}_k := R_{k+1,k+1} - v_k R_k^{-1} u_k$

By assumption $R_k^{-1} \in R_+^{k \times k}$. From (32) it follows that $R_{k+1}^{-1} \in R_+^{(k+1) \times (k+1)}$ if $\overline{R}_k > 0$. It is well-known that R_{k+1} is a positive defined and $\det R_{k+1} > 0$. Using (31) we may write

$$
\det R_{k+1} = \det \begin{bmatrix} R_k & u_k \\ v_k & R_{k+1,k+1} \end{bmatrix} = \det \begin{bmatrix} R_k & , & u_k \\ 0 & , & R_{k+1,k+1} - v_k R_k^{-1} u_k \end{bmatrix}
$$

$$
= \det R_k \det \left(R_{k+1,k+1} - v_k R_k^{-1} u_k \right) = \overline{R}_k \det R_k
\tag{33}
$$

From (33) it follows that $\det R_{k+1} > 0$ and $\det R_k > 0$ implies $\overline{R}_k > 0$. $\Box$

From Theorem 2 and Lemma 2 we have the following

Collorary 2. The electrical circuits consisting of resistances, inductances (capacetances) and source voltages can be reduced to positive standard systems.

Theorem 4. If the matrices E and A have the form (29) then

$$
\Phi_{-1} = \begin{bmatrix} 0 & 0 \\ 0 & A_4^{-1} \end{bmatrix} \in R_+^{n \times n}
\tag{34a}
$$

$$
\Phi_0 = \begin{bmatrix} I_r & A_2 A_4^{-1} \\ A_4^{-1} A_3 & A_4^{-1} A_3 A_2 A_4^{-1} \end{bmatrix} \in R_+^{n \times n}
\tag{34b}
$$

Proof. From (3) we have

$$
\Phi_{-1} = \lim_{s \to \infty} [Es - A]^{-1} = \lim_{s \to \infty} \begin{bmatrix} I_r s - A_1, & -A_2 \\ -A_3, & A_4 \end{bmatrix}^{-1} =
$$

$$
= \lim_{s \to \infty} \begin{bmatrix} [I_r s - A_1]^{-1} \left(I_r + A_2 H^{-1} A_2 [I_r s - A_1]^{-1} \right), \\ H^{-1} A_3 [I_r s - A_1]^{-1} \end{bmatrix}
$$

$$
\begin{matrix} [I_r s - A_1]^{-1} A_2 H^{-1} \\ H^{-1} \end{matrix} = \begin{bmatrix} 0 & 0 \\ 0 & A_4^{-1} \end{bmatrix}
$$

since $\lim_{s \to \infty}\left[I_r s - A_1\right]^{-1} = 0$ and $\lim_{s \to \infty} H^{-1} = \lim_{s \to \infty}\left[A_4 - A_3\left[I_r s - A_1\right]^{-1} A_2\right]^{-1} = A_4^{-1}$

By Lemma $A_4^{-1} \in R_+^{(n-r) \times (n-r)}$ and (34a) holds.

Using the formula

$$\Phi_0 = \lim_{s \to \infty}\left\{\left(s\left[Es - A\right]^{-1} E\right)^{\mu} s\left[Es - A\right]^{-1}\right\}$$

for $\mu = 1$ and (29), (24) we obtain

$$\Phi_0 = \lim_{s \to \infty}\left\{s\left[Es - A\right]^{-1} Es\left[Es - A\right]^{-1}\right\} =$$

$$= \lim_{s \to \infty}\left\{s^2\begin{bmatrix} I_r s - A_1, & -A_2 \\ -A_3, & A_4 \end{bmatrix}^{-1}\begin{bmatrix} I_r & 0 \\ 0 & 0 \end{bmatrix}\begin{bmatrix} I_r s - A_1, & -A_2 \\ -A_3, & A_4 \end{bmatrix}^{-1}\right\} = \begin{bmatrix} I_r & A_2 A_4^{-1} \\ A_4^{-1} A_3, & A_4^{-1} A_3 A_2 A_4^{-1} \end{bmatrix}$$

in a similar way as in proof of formula (20).

Note that $A_2 \in R_+^{r \times (n-r)}, A_3 \in R_+^{(n-r) \times r}$. By Lemma $A_4^{-1} \in R_+^{(n-r) \times (n-r)}$ and (34b) holds. $\square$

5   Conclusions

A new class of weakly positive singular continuous-time systems has been introduced. It has been shown that the equations (1a) and (7) have the same solution (8). Necessary and sufficient conditions have been established (Theorem 2) under which a weakly positive singular system (1) can be reduced by the strict equivalence transformation (10), (12) to the positive standard system (15), (15). It has been shown that linear electrical circuits consisting of resistances, inductances (capacitances) and source voltages are examples of positive singular continuous-time linear systems and they can be reduced to positive standard systems.

References

[1] Cobb D 1984 Controllability, observability and duality in singular systems. *IEEE Trans Automat Contr* 29(12):1076-1082

[2] Dai L 1989 *Singular Control Systems* (Lectures Note in Control and Information Sciences) Springer-Verlag

[3] Fanti M P, Maione B, Turchiano B 1990 Controllability of multi-input positive discrete-time systems. *Int. J. Control* 51(6):1295-1308

[4] Kaczorek T 1998 Computation of fundamental matrices and reachability of positive singular discrete linear systems *Bull. Pol. Acad. Techn. Sci.*46(4) (in press)

[5] Kaczorek T 1993 *Linear Control Systems, vol. 2*, Research Studies Press and J.Wiley, New York

[6] Kaczorek T 1997 *Positive linear systems and their relationship with electrical circuits*, XX – SPETO : 33-41.

[7] Kaczorek T 1998 Positive descriptor discrete-time linear systems *Problems of Nonlinear Analysis in Engineering Systems* (in press)

[8] Kaczorek T 1997 Positive singular discrete linear systems *Bull. Pol. Acad Techn. Sci.*, 45(4) : 619-631

[9] Klamka J 1991 *Controllability of Dynamical Systems,* Kluwer Academic Publ, Dordecht

[10] Lewis F L 1986 A survey of linear singular systems *Circuits Systems Signal Process,* 5(1):1-36.

[11] Lewis F L 1984 Descriptor systems: Decomposition into forward and backward subsystems *IEEE Trans Automat Contr* 29:167-170

[12] Luenberger D G 1978 Time-invariant descriptor systems *Automatica,* 14:473-480

[13] Luenberger G 1977 Dynamic equations in descriptor *IEEE Trans Automat Contr* 22(3):312-321

[14] Mertzios B G, Lewis F L 1989 Fundamental matrix of discrete singular systems *Circuits, Syst., Signal Processing,* 8(3):341-355

[15] Ohta Y, Madea H, Kodama S 1984 Reachability, observability and realizability of continuous-time positive systems *SIAM J. Control and Optimization* 22(2):171-180

2

Harmonic Analysis of Linear Sampled-Data Systems

J.J. Yamé and R. Hanus

1   Introduction

A sampled-data system arises when an analog process is controlled by a micro-processor based controller and therefore such a system combines both continuous time and discrete-time dynamic subsystems which are interconnected *via* periodic samplers and hold devices. As a result sampled-data systems are hybrid systems and their continuous-time dynamics is described by periodically time-varying operators. The system theoretic concepts such as transfer functions and frequency responses are no longer applicable to such systems, that is, in the classical frequency domain they do not have the form of multiplication operators. Due to the importance of the concept of frequency response as a powerful tool in control system analysis and design, it has been highly desirable to extend this notion to sampled-data systems. Recently, different frameworks dealing with the concept of frequency response for sampled-data systems have been introduced all taking into account their continuous-time behavior. One of these frameworks is the time-domain lifting technique proposed by [1] and [2]. It is shown that a linear sampled-data system can be viewed as a linear time-invariant (LTI) discrete-time system with infinite-dimensional input and output signals which capture the intersample behavior of the original signals. This technique allows for the use of a compatible transform to define notions such as transfer functions and by extension the concept of frequency response [3]. Another framework related to the frequency domain and dealing with the intersample behavior of sampled-data systems is based on their steady state response to a sinusoidal input in terms of impulse modulation [4]. In [4], the notion of frequency response is defined as a mapping from a certain signal space into itself. These frameworks offers two different but equivalent viewpoints on frequency responses for sampled-data systems as shown in [5]. In this chapter, we further investigate the concept of frequency response of sampled-data systems using a point of view based on a symmetry argument which we exploit *via* a group theoretic approach. More precisely, the method consists of expansion of sampled-data systems with respect to their "harmonics". On the function space under consideration, "harmonics" stands for the "elementary subspaces" which exhibit the simplest behavior in presence of some action, here a group of transformations which leave invariant sampled-data systems. It is shown that the restriction of a bounded sampled-data system to the "harmonics" defines an operator-valued function of the frequency variable which has the nice properties of the frequency response of LTI systems.

The chapter is organized as follows. The setup of sampled-data systems and some necessary mathematical background are given in section 2. In section 3, we review the notion of frequency response of LTI systems and we exhibit the nature of the spectral analysis problem of sampled-data systems. Section 4 derives the decomposition of the function space under consideration into elementary parts. In section 5 this decomposition is used to state the fundamental structure of sampled-data systems and to define the concept of frequency response for such systems. Section 6 is the conclusion, where the results of this chapter are summarized and some comments are given on its relationship to other related contributions.

2 Preliminaries

The sets of integers, real and complex numbers are respectively denoted by $\mathbb{Z}$, $\mathbb{R}$ and $\mathbb{C}$. Let $\mathcal{H}$ be a Hilbert space with inner product $\langle ., .\rangle_{\mathcal{H}}$ and norm $\|.\|_{\mathcal{H}}$, the set of all linear bounded operators on $\mathcal{H}$ is a Banach algebra denoted by $\mathcal{L}(\mathcal{H})$ with norm $\|.\|_{\mathcal{L}(\mathcal{H})}$. The spaces $l^2(\mathbb{Z})$, $L^2(\mathbb{R})$ and $L^2_{loc}(\mathbb{R})$ denote respectively the Hilbert space of square-summable sequences defined on $\mathbb{Z}$, the Hilbert space of square-integrable functions on $\mathbb{R}$ and the space of locally square-integrable functions on $\mathbb{R}$. The identity operator on a space is denoted by I and $\mathcal{A}^*$ stands for the adjoint of the operator $\mathcal{A}$ on a Hilbert space.

2.1 Sampled-data systems

A standard setup of a single-input, single-output sampled-data system with sampling period T is depicted in figure 1.

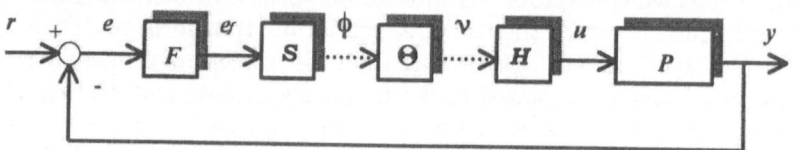

Figure 1: Sampled-data system

The reference input r, the plant input u and output y, the filter input e and output e_f are all continuous-time signals whereas the control signal ν and the sampler output ϕ are discrete-time signals . We assume that the continuous-time signals are elements of the function space $L^2(\mathbb{R})$ and the discrete-time signals belong to $l^2(\mathbb{Z})$. The plant P and the filter F are LTI continuous-time systems, moreover F is strictly proper[1] and the digital control algorithm Θ is a discrete-time shift-invariant system. The sampler is the hybrid operator $\mathbf{S} : L^2(\mathbb{R}) \supset Ran(F) \to l^2(\mathbb{Z})$ defined by $\mathbf{S}e_f = \phi = \{\phi[k]\}_{k=-\infty}^{\infty}$ where the sequence $\{\phi[k]\}_{k=-\infty}^{\infty}$ is the digital signal with $\phi[k] = e_f(kT), k \in \mathbb{Z}$ and $Ran(F)$ denotes the range of F. The hold device is also a hybrid system $\mathbf{H} : l^2(\mathbb{Z}) \to L^2(\mathbb{R})$ which is defined by $\mathbf{H}\nu = u$ with the continuous-time signal u given by $u(kT + t) = (\mathbf{H}\nu)(t) = h(t)\nu[k]$ for $0 \le t < T, k \in \mathbb{Z}$. The hold function $h(t)$ is assumed to be a piecewise continuous function on $[0, T)$. It is straightforward to show that the operator $\mathcal{K} = \mathbf{H}\Theta\mathbf{S}F$ on $L^2(\mathbb{R})$ is a T-periodically time-varying operator and that the feedback map on $L^2(\mathbb{R})$, $\mathcal{T} : r \to y$ which is given by $\mathcal{T} = (1 + P\mathcal{K})^{-1} P\mathcal{K}$, is also T-periodic provided it is a bounded operator on $L^2(\mathbb{R})$. In the sequel, we assume the boundedness of the sampled-data operator on $L^2(\mathbb{R})$.

2.2 Groups and their representations

We recall the notion of a group and introduce the related concept of group representation, then we indicate in the next section how these notions arise naturally in the spectral analysis problem of sampled-data systems. Background material on groups and their representations may be found in [6].

A group G is a set of elements $g_1, g_2, \ldots$ endowed with a composition law written $g_1 g_2$ for every pair of elements g_1, g_2 and defined as an element of G such that it satisfies associativity, the existence of an identity element and the existence of inverse elements in G. If composition is commutative, i.e. every pair of elements of G obeys the rule $g_1 g_2 = g_2 g_1$, then G is said to be *Abelian* or commutative. A *homomorphism* is a mapping ϕ of a group G into another group G' such that $\phi(g_1 g_2) = \phi(g_1)\phi(g_2)$ for all $g_1, g_2 \in G$. A class of homomorphisms is formed by the so-called realizations or representations of groups. By a representation of an abstract group we mean in general any group composed of concrete mathematical entities which is homomorphic to the original group. An important class of representations is the so-called unitary representation defined below.

Definition 1 *A unitary representation of G consists in associating with each element $g \in G$ a unitary operator $U(g)$ of a Hilbert space $\mathcal{H}$*

$$\begin{aligned} U : \quad G \quad &\longrightarrow \quad \mathcal{L}(\mathcal{H}) \\ g \quad &\longmapsto \quad U(g) \end{aligned} \tag{1}$$

in such a way that these transformations obey the law of composition $U(g_1 g_2) = U(g_1)U(g_2)$ for every $g_1, g_2 \in G$. The Hilbert space $\mathcal{H}$ is called the representation space and is denoted by $\mathcal{H}(U)$.

The dimension $d(U)$ of $\mathcal{H}$ is called the degree of the representation U. If $d(U) < \infty$, then representation U is said to be *finite dimensional* otherwise it

---

[1]This assumption guarantees that the sampling operation is well defined

is *infinite dimensional*. In a sense, the "building blocks" or elementary parts of unitary representations are the irreducible representations which are defined as follows.

Definition 2 *Let U be a unitary representation of a group G. A closed linear manifold, i.e. a subspace, $\mathcal{H}_0$ of the representation space $\mathcal{H}(U)$ is called invariant under U or U-invariant if we have $U(g)\mathcal{H}_0 \subset \mathcal{H}_0$ for all $g \in G$. A unitary representation U is called irreducible if $\mathcal{H}(U) \neq \{0\}$ and if $\mathcal{H}(U)$ and $\{0\}$ are the only invariant subspaces of $\mathcal{H}(U)$ under U.*

Non-irreducible unitary representations are decomposed as the "sum" of irreducible representations. Here the meaning of "sum" should be clarified because some difficulties arise for infinite-dimensional unitary representations mainly due to the presence of continuous spectrum. In this case, the notion of direct sum decomposition which is related to the classical notion of direct sum of Hilbert spaces suggests that this must be generalized to that of direct integral decomposition via the concept of continuous sum or direct integral of Hilbert spaces [6] which will be introduced in section 4. The classical notion of direct sum decomposition of a unitary representation is introduced as follows. Suppose the representation space $\mathcal{H}(U)$ of a unitary representation U of the group G is decomposed into a direct sum of U-invariant closed subspaces $\{\mathcal{H}_\alpha\}_{\alpha \in A}$, that is $\mathcal{H}(U) = \bigoplus_{\alpha \in A} \mathcal{H}_\alpha$. Then the unitary representation U is called a direct sum of unitary representations $\{U_\alpha\}_{\alpha \in A}$ where U_α is the restriction of U on $\mathcal{H}_\alpha$. This means that the Hilbert space $\mathcal{H}(U)$ has an orthogonal decomposition whose summands are minimal invariant subspaces under the action of U. Minimality is understood in the sense that no proper subspace of $\mathcal{H}_\alpha$ is mapped into itself by the collection $\{U_\alpha(g)\}$ for all $g \in G$. Actually these invariant subspaces are reducing subspaces.

3 Nature of the problem

In order to motivate our study we first examine some aspects of the frequency response of LTI continuous-time systems. Roughly speaking, such systems are convolution operators, that is, systems described in an input/output setting by the convolution $y(t) = (Hr)(t) = \int_{-\infty}^{\infty} h(\tau) r(t-\tau) d\tau$ defined in terms of the kernel $h(t)$ called the impulse response. An important feature of such systems which are stable is that they transform complex exponential inputs into complex exponential outputs and, this leads to the notion of frequency response. Indeed consider the input signal $r(t) = e^{j\omega t}$, then the steady state output is $y(t) = \left(\int_{-\infty}^{\infty} h(\tau) e^{-j\omega\tau} d\tau \right) e^{j\omega t} = \widehat{h}(\omega) e^{j\omega t}$ where $\widehat{h}$, the Fourier transform of h, is called the frequency response of the system. Thus, the action of a LTI system on a complex exponential is simply *multiplication* by a scaling factor (the frequency response) which depends only on the frequency ω of the input signal. This fact allows to describe in "Fourier coordinates" the effect of such systems on more general inputs as pointwise multiplication

$$\widehat{y}(\omega) = \widehat{h}(\omega).\widehat{r}(\omega) \tag{2}$$

where $\widehat{y}$ and $\widehat{r}$ are the Fourier transforms of y and r. Mathematically, the basic task of frequency response analysis of a LTI continuous-time system H can be described by the following "eigenvalue problem"[2] $(He_\omega)(t) = \lambda_\omega.e_\omega(t)$ to determine the scaling factors $\lambda_\omega = \widehat{h}(\omega)$ corresponding to the formal eingenfunctions $e_\omega(t) = e^{j\omega t}$. The one-dimensional vector spaces (the "harmonics")

$$\mathcal{V}_\omega = \{u : u(t) = \mathbf{u}_m.e_\omega(t), \ \mathbf{u}_m \in \mathbb{C}\} \tag{3}$$

are thus invariant under H, that is, $u \in \mathcal{V}_\omega$ implies $Hu \in \mathcal{V}_\omega$ and $Hu = \lambda_\omega u$. The family $\{e_\omega(t)\}_{\omega=-\infty}^{\infty}$ forms a "basis" which clearly reduces H to a diagonal form. Therefore, frequency domain analysis of LTI systems amounts to looking at the action of such systems on these reducing one-dimensional spaces. Actually, the fact that exponentials are "eigenfunctions" of an arbitrary LTI system H is a direct consequence of the time invariance property which may be defined by the commutation rule

$$D_\tau H = HD_\tau, \quad \text{for all } \tau \in \mathbb{R} \tag{4}$$

and where $D_\tau : L^2(\mathbb{R}) \to L^2(\mathbb{R})$ is the time translation operator

$$(D_\tau f)(t) = f(t+\tau) \tag{5}$$

Indeed, each exponential $e_\omega(t)$ is an "eigenfunction" of *all* translations D_τ with *distinct* eigenvalues $e_\omega(\tau)$ and therefore they provide a common spectral decomposition for the operators that commute with these translations. Let us clarify this by assuming that H verifies the commutation rule (4) and set $y(t) = (He_\omega)(t)$, then $(D_\tau y)(t) = H(D_\tau e_\omega)(t) = e_\omega(\tau)y(t)$ that is $y(t)$ is also an "eigenfunction" of D_τ. Letting $t = 0$, we obtain $y(\tau) = e_\omega(\tau)y(0)$ and consequently $(He_\omega) = y(0)e_\omega$ which means that "eigenfunctions" of D_τ are exactly the "eigenfunctions" of H. The significance of this result is that $\{D_\tau\}_{-\infty \leq \tau \leq \infty}$ being a set of mutually commuting operators, all of which commute with H, their simultaneous eigenfunctions can be chosen as a basis which diagonalize H. The Fourier transform which is an "expansion" onto these "eigenfunctions" has in some sense the effect to simultaneously diagonalize all those commuting operators, that is, to express them as multiplication operators thereby simplifying the analysis of LTI systems. Conversely, we can look upon LTI systems as being obtained by synthesis of very simple operators, i.e. the scalars $\widehat{h}(\omega)$, on the irreducible representation spaces $\mathcal{V}_\omega$ into which the translations D_τ decompose. Now, if we input an exponential signal to a sampled-data system, it is straightforward to show that its output is given by $y(t) = (\mathcal{T}e_\omega)(t) = \sigma(t,\omega)e^{j\omega t}$, i.e., an exponential signal modulated by the function $\sigma(t,\omega)$ called the *symbol* of the operator $\mathcal{T}$. It is a function defined on

---

[2] Complex exponentials do no belong to the space $L^2(\mathbb{R})$ of finite energy signals.

$\mathbb{R}^2$ which is $T-$periodic in t for every fixed ω and the one-dimensional spaces $\mathcal{V}_\omega$ are no longer invariant under the action of sampled-data systems. In the search for extending the frequency response concept to linear sampled-data systems, we need an analog of the family of the reducing spaces $\mathcal{V}_\omega$, and for this, we will resort to the theory of group representation. The key property of sampled-data systems which will be relevant in our analysis is their T-periodic structure which is usually expressed on $L^2(\mathbb{R})$ as $D_T \mathcal{T} = \mathcal{T} D_T$ with T being the sampling period. But, instead of one commutation rule for characterizing the T-periodicity of a sampled-data system, we consider the following set of commutation rules with the family of $\gamma-$translation operators on $L^2(\mathbb{R})$

$$D_\gamma \mathcal{T} = \mathcal{T} D_\gamma \qquad \text{for all } \gamma \in \Gamma \tag{6}$$

with $\Gamma = \{..., -2T, -T, 0, T, 2T, ...\}$. The commutation rules (6) means that the system looks the same after a coordinate transformation γ time units later, in other words, the operator $\mathcal{T}$ is invariant under the transformation $\iota_\gamma : t \mapsto t + \gamma$ of $\mathbb{R}$ into itself. Therefore, such a transformation is a symmetry operator for the system and the following composition rule holds $\iota_{\gamma_1} . \iota_{\gamma_2} = \iota_{\gamma_1 + \gamma_2}$. Clearly, the set of coordinate transformations $\{\iota_\gamma\}$ forms an Abelian group which is isomorphic to the index set Γ considered as an additive group which itself is isomorphic to the additive group $\mathbb{Z}$. From an abstract group point of view, the sets $\{\iota_\gamma\}$ and Γ are identical. Furthermore, it is easily verified that the $\gamma-$translation operators D_γ which act on functions satisfy the same group law as the elements of $\{\iota_\gamma\}$. Namely

$$D_{\gamma_1} D_{\gamma_2} = D_{\gamma_1 + \gamma_2} \tag{7}$$

and the set of operators $\{D_\gamma\}$ is a group isomorphic to the group $\{\iota_\gamma\}$. That is, the action of ι_γ on the *time* $t \in \mathbb{R}$ is now represented by an induced action of D_γ on *signals* $f \in L^2(\mathbb{R})$. The correspondence (mapping the action on time to the action on signals)

$$\begin{array}{cccc} D: & \Gamma & \to & \mathcal{L}\left(L^2(\mathbb{R})\right) \\ & \gamma & \mapsto & D_\gamma \end{array} \tag{8}$$

is a representation of Γ in $L^2(\mathbb{R})$. It is easily verified that the D_γ's are unitary operators on $L^2(\mathbb{R})$ and hence the representation D is an infinite-dimensional unitary representation of the symmetry group Γ on the Hilbert space $\mathcal{H}(D) = L^2(\mathbb{R})$. From the above considerations and adopting a continuous-time point of view, a sampled-data system may be characterized as follows: the time configuration of the system is the real line $\mathbb{R}$, the signals on which it acts belong to $L^2(\mathbb{R})$ and the fundamental time property of its evolution is described by the group Γ of one-to-one transformations of $\mathbb{R}$ into itself. From the commuting rules (6), it is clear that the $\gamma-$translation operators D_γ transform the space of solutions of the equation $(\mathcal{T} - \lambda I) u = 0$ into itself. Now by analogy with the LTI case, the harmonic analysis problem of sampled-data systems is that of decomposing the function space $L^2(\mathbb{R})$ into irreducible, i.e. minimal invariant, submanifolds under the action of this representation described by the family of $\gamma-$translation operators D_γ.

4 Harmonic decomposition

In order to determine these submanifolds, we set a formal eigenvalue problem for the operators D_γ

$$(D_\gamma \psi)(t) = \lambda(\gamma)\psi(t) \quad \Leftrightarrow \quad \psi(t+\gamma) = \lambda(\gamma)\psi(t) , \quad \lambda(\gamma) \in \mathbb{C} \quad (9)$$

The following properties are easily obtained from equation (7)

$$(a)\,\lambda(0) = 1, (b)\,\lambda(\gamma_1)\lambda(\gamma_2) = \lambda(\gamma_1 + \gamma_2), (c)\,|\lambda(\gamma)| = 1$$

Properties (a) and (b) means that λ is a homomorphism from Γ into the multiplicative group $\mathbb{C}^* = \mathbb{C} \setminus \{0\}$, and it is well-known that the only complex functions satisfying the above properties are the exponentials

$$\lambda(\gamma) = e_\omega(\gamma) = e^{j\omega\gamma} , \quad \gamma \in \Gamma \quad (10)$$

where $\omega \in \mathbb{R}$ and is independent of γ. From equation (9) , we have $\psi(t) = e^{-j\omega\gamma}\psi(t+\gamma)$ and multiplying the two sides of this last identity by $e^{-j\omega t}$, we obtain $e^{-j\omega t}\psi(t) = e^{-j\omega(t+\gamma)}\psi(t+\gamma)$, i.e., $D_\gamma(e_{-\omega}\psi)(t) = (e_{-\omega}\psi)(t)$ which implies that the function $u = e_{-\omega}\psi$ is γ-periodic. We have therefore

$$\psi = e_\omega.u \Leftrightarrow \psi(t) = e_\omega(t).u(t) \quad (11)$$

that is, the "eigenfunctions" of D_γ are exponential signals modulated by γ-periodic functions, and in the sequel we write them as $\psi_\omega(t)$ and we call the label ω the *pseudofrequency*. This terminology is motivated by the following observation. The exponential $e_\omega(t)$ is γ-periodic in t if and only if $\omega\gamma \in 2\pi\mathbb{Z} = \{2\pi n : n \in \mathbb{Z}\}$ for any $\gamma \in \Gamma$. The points ω that satisfy this condition form a discrete set $\Gamma^\perp$ which consists of all points $n\omega_s$ ($\omega_s = 2\pi/T$ is the sampling frequency), $n \in \mathbb{Z}$, i.e., $\Gamma^\perp = \{..., -2\omega_s, -\omega_s, 0, \omega_s, 2\omega_s, ...\}$. Setting $\omega_n = \omega + n\omega_s$, $n \in \mathbb{Z}$ and operating on ψ_{ω_n} with D_γ then gives

$$\left(D_\gamma \psi_{\omega_n}\right)(t) = \psi_{\omega_n}(t+\gamma) = e^{-j(\omega+n\omega_s)\gamma}\psi_{\omega_n}(t) = e^{-j\omega\gamma}\psi_{\omega_n}(t) \quad (12)$$

which shows that ω is actually not uniquely specified. Thus it is not possible to distinguish between ω and ω_n, all these points in the ω-space being equivalent. This is the well-known aliasing or frequency folding phenomena. We can therefore assume that $\omega \in \Omega$, where Ω is a fundamental domain of the action of $\Gamma^\perp$ on $\mathbb{R}$ by ω_s-translations and take it as that part of the ω-space which contains the origin, i.e., $\Omega = [-\omega_s/2, \omega_s/2)$. We denote the space of all eigenfunctions of D_γ belonging to $L^2_{loc}(\mathbb{R})$ and having the *same* pseudofrequency ω by $\mathcal{H}_\omega$

$$\mathcal{H}_\omega = \left\{\psi \in L^2_{loc}(\mathbb{R}) : \psi(t) = e^{j\omega t}u(t) \; ; \; u(t) \text{ is } \gamma - \text{periodic}\right\} \quad (13)$$

which is a Hilbert space with inner product $\langle\psi, \phi\rangle_{\mathcal{H}_\omega} = \frac{1}{T}\int_0^T \psi(t)\overline{\phi}(t)\,dt$. It is verified that the exponentials $\left\{e^{j(\omega+\varpi)t}, \varpi \in \Gamma^\perp\right\}$ form an orthonormal basis in $\mathcal{H}_\omega$ and this space is a space of a new representation D^ω of Γ, i.e., $D^\omega : \gamma \mapsto D_\gamma^\omega$ maps Γ into $\mathcal{L}(\mathcal{H}_\omega)$ with the operator $D_\gamma^\omega : \mathcal{H}_\omega \to \mathcal{H}_\omega$ defined by

$$D_\gamma^\omega \psi_\omega = e_\omega(\gamma)\psi_\omega \qquad (14)$$

that is D_γ^ω is the operator of multiplication by $e_\omega(\gamma)I$ on $\mathcal{H}_\omega$, it is a diagonal operator on $\mathcal{H}_\omega$ called a scalar operator.

Now, we introduce the notion of *continuous sum* or *direct integral* of a family of Hilbert spaces $\{\mathcal{H}_\omega\}_{\omega \in \Omega}$ with respect to the ordinary Lebesgue measure of the parameter space Ω ([6]). Roughly speaking, this space denoted by $\mathcal{H}^{cont}$ should consist of functions $\tilde{f}$ on Ω such that $\tilde{f}_\omega = \tilde{f}(\omega) \in \mathcal{H}_\omega$ for each ω and the $\mathcal{H}^{cont}$-norm of $\tilde{f}$ is bounded. Such a continuous sum is defined by

$$\mathcal{H}^{cont} \triangleq \int_\Omega \mathcal{H}_\omega d\omega \triangleq \left\{ \tilde{f} \mid \tilde{f} : \Omega \to \mathcal{H}_\omega \text{ with } \left\| \tilde{f} \right\|_{\mathcal{H}^{cont}}^2 = \int_\Omega \left\| \tilde{f}(\omega) \right\|_{\mathcal{H}_\omega}^2 d\omega < \infty \right\} \qquad (15)$$

The space $\mathcal{H}^{cont}$ is the Hilbert space of square integrable $\mathcal{H}_\omega$−valued functions on Ω, i.e., $\mathcal{H}^{cont} = L^2(\Omega, \mathcal{H}_\omega)$ and this may be viewed also as a space of bundle where each fiber is $\mathcal{H}_\omega$. The following equivalence lemma is crucial for the statement of the main result on the structure of sampled-data systems.

Lemma 3 $\mathcal{H}^{cont}$ *is a realization of* $L^2(\mathbb{R})$

Proof. We quote the following result from ([7],p. 506) : $L^2(\Omega, \mathcal{H}_\omega)$ is isomorphic to the direct sum of countably many copies of $L^2(\Omega)$ indexed by the set of all integers, i.e., $L^2(\Omega, \mathcal{H}_\omega) \cong l^2(\mathbb{Z}, L^2(\Omega))$. Denote by $\Omega + n\omega_s$ the set $\{\omega' : \omega' = \omega + n\omega_s, \ \omega \in \Omega, n \in \mathbb{Z}\}$. Clearly $L^2(\Omega) \cong L^2(\Omega + n\omega_s)$ for all $n \in \mathbb{Z}$, and the following isomorphisms hold $l^2(\mathbb{Z}, L^2(\Omega)) \cong \bigoplus_{n=-\infty}^{\infty} L^2(\Omega + n\omega_s)$ $\cong L^2\left(\bigcup_{n=-\infty}^{\infty}(\Omega + n\omega_s)\right)$. Since $\bigcup_{n=-\infty}^{\infty}(\Omega + n\omega_s) = \mathbb{R}$, this completes the proof. ∎

This lemma states that $L^2(\mathbb{R})$ and $\mathcal{H}^{cont}$ are in a sense "equal"

$$L^2(\mathbb{R}) \cong \mathcal{H}^{cont} = \int_\Omega \mathcal{H}_\omega d\omega = L^2(\Omega, \mathcal{H}_\omega) \qquad (16)$$

The identification of $L^2(\mathbb{R})$ with $\mathcal{H}^{cont}$ means that the space $L^2(\mathbb{R})$ is the continuous sum of the submanifolds $\mathcal{H}_\omega$ and thus under this identification, every function f in $L^2(\mathbb{R})$ can be viewed as an element $\tilde{f}$ in $\mathcal{H}^{cont}$, with its "components" $\tilde{f}_\omega$ given by

$$\tilde{f}_\omega(t) = \sum_{\varpi \in \Gamma^\perp} e^{j(\omega + \varpi)t} \hat{f}(\omega + \varpi) \qquad (17)$$

where $\hat{f}$ is the Fourier transform of f. This expression is obtained from the fact that the family of exponentials $\left\{e^{j(\omega+\varpi)t}, \varpi \in \Gamma^{\perp}\right\}$ is an orthonormal basis in $\mathcal{H}_{\omega}$. Equation (17) can be transformed to the following explicit expression using Poisson summation formula ([8], p. 149)

$$\tilde{f}_{\omega}(t) = \sum_{\gamma \in \Gamma} e^{-j\omega\gamma} f(t+\gamma) = \sum_{k \in \mathbb{Z}} e^{-j\omega kT} f(t+kT) \qquad (18)$$

Equation (18) expresses the decomposition of a signal f of $L^2(\mathbb{R})$ into its components $\tilde{f}_{\omega} \in \mathcal{H}_{\omega}$, this decomposition defines a generalized Fourier transform $\mathcal{F}_G$ when an element $f \in L^2(\mathbb{R})$ is viewed as an element $\tilde{f} \in L^2(\Omega, \mathcal{H}_{\omega})$

$$\begin{array}{cccc} \mathcal{F}_G: & L^2(\mathbb{R}) & \rightarrow & L^2(\Omega, \mathcal{H}_{\omega}) \\ & f & \mapsto & \tilde{f} \end{array} \qquad (19)$$

Moreover the Parseval identity holds, i.e. $\left\|\tilde{f}\right\|_{L^2(\Omega, \mathcal{H}_{\omega})} = \|f\|_{L^2(\mathbb{R})}$. The operator $\mathcal{F}_G$ is unitary so that the inverse map $\mathcal{F}_G^{-1}$ is the adjoint operator $\mathcal{F}_G^*: L^2(\Omega, \mathcal{H}_{\omega}) \rightarrow L^2(\mathbb{R})$ mapping $\tilde{f}$ to f and given by the formal expression

$$f = \mathcal{F}_G^* \tilde{f} = \int_{\Omega} \tilde{f}(\omega)\, d\omega \qquad (20)$$

To give a sense to this formal expression, we need a concrete realization of $L^2(\Omega, \mathcal{H}_{\omega})$ given by the following lemma

Lemma 4 *The Hilbert space $L^2(\Omega \times [0,T))$ is a realization of $L^2(\Omega, \mathcal{H}_{\omega})$*

Proof. From (18), $\tilde{f}_{\omega}(t)$ is known for every $t \in \mathbb{R}$ if and only if it is known for every t in the interval $[0,T)$. The Hilbert spaces identification is done by defining the two-variables complex function $F : \Omega \times [0,T) \rightarrow \mathbb{C}$, and setting $F(\omega, t) = \tilde{f}(\omega)$ with $\int_{\Omega} \left\|\tilde{f}(\omega)\right\|_{\mathcal{H}_{\omega}}^2 d\omega = \int_{\Omega} \int_0^T |F(\omega, t)|^2 dt d\omega$. ∎

Expression (20) is then defined by $f(t) = \left(\mathcal{F}_G^* \tilde{f}\right)(t) = \int_{\Omega} F(\omega, t)\, d\omega$.

Remark 1 *The spectral decomposition of a signal f of $L^2(\mathbb{R})$ into its spectral components $f_{\omega} \in \mathcal{H}_{\omega}$ is peculiar to the action of the symmetry group Γ. Note that this spectral components are time-dependent, so that the transformation (18) is a mixed time/frequency representation of the original time-domain signal. One may have recognized in equation (18), the discrete-time Fourier transform of the sequence $\{f(t+\gamma)\}_{\gamma \in \Gamma}$ and (18) can be reinterpreted as a transform which first chops the continuous-time signal into the set of pieces $\{..., f_{-1}(t), f_0(t), f_1(t), ...\}$ where $f_k(t) = f(t+kT), 0 \leq t < T, k \in \mathbb{Z}$ and then takes the classical Fourier transform of this sequence. This is actually the idea of lifting signals in the framework proposed in [1], [2].*

5 Frequency response of sampled-data systems

The basic idea generalizing the concept of frequency response for sampled-data systems is to synthesize such systems, on the space $L^2(\mathbb{R})$ on which the symmetry group Γ has a unitary representation, from operators on irreducible representation subspaces, i.e., the "harmonics". Returning back to the unitary representations D^ω as given by (14), we can form the continuous sum $\widetilde{D}$ of the representations D^ω (see [6]), that is, the operator-valued function $\widetilde{D} : \gamma \mapsto \widetilde{D}_\gamma$ which maps Γ into $\mathcal{L}(\mathcal{H}^{cont})$ with $\widetilde{D}_\gamma \mid_{\mathcal{H}_\omega} = D^\omega_\gamma$ or explicitly $\left(\widetilde{D}_\gamma \widetilde{f}\right)(\omega) = d_\gamma(\omega).\widetilde{f}(\omega)$ where $d_\gamma(\omega) = e_\omega(\gamma)$. Under the unitary map $\mathcal{F}_G$, it is not difficult to show that the representation $\widetilde{D}$ in $\mathcal{H}^{cont}$ is equivalent to the representation D in $L^2(\mathbb{R})$. The operator $\widetilde{D}_\gamma$ which is a continuous sum of the operators D^ω_γ is said to be decomposable. A bounded operator $\mathcal{A}$ on $\mathcal{H}^{cont} = \int_\Omega \mathcal{H}_\omega d\omega$ is called *decomposable* if there exists a family $\{A(\omega)\}_{\omega \in \Omega}$ of operators on the spaces $\mathcal{H}_\omega$ such that $(\mathcal{A}f)(\omega) = A(\omega) f(\omega)$ almost everywhere on Ω and the operator norm of $\mathcal{A}$ in $\mathcal{L}(\mathcal{H}^{cont})$ is given by

$$\|\mathcal{A}\|_{\mathcal{L}(\mathcal{H}^{cont})} = \underset{\omega \in \Omega}{ess \sup} \|A(\omega)\|_{\mathcal{L}(\mathcal{H}_\omega)} \tag{21}$$

We state the main structure theorem of sampled-data systems.

Theorem 5 *Every bounded sampled-data operator $\mathcal{T}$ on $L^2(\mathbb{R})$ is a decomposable operator.*

Proof. The proof is based on ([6], theorem 1, §4.5, p. 58) and is a direct consequence of the commutation rules (6). The space $L^2(\mathbb{R}) \cong \mathcal{H}^{cont}$ is the continuous sum of $D_\gamma - invariant$ submanifolds $\mathcal{H}_\omega$ and every bounded sampled-data operator $\mathcal{T}$ on $L^2(\mathbb{R})$ commutes with all the operators D_γ. From [6], the operator $\mathcal{T}$ is decomposable, i.e., there exists a family $\{T(\omega)\}_{\omega \in \Omega}$ of operators on the submanifolds $\mathcal{H}_\omega$ such that $(\mathcal{T}f)(\omega) = T(\omega) f(\omega)$ and equation (21) holds for $\mathcal{T}$. ∎

Theorem 5 asserts the existence of "simplest operators" from which $\mathcal{T}$ on $L^2(\mathbb{R})$ is built. These "simplest operators" act on the fibers $\mathcal{H}_\omega$ of $L^2(\mathbb{R})$ as pointwise multiplication operators $T(\omega) : f(\omega) \mapsto T(\omega).f(\omega)$. The "harmonics" of sampled-data systems are the fibers $\mathcal{H}_\omega$ which are the irreducible representation spaces $\mathcal{H}_\omega$ under the unitary representation D of the symmetry group Γ. The isometric isomorphism $\mathcal{F}_G$ is actually a "basis changing" transformation and allows us to view alternatively the operator $\mathcal{T} \in \mathcal{L}(L^2(\mathbb{R}))$ as an element $\widetilde{\mathcal{T}}$ of $\mathcal{L}(L^2(\Omega, \mathcal{H}_\omega))$ given by the identity

$$\widetilde{\mathcal{T}} = \mathcal{F}_G \mathcal{T} \mathcal{F}_G^* \tag{22}$$

Under the identification (22), $\mathcal{T}$ and $\widetilde{\mathcal{T}}$ are in a sense the same operator and their restrictions to the submanifold $\mathcal{H}_\omega$ are identical, i.e., $\mathcal{T} \mid_{\mathcal{H}_\omega} = \widetilde{\mathcal{T}} \mid_{\mathcal{H}_\omega} =$

$T(\omega)$. We call the operator-valued function on Ω

$$\widetilde{T}: \quad \Omega \quad \rightarrow \quad \mathcal{L}(\mathcal{H}_\omega) \tag{23}$$
$$\omega \quad \mapsto \quad T(\omega)$$

the *frequency response operator* of the sampled-data system T. The frequency response of the sampled-data system *at pseudofrequency* ω, written $\widetilde{T}(\omega)$, is the restriction of the operator $\widetilde{T}$ on the invariant submanifold $\mathcal{H}_\omega$, that is, the operator of multiplication $T(\omega)$ which is directly obtained from the original operator T as

$$\widetilde{T}(\omega) = T(\omega) = (\mathcal{F}_G T \mathcal{F}_G^*)|_{\mathcal{H}_\omega} \tag{24}$$

From (21) and (22), the $L^2(\mathbb{R})$ –induced gain of T is precisely the supremum over all ω of the frequency gain function $\|T(\omega)\|_{\mathcal{L}(\mathcal{H}_\omega)}$, that is,

$$\|T\|_{\mathcal{L}(L^2(\mathbb{R}))} = \left\|\widetilde{T}\right\|_{\mathcal{L}(L^2(\Omega,\mathcal{H}_\omega))} = \underset{\omega \in \Omega}{ess\,sup}\, \|T(\omega)\|_{\mathcal{L}(\mathcal{H}_\omega)}$$

For every $x \in L^2(\mathbb{R})$, the effect of the sampled-data operator T is equivalently described by the *pointwise* multiplication operator $\widetilde{T}$ on $L^2(\Omega, \mathcal{H}_\omega)$

$$y(t) = (Tx)(t) \Leftrightarrow \widetilde{y}(\omega) = \left(\widetilde{T}\widetilde{x}\right)(\omega) = \widetilde{T}(\omega)\,\widetilde{x}(\omega)$$

The following algebraic properties holds almost everywhere on Ω

$$\widetilde{(T_1 + T_2)}(\omega) = \widetilde{T_1}(\omega) + \widetilde{T_2}(\omega) \qquad \widetilde{(T_1 T_2)}(\omega) = \widetilde{T_1}(\omega)\,\widetilde{T_2}(\omega)$$
$$\widetilde{(T^*)}(\omega) = \left(\widetilde{T}(\omega)\right)^* \qquad \widetilde{(\lambda T)}(\omega) = \lambda\widetilde{T}(\omega) \ , \ \lambda \in \mathbb{C}$$

and if T is invertible, then $\widetilde{(T^{-1})}(\omega) = \left(\widetilde{T}(\omega)\right)^{-1}$. Hence the unitary transformation $\mathcal{F}_G$ is a homomorphism of the algebra of all bounded linear periodic (and in particuliar of sampled-data) systems on $L^2(\mathbb{R})$ into the algebra of decomposable operators on $\mathcal{H}^{cont}$ with the *usual pointwise* operations.

Remark 2 *Araki and co-workers [4] have defined a concept of frequency response for sampled-data systems as a mapping of a signal space into itself. Their approach is based on the impulse modulation formula and their signal space is similar to the submanifold $\mathcal{H}_\omega$. They do not relate their signal space to $L^2(\mathbb{R})$ but make it isomorphic to the sequence space $l^2(\mathbb{Z})$. The frequency response operator is then defined on $l^2(\mathbb{Z})$ and is expressed as an infinite-dimensional matrix. However, the concept of frequency response as defined in [4] is in fact identical to that defined in this section. Our approach has the advantage to provide an understanding of the structure of sampled-data systems as decomposable operators and to show in a natural way the deep connection between [1], [2], [3] and [4]. Actually these frameworks rely on a simple motivating principle, i.e. a symmetry property, on which they are raised.*

6 Conclusion

This chapter has presented a group theoretic approach to the concept of frequency response of sampled-data systems. Specifically, we have considered sampled-data systems as operators on the space of finite-energy signals defined on the real line where a specific group of transformations acts. Under the action of this group the signal space decomposes into certain infinitesimal invariant submanifolds. The frequency response is defined as the restriction of the sampled-data operator to these invariant submanifolds. This decomposition yields a unitary transformation which is a generalized Fourier transform matched to sampled-data signal and system analysis and its relation with the lifting technique has been pointed out. The approach proposed in this chapter provides a unified framework for the concept of frequency response of sampled-data systems and shows that recently introduced approaches for frequency domain analysis of sampled-data systems follow actually from one mathematical core, i.e. a group structure describing a time domain symmetry.

References

[1] Bamieh B.A., Pearson J.B. 1992 A General Framework for Linear Periodic Systems with Applications to $\mathcal{H}^\infty$ Sampled-Data Control. *IEEE Transactions on Automatic Control*, Vol. 37, N°. 4, pp. 418-435

[2] Yamamoto Y. 1994 A Function Space Approach to Sampled Data Control Systems and Tracking Problems. *IEEE Transactions on Automatic Control*, Vol. 39, N°. 4, pp. 703-713

[3] Yamamoto Y., Khargonekar P.P. 1996 Frequency Response of Sampled-Data Systems. *IEEE Transactions on Automatic Control*, Vol. 41, N°. 2, pp. 166-176

[4] Araki M., Ito Y. and Hagiwara T. 1995 Frequency Response of Sampled-data Systems. *Automatica*, Vol. 32, N°. 4, pp. 483-497

[5] Yamamoto Y., Araki M. 1994 Frequency Responses for Sampled-Data Systems : Their Equivalence and Relationships. *Linear Algebra and Its Applications*, Vol. 205-206, pp. 1319-1339

[6] Kirillov A.A. 1976 *Elements of the Theory of Representations*. Springer-Verlag, berlin

[7] Naimark, M.A. 1972 *Normed Algebras*. Wolters-Noordhoff Publishing, Groningen

[8] Yosida, K. 1995 *Functional Analysis*. Springer-Verlag, Berlin

3

Singular Value Properties of the Discrete-Time LQ Optimal Regulator

K.G. Arvanitis , G. Kalogeropoulos and T.G. Koussiouris

1 Introduction

The stability robustness as well the singular value properties of the continuous-time LQ state feedback regulator have received much attention in the past [1]-[4]. From the so far reported results on the subject, it is well recognized that continuous-time LQ regulators ensure considerably large guaranteed stability margins, i.e. guaranteed gain margin in the interval $\left(\frac{1}{2} , \infty \right)$ and phase margin of at least 60 degrees. Furthermore, the singular values of the closed-loop transfer function of the continuous-time LQ regulator are no greater than the singular values of the open-loop transfer function. Finally, in the case of the output-weighted cost function, the singular values of the closed-loop transfer function of the continuous-time LQ regulator are no greater than the output-weighting parameter.

Stability robustness properties of the discrete-time LQ regulator have been studied in the past in [5], wherein it has been confirmed that the stability margins of the discrete regulator are inferior to those of the continuous-time one. Moreover, in [5], guaranteed stability margins for the discrete LQ regulator, are suggested for some particular cases. Finally, to the authors' best knowledge, an analysis of the singular value properties of the discrete LQ regulator, similar to that reported in [4] for the continuous-time case, has not yet been reported in the literature.

In this respect, in the present paper, a series of useful singular value properties for the state feedback discrete LQ optimal regulator, is established. These properties are obtained by extensively investigating the behavior: (a) of the minimum singular values of the regulator's return difference matrix, which is a fundamental transfer function interwoven with the study of the robustness properties of the regulator, and (b) of the singular values of the regulator's closed-loop transfer function compared to the behavior of the singular values of its open-loop transfer function. Our investigation allows to suggest new lower bounds for the minimum singular value of the regulator's return difference matrix. On the basis of these bounds, we establish new guaranteed stability margins for such a type of LQ regulators. These margins are, in many cases, sharper than the guaranteed stability margins, proposed in the literature. Furthermore, our investigation provides guaranteed stability margins in cases where known techniques fail. The proposed stability margins are obtained on the basis of a fundamental spectral factorization equality, called the Return Difference Equality, and are expressed directly in terms of the elementary cost and system matrices. On the other hand, it is verified in the paper that, in contrast to what happens in the continuous-time case, the singular values of the closed-loop transfer function of the discrete LQ regulator, can be, in general, greater than the singular values of the open-loop transfer function. Moreover, in the case of the output-weighted cost function, the singular values of the closed-loop transfer function of the discrete LQ regulator can be, in general, greater than the output-weighting parameter. This difference between the discrete and the continuous-time regulator, is mainly due to the entanglement of the solution of the discrete Riccati equation to the aforementioned Return Difference Equality. However, relatively simple bounds for the singular values of the closed-loop transfer function of the discrete-time LQ regulator can still be established. These bounds, which relate the singular values of the closed-loop and open-loop transfer functions of the discrete LQ regulator are suggested in the paper, for the first time.

2   Preliminaries

Consider the linear discrete-time, detectable and stabilizable plant of the form
$$\mathbf{x}(k+1) = \mathbf{A}\mathbf{x}(k) + \mathbf{B}\mathbf{u}(k) \ , \ \mathbf{y}(k) = \mathbf{C}\mathbf{x}(k) \tag{1}$$
where $\mathbf{x}(k) \in \mathbf{R}^n$, $\mathbf{u}(k) \in \mathbf{R}^m$, $\mathbf{y}(k) \in \mathbf{R}^p$ and where $\mathbf{A}$, $\mathbf{B}$ and $\mathbf{C}$ have appropriate dimensions. Suppose that system (1) is controlled by the state feedback law
$$\mathbf{u}(k) = -\mathbf{F}\mathbf{x}(k)$$
in order to minimize the following index of performance
$$J = \sum_{k=0}^{\infty} \left\{ \mathbf{y}^{\mathrm{T}}(k)\mathbf{Q}\mathbf{y}(k) + \mathbf{u}^{\mathrm{T}}(k)\mathbf{R}\mathbf{u}(k) \right\}$$
for the nominal plant, where $\mathbf{Q} \in \mathbf{R}^{p \times p}$, $\mathbf{R} \in \mathbf{R}^{m \times m}$ are positive definite matrices.

The solution of the above optimal regulation problem is well known to be [5]
$$\mathbf{F} = \left(\mathbf{R} + \mathbf{B}^{\mathrm{T}}\mathbf{P}\mathbf{B} \right)^{-1} \mathbf{B}^{\mathrm{T}}\mathbf{P}\mathbf{A}$$
where $\mathbf{P}$ is the positive definite solution of the discrete algebraic Riccati equation
$$\mathbf{P} = \mathbf{A}^{\mathrm{T}}\mathbf{P}\mathbf{A} + \mathbf{C}^{\mathrm{T}}\mathbf{Q}\mathbf{C} - \mathbf{A}^{\mathrm{T}}\mathbf{P}\mathbf{B}\left(\mathbf{R} + \mathbf{B}^{\mathrm{T}}\mathbf{P}\mathbf{B} \right)^{-1} \mathbf{B}^{\mathrm{T}}\mathbf{P}\mathbf{A} \tag{2}$$

Let $, \Omega(z) = \mathbf{I} + \mathbf{T}(z) \equiv \mathbf{I} + \mathbf{F}(z\mathbf{I} - \mathbf{A})^{-1}\mathbf{B}$. Matrix $\Omega(z)$ satisfies the following

Return Difference Equality [5]

$$\Omega^T(z^{-1})\left(\mathbf{R} + \mathbf{B}^T\mathbf{PB}\right)\Omega(z) = \mathbf{R} + \mathbf{B}^T\left(z^{-1}\mathbf{I} - \mathbf{A}^T\right)^{-1}\mathbf{C}^T\mathbf{QC}(z\mathbf{I} - \mathbf{A})^{-1}\mathbf{B} \qquad (3)$$

In the sequel, let $\sigma_{max}(\mathbf{M})$ and $\sigma_{min}(\mathbf{M})$ be the maximum and minimum singular values of $\mathbf{M}$. In [5], lower bounds for $\sigma_{min}[\Omega(z)]$ and guaranteed stability margins for the discrete LQ regulator, have been proposed for the cases where $\mathbf{A}$, either is asymptotically stable, with $\sigma_{max}(\mathbf{A}) < 1$, or is asymptotically stable with distinct eigenvalues and $\sigma_{max}(\mathbf{A}) \geq 1$, or finally has distinct eigenvalues, which do not lie on the unit circle, but may be lie outside this circle. However, it seems that, the lower bounds for $\sigma_{min}[\Omega(z)]$, proposed in [5], are quite tight, thus leading to small guaranteed stability margins. Moreover, there are several cases, in which the results of [5] cannot be applied; see for example the case where

$$\mathbf{A} = \begin{bmatrix} 1 & 0 \\ 1 & -1 \end{bmatrix}, \mathbf{B} = \begin{bmatrix} 1 & 1 \\ -1 & 1 \end{bmatrix}, \mathbf{C} = \begin{bmatrix} 0.5 & -0.7 \\ 0.6 & -0.25 \end{bmatrix}, \mathbf{Q} = \begin{bmatrix} 4 & 1 \\ 1 & 4 \end{bmatrix}, \mathbf{R} = \begin{bmatrix} 2 & 0 \\ 0 & 2 \end{bmatrix}.$$

On the other hand, in the case of the continuous-time LQ regulator with $\mathbf{R}=\mathbf{I}$, it has been confirmed in [4], that the singular values of the closed-loop transfer function of the regulator are no greater than the singular values of the open-loop transfer function. Moreover, in the case of the output-weighted cost function, the singular values of the closed-loop transfer function of the regulator are no greater than the output-weighting parameter.

These two properties do not hold for the discrete LQ regulator. As an example, consider the case where $\mathbf{A} = \begin{bmatrix} 0.3 & -0.8 & 0 \\ -0.6 & 0 & 0.1 \\ 0 & -1 & 1 \end{bmatrix}$, $\mathbf{B} = \begin{bmatrix} 2 & 0 \\ 0 & 1 \\ -1 & 2 \end{bmatrix}$, $\mathbf{Q} = \begin{bmatrix} 4 & 0 & 1 \\ 0 & 3 & 0 \\ 1 & 0 & 2 \end{bmatrix}$,

$\mathbf{C} = \mathbf{I}_{3\times3}$, $\mathbf{R} = \mathbf{I}_{2\times2}$. The discrete regulator is $\mathbf{F} = \begin{bmatrix} 0.2465 & -0.4880 & 0.0620 \\ -0.0825 & -0.4860 & 0.3694 \end{bmatrix}$. It is not difficult to check that the minimum singular value of the regulator's closed-loop transfer function is greater than the minimum singular value of the open-loop system. Moreover, observe that $\mathbf{Q} \geq \dfrac{1}{0.7942^2}\mathbf{I}_3$. Then, $\xi = 0.7942$. Note that here,

$\sigma_j(\mathbf{G}_c(z)) > \xi$, for $|z| = 1$ and $\forall j$.

From the previous analysis, it becomes clear that new estimates of the stability margins, as well as a thorough study of the singular value properties of the discrete LQ regulator are necessary. These two tasks are accomplished in sequel.

3   Stability Margins of the Discrete Regulator

We first establish the following two fundamental results.

Theorem 3.1. Suppose that matrix $\hat{\mathbf{R}} = \mathbf{BR}^{-1}\mathbf{B}^T$ is nonsingular. Then, the guaranteed stability margins of the discrete LQ optimal regulator, are given by

$$GM_\beta = (1 \pm \beta_\Omega)^{-1} \quad , \quad PM_\beta = \pm \arccos\left(1 - \frac{\beta_\Omega^2}{2}\right) \tag{4}$$

$$\beta_\Omega^2 = \frac{\sigma_{min}(\mathbf{R})}{\sigma_{max}(\mathbf{R}) + \sigma_{max}^2(\mathbf{B})\vartheta}\left[1 + \frac{\sigma_{min}^2(\mathbf{B})\sigma_{min}^2(\mathbf{C}^T)\sigma_{min}(\mathbf{Q})}{\sigma_{max}(\mathbf{R})[1 + \sigma_{max}(\mathbf{A})]^2}\chi\right]$$

$$\vartheta = \frac{\sigma_{max}^2(\mathbf{A})\sigma_{max}(\mathbf{R})}{\sigma_{min}^2(\mathbf{B})} + \sigma_{max}^2(\mathbf{C})\sigma_{max}(\mathbf{Q})$$

Proof: From the analysis reported in [5], we have

$$\sigma_{min}^2[\Omega(z)] \geq \frac{\sigma_{min}(\mathbf{R})}{\sigma_{max}(\mathbf{R}) + \sigma_{max}^2(\mathbf{B})\sigma_{max}(\mathbf{P})}\left[1 + \frac{\sigma_{min}^2(\mathbf{B})\sigma_{min}^2(\mathbf{C}^T)\sigma_{min}(\mathbf{Q})}{\sigma_{max}(\mathbf{R})[1 + \sigma_{max}(\mathbf{A})]^2}\chi\right] \tag{5}$$

Observe now that equation (2) can also be written as

$$\mathbf{P} = \mathbf{A}^T\mathbf{P}\mathbf{A} + \mathbf{C}^T\mathbf{Q}\mathbf{C} - \mathbf{A}^T\mathbf{P}\mathbf{B}_R\left(\mathbf{I} + \mathbf{B}_R^T\mathbf{P}\mathbf{B}_R\right)^{-1}\mathbf{B}_R^T\mathbf{P}\mathbf{A}$$

where $\mathbf{B}_R = \mathbf{B}\mathbf{R}^{-\frac{1}{2}}$. Suppose now that $\hat{\mathbf{R}}$ is nonsingular. Then, from [6] we obtain

$$\mathbf{P} \leq \mathbf{A}^T\hat{\mathbf{R}}^{-1}\mathbf{A} + \mathbf{C}^T\mathbf{Q}\mathbf{C} \tag{6}$$

Therefore,

$$\sigma_{max}(\mathbf{P}) \leq \sigma_{max}\left(\mathbf{A}^T\hat{\mathbf{R}}^{-1}\mathbf{A} + \mathbf{C}^T\mathbf{Q}\mathbf{C}\right) \leq \sigma_{max}^2(\mathbf{A})\sigma_{max}\left(\hat{\mathbf{R}}^{-1}\right) + \sigma_{max}\left(\mathbf{C}^T\mathbf{Q}\mathbf{C}\right)$$

$$\leq \frac{\sigma_{max}^2(\mathbf{A})}{\sigma_{min}\left(\hat{\mathbf{R}}\right)} + \sigma_{max}^2(\mathbf{C})\sigma_{max}(\mathbf{Q}) = \frac{\sigma_{max}^2(\mathbf{A})}{\sigma_{min}\left(\mathbf{B}\mathbf{R}^{-1}\mathbf{B}^T\right)} + \sigma_{max}^2(\mathbf{C})\sigma_{max}(\mathbf{Q})$$

$$\leq \frac{\sigma_{max}^2(\mathbf{A})}{\sigma_{min}^2(\mathbf{B})\sigma_{min}\left(\mathbf{R}^{-1}\right)} + \sigma_{max}^2(\mathbf{C})\sigma_{max}(\mathbf{Q}) = \frac{\sigma_{max}^2(\mathbf{A})\sigma_{max}(\mathbf{R})}{\sigma_{min}^2(\mathbf{B})} + \sigma_{max}^2(\mathbf{C})\sigma_{max}(\mathbf{Q})$$

and hence $\sigma_{max}(\mathbf{P}) \leq \vartheta$. Then, $\sigma_{min}[\Omega(z)] \geq \beta_\Omega$ and (4) follows. $\square$

Theorem 3.2. Suppose that $\hat{\mathbf{R}}$ is singular and that

$$\mathbf{C}^T\mathbf{Q}\mathbf{C} > 0 \quad \text{and} \quad \sigma_{max}^2(\mathbf{A}) < 1 + \sigma_{min}^2(\mathbf{B}_R)\eta \tag{7a}$$

$$\eta = \lambda_{max}\left[\mathbf{A}^T\left[\left(\mathbf{C}^T\mathbf{Q}\mathbf{C}\right)^{-1} + \hat{\mathbf{R}}\right]^{-1}\mathbf{A} + \mathbf{C}^T\mathbf{Q}\mathbf{C}\right] \tag{7b}$$

Then, the guaranteed stability margins of the discrete LQ regulator are given by

$$GM_\gamma = (1 \pm \gamma_\Omega)^{-1} \quad , \quad PM_\gamma = \pm\arccos\left(1 - \frac{\gamma_\Omega^2}{2}\right) \tag{8}$$

$$\gamma_\Omega^2 = \frac{\sigma_{min}(\mathbf{R})}{\sigma_{max}(\mathbf{R}) + \sigma_{max}^2(\mathbf{B})\mu}\left[1 + \frac{\sigma_{min}^2(\mathbf{B})\sigma_{min}^2(\mathbf{C}^T)\sigma_{min}(\mathbf{Q})}{\sigma_{max}(\mathbf{R})[1 + \sigma_{max}(\mathbf{A})]^2}\chi\right]$$

$$\mu = \frac{\lambda_{\max}\left(\mathbf{C}^T\mathbf{QC}\right)}{1+\sigma_{\min}^2\left(\mathbf{B}_R\right)\eta-\sigma_{\max}^2\left(\mathbf{A}\right)}\sigma_{\max}^2\left(\mathbf{A}\right)+\sigma_{\max}^2\left(\mathbf{C}\right)\sigma_{\max}\left(\mathbf{Q}\right)$$

Proof: If (7a) holds, then $\mathbf{P}$ obeys the following inequality [7]

$$\mathbf{P} \le \frac{\lambda_{\max}\left(\mathbf{C}^T\mathbf{QC}\right)}{1+\sigma_{\min}^2\left(\mathbf{B}_R\right)\eta-\sigma_{\max}^2\left(\mathbf{A}\right)}\mathbf{A}^T\mathbf{A}+\mathbf{C}^T\mathbf{QC} \tag{9}$$

Therefore,

$$\sigma_{\max}\left(\mathbf{P}\right) \le \sigma_{\max}\left(\frac{\lambda_{\max}\left(\mathbf{C}^T\mathbf{QC}\right)}{1+\sigma_{\min}^2\left(\mathbf{B}_R\right)\eta-\sigma_{\max}^2\left(\mathbf{A}\right)}\mathbf{A}^T\mathbf{A}+\mathbf{C}^T\mathbf{QC}\right)$$

$$\le \frac{\lambda_{\max}\left(\mathbf{C}^T\mathbf{QC}\right)}{1+\sigma_{\min}^2\left(\mathbf{B}_R\right)\eta-\sigma_{\max}^2\left(\mathbf{A}\right)}\sigma_{\max}^2\left(\mathbf{A}\right)+\sigma_{\max}^2\left(\mathbf{C}\right)\sigma_{\max}\left(\mathbf{Q}\right)=\mu \tag{10}$$

Taking into account (10), we finally obtain $\sigma_{\min}\left[\Omega(z)\right]\ge\gamma_\Omega$ and (8) follows. $\square$

When $\mathbf{A}$ is asymptotically stable, the following results can be established.

Theorem 3.3. Let $\mathbf{A}$ be asymptotically stable with $\sigma_{\max}\left(\mathbf{A}\right)<1$. Then, the guaranteed stability margins of the discrete LQ regulator are given by

$$\mathrm{GM}_\delta = \left(1\pm\delta_\Omega\right)^{-1}\ ,\quad \mathrm{PM}_\delta = \pm\arccos\left(1-\frac{\delta_\Omega^2}{2}\right) \tag{11}$$

$$\delta_\Omega^2 = \frac{\sigma_{\min}\left(\mathbf{R}\right)}{\sigma_{\max}\left(\mathbf{R}\right)+\sigma_{\max}^2\left(\mathbf{B}\right)\pi}\left[1+\frac{\sigma_{\min}^2\left(\mathbf{B}\right)\sigma_{\min}^2\left(\mathbf{C}^T\right)\sigma_{\min}\left(\mathbf{Q}\right)}{\sigma_{\max}\left(\mathbf{R}\right)\left[1+\sigma_{\max}\left(\mathbf{A}\right)\right]^2}\chi\right]$$

$$\pi = \frac{\lambda_{\max}\left(\mathbf{C}^T\mathbf{QC}\right)}{1-\sigma_{\max}^2\left(\mathbf{A}\right)}\sigma_{\max}^2\left(\mathbf{A}\right)+\sigma_{\max}\left(\mathbf{Q}\right)\sigma_{\max}^2\left(\mathbf{C}\right)$$

Proof: Let $\mathbf{P}_L$ be the positive definite solution of the Lyapunov equation

$$\mathbf{P}_L = \mathbf{A}^T\mathbf{P}_L\mathbf{A}+\mathbf{C}^T\mathbf{QC} \tag{12}$$

If matrix $\mathbf{A}$ is asymptotically stable, then, it is evident that $0\le\mathbf{P}\le\mathbf{P}_L$. Therefore,

$$\sigma_{\max}\left(\mathbf{P}\right)\le\sigma_{\max}\left(\mathbf{P}_L\right) \tag{13}$$

If $\sigma_{\max}\left(\mathbf{A}\right)<1$, then according to the results in [8] we have

$$\mathbf{P}_L \le \frac{\lambda_{\max}\left(\mathbf{C}^T\mathbf{QC}\right)}{1-\sigma_{\max}^2\left(\mathbf{A}\right)}\mathbf{A}^T\mathbf{A}+\mathbf{C}^T\mathbf{QC}$$

Therefore,

$$\sigma_{\max}\left(\mathbf{P}_L\right)\le\sigma_{\max}\left[\frac{\lambda_{\max}\left(\mathbf{C}^T\mathbf{QC}\right)}{1-\sigma_{\max}^2\left(\mathbf{A}\right)}\mathbf{A}^T\mathbf{A}+\mathbf{C}^T\mathbf{QC}\right]$$

$$\le \frac{\lambda_{\max}\left(\mathbf{C}^T\mathbf{QC}\right)}{1-\sigma_{\max}^2\left(\mathbf{A}\right)}\sigma_{\max}\left(\mathbf{A}^T\mathbf{A}\right)+\sigma_{\max}\left(\mathbf{C}^T\mathbf{QC}\right)$$

$$\leq \frac{\lambda_{max}\left(\mathbf{C}^{T}\mathbf{Q}\mathbf{C}\right)}{1-\sigma^2_{max}(\mathbf{A})}\sigma^2_{max}(\mathbf{A})+\sigma_{max}(\mathbf{Q})\sigma^2_{max}(\mathbf{C})=\pi$$

Hence, $\sigma_{min}[\Omega(z)]\geq\delta_\Omega$ and (11) follows. $\qquad\square$

Theorem 3.4. Let **A** be an asymptotically stable matrix with distinct eigenvalues. Then, the guaranteed stability margins of the discrete LQ regulator are given by

$$GM_\varepsilon=\left(1\pm\varepsilon_\Omega\right)^{-1}\ ,\ PM_\varepsilon=\pm\arccos\left(1-\frac{\varepsilon^2_\Omega}{2}\right) \qquad (14)$$

$$\varepsilon^2_\Omega=\frac{\sigma_{min}(\mathbf{R})}{\sigma_{max}(\mathbf{R})+\sigma^2_{max}(\mathbf{B})\phi}\left[1+\frac{\sigma^2_{min}(\mathbf{B})\sigma^2_{min}\left(\mathbf{C}^T\right)\sigma_{min}(\mathbf{Q})}{\sigma_{max}(\mathbf{R})\left[1+\sigma_{max}(\mathbf{A})\right]^2}\chi\right]$$

$$\phi=\frac{\lambda_{max}\left(\mathbf{M}^T\mathbf{C}^T\mathbf{Q}\mathbf{C}\mathbf{M}\right)}{1-v^2_{max}(\mathbf{A})}\sigma^{-2}_{min}(\mathbf{M}),\ \mathbf{M}^{-1}\mathbf{A}\mathbf{M}=\mathbf{L},\ \mathbf{L}=\mathrm{diag}\{\lambda_i(\mathbf{A})\},\ i=1,...,n \ (15)$$

Proof: If **A** is asymptotically stable and diagonalizable, then according to [9] we have

$$\mathbf{P}_L\leq\lambda_{max}\left(\mathbf{M}^T\mathbf{C}^T\mathbf{Q}\mathbf{C}\mathbf{M}\right)\left(\mathbf{M}^{-1}\right)^T\ \underset{i=1,2,...,n}{\mathrm{diag}}\left\{\frac{1}{1-\left|\lambda_i(\mathbf{A})\right|^2}\right\}\mathbf{M}^{-1}$$

Therefore,

$$\sigma_{max}\left(\mathbf{P}_L\right)\leq\sigma_{max}\left[\lambda_{max}\left(\mathbf{M}^T\mathbf{C}^T\mathbf{Q}\mathbf{C}\mathbf{M}\right)\left(\mathbf{M}^{-1}\right)^T\ \underset{i=1,2,...,n}{\mathrm{diag}}\left\{\frac{1}{1-\left|\lambda_i(\mathbf{A})\right|^2}\right\}\mathbf{M}^{-1}\right]$$

$$=\lambda_{max}\left(\mathbf{M}^T\mathbf{C}^T\mathbf{Q}\mathbf{C}\mathbf{M}\right)\sigma_{max}\left[\left(\mathbf{M}^{-1}\right)^T\ \underset{i=1,2,...,n}{\mathrm{diag}}\left\{\frac{1}{1-\left|\lambda_i(\mathbf{A})\right|^2}\right\}\mathbf{M}^{-1}\right]$$

$$\leq\lambda_{max}\left(\mathbf{M}^T\mathbf{C}^T\mathbf{Q}\mathbf{C}\mathbf{M}\right)\sigma^2_{max}\left(\mathbf{M}^{-1}\right)\sigma_{max}\left[\underset{i=1,2,...,n}{\mathrm{diag}}\left\{\frac{1}{1-\left|\lambda_i(\mathbf{A})\right|^2}\right\}\right]$$

$$=\frac{\lambda_{max}\left(\mathbf{M}^T\mathbf{C}^T\mathbf{Q}\mathbf{C}\mathbf{M}\right)}{1-v^2_{max}(\mathbf{A})}\sigma^{-2}_{min}(\mathbf{M})=\phi \qquad (16)$$

Combining (13) and (16) yields

$$\sigma_{max}(\mathbf{P})\leq\phi \qquad (17)$$

Combining (5) and (17), readily yields (14). $\qquad\square$

Theorem 3.5. Let **A** be asymptotically stable and normal. Then, the guaranteed stability margins of the discrete LQ regulator are given by

$$GM_\zeta=\left(1\pm\zeta_\Omega\right)^{-1}\ ,\ PM_\zeta=\pm\arccos\left(1-\frac{\zeta^2_\Omega}{2}\right) \qquad (18)$$

$$\zeta_{S\Omega}^2 = \frac{\sigma_{min}(\mathbf{R})}{\sigma_{max}(\mathbf{R}) + \sigma_{max}^2(\mathbf{B})\psi}\left[1 + \frac{\sigma_{min}^2(\mathbf{B})\sigma_{min}^2(\mathbf{C}^T)\sigma_{min}(\mathbf{Q})}{\sigma_{max}(\mathbf{R})[1 + \sigma_{max}(\mathbf{A})]^2}\chi\right]$$

$$\psi = \frac{\lambda_{max}(\mathbf{C}^T\mathbf{QC})}{\sigma_{min}(\mathbf{I} - \mathbf{A}^T\mathbf{A})}$$

Proof: If $\mathbf{A}$ is asymptotically stable and normal, then we have [9]

$$\mathbf{P_L} \le \lambda_{max}(\mathbf{C}^T\mathbf{QC})(\mathbf{I} - \mathbf{A}^T\mathbf{A})^{-1}$$

Therefore,

$$\sigma_{max}(\mathbf{P_L}) \le \sigma_{max}\left[\lambda_{max}(\mathbf{C}^T\mathbf{QC})(\mathbf{I} - \mathbf{A}^T\mathbf{A})^{-1}\right] = \lambda_{max}(\mathbf{C}^T\mathbf{QC})\sigma_{max}\left[(\mathbf{I} - \mathbf{A}^T\mathbf{A})^{-1}\right]$$

$$= \frac{\lambda_{max}(\mathbf{C}^T\mathbf{QC})}{\sigma_{min}(\mathbf{I} - \mathbf{A}^T\mathbf{A})} = \psi \tag{19}$$

Combining (13) and (19) yields

$$\sigma_{max}(\mathbf{P}) \le \psi \tag{20}$$

Combining (5) and (20), readily yields (18). $\qquad\square$

4 Singular Value Properties of the Discrete Regulator

In the sequel, let $\mathbf{R}=\mathbf{I}$, $\mathbf{G}_o(z) = \mathbf{C}(z\mathbf{I} - \mathbf{A})^{-1}\mathbf{B}$, $\mathbf{G}_c(z) = \mathbf{C}(z\mathbf{I} - \mathbf{A} + \mathbf{BF})^{-1}\mathbf{B}$.
Then, the following results, can be established.

Theorem 4.1. Suppose that matrix $\mathbf{BB}^T$ is nonsingular. Also, let

$$\varepsilon_1 = 1 + \left\{\lambda_{max}\left[\mathbf{A}^T(\mathbf{BB}^T)^{-1}\mathbf{A}\right] + \lambda_{max}(\mathbf{Q})\sigma_{max}^2(\mathbf{C})\right\}^{1/2}\sigma_{max}(\mathbf{B})$$

Then, for any $\mathbf{Q} \ge 0$ and for $|z| = 1$

$$\frac{\sigma_j(\mathbf{G}_c(z))}{\sigma_j(\mathbf{G}_o(z))} \le \varepsilon_1 \ , \ \forall j$$

Proof: Let $S(z) = \Omega^{-1}(z)$ be the sensitivity function. As it can be easily shown, the following relation holds

$$\mathbf{G}_c(z) = \mathbf{G}_o(z)S(z) \tag{21}$$

On the other hand from (3) and for $\mathbf{R}=\mathbf{I}$ we obtain

$$\Omega^T(z^{-1})(\mathbf{I} + \mathbf{B}^T\mathbf{PB})\Omega(z) = \mathbf{I} + \mathbf{G}_o^T(z^{-1})\mathbf{Q}\mathbf{G}_o(z) \tag{22}$$

Multiplying (22) left and right by $S^T(z^{-1})$ and $S(z)$, respectively, yields

$$\mathbf{I} + \mathbf{B}^T\mathbf{PB} = S^T(z^{-1})S(z) + S^T(z^{-1})\mathbf{G}_o^T(z^{-1})\mathbf{Q}\mathbf{G}_o(z)S(z) \tag{23}$$

Since $\mathbf{Q} \ge 0$, then $S^T(z^{-1})\mathbf{G}_o^T(z^{-1})\mathbf{Q}\mathbf{G}_o(z)S(z) \ge 0$, for $|z| = 1$. Hence, for $|z| = 1$,

$$\mathbf{S}^{\mathrm{T}}(z^{-1})\mathbf{S}(z) \leq \mathbf{I} + \mathbf{B}^{\mathrm{T}}\mathbf{P}\mathbf{B} = \begin{bmatrix} \mathbf{I} & \mathbf{B}^{\mathrm{T}}\mathbf{P}^{\frac{1}{2}} \end{bmatrix}\begin{bmatrix} \mathbf{I} \\ \mathbf{P}^{\frac{1}{2}}\mathbf{B} \end{bmatrix} \text{ or } \sigma_{\max}(\mathbf{S}(z)) \leq \sigma_{\max}\left(\begin{bmatrix} \mathbf{I} \\ \mathbf{P}^{\frac{1}{2}}\mathbf{B} \end{bmatrix}\right) \quad (24)$$

Observe now that

$$\sigma_{\max}\left(\begin{bmatrix} \mathbf{I} \\ \mathbf{P}^{\frac{1}{2}}\mathbf{B} \end{bmatrix}\right) \leq 1 + \sigma_{\max}\left(\mathbf{P}^{\frac{1}{2}}\mathbf{B}\right) \leq 1 + \sigma_{\max}\left(\mathbf{P}^{\frac{1}{2}}\right)\sigma_{\max}(\mathbf{B}) = 1 + \lambda_{\max}^{\frac{1}{2}}(\mathbf{P})\sigma_{\max}(\mathbf{B}) \quad (25)$$

Combining (24) and (25) yields

$$\sigma_{\max}(\mathbf{S}(z)) \leq 1 + \lambda_{\max}^{\frac{1}{2}}(\mathbf{P})\sigma_{\max}(\mathbf{B}) \quad , \text{ for } |z| = 1 \quad (26)$$

For $\mathbf{R}=\mathbf{I}$, from (6) we obtain

$$\lambda_{\max}(\mathbf{P}) \leq \lambda_{\max}\left[\mathbf{A}^{\mathrm{T}}(\mathbf{B}\mathbf{B}^{\mathrm{T}})^{-1}\mathbf{A} + \mathbf{C}^{\mathrm{T}}\mathbf{Q}\mathbf{C}\right]$$

From Weyl's inequality, we have

$$\lambda_{\max}(\mathbf{P}) \leq \lambda_{\max}\left[\mathbf{A}^{\mathrm{T}}(\mathbf{B}\mathbf{B}^{\mathrm{T}})^{-1}\mathbf{A}\right] + \lambda_{\max}\left(\mathbf{C}^{\mathrm{T}}\mathbf{Q}\mathbf{C}\right)$$

$$= \lambda_{\max}\left[\mathbf{A}^{\mathrm{T}}(\mathbf{B}\mathbf{B}^{\mathrm{T}})^{-1}\mathbf{A}\right] + \sigma_{\max}^2\left(\mathbf{Q}^{1/2}\mathbf{C}\right)$$

$$\leq \lambda_{\max}\left[\mathbf{A}^{\mathrm{T}}(\mathbf{B}\mathbf{B}^{\mathrm{T}})^{-1}\mathbf{A}\right] + \sigma_{\max}^2\left(\mathbf{Q}^{1/2}\right)\sigma_{\max}^2(\mathbf{C})$$

$$= \lambda_{\max}\left[\mathbf{A}^{\mathrm{T}}(\mathbf{B}\mathbf{B}^{\mathrm{T}})^{-1}\mathbf{A}\right] + \lambda_{\max}(\mathbf{Q})\sigma_{\max}^2(\mathbf{C}) \quad (27)$$

Introducing (27) into (26), yields

$$\sigma_{\max}(\mathbf{S}(z)) \leq 1 + \left\{\lambda_{\max}\left[\mathbf{A}^{\mathrm{T}}(\mathbf{B}\mathbf{B}^{\mathrm{T}})^{-1}\mathbf{A}\right] + \lambda_{\max}(\mathbf{Q})\sigma_{\max}^2(\mathbf{C})\right\}^{1/2}\sigma_{\max}(\mathbf{B}) = \varepsilon_1$$

which leads to

$$\mathbf{S}(z)\mathbf{S}^{\mathrm{T}}(z^{-1}) \leq \varepsilon_1^2\mathbf{I} \quad , \text{ for } |z| = 1 \quad (28)$$

Multiplying (28) left and right by $\mathbf{G}_o(z)$ and $\mathbf{G}_o^{\mathrm{T}}(z^{-1})$ (for $|z| = 1$), respectively and taking into account (21), we have

$$\mathbf{G}_c(z)\mathbf{G}_c^{\mathrm{T}}(z^{-1}) \leq \varepsilon_1^2\mathbf{G}_o(z)\mathbf{G}_o^{\mathrm{T}}(z^{-1}) \quad , \text{ for } |z| = 1$$

and hence, since $\mathbf{G}_c(z)\mathbf{G}_c^{\mathrm{T}}(z^{-1}) \geq 0$ and $\mathbf{G}_o(z)\mathbf{G}_o^{\mathrm{T}}(z^{-1}) \geq 0$, provided that the eigenvalues are in descending order, we obtain

$$\lambda_j\left(\mathbf{G}_c(z)\mathbf{G}_c^{\mathrm{T}}(z^{-1})\right) \leq \lambda_j\left(\varepsilon_1^2\mathbf{G}_o(z)\mathbf{G}_o^{\mathrm{T}}(z^{-1})\right) \quad , \quad \forall j \text{ and for } |z| = 1$$

or equivalently

$$\sigma_j\left(\mathbf{G}_c(z)\right) \leq \varepsilon_1\sigma_j\left(\mathbf{G}_o(z)\right) \quad , \quad \forall j \text{ and for } |z| = 1 \qquad \square$$

Proposition 4.1. Suppose that $\mathbf{C}^{\mathrm{T}}\mathbf{Q}\mathbf{C} > 0$ and $\sigma_{\max}^2(\mathbf{A}) < 1 + \sigma_{\min}^2(\mathbf{B})\eta_1$, with

$$\eta_1 = \lambda_{\max}\left[\mathbf{A}^{\mathrm{T}}\left[(\mathbf{C}^{\mathrm{T}}\mathbf{Q}\mathbf{C})^{-1} + \mathbf{B}\mathbf{B}^{\mathrm{T}}\right]^{-1}\mathbf{A} + \mathbf{C}^{\mathrm{T}}\mathbf{Q}\mathbf{C}\right]$$

Also let

$$\varepsilon_2 = 1 + \left[\frac{1 + \sigma_{min}^2(\mathbf{B})\eta_1}{1 + \sigma_{min}^2(\mathbf{B})\eta_1 - \sigma_{max}^2(\mathbf{A})} \lambda_{max}(\mathbf{Q}) \right]^{1/2} \sigma_{max}(\mathbf{C}) \, \sigma_{max}(\mathbf{B})$$

Then, for any $\mathbf{Q} \geq 0$ and for $|z| = 1$

$$\frac{\sigma_j(\mathbf{G}_c(z))}{\sigma_j(\mathbf{G}_o(z))} \leq \varepsilon_2 \quad , \quad \forall j \tag{29}$$

Proof: From (9), and for $\mathbf{R} = \mathbf{I}$, we have

$$\lambda_{max}(\mathbf{P}) \leq \lambda_{max} \left[\frac{\lambda_{max}(\mathbf{C}^T\mathbf{Q}\mathbf{C})}{1 + \sigma_{min}^2(\mathbf{B})\eta_1 - \sigma_{max}^2(\mathbf{A})} \mathbf{A}^T\mathbf{A} + \mathbf{C}^T\mathbf{Q}\mathbf{C} \right]$$

$$\leq \left[1 + \frac{\sigma_{max}^2(\mathbf{A})}{1 + \sigma_{min}^2(\mathbf{B})\eta_1 - \sigma_{max}^2(\mathbf{A})} \right] \lambda_{max}(\mathbf{C}^T\mathbf{Q}\mathbf{C})$$

$$\leq \frac{1 + \sigma_{min}^2(\mathbf{B})\eta_1}{1 + \sigma_{min}^2(\mathbf{B})\eta_1 - \sigma_{max}^2(\mathbf{A})} \lambda_{max}(\mathbf{Q})\sigma_{max}^2(\mathbf{C}) \tag{30}$$

Introducing (30) to (26) we readily obtain

$$\sigma_{max}(\mathbf{S}(z)) \leq 1 + \left[\frac{1 + \sigma_{min}^2(\mathbf{B})\eta_1}{1 + \sigma_{min}^2(\mathbf{B})\eta_1 - \sigma_{max}^2(\mathbf{A})} \lambda_{max}(\mathbf{Q}) \right]^{1/2} \sigma_{max}(\mathbf{C}) \, \sigma_{max}(\mathbf{B}) = \varepsilon_2$$

for $|z| = 1$, which, according to the proof of Theorem 4.1, leads to (29). □

Proposition 4.2. Let $\mathbf{A}$ be asymptotically stable with $\sigma_{max}(\mathbf{A}) < 1$. Define

$$\varepsilon_3 = 1 + \left[\frac{\lambda_{max}(\mathbf{Q})}{1 - \sigma_{max}^2(\mathbf{A})} \right]^{1/2} \sigma_{max}(\mathbf{C})\sigma_{max}(\mathbf{B})$$

Then, for any $\mathbf{Q} \geq 0$ and for $|z| = 1$

$$\frac{\sigma_j(\mathbf{G}_c(z))}{\sigma_j(\mathbf{G}_o(z))} \leq \varepsilon_3 \quad , \quad \forall j \tag{31}$$

Proof: As already mentioned $0 \leq \mathbf{P} \leq \mathbf{P}_L$ where, $\mathbf{P}_L$ satisfies (12). Therefore,

$$\lambda_{max}(\mathbf{P}) \leq \lambda_{max}(\mathbf{P}_L) \tag{32}$$

If $\sigma_{max}(\mathbf{A}) < 1$, then according to the results in [8] we have

$$\lambda_{max}(\mathbf{P}_L) \leq \frac{\lambda_{max}(\mathbf{C}^T\mathbf{Q}\mathbf{C})}{1 - \sigma_{max}^2(\mathbf{A})} \leq \frac{\lambda_{max}(\mathbf{Q})\sigma_{max}^2(\mathbf{C})}{1 - \sigma_{max}^2(\mathbf{A})} \tag{33}$$

Combining (32) and (33) yields

$$\lambda_{max}(\mathbf{P}) \leq \frac{\lambda_{max}(\mathbf{Q})\sigma_{max}^2(\mathbf{C})}{1 - \sigma_{max}^2(\mathbf{A})} \tag{34}$$

Introducing (34) into (26), yields

$$\sigma_{max}(S(z)) \le 1 + \left[\frac{\lambda_{max}(Q)}{1 - \sigma_{max}^2(A)}\right]^{1/2} \sigma_{max}(C)\sigma_{max}(B) = \varepsilon_3$$

which, according to the proof of Theorem 4.1, leads to (31).   □

Proposition 4.3. Let A be asymptotically stable with distinct eigenvalues, M be defined by (15) and

$$\varepsilon_4 = 1 + \left[\frac{\lambda_{max}(Q)}{1 - v_{max}^2(A)}\right]^{1/2} \frac{\sigma_{max}(M)}{\sigma_{min}(M)} \sigma_{max}(C)\sigma_{max}(B)$$

Then, for any $Q \ge 0$ and for $|z| = 1$

$$\frac{\sigma_j(G_c(z))}{\sigma_j(G_o(z))} \le \varepsilon_4 \quad , \quad \forall j \tag{35}$$

Proof: If A is an asymptotically stable and diagonalizable matrix, then according to the results reported in [9] we obtain

$$\lambda_{max}(P_L) \le \lambda_{max}\left(\lambda_{max}(M^T C^T QCM)(M^{-1})^T \underset{i=1,2,...,n}{\text{diag}}\left\{\frac{1}{1 - |\lambda_i(A)|^2}\right\}M^{-1}\right)$$

$$= \lambda_{max}(M^T C^T QCM)\lambda_{max}\left((M^{-1})^T \underset{i=1,2,...,n}{\text{diag}}\left\{\frac{1}{1 - |\lambda_i(A)|^2}\right\}M^{-1}\right)$$

$$= \sigma_{max}^2(Q^{1/2}CM)\sigma_{max}^2\left(\left[\underset{i=1,2,...,n}{\text{diag}}\left\{\frac{1}{1 - |\lambda_i(A)|^2}\right\}\right]^{1/2}M^{-1}\right)$$

$$\le \lambda_{max}(Q)\sigma_{max}^2(C)\sigma_{max}^2(M)\lambda_{max}\left(\underset{i=1,2,...,n}{\text{diag}}\left\{\frac{1}{1 - |\lambda_i(A)|^2}\right\}\right)\sigma_{max}^2(M^{-1})$$

$$= \frac{\lambda_{max}(Q)\sigma_{max}^2(C)\sigma_{max}^2(M)}{\left[1 - |\lambda_{max}(A)|^2\right]\sigma_{min}^2(M)} \tag{36}$$

Combining (32) and (36) yields

$$\lambda_{max}(P) \le \frac{\lambda_{max}(Q)\sigma_{max}^2(C)\sigma_{max}^2(M)}{\left[1 - v_{max}^2(A)\right]\sigma_{min}^2(M)} \tag{37}$$

Introducing (37) into (26) yields

$$\sigma_{max}(S(z)) \le 1 + \left[\frac{\lambda_{max}(Q)}{1 - v_{max}^2(A)}\right]^{1/2} \frac{\sigma_{max}(M)}{\sigma_{min}(M)} \sigma_{max}(C)\sigma_{max}(B) = \varepsilon_4$$

which, according to the proof of Theorem 4.1, leads to (35).   □

Proposition 4.4. Let A be an asymptotically stable and normal matrix and define

$$\varepsilon_5 = 1 + \left[\frac{\lambda_{max}(\mathbf{Q})}{\lambda_{min}(\mathbf{I} - \mathbf{A}^T\mathbf{A})} \right]^{1/2} \sigma_{max}(\mathbf{C})\sigma_{max}(\mathbf{B})$$

Then, for any $\mathbf{Q} \geq 0$ and for $|z| = 1$

$$\frac{\sigma_j(\mathbf{G}_c(z))}{\sigma_j(\mathbf{G}_o(z))} \leq \varepsilon_5 \quad , \quad \forall j \tag{38}$$

Proof: If $\mathbf{A}$ is asymptotically stable and normal, then we have [9]

$$\lambda_{max}(\mathbf{P}_L) \leq \lambda_{max}(\mathbf{C}^T\mathbf{Q}\mathbf{C})\lambda_{max}\left((\mathbf{I} - \mathbf{A}^T\mathbf{A})^{-1}\right) \leq \frac{\lambda_{max}(\mathbf{Q})\sigma_{max}^2(\mathbf{C})}{\lambda_{min}(\mathbf{I} - \mathbf{A}^T\mathbf{A})} \tag{39}$$

Combining (32) and (39) yields

$$\lambda_{max}(\mathbf{P}) \leq \frac{\lambda_{max}(\mathbf{Q})\sigma_{max}^2(\mathbf{C})}{\lambda_{min}(\mathbf{I} - \mathbf{A}^T\mathbf{A})} \tag{40}$$

Introducing (40) into (26) yields

$$\sigma_{max}(\mathbf{S}(z)) \leq 1 + \left[\frac{\lambda_{max}(\mathbf{Q})}{\lambda_{min}(\mathbf{I} - \mathbf{A}^T\mathbf{A})} \right]^{1/2} \sigma_{max}(\mathbf{C})\sigma_{max}(\mathbf{B}) = \varepsilon_5$$

which, according to the proof of Theorem 4.1, leads to (38). □

In the case where $\mathbf{Q} \geq \frac{1}{\xi^2}\mathbf{I}$, $\xi > 0$, the following result can be obtained.

Theorem 4.2. Suppose that $\mathbf{B}\mathbf{B}^T$ is nonsingular. Then, for $\mathbf{Q} \geq \frac{1}{\xi^2}\mathbf{I}$, $\xi > 0$

$$\sigma_j(\mathbf{G}_c(z)) \leq \xi\varepsilon_1 \quad , \quad \forall j \text{ and for } |z| = 1 \tag{41}$$

Proof: If $\mathbf{Q} \geq \frac{1}{\xi^2}\mathbf{I}$, relation (23) implies that

$$\mathbf{I} + \mathbf{B}^T\mathbf{P}\mathbf{B} \geq \mathbf{S}^T(z^{-1})\mathbf{S}(z) + \frac{1}{\xi^2}\mathbf{S}^T(z^{-1})\mathbf{G}_o^T(z^{-1})\mathbf{G}_o(z)\mathbf{S}(z)$$

$$= \mathbf{S}^T(z^{-1})\mathbf{S}(z) + \frac{1}{\xi^2}\mathbf{G}_c^T(z^{-1})\mathbf{G}_c(z)$$

or $\xi^2(\mathbf{I} + \mathbf{B}^T\mathbf{P}\mathbf{B}) \geq \xi^2\mathbf{S}^T(z^{-1})\mathbf{S}(z) + \mathbf{G}_c^T(z^{-1})\mathbf{G}_c(z)$. Considering $\mathbf{S}^T(z^{-1})\mathbf{S}(z) \geq 0$, we obtain

$$\mathbf{G}_c^T(z^{-1})\mathbf{G}_c(z) \leq \xi^2(\mathbf{I} + \mathbf{B}^T\mathbf{P}\mathbf{B}) = \xi\begin{bmatrix} \mathbf{I} & \mathbf{B}^T\mathbf{P}^{1/2} \end{bmatrix}\begin{bmatrix} \mathbf{I} \\ \mathbf{P}^{1/2}\mathbf{B} \end{bmatrix}\xi$$

which, on the basis of (25), leads to

$$\sigma_{max}(\mathbf{G}_c(z)) \leq \xi\left[1 + \lambda_{max}^{1/2}(\mathbf{P})\sigma_{max}(\mathbf{B})\right] \quad , \quad \text{for } |z| = 1$$

If $\mathbf{B}\mathbf{B}^T$ is nonsingular, then from the proof of Theorem 4.1, we finally obtain

$$\sigma_{max}(\mathbf{G}_c(z)) \leq \xi\varepsilon_1 \quad , \quad \text{for } |z| = 1$$

which leads to

$$\mathbf{G}_c(z)\mathbf{G}_c^T\left(z^{-1}\right) \leq \xi^2 \varepsilon_1^2 \mathbf{I}$$

Provided the eigenvalues are in descending order, we obtain

$$\lambda_j\left(\mathbf{G}_c(z)\mathbf{G}_c^T\left(z^{-1}\right)\right) \leq \xi^2 \varepsilon_1^2 \quad, \quad \forall j \text{ and for } |z| = 1$$

or equivalently (41). □

Alternative upper bounds for $\sigma_j\left(\mathbf{G}_c(z)\right)$, whose derivation is left to the reader, can be easily obtained by taking into account the proof of Theorem 4.2 and the upper bounds for $\lambda_{max}(\mathbf{P})$, given in Propositions 4.1-4.4.

5   Conclusions

In this paper, useful singular value properties for the discrete LQ regulator have been proposed. On the basis of these properties, new guaranteed stability margins for such a type of LQ regulators have been established. Moreover, new results relating the singular values of the closed-loop and the open-loop transfer functions of the discrete LQ regulator, have also been proposed.

References

[1]. Safonov M.G., Athans M. 1977 Gain and phase margin for multiloop LQG regulators. *IEEE Trans. Autom. Control* AC-22, 173-179.

[2]. Lehtomaki N.A., Sandell N.R., Athans M. 1981 Robustness results in linear quadratic Gaussian based multivariable control designs. *IEEE Trans. Autom. Control* AC-26, 75-93.

[3]. Soroka E., Shaked U. 1984 Stability margins of LQ regulators. *IEEE Trans. Autom. Control* AC-29, 664-665.

[4]. Safonov M.G., Wang W. 1992 Singular value properties of LQ regulators. *IEEE Trans. Autom. Control* AC-37, 1210-1211.

[5]. Shaked U. 1986 Guaranteed stability margins for the discrete-time linear quadratic optimal regulator. *IEEE Trans. Autom. Control* AC-31, 162-165.

[6]. Komaroff N. 1994 Iterative matrix bounds and computational solutions to the discrete algebraic Riccati equation. *IEEE Trans. Autom. Control* AC-39, 1676-1678.

[7]. Lee C.-H. 1997 Upper matrix bound of the solution for the discrete Riccati equation. *IEEE Trans. Autom. Control* AC-42, 840-842.

[8]. Lee C.-H., Kung F.-C. 1997 Upper and lower matrix bounds of the solutions for the continuous and discrete Lyapunov equations. *J. Franklin Inst.* 334B, 539-546.

4

Identification of Non-Minimum Phase Finite Impulse Response Systems Using the Fourth-Order Cumulants

K. Abderrahim, R. Ben Abdennour, F. Msahli, M. Ksouri and G. Favier

1 Introduction

The NMP FIR system parameters estimation based on higher-order statistics, has received great attention over the ten last years. In fact, several methods using cumulant statistics for the identification of NMP FIR systems have been developed by many authors [1-20]. These methods can be classified as closed-form solutions, linear algebra solutions and nonlinear optimisation solutions [14]. Only the linear algebra solutions are considered in this chapter because they yield consistent estimates and their computationally complexity are not expensive [14]. They consist of constructing overdetermined systems of equations relating correlation with either third or fourth order cumulants to the impulse response coefficients and resolving these systems by using the least squares approaches. The performance of these methods degrades when the data are contaminated by white or colored Gaussian noise because the correlation is not blind to it [14]. Moreover, most of these methods propose a redundant vector of unknown parameters. In fact, the obtained systems are nonlinear, hence requiring nonlinear algorithms to be solved. To overcome these problems, we propose in this chapter linear algebra methods that can be used to identify the parameters of NMP FIR systems using higher-order statistics. The developed methods use only the fourth-order cumulants of the output process, thereby giving consistent estimates in the presence of colored noise.

The rest of the chapter is organised as follows: In section 2, the model and its assumptions are presented. In section 3, least squares methods are developed that allow us to estimate the parameters of NMP FIR systems. Some simulations results showing the performance of the proposed methods are presented in section 4. This chapter is finally concluded in section 5.

2 Model and assumptions

We consider a Single-Input Single-Output LTI FIR system depicted in figure 1.

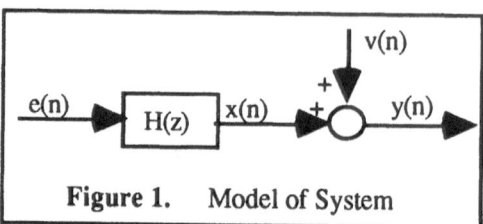

Figure 1.   Model of System

where $\{x(n)\}$ is the true output sequence, $h(n)$ is the impulse response of the FIR LTI system, $\{e(n)\}$ is the input sequence, $\{v(n)\}$ is the noise sequence and $\{y(n)\}$ is the observable output sequence.

We associate to the figure 1. the following relations:

$$x(n) = \sum_{i=0}^{q} h(i).e(n-i) \tag{1}$$

$$y(n) = x(n) + v(n) \tag{2}$$

The following conditions are assumed to be hold:

   C1. The driving noise sequence $\{e(n)\}$ is zero mean independent and identically distributed (i.i.d), non-Gaussian and its mth order cumulant given by:

$$C_{me}(\tau_1,\ldots,\tau_{m-1}) = \begin{cases} \gamma_{m,e} & \text{if } \tau_1 =\ldots= \tau_{m-1} = 0 \\ 0 & \text{otherwise} \end{cases}$$

   C2. The measurement noise sequence $\{v(n)\}$ is assumed to be zero mean, i.i.d Gaussian sequence and independent of $\{e(n)\}$ with unknown variance.

   C3. The order q of the model is known.

   C4. The transfer function $H(z)$ is stable and causal and $h(0)=1$.

3 The proposed methods

The developed methods are based on the following equation which relates different slices of same order cumulants [18]:

$$\sum_{j=0}^{q} h(j)\left[\prod_{k=1}^{s} h(j+\alpha_k)\right] C_{mx}(\tau_1,\tau_2,\ldots,\tau_s,j+\beta_{s+1},\ldots,j+\beta_{m-1}) =$$

$$\sum_{i=0}^{q} h(i)\left[\prod_{k=1}^{s} h(i+\tau_k)\right] C_{mx}(\alpha_1,\alpha_2,\ldots,\alpha_s,i+\beta_{s+1},\ldots,i+\beta_{m-1}) \tag{3}$$

where $1 \le s \le m-1$ (for proof see [18])

For m=4 and s=2, we have:

$$\sum_{j=0}^{q} h(j)h(j+\alpha_1)h(j+\alpha_2)C_{4x}(\tau_1,\tau_2,j+\beta_3) =$$

$$\sum_{i=0}^{q} h(i)h(i+\tau_1)h(i+\tau_2)C_{4x}(\alpha_1,\alpha_2,i+\beta_3)$$

(4)

If we let $\alpha_1 = \alpha_2 = 0$ and $\tau_1 = q$ in (4), we obtain:

$$\sum_{j=1}^{q} h^3(j)\, C_{4x}(q,\tau_2,j+\beta_3) -$$

$$h(\tau_2)h(q)\, C_{4x}(0,0,\beta_3) = -C_{4x}(q,\tau_2,\beta_3)$$

(5)

The relation (5) is treated as a system of equations of (2q+1) unknowns: $\{h(i)h(q)\}_{i=0,\dots,q}$ and $\{h^3(i)\}_{i=1,\dots,q}$. The solution can be obtained either by using a closed-form solution or by solving a linear system of equations. From these estimates, we can compute the values of $\{h(i)\}_{i=1,\dots,q}$.

3.1. Closed-form recursive solution

In this section, we show that the unknown parameters $\{h(i)h(q)\}_{i=0,\dots,q}$ and $\{h^3(i)\}_{i=1,\dots,q}$ can be computed from (5) using a closed-form recursive algorithm.

Taking (5) for $\beta_3 = q$, we have:

$$h(\tau_2)h(q) = \frac{C_{4x}(q,\tau_2,q)}{C_{4x}(q,0,0)}$$

(6)

by setting $\tau_2 = 0, 1,\dots,q$, we can compute the values $\{h(i)h(q)\}_{i=0,\dots,q}$.

Again, for $\tau_2 = 0$, if we take $\beta_3 = q - j$ for j=1...,q, we have the following recursive equation:

$$h^3(j) = \frac{h(q).C_{4x}(0,0,q-j) - \sum_{i=0}^{j-1} h^3(i).C_{4x}(q,0,q-j+i)}{C_{4x}(q,0,q)}$$

(7)

$h(0)h(q)$ is computed from (6).

3.2. Least squares solution

A least squares method is used here to determine the unknown parameters of the relation (5): $\{h(i)h(q)\}_{i=0,\dots,q}$ and $\{h^3(i)\}_{i=1,\dots,q}$. This relation is characterised by the parameters: β_3 and τ_2 and the cumulants $\left\{C_{4x}(q,\tau_2,j+\beta_3)\right\}_{j=0,\ldots,q}$. In fact, it is important to determine the set of these parameters giving a nonredundant system of equations. Moreover, the last cumulants of each equation are not all zero. It is very easy to verify that the cumulants $\left\{C_{4x}(q,\tau_2,j+\beta_3)\right\}_{j=0,\ldots,q}$ are not all zero when the coefficients q, τ_2 and $\left\{j+\beta_3\right\}_{j=0,\ldots,q}$ have the same sign. Thus, this set is defined as follows:

$$S = \left\{\tau_2, \beta_3 \ / \ (0 \le \tau_2 \le q) \text{ and } (-q \le \beta_2 \le q)\right\}$$

Concatenating (5) for τ_2 and $\beta_3 \in S$, we obtain the following system of equations:

$$A\theta = b \tag{8}$$

where

- $\theta = \left[h^3(1) \ \ldots \ h^3(q) \ h(0)h(q) \ h(1)h(q) \ \ldots \ h(q)h(q)\right]^T$ is a vector of $(2q+1)$.

- $b = \left[b_0 \ b_1 \ \ldots \ b_q\right]^T$ is a vector of $(q+1)(2q+1)$ elements.

$$b_i = \left[0 \ \ldots \ 0 \ -C_{4x}(q,i,0) \ -C_{4x}(q,i,1) \ \ldots \ -C_{4x}(q,i,q)\right]^T$$

- A is a $[(q+1)(2q+1), (2q+1)]$ matrix having the following form:

$$A = \begin{bmatrix} G_0 & H_0 \\ G_1 & H_1 \\ . & . \\ G_q & H_q \end{bmatrix}$$

where $\left\{G_i\right\}_{i=0,\ldots,q}$ and $\left\{H_i\right\}_{i=0,\ldots,q}$ are matrix of the sizes $[(2q+1),q]$ and $[(2q+1),q+1]$ respectively. They have the following form:

$$G_i = \begin{bmatrix} 0 & . & . & 0 & C_{4x}(q,i,0) \\ . & . & . & C_{4x}(q,i,0) & C_{4x}(q,i,1) \\ . & . & . & . & . \\ 0 & . & . & . & . \\ C_{4x}(q,i,0) & . & . & . & C_{4x}(q,i,q-1) \\ . & . & . & . & C_{4x}(q,i,q) \\ . & . & . & C_{4x}(q,i,q) & 0 \\ C_{4x}(q,i,q) & 0 & . & 0 & . \\ 0 & . & . & . & 0 \end{bmatrix}$$

$$H_i = \begin{bmatrix} V_0 & V_1 & \cdots & V_j & \cdots & V_q \end{bmatrix}$$

$\{V_j\}_{j=0,\ldots,q}$ are vectors of $(2q+1)$ elements defined by:

$$V_j = \begin{cases} \begin{bmatrix} -C_{4x}(0,0,-q) & -C_{4x}(0,0,-q+1) & \cdots & -C_{4x}(0,0,-q) \end{bmatrix}^T & \text{if } i = j \\ \\ 0 \ (\text{null vector}) \ \text{otherwise} \end{cases}$$

The matrix A has full rank as proved in the following theorem.

Theorem
Under the conditions C1, C2, C3 and C4, given the true output statistics $C_{4x}(q,i,j)$ and $C_{4x}(0,0,k)$ of the model (1) for $(0 \le i \le j \le q)$ and $(-q \le k \le q)$ respectively, the matrix A has full rank $(2q+1)$.

Proof
The theorem can be proved by contradictions. Suppose that the rank of A is less than $(2q+1)$. Then, there is more than one solution to the system of equations (1). However, all the solutions should satisfy (6) and (7), hence all the solutions must be identical. This is a contradiction, hence the rank of A is $(2q+1)$.

The matrix A has full rank, hence the obtained system has a unique least squares solution defined by:

$$\theta = \left(A^T A \right)^{-1} A^T b \tag{9}$$

3.3. Determination of the unknown parameters
The impulse response coefficients $\{h(i)\}_{i=1,\ldots,q}$ are obtained by using one of the following equations:

for $i = 1,\ldots,q$

$$h(i) = \frac{\sqrt[3]{\theta(i)} + \dfrac{\theta(q+1+i)}{\theta(q+1)}}{2} \tag{10}$$

for $i = 1,\ldots,q$

$$h(i) = \frac{\theta(q+1+i)}{\theta(q+1)} \tag{11}$$

for $i = 1,\ldots,q$

$$h(i) = \sqrt[3]{\theta(i)} \tag{12}$$

Remark

The developed method defines a redundant vector of unknowns parameters (as the methods described in [2, 11, 15, 18]), although, the authors of [12] have critiqued this approach. But, we can avoided this redundancy by using the following proposition.

3.4. A proposition for avoiding the redundancy

In this section, we present a second method allowing to avoid the redundancy appearing in the vector containing the unknown parameters. This proposition is a particular case of the developed method. It consists in concatenating (5) for $(\tau_2 = 0)$ and $(-q \leq \beta_2 \leq q)$, we obtain a system of equations having the form defined by (8) where:

$$\theta = \begin{bmatrix} h^3(1) & \dots & h^3(q) & h(q) \end{bmatrix}^T$$

$$b = \begin{bmatrix} 0 & \dots & 0 & -C_{4x}(q,0,0) & -C_{4x}(q,0,1) & \dots & -C_{4x}(q,0,q) \end{bmatrix}^T$$

$$A = \begin{bmatrix} 0 & . & . & 0 & C_{4x}(q,0,0) & -C_{4x}(0,0,-q) \\ . & . & . & C_{4x}(q,0,0) & C_{4x}(q,0,1) & -C_{4x}(0,0,-q+1) \\ . & . & . & . & . & . \\ 0 & . & . & . & . & . \\ C_{4x}(q,0,0) & . & . & . & C_{4x}(q,0,q-1) & . \\ . & . & . & . & C_{4x}(q,0,q) & . \\ . & . & . & C_{4x}(q,0,q) & 0 & . \\ C_{4x}(q,0,q) & . & . & 0 & . & . \\ 0 & . & . & . & 0 & -C_{4x}(0,0,q) \end{bmatrix}$$

A, θ and b have the following sizes $[2q+1, q+1]$, $[q+1,1]$ and $[2q+1,1]$ respectively.

Using the above theorem, we can shown that the rank of the matrix A is $(q+1)$.

The impulse response coefficients $\{h(i)\}_{i=1,\dots,q}$ are obtained by using (11).

3.5. Proprieties of the proposed method

The proposed method has several interesting points which can be summarised as follows:

- The proposed method uses the fourth order cumulants and not the second order statistics.
- The method is blind to Gaussian noise (colored or white) and it is useful when the characteristics of noise are not known or when the SNR value is low.
- The proposed method requires an input having a symmetric distribution and consequently it is useful in the case of blind equalisation problems.
- The second contribution proposes a nonredundant vector of unknown parameters.

4 Simulation examples

The objective of the simulations is to compare the performances of our contributions (Prop 2.: section 3.3 and Prop 1.: Section 3.4) with the following existing methods: C(q,k) [10], Giannakis and Mendel modified by Tugnait [20], Comon [3] and Dembele [5]. The C(q,k) algorithm was first introduced by Giannakis [10]. It uses third order cumulants. An extension to the fourth order was derived by Dianat and Raghuveer [6]. The Giannakis and Mendel method was the first linear algebra method proposed in the literature. It uses both second and third or fourth order cumulants. Several deficiencies have been affected to this method by many authors [12, 19]. To overcome certain problems, Tugnait proposed two modifications [19, 20]. The first one consists in increasing the number of equations in order to obtain a system of equations having a unique solution, while the second one is to avoid numerical ill conditioning. Only the second modification is implemented here. Furthermore, Tugnait proposed another solution allowing to avoid redundancy appearing of the vector containing the unknown parameters [20]. The Comon method uses only fourth order cumulants and it proposes a nonredundant vector of unknown parameters. The method proposed by Dembele consists in constructing a matrix of fourth order cumulants and extracting the coefficients after Cholesky type factorisation of the obtained matrix.

The model that we consider in our simulations is described by the following equation [15]:

$$x(n) = e(n) + 0.9e(n-1) + 0.79e(n-2) - 0.745e(n-3) \qquad (13)$$

with zeros at 0.5 and -0.7±j.

The colored gaussian noise is generated by [15]:

$$v(n) = w(n) + 0.5w(n-1) - 0.25w(n-2) + 0.5w(n-3) \qquad (14)$$

where w(k) is i.i.d. Gaussian noise.

The input sequence {e(n)} is zero mean Pseudo-Random Binary Sequence.

We define the Signal-to-Noise Ratio (SNR) as:

$$SNR(dB) = 10. Log\left[\frac{E\{x^2(n)\}}{E\{v^2(n)\}}\right] \qquad (15)$$

To reduce the realisation dependency, the parameter estimates were averaged over 20 Monte Carlo runs. The accuracy of system identification is assessed by computing the mean square error (mse). The mean (μ), the standard deviation (σ) and the mse values are presented in Tables 1 and 2.

Interpretation

From tables 1 and 2, we can have the following observations:

- For all values of SNR considered, the standard deviation and mse values of the proposed method are lower than the other methods.

- The performance of the GMT and Tug. methods degrades when the data are contaminated by colored Gaussian noise because the order of the filter which generated the noise does not satisfy the restrictions imposed by these methods. On the contrary, the proposed method yields consistent estimates in this case also.

- The mse value of the estimates of the Prop. 2 are lower than the Prop. 1. However, the last defines a nonredundant vector of unknown parameters, but it uses a less cumulant slices than the Prop. 2. Hence, we can concluded that the system identiafibility improves when we use a large cumulant slices.

From the above observations and the results presented in tables 1 and 2, it is to easy to remark that the proposed method shows a significant improvement to the estimation of LTI NMP FIR systems.

Table 1. Simulation results (white noise).

	C(q,k)	GMT	Tug	Com	Dem.	Prop1	Prop.2
SNR = 50 dB							
h1: μ	0.8939	0.8882	0.8894	0.8970	0.8918	0.8982	0.8988
σ	0.0203	0.0253	0.0252	0.0111	0.0165	0.0064	0.0063
mse	0.0004	0.0007	0.0007	0.0001	0.0003	0.0000	0.0000
h2: μ	0.7787	0.7683	0.7697	0.7853	0.7703	0.7851	0.7870
σ	0.0193	0.0412	0.0415	0.0111	0.0376	0.0094	0.0078
mse	0.0005	0.0021	0.0020	0.0001	0.0017	0.0001	0.0001
h3: μ	-0.7486	-0.7319	-0.7325	-0.7471	-0.7784	-0.7443	-0.7466
σ	0.0176	0.0255	0.0259	0.0078	0.0785	0.0074	0.0059
mse	0.0003	0.0008	0.0008	0.0001	0.0070	0.0001	0.0000
SNR = 10 dB							
h1: μ	0.8608	0.6626	0.7055	0.8046	0.7805	0.8465	0.8567
σ	0.2356	0.2079	0.2260	0.0860	0.0655	0.0386	0.0699
mse	0.0543	0.0974	0.0864	0.0161	0.0184	0.0043	0.0065
h2: μ	0.7488	0.5453	0.6131	0.7120	0.6537	0.7061	0.7471
σ	0.2557	0.1708	0.2397	0.0895	0.0795	0.0939	0.0950
mse	0.0638	0.0876	0.0859	0.0137	0.0246	0.0154	0.0104
h3: μ	-0.8950	-0.5393	-0.6035	-0.7397	-1.0566	-0.6711	-0.7457
σ	0.2522	0.1347	0.1640	0.0688	0.1965	0.0630	0.0573
mse	0.0830	0.0595	0.0456	0.0045	0.1338	0.0092	0.0031
SNR = 5 dB							
h1: μ	0.8505	0.6221	0.6504	0.7604	0.7430	0.7635	0.8346
σ	0.1794	0.2700	0.3199	0.0893	0.0919	0.0820	0.0886
mse	0.0330	0.1465	0.1595	0.0271	0.0327	0.0250	0.0117
h2: μ	0.6425	0.4450	0.5076	0.6273	0.5896	0.6320	0.7034
σ	0.2987	0.1984	0.1863	0.1410	0.0972	0.1529	0.1278
mse	0.1065	0.1564	0.1127	0.0454	0.0492	0.0472	0.0230
h3: μ	-1.0051	-0.4770	-0.5263	-0.7412	-1.1661	-0.6496	-0.7667
σ	0.3709	0.1716	0.1662	0.1155	0.2516	0.1062	0.1079
mse	0.1983	0.0998	0.0741	0.0127	0.2374	0.0198	0.0115

5 Conclusions

We have presented two contributions for identification of LTI NMP FIR systems. They use the fourth order cumulants of the noisy observations of the system output and consequently yield consistent parameters estimation in the presence of additive Gaussian noise. Both recursive closed-form and batch least squares solutions of the parameters estimation are proposed for each contribution. The second contribution allows to reduce the redundancy of the vector of unknown parameters is developed. Finally, the simulation results showed that the performance of our contributions is better than the other methods.

Table 2. Simulation results (colored noise).

	C(q,k)	GMT	Tug	Com	Dem.	Prop1	Prop.2
SNR = 50 dB							
h1: μ	0.8930	0.8933	0.8954	0.8968	0.8875	0.8977	-0.8980
σ	0.0177	0.0129	0.0140	0.0083	0.0153	0.0058	0.0041
mse	0.0003	0.0002	0.0002	0.0001	0.0004	0.0000	0.0000
h2: μ	0.7848	0.7712	0.7734	0.7872	0.7669	0.7875	0.7891
σ	0.0223	0.0321	0.0320	0.0092	0.0364	0.0086	0.0070
mse	0.0005	0.0013	0.0012	0.0001	0.0018	0.0001	0.0000
h3: μ	-0.7523	-0.7305	-0.7318	-0.7483	-0.7944	-0.7419	-0.7465
σ	0.0159	0.0193	0.0197	0.0078	0.0805	0.0078	0.0059
mse	0.0003	0.0006	0.0005	0.0001	0.0086	0.0001	0.0000
SNR = 10 dB							
h1: μ	0.9775	0.9754	1.0329	0.8600	0.7890	0.8165	0.8767
σ	0.2743	0.2267	0.2850	0.0815	0.0748	0.0687	0.0623
mse	0.0775	0.0545	0.0948	0.0079	0.0176	0.0115	0.0042
h2: μ	0.7801	0.4154	0.5008	0.7130	0.6430	0.7215	0.7539
σ	0.2974	0.2017	0.2097	0.0910	0.0687	0.0911	0.0988
mse	0.0841	0.1790	0.1254	0.0138	0.0261	0.0126	0.0106
h3: μ	-1.0099	-0.3282	-0.3990	-0.7848	-1.1272	-0.6759	-0.7833
σ	0.3302	0.1058	0.1192	0.0976	0.2275	0.0747	0.0880
mse	0.1738	0.1844	0.1332	0.0106	0.1953	0.0101	0.0088
SNR = 5 dB							
h1: μ	1.0628	1.2205	1.3355	0.8354	0.7645	0.8181	0.9033
σ	0.3794	0.4700	0.5739	0.1301	0.1143	0.0797	0.0899
mse	0.1632	0.3126	0.5026	0.0203	0.0308	0.0128	0.0077
h2: μ	0.8444	0.3528	0.4320	0.7156	0.6545	0.7271	0.7878
σ	0.3046	0.2036	0.2934	0.1266	0.0843	0.1108	0.0902
mse	0.0911	0.2306	0.2099	0.0208	0.0251	0.0156	0.0077
h3: μ	-1.2056	-0.1705	-0.2330	-0.7814	-1.2746	-0.6024	-0.8186
σ	0.4550	0.0934	0.1494	0.1443	0.2753	0.2326	0.1145
mse	0.4088	0.3383	0.2833	0.0211	0.3524	0.0717	0.0179

References

1. K. Abderrahim, R. Ben Abdennour, M. Ksouri and G. Favier, MA model parameters estimation using both third and fourth order cumulants, Proceedings of CESA'98 IMACS Multiconference, pp. 1100-1105, Nabeul-Hammamet, April, 1-4, 1998, Tunisia.
2. S. A. Alshebeili, A.N. Venetsanopoulos, A.E. Cetin, Cumulant based identification approaches for nonminimum phase FIR systems, IEEE SP, 41(4) pp:1576-1588, April 1993.

3. P. Comon, MA identification using fourth order cumulants, Signal Processing, 26(3) pp:381-388, March 1992.
4. D. Dembélé, Identification de modèles ARMA linéaires à l'aide de statistiques d'ordre élevé: application à l'égalisation aveugle, Thèse de Doctorat, Université de Nice Sophia Antipolis, Juillet 1995.
5. D. Dembélé and G. Favier, Une nouvelle méthode d'identification de modèles RIF à l'aide des cumulants d'ordre quatre, application à l'égalisation de canal, GRETSI, September 1995, Juan-les-Pins, France.
6. S. A. Dianat and M.R. Raghuveer, Polyspectral factorisation: necessary and sufficient condition for finite extent cumulant sequences, Proceedings of ICASSP, 2322-2324.
7. G. Favier, Statistiques d'ordre élevé: Applications à l'identification et aux télécommunications, Proceeding of Ecole de Printemps sur le traitement du signal et de l'image et ses applications, edited by G. Favier & M. Ksouri, Douz March 25-29, 1996, Tunisia.
8. G. Favier, D. Dembélé and J. L. Peyre, Identification de modèles AR, MA et ARMA à l'aide des statistiques d'ordre supérieur: comparaison des méthodes et analyse de performance, GRETSI, pp. 137-140, September 13-16, 1993.
9. J. A. R. Fonollosa and J. Vidal, System identification using a linear combination of cumulant slices, IEEE Trans. Signal Processing, 41(7), pp: 2405-2412, July 1993.
10. G. B. Giannakis, Apowerful tool in signal processing, Proc. of the IEEE, 75(9) pp: 1333-1334, September 1987.
11. G. B. Giannakis, J.M. Mendel, Identification of nonminimum phase systems using higher order statistics, IEEE SP 37(3) pp: 360-377, March 1989.
12. B. Jelonnek and K.D. Kammeyer, Improves methods for the blind system identification using higher order statistics, IEEE SP 40(12) pp: 2947-2960, December, 1992.
13. K. S. Lii and M. Rosenblatt, Deconvolution and estimation of transfert function phase and coefficients for non gaussian linear processes, The annals of statistics volume 10, pp. 1195-1208, April, 1982.
14. J. M. Mendel, Tutorial on Higher Order Statistics (Spectra) in signal processing and system theory: Theoretical results and some applications, Proc. of the IEEE, 79(3),pp:278-305, March 1991.
15. Y. J. Na, K. S. Kim, I. Song, T. Kim, Identification of nonminimum phase Systems using the third and fourth order cumulants, IEEE SP 43(8), pp: 2018-2022, August 1995.
16. C. L. Nikias, A. P. Petropulu, Higher-Order spectra analysis, PTR Prentice Hall, Englewood Cliffs, New Jersey, 07632, 1993.
17. J. L. Peyre, Statistiques d'ordre élevé appliquées à l'identification de signaux analogiques stationnaires à bande limitée, Thèse de Doctorat, Université de Nice Sophia Antipolis, Mars 1996.
18. A. G. Stogioglou, S. McLaughlin, MA parameter estimation and cumulant enhancement, IEEE SP, 44(7) pp:1704-1718, July 1996.
19. J. K. Tugnait, Approaches to FIR system identification with noisy data using higher order statistics, IEEE ASSP 38(7) pp:1307-1317, July 1990.
20. J. K. Tugnait, New results on FIR system identification using higher order statistics, IEEE SP 39(10) pp: 2216-2221, October 1991.

5

A Robust Frequency Domain Identification Method Revisited: Application in Steel Casting

R. DeKeyser and S. Zhang

1 Introduction

System identification is very important for designing controllers since the results of the identification directly effect the dynamic performance of the control system. Parametric identification algorithms, such as the least-squares methods, maximum-likelihood method and stochastic approximation methods are widely used in advanced control systems [1].

In recent years, identification techniques based on frequency domain methods [2], the orthonormal basis functions [3], the time-frequency analysis (or wavelet analysis) [4,5] are rapidly developing and widely used in many areas such as filtering, prediction, non-stationary processes, on-line detection and time-varying system control [6].

The method presented in this paper is mainly based on a frequency domain technique as a first step. A parametric model is obtained in a second step. In the first step the orthonormal basis analysis, or orthogonal detection, is used so that the method has very powerful ability to overcome the noise effects in system identification and a simple and stable algorithm can offer on-line detection and identification.

The method is applied to real-life data obtained in a steel factory. The resulting models have been the basis for a model based predictive controller leading to satisfactory results [7].

2 Process Description

Continuous bloom casting is the process of moulding molten steel into solid slabs. A diagrammatic picture of the process is shown in Fig. 1 with:
- the tundish acting as a reservoir, feeding molten steel to the mould
- the tundish valve ('stopper') being the actuator to control the steel flow to the mould.

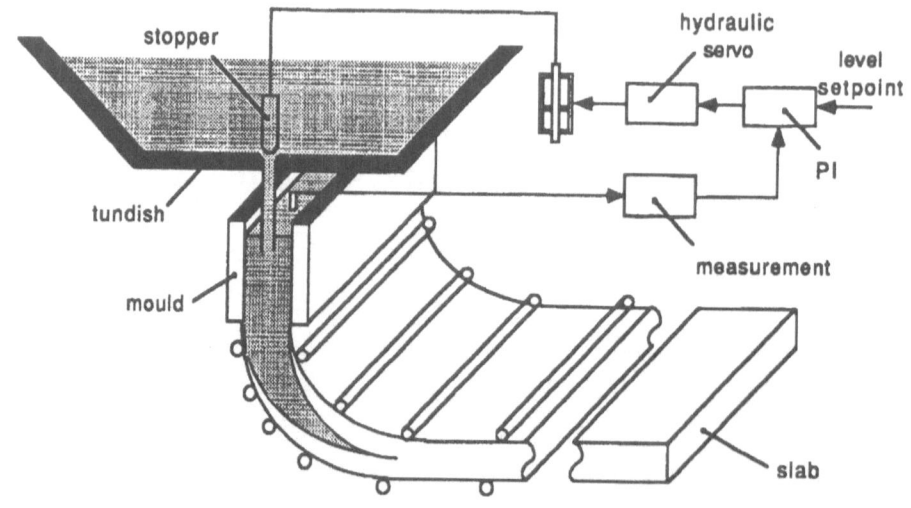

Fig. 1: The continuous casting process.

The cast metal undergoes two cooling processes. Primary cooling occurs in the mould and produces a supporting shell around the still molten centre. This semi-elastic strand is continuously withdrawn from the mould onto a series of supporting rolls containing the secondary cooling stage. This is a continuous production process. After solidification, the newly cast slab goes to the rolling mills. It is generally assumed that the surface quality of the rolled steel is directly correlated to the stability of the molten steel level in the mould [7].

In order to develop a model based predictive controller for this system, a process model is required. The process of interest (as seen by the controller, i.e. between manipulated and controlled variable) is defined by :
- u: manipulated variable = stopper position in %
- y: controlled variable = measured mould level in mm.

The integrating effect of the mould (the level is the integral of the in-flow, the out-flow being constant) can be represented by a transfer function $1/Cs$. Its parameter C is the cross-sectional area of the mould. The transfer characteristic between stopper position and in-flow is unknown and strongly time-varying; it has to be identified from experimental data.

Extensive test campaigns have been done in the industrial casting unit in order to obtain experimental data for modeling the process. Methods based on random steps, PRBS and stochastic test inputs were not successful [7]. In this paper a method is described which is based on the use of sine test-inputs for identification. A typical data sequence obtained at the real process is given in Fig. 2 with a blow-out in Fig. 3.

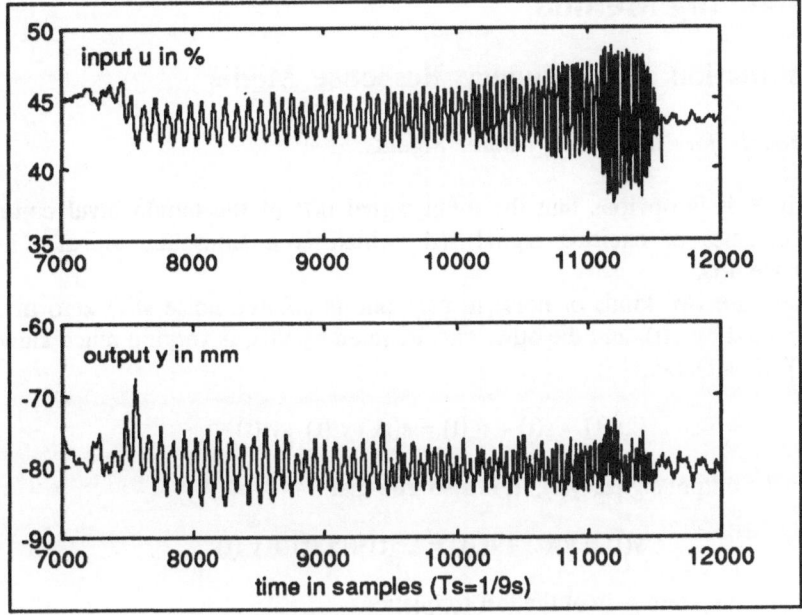

Fig. 2: Typical input-output data for identification

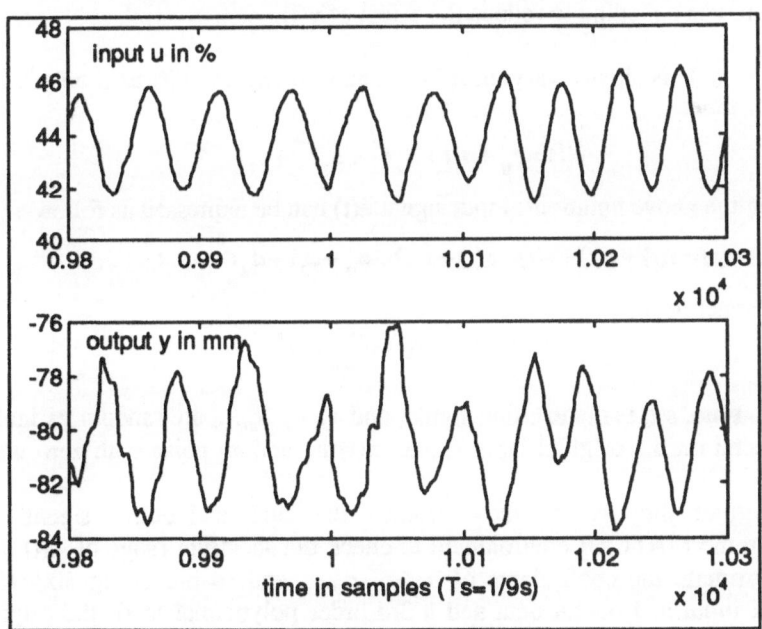

Fig. 3: Detail of time-series in Fig. 2

3 Modeling Method

3.1 Estimation of Frequency Response Model

3.1.1 Models for Input and Output Signals.

From Fig.2, it is obvious that the input signal u(t) of the mould level control system consists of exciting signal s(t), which is a sinusoidal signal, and interference n(t).

There are two kinds of noise in n(t): one is colored noise with zero mean value denoted by x(t), and the other one, denoted by v(t), is varying much slower than x(t) and s(t), i.e.:

$$u(t) = s(t) + n_a(t) = s(t) + x_a(t) + v_a(t)$$

For the output signal y(t), it is also true that

$$y(t) = ks(t) + n_y(t) = ks(t) + x_y(t) + v_y(t)$$

where k is the system gain at the test frequency.

Because v(t) is slowly varying the interference effect is difficult to remove by using the classical filtering method when detection is on-line performed. It is necessary to use a method to reduce the effect of v(t).

Since v(t) is slowly varying it is possible to express v(t) as a polynomial function of time:

$$v(t) = a_0 + a_1 t + a_2 t^2 + a_3 t^3 + ...$$

Using the above notations, input signal u(t) can be expressed as follows:

$$u(t) = s(t) + x(t) + v(t) = s(t) + x(t) + a_0 + a_1 t + a_2 t^2 + a_3 t^3 + ...$$

or in a compact form:

$$u(t) = s(t) + x(t) + \sum_k a_k t^k$$

where t is time, s(t) is the exciting signal, and a_0, a_1, a_2,..., are random variables, which depend on the original input signal, x(t) is random noise with zero mean value.

To further simplify the above result, the input and output signals are analyzed in the MATLAB environment to check the spectrum range of s(t) and x(t), to estimate the coefficients of a_0, a_1, a_2,..., and so on. Using 300~1500 samples of input and output data and a 3rd-order polynomial to fit the original input and output signals, it is shown that a_0 is in the range from 39 to 43, a_1 is quite small, from about -0.002 to 0.018, a_2 is much smaller than a_1, from about -0.0002 to 0.00001 and a_3, is 0.

From these results, the following signal model can be built:

$$u(t) = s(t) + x(t) + a_0 + a_1 t$$

and similarly for y(t). If s(t) and ks(t) can be estimated from u(t) and y(t) then the system gain k can be solved.

3.1.2 Orthogonal Detection

The frequency response of a system is the basic knowledge of the system. For a system excited by sinusoidal signals, the frequency response is easy to estimate, because in this case the orthogonal detection method is very powerful.

If there is a set of functions, $\{ u_i(t), \ |i = 0,1,2,... \}$, and all the dot products of $u_k(t)$ with $u_p(t)$ (k,p = 0,1,2,...) in the function space satisfy the following conditions:

$$\int_a^b u_k(t)u_p(t)dt = \delta_{kp}$$

and

$$\delta_{kp} = \begin{cases} d_k, & k = p \\ 0, & \text{otherwise} \end{cases}$$

where a,b and d_k ($k = 0,1,2,...$) are constant, then the sets of functions are called orthogonal basis function set and the arbitrary linear combinations of all these orthogonal basis functions, $u_i(t)$ (i=0,1,2,...), generate an orthogonal function space.

For example, the sinusoidal functions, sin(kω t) and cos(kω t), (k= 0,1,2,...) are orthogonal basis function set and all the linear combinations of sin(kω t) and cos(kω t) (k= 0,1,2,...) generate an orthogonal function space.

Using the concept of orthogonal basis function set and the orthogonal function space, it is possible to build an orthogonal detection method for detecting a certain orthogonal function basis component from a compound signal.

For example, c(t) is a compound signal in the orthogonal function space, and we want to check if there exists a certain component of the orthogonal basis $u_p(t)$ in c(t).

As the statement above, c(t) can be expanded into the following form:
$$c(t) = \sum c_i u_i(t) \qquad (i=0,1,2,...),$$

where c_i (i =0,1,2,...) are the coefficients of the orthogonal basis $u_i(t)$.

On one hand, performing the dot product of $c(t)$ with $u_p(t)$ we can get

$$\int_a^b c(t)u_p(t)dt = l$$

where l is a constant.

On the other hand, according to the orthogonal condition, it is held that

$$\int_a^b c(t)u_p(t)dt = \int_a^b (\Sigma c_i u_i(t))u_p(t)dt = c_p \int_a^b u_p(t)u_p(t)dt = c_p d_p$$

So it is clear that if $c_p d_p \neq 0$ then there exists the p-th orthogonal basis function component and the coefficient c_p is $1/d_p$; if $c_p d_p = 0$ then there is no such component of $u_p(t)$ in $c(t)$.

For another example,

$$z(t) = z_1 \cos(\omega t) + z_3 \cos(3\omega t)$$

where z_1 and z_3 are constant. After operating the dot product of $z(t)$ with $\sin(k\omega t)$ and $\cos(k\omega t)$ ($k = 0,1,2,...$), the result shows only two nonzero terms of the dot product with components $\cos(\omega t)$ and $\cos(3\omega t)$.

The orthogonal basis functions used in the orthogonal detection, such as $u_p(t)$, $\cos(\omega t)$ and $\cos(3\omega t)$, are called reference signals in this paper. It is clear that the reference signals are sinusoidal signals if a linear system is excited by sinusoidal signals.

Orthogonal detection has very strong ability to remove the effect of noise so that it is widely used for detecting signal from noise.

3.1.3 A Novel Method

It can be shown that the orthogonal detection is not valid in our case because the time-varying interference $v(t)$ and sinusoidal function do not belong to the same orthogonal function set. In detail, performing the dot product, it is easy to show

$$\int_0^T (a_0 + a_1 t)\sin\omega t\, dt = -\frac{a_1 T^2}{2\pi} \neq 0$$

Because a_1 is a random variable the expectation is

$$E[\int_0^T v(t)\sin\omega t\, dt] = -\frac{T^2 E[a_1]}{2\pi}$$

It means that if the orthogonal detection is used in our case then following two problems exist.

First, although the result is independent of a_0, it still depends on the random variable a_1. Even if a_1 could be estimated well the expectation $E[a_1]$ depends on the distribution of a_1, which is very difficult to estimate. So the above result only can offer a rough average contribution of $v(t)$ on the range $(0, T)$.

Second, the result also depends on the integral time T, or the measuring time. On one hand, since the expectation of $v(t)$ is over the interval $(0, T)$, a smaller T is better for a higher resolution. On the other hand, a larger T is better in order to average the effect of random noise $x(t)$. This conflict is hard to trade off in our point of view.

In order to cancel the effect of time-varying interference, it is necessary to choose another reference signal, denoted g(t), to satisfy the following conditions instead of the orthogonal condition:

$$\int_a^b v(t)g(t)dt \approx 0$$

and g(t) can be expanded in the orthogonal function space which contains s(t), i.e.,

$$g(t) = \sum g_i u_i(t) \qquad (i=0,1,2,...)$$

Operating the dot product of s(t) with g(t), we can get

$$\int_a^b s(t)g(t)dt = \int_a^b (\sum s_i u_i(t))(\sum g_i u_i(t))dt = \sum s_i g_i d_i$$

But it is not sure this kind of reference signal g(t) always exists. Fortunately, g(t) exists for our task, being g(t) = cos(2πt/T) where T is the period of exciting signal.

It is easy to verify that g(t) satisfies the condition we suppose above because

$$\int_0^T v(t)g(t)dt = \int_0^T (a_0 + a_1 t)\cos(\frac{2\pi t}{T})dt = 0$$

This means the effect of time-varying interference is exactly canceled. Performing the dot product of u(t) with g(t)

$$DP_u = \int_0^T u(t)g(t)dt = \int_0^T (a_0 + a_1 t)g(t)dt + \int_0^T (x(t)+s(t))g(t)dt$$

$$= \int_0^T s(t)g(t)dt = \frac{s_0 g_0 T}{2}$$

where s_0 and g_0 are magnitudes of s(t) and g(t), T is the period of g(t). Then the gain of the signal s(t) in u(t) to the reference signal g(t), k_u is

$$k_u = \left|\frac{g_0}{s_0}\right| = \left|\frac{g_0^2 T}{2DP_u}\right|$$

In the same way the gain of the signal ks(t), corresponding to the exciting signal in y(t) to the reference signal g(t), k_y is

$$k_y = \left| \frac{g_0^2 T}{2DP_u} \right|$$

And the system gain k is

$$k = \frac{k_y}{k_u} = \left| \frac{DP_u}{DP_y} \right|$$

In the above statement, it is implied that the phase delay between the reference signal g(t) to y(t) and u(t) is zero. So the problem left is how to estimate the phase delay of the system. If the phase delay is unknown then the new method cannot be directly applied, or in other words, it is neccesary to estimate the phase delay between g(t) to y(t) and u(t), D_y and D_u.
Computing the correlation function

$$R_u(\tau) = \int u(t) g(t + \tau) d\tau$$

and according to the property that $R_u(\tau)$ reaches its maximum value when $(\tau = \tau_u)$ makes $u(t)$ and $g(t + \tau_u)$ having the same phase, it is clear that

$$D_u = \tau_u \cdot \frac{360}{P}$$

where P is the number of samples in one period, and D_u is in degree.
Similarly

$$D_y = \tau_u \cdot \frac{360}{P}$$

and the phase shift is: $D = D_y - D_u = (\tau_y - \tau_u) \cdot \frac{360}{P}$

3.2 Estimation of Parametric Model

A parametric process model is needed in the predictive control algorithm. This model has the structure:

$$A(q^{-1}) y(t) = \frac{B(q^{-1})}{F(q^{-1})} u(t)$$

with $A(q^{-1}) \equiv 1 - q^{-1}$ representing the mould integrator and $\{B(q^{-1}), F(q^{-1})\}$ polynomials in the backward shift operator q^{-1}. The parameters in $\{B(q^{-1}), F(q^{-1})\}$ are unknown and have to be estimated.

The parametric model is obtained from the triplets $\{f_i, M_i, \phi_i\}$, $i = 1...n$, using the following procedure (f_i denotes frequency of the exciting sine signal; M_i denotes modulus, being the result k of section 3.1; ϕ_i denotes phase shift, being the result D of section 3.1):

- generate the signal $u^*(t) = \sum_{i=1}^{n} \sin(2\pi f_i t)$

- generate the signal $y^*(t) = \sum_{i=1}^{n} M_i \sin(2\pi f_i t + \phi_i)$

- estimate the parameters in the model $y^*(t) - y^*(t-1) = \dfrac{B(q^{-1})}{F(q^{-1})} u^*(t)$

with $B(q^{-1}) \Big/ F(q^{-1})$ identified from *the computer-generated* (thus noise free) signals $u^*(t)$ and $y^*(t)$ with a standard least-squares parameter estimator [1].

4 Experimental Results

4.1 Results of Frequency Response Modeling

Two sets of input and output data of the mould level system acquired from the steel factory are used to build the model. The first data set is shown in Fig. 2. There are about 4000 samples of the input and output signals in each data set. In each signal, there are 6 segments with different exciting frequencies, which are 1/10, 1/8, 1/6, 1/5, 1/4, 1/3 Hz respectively. Each signal segment lasts 10 periods of the exciting frequency, and the sampling period is 1/9 seconds, so it means that the first segment, corresponding to exciting frequency 1/10Hz, contains 900 samples; the second segment, corresponding to exciting frequency 1/8Hz, has 720 samples; the last segment, corresponding to exciting frequency 1/3Hz, has 270 samples.

The novel method is used to estimate the gain and phase delay of the mould level system. For the first data set, the results are as given in table I:

Hz	1/10	1/8	1/6	1/5	1/4	1/3
G_i	1.0	0.6819	0.4233	0.3646	0.3036	0.2007
D_i	-132	-135	-146	-160	-170	-173
G_{10}			3.144			

Table I: Results of frequency response modeling for first data sequence

where G_{10} is the gain at the frequency 1/10Hz, G_1 is the gain normalized by G_{10}, D_1 is the phase delay (in degree).

For the second data set, the results are given in table II:

Hz	1/10	1/8	1/6	1/5	1/4	1/3
G_2	1.0	0.6862	0.5514	0.4341	0.36	0.2041
D_2	-124	-140	-140	-152	-170	-186
G_{20}			3.0425			

Table II: Results of frequency response modeling for second data sequence

where G_{20} is the gain at the frequency 1/10Hz, G_2 is the gain normalized by G_{20}, D_2 is the phase delay (in degree).

To check the validity of the novel method to different data lengths or measuring time, 3 periods of each segment in the first and second data sets are used to estimate the gain and phase delay. The results are as follows (table III and table IV):

Hz	1/10	1/8	1/6	1/5	1/4	1/3
G_1	1.0	0.5863	0.4136	0.3676	0.3023	0.1799
D_1	-132	-135	-146	-160	-160	-173
G_{10}			3.2733			

Table III: Results using only 3 periods of first data set

Hz	1/10	1/8	1/6	1/5	1/4	1/3
G_2	1.0	0.6831	0.5725	0.4279	0.3729	0.1818
D_2	-124	-135	-140	-152	-160	-173
G_{20}			3.0101			

Table IV: Results using only 3 periods of second data set

From these results, it is clear that the novel method is powerful to estimate the frequency response from the input and output signals with very slowly time-varying interference and high stochastic disturbances. Moreover the method is easy to perform as an on-line estimator because only the dot product is needed.

It is clear that results are also quite stable w.r.t the measuring time. In fact if the measuring time is longer than (2~3) periods of the exciting signal then the method can offer good results. It shows that the novel method is robust.

4.2 Results of Parametric Modeling

In Fig. 4 we illustrate the method using the 6 frequency response results obtained in the previous section for the first data sequence. The estimated results from table I are indicated by crosses, while the solid lines correspond to the frequency response of the parametric model obtained using the procedure of section 3.2.

The following structure for the parametric model resulted in a good balance between complexity and accuracy:

$$\begin{cases} A(q^{-1}) = 1 - q^{-1} \quad \text{(mould integrator)} \\ B(q^{-1}) = bq^{-d} \quad \text{(a pure time-delay } d > 1 \text{ is obvious from the phase-delay diagram)} \\ F(q^{-1}) = 1 + fq^{-1} \end{cases}$$

resulting in the model:
$$y(t) = \frac{0.029q^{-3}}{(1-q^{-1})(1-0.885q^{-1})}u(t)$$

These results are very well reproducible, indicating that the method is quite robust.

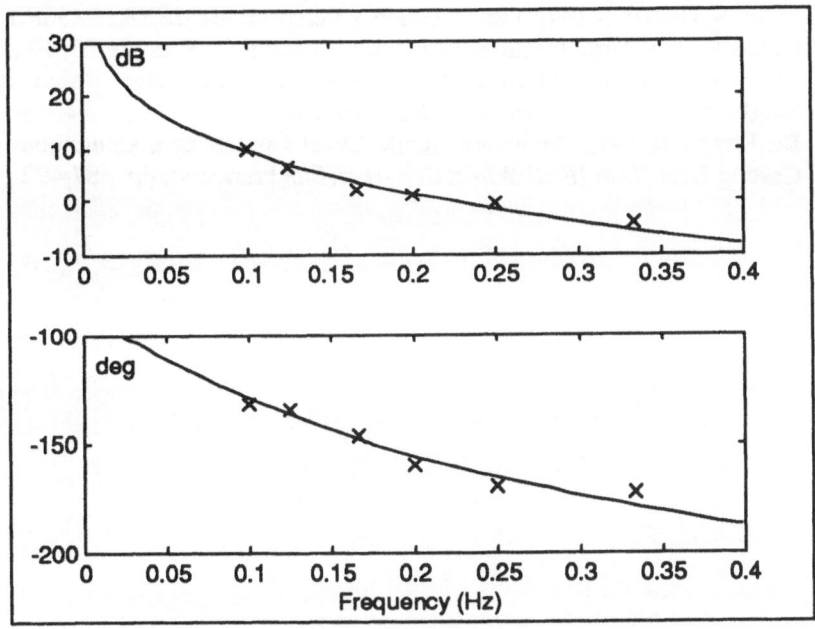

Figure 4: Identification results for real-life casting line.

5 Conclusions

The mould level process in continuous steel casting is a very tough system because of nonlinearities, dead-time, integrating effects, time-varying characteristics and a very high level of stochastic disturbances. Consequently the model identification is the most difficult part in the design of a model based predictive controller for this process. A method based on frequency domain identification concepts has been presented in this paper. It is quite attractive for practical applications thanks to its simplicity and robustness. It can be implemented by applying a low-frequency sine test-input which does not disturb the normal process operation. The experiments on the real casting line indicate that this is a realistic approach: it leads to good mould level control and the procedure is well accepted by the process operators.

References

1. Ljung L 1987 *System Identification: Theory for the User*. Prentice Hall, Englewood Cliffs
2. Papoulis A 1977 *Signal Analysis*. McGraw-Hill, New York
3. Ahmed N Rao K R 1975 *Orthogonal Transforms of Digital Signal Processing*. Wiley, New York.
4. Mann S Haykin S 1992 Time-frequency Perspectives. *IEEE ICASSP.1*
5. Cohen L 1989 Time-frequency Analysis - A Review. *Proc. IEEE, 77 (7)*
6. Zhang S He Y M 1993 Unstable Position of Bayes Receiver Based on the Single Observation. *Journal of HSEI* 14, No.3
7. De Keyser R 1996 Predictive Mould Level Control in a Continuous Steel Casting Line. *13th IFAC World Congress*, San Francisco, pp. 487-492

6

Fault Detection in Flight Control Systems via Innovation Sequence of Kalman Filter

C.M. Hajiyev and F. Caliskan

1 Introduction

The use of automatic flight control systems (AFCS) to control the motion of an aircraft is a very wide and well- established application. The motion of the aircraft is normally measured by sensors, such as gyroscopes and accelerometers. Since these sensors are subject to unexpected changes the affected feedback signals influence both handling and flying qualities of the aircraft. When faults occur, the AFCS is still expected to function as nearly normally as possible. To maintain such performance in the presence of sensor faults needs two essential stages:
- Any sensor fault must be detected and isolated, and
- Appropriate remedies must be applied.

This paper deals with the detection of sensor faults, and can be augmented for actuator and control surface faults.

Redundancy techniques are closely related to fault tolerance. There are mainly two redundancy techniques; hardware redundancy, and analytical (functional) redundancy.

Repeated hardware elements are usually distributed spatially around the system to provide protection against localized damage. The major problems encountered with hardware redundancy are the extra cost, and software, and furthermore the additional space required to accommodate the additional sensors and other hardware components.

The analytical redundancy techniques are basically signal processing techniques employing state estimation, parameter identification, statistical decision theory etc. [1].

The diagnostic Kalman filter used in diagnosis problems has some drawbacks. The order of the diagnostic filter is generally high, and consequently the computational load is high. The order of the filter is often equal to the order of the dynamical system to be monitored. Birdin diagnosis filter is one of the examples [2-5]. The order of the filter is even higher than the order of the system as in the approaches using multi-models.

The approaches above requires a priori statistical characteristics of the faults occurred in the system. Such information, in practice, is mostly not available, and using those approaches have limited applications for Fault Detection and Isolation (FDI). One of the diagnosis approaches based on Kalman filtering is the analysis of the innovation sequence [6-8]. These approaches do not require a priori statistical characteristics of the faults, and the computational burden is not very heavy.

Faults in dynamical systems can be detected with the aid of an innovation sequence that has the property that if the system operates normally the normalized innovation sequence in a correlated Kalman filter is a Gaussian white noise with a zero mean and with a unit covariance matrix. Faults that change the system dynamics by causing surges of drifts of the state vector components, abnormal measurements, sudden shifts in the measurement channel, and other difficulties such a decrease of instrument accuracy, an increase of background noise, etc., affect the characteristics of the normalized innovation sequence by changing its white noise nature, displacing its zero mean, and varying unit covariance matrix. Thus, the problem is how to detect as quickly as possible any change of these parameters from their nominal value. Methods of testing the agreement between the innovation sequence and white noise, and the detection of any change of its mathematical expectation have been discussed in [6,9].

In this paper a novel approach based on the innovation sequence of Kalman filter is introduced for fault detection, and applied to an aircraft model. The faults are assumed as changes in the mean value or covariance matrix of the sensor measurements of the aircraft. There are quite a lot of methods detecting faults affecting the mean value, but there are small number of methods detecting faults affecting the covariance matrix. The method presented in this paper can quickly detect the latter sort of faults. The detection time is less than filtering time.

2 Verification of the covariance matrix of innovation sequence

Let the discussed linear dynamic system be specified by the state equation

$$\mathbf{x}(k + 1) = \Phi(k + 1, k)\mathbf{x}(k) + \mathbf{G}(k + 1, k)\mathbf{w}(k) \tag{1}$$

and measurement equation

$$\mathbf{z}(k) = \mathbf{H}(k)\mathbf{x}(k) + \mathbf{v}(k) \qquad (2)$$

where $\mathbf{x}$(k) is an n-dimensional system state vector, ϕ(k+1,k) is an nxn system matrix, $\mathbf{w}$(k) is a random n-dimensional vector, $\mathbf{G}$(k+1,k) is an nxn perturbation-noise transition matrix, $\mathbf{z}$(k) is an s-dimensional measurement vector, $\mathbf{H}$(k) is an sxn system measurement matrix, and $\mathbf{v}$(k) is an s-dimensional measurement noise vector.

The random vectors $\mathbf{w}$(k) and $\mathbf{v}$(k) are both assumed to represent Gaussian white noise. Their mean values and covariances are

$$E\big[\mathbf{w}(k)\big] = 0;\; E\big[\mathbf{w}(k)\mathbf{w}^T(j)\big] = \mathbf{Q}(k)\delta(kj)$$
$$E\big[\mathbf{v}(k)\big] = 0;\; E\big[\mathbf{v}(k)\mathbf{v}^T(j)\big] = \mathbf{R}(k)\delta(kj) \qquad (3)$$
$$E\big[\mathbf{w}(k)\mathbf{v}^T(j)\big] = 0$$

where E is a statistical averaging operator, and $\delta(kj)$ is the Kronecker delta function:

$$\delta(k,j) = \begin{cases} 1, k = j \\ 0, k \neq j \end{cases} \qquad (4)$$

The estimated state vector $\mathbf{xe}$(k/k) and the covariance matrix of estimate errors $\mathbf{P}$(k/k) can be found with the aid of a Kalman filter of the form [10]:

$$\mathbf{xe}(k\,/\,k) = \mathbf{xe}(k\,/\,k-1) + \mathbf{K}(k)\Delta(k)$$
$$\Delta(k) = \mathbf{z}(k) - \mathbf{H}(k)\mathbf{xe}(k\,/\,k-1)$$
$$\mathbf{K}(k) = \mathbf{P}(k\,/\,k-1)\mathbf{H}^T(k)$$
$$\big[\mathbf{H}(k)\mathbf{P}(k\,/\,k-1)\mathbf{H}^T(k) + \mathbf{R}(k)\big]^{-1}$$
$$\mathbf{xe}(k\,/\,k-1) = \Phi(k\,/\,k-1)\mathbf{xe}(k-1\,/\,k-1) \qquad (5)$$
$$\mathbf{P}(k\,/\,k) = \big[\mathbf{I} - \mathbf{K}(k)\mathbf{H}(k)\big]\mathbf{P}(k\,/\,k-1)$$
$$\mathbf{P}(k\,/\,k-1) = \Phi(k\,/\,k-1)\mathbf{P}(k-1\,/\,k-1)$$
$$\Phi^T(k,k-1) + \mathbf{G}(k,k-1)\mathbf{Q}(k-1)\mathbf{G}^T(k,k-1)$$

where $\mathbf{P}$(k-1/k-1) is a covariance matrix of estimate errors at the preceding step, $\mathbf{K}$(k) is the gain matrix of the Kalman filter, and $\mathbf{I}$ is a unit matrix.

To detect faults it is convenient to use normalized innovation sequence [10]:

$$\tilde{\Delta}(k) = \left[\mathbf{H}(k)\mathbf{P}(k \,/\, k-1)\mathbf{H}^T(k) + \mathbf{R}(k) \right]^{-1/2} \Delta(k) \tag{6}$$

since then $E\left[\tilde{\Delta}(k)\tilde{\Delta}^T(j) \right] = \mathbf{P}_\Delta = \mathbf{I}\delta(kj)$. To test the covariance matrix of the innovation sequence $\tilde{\Delta}(k)$, the trace of the following sample covariance matrix may be used:

$$\mathbf{S}(k) = \frac{1}{M-1} \sum_{j=k-M+1}^{k} \left[\tilde{\Delta}(j) - \overline{\tilde{\Delta}}(k) \right]\left[\tilde{\Delta}(j) - \overline{\tilde{\Delta}}(k) \right]^T \tag{7}$$

where

$$\overline{\tilde{\Delta}}(k) = \frac{1}{M} \sum_{j=k-M+1}^{k} \tilde{\Delta}(j)$$

is the sample mean, and M is the number of the implementation having X^2 distribution. However, the fact that the nondiagonal elements of matrix S do not participate in, the test can lead to incorrect conclusions in the detection of the faults in dynamic systems.

In this paper a method of testing the covariance matrix of innovation sequences that is free of the above drawback is proposed, and how to detect faults as fast as possible is shown. For this purpose the following statistics is used:

$$\lambda = \frac{\mathbf{L}^T \mathbf{P}_\Delta^{-1} \mathbf{L}}{\mathbf{L}^T \mathbf{A}^{-1} \mathbf{L}} \tag{8}$$

where $\mathbf{A} = (M-1)\mathbf{S}$ is a random Whishart matrix, and $\mathbf{L}$ is any fixed vector.

Since matrices $\mathbf{P}_\Delta$ and $\mathbf{A}$ are positive definite and since $\mathbf{A}$ has a Whishart distribution $[\mathbf{A} \sim \mathbf{W}(\mathbf{P}_\Delta, M)]$, λ has a X^2_{M-s+1} distribution for any fixed vector $\mathbf{L}$ [11]. This result allows the analysis of a multivariable Whishart distribution to be reduced to the analysis of a univariate X^2 distribution. The detectability of a fault depends on λ and thus also on the choice of vector $\mathbf{L}$ [8]. To ensure that fault detection is as fast as possible, a nonzero vector $\mathbf{L}$ must be found from the condition that ratio of two quadratic forms (8) is maximum.

Theorem: The ratio of two quadratic forms (8) is maximum when vector **L** (argument of the quadratic forms) is the eigenvector corresponding to the maximum eigenvalue of the matrix $\mathbf{AP_\Delta^{-1}}$.

Proof: The necessary conditions for an $\lambda(\mathbf{L})$ extremum are $\lambda(\mathbf{L}): \partial\lambda(\mathbf{L}) / \partial\mathbf{L} = 0$. As known well [12], the inverse matrix $\mathbf{Y^{-1}}$ of a symmetrical nonsingular matrix **Y** is also symmetrical. If **Y** is positive definite the same is true for $\mathbf{Y^{-1}}$.

From the above it follows that since **A** and $\mathbf{P_\Delta}$ are symmetrical and positive definite, so are the matrices $\mathbf{A^{-1}}$ and $\mathbf{P_\Delta^{-1}}$. Then differentiating (8) with respect to **L**, and considering that $\mathbf{A^{-1}}$ and $\mathbf{P_\Delta^{-1}}$ are symmetrical yield;

$$\frac{2\mathbf{DP_\Delta^{-1}L} - 2\mathbf{CA^{-1}L}}{\mathbf{D^2}} = 0$$

where $\mathbf{L^T A^{-1} L = D}$, and $\mathbf{L^T P_\Delta^{-1} L = C}$.

Hence

$$\mathbf{DP_\Delta^{-1}L - CA^{-1}L} = 0 \tag{9}$$

Considering that $\mathbf{D} \neq 0$, (9) can be rewritten as an equation:

$$\mathbf{P_\Delta^{-1}L - (C/D)A^{-1}L} = 0 \tag{10}$$

Expression (10) which is simultaneously sufficient condition defines an optimal vector $\mathbf{L}_{opt}$ for which

$$\lambda(\mathbf{L}_{opt}) = \lambda_{max} = \frac{\mathbf{L_{opt}^T P_\Delta^{-1} L_{opt}}}{\mathbf{L_{opt}^T A^{-1} L_{opt}}} = \left(\frac{\mathbf{C}}{\mathbf{D}}\right)_{max} \tag{11}$$

Equation (11) can then be written as

$$\mathbf{P_\Delta^{-1}L} = \lambda\mathbf{A^{-1}L} \quad , \quad \lambda = \lambda_{max} \tag{12}$$

Equation (12) can be written in the equivalent form as

$$\mathbf{AP_\Delta^{-1}L} = \lambda\mathbf{L} \tag{13}$$

The system of homogeneous equations (13) has a nontrivial solution for

$$\left| \mathbf{A}\mathbf{P}_\Delta^{-1} - \lambda \right| = 0 \tag{14}$$

The roots of (14) $\lambda_1, \ldots \ldots \lambda_s$ are eigenvalues of the matrix $\mathbf{A}\mathbf{P}_\Delta^{-1}$ [13]. Using (12) it is possible for every eigenvalue $\lambda_1, \ldots \ldots \lambda_s$ to find an eigenvector $\mathbf{L}_1, \ldots \ldots \mathbf{L}_s$ where

$$\mathbf{L}_i^T \mathbf{P}_\Delta^{-1} \mathbf{L}_i = \lambda_i \mathbf{L}_i^T \mathbf{A}^{-1} \mathbf{L}_i \tag{15}$$

The best vector $\mathbf{L}_{opt}$ is the eigenvector of matrix $\mathbf{A}\mathbf{P}_\Delta^{-1}$ (in this case $\mathbf{A}$ since $\mathbf{P}_\Delta^{-1} = \mathbf{I}$.) that corresponds the maximum eigenvalue of this matrix. Since $\mathbf{P}_\Delta$ and $\mathbf{A}$ are symmetrical and positive definite, the matrix $\mathbf{A}\mathbf{P}_\Delta^{-1}$ has real eigenvectors and real eigenvalues [13]. Hence, (12)-(14) can be used to find analytically or numerically the vector $\mathbf{L}_{opt}$ for which the ratio of two quadratic forms (8) is maximum.

3 Simulation of aircraft via Kalman filter

The continuous-time longitudinal dynamics of an aircraft is used in the simulation[14]. The state-space model of the aircraft is

$$\dot{\mathbf{x}} = \mathbf{A}_s \mathbf{x} + \Gamma \mathbf{w}$$
$$\mathbf{z} = \mathbf{H}\mathbf{x} + \mathbf{v} \tag{16}$$

where

$$\mathbf{x} = \left[u, w', q, \theta \right]^T$$

u : perturbation velocity along Ox
w' : perturbation velocity along Oz
q : perturbation pitch rate
θ : perturbation pitch angle

$$\mathbf{A}_s = \begin{bmatrix} -0.033 & 0.0001 & 0.0 & -9.81 \\ 0.168 & -0.387 & 260.0 & 0.0 \\ 0.005 & -0.0064 & -0.55 & 0.0 \\ 0.0 & 0.0 & 1.0 & 0.0 \end{bmatrix}$$

$\mathbf{H} = \mathbf{I} \ (4 \times 4)$

From the graphs in Fig. 1 it is seen that the Kalman filter can appropriately estimates the actual aircraft states. The simulation results are given in Fig. 1.

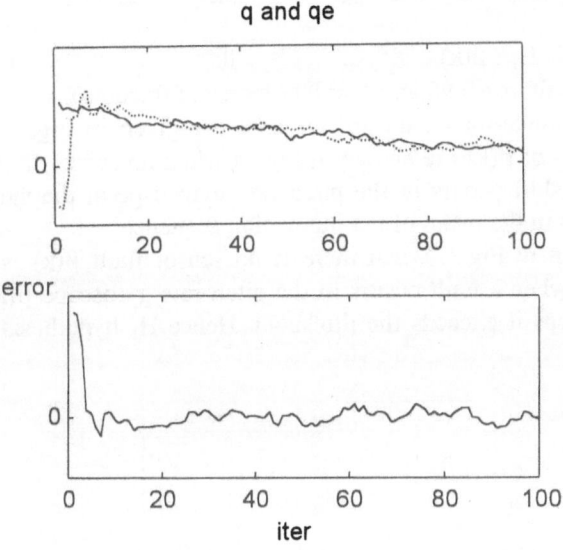

Fig 1. Actual pitch rate (q)(solid line), estimated pitch rate (q_e)(dotted line), and error between two.

4 Simulation of Fault Detection Algorithm

a) Detection of faults changing the mean of the innovation sequence: To detect faults changing the mean of the innovation sequence the following statistical function is used [6].

$$\beta(k) = \sum_{j=k-M+1}^{k} \widetilde{\Delta}^T(j)\widetilde{\Delta}(j) \qquad (17)$$

This statistical function has X^2 distribution with Ms degree of freedom. Now consider the following two hypotheses:

H_0 : System operates normally, and
H_1 : Fault occurs in the system.

If the hypothesis H_1 is correct then X^2 level for a confidence probability α will be greater than X^2 level found for the hypothesis H_0, i.e.:

$$H_0 : \beta(k) \leq X^2_{\alpha,Ms} \qquad \forall k$$

$$H_1 : \beta(k) > X^2_{\alpha,Ms} \qquad \exists k$$

$X^2_{\alpha,Ms} = 101.88$ for $\alpha=0.95$ and $Ms=80$ (degree of freedom).

The simulation results in this case are given in Fig. 2. The graphs of statistical values of $\beta(k)$ are shown in Fig.2a when no sensor fault occurs, and in Fig.2b when a shift occurs in the pitch rate gyroscope at the step 30. This fault causes a change in the mean of the innovation sequence.

As seen in Fig.2, when there is no sensor fault $\beta(k)$ is lower than the threshold, and when a fault occurs in the pitch rate gyroscope $\beta(k)$ grows rapidly and after 14 steps it exceeds the threshold. Hence H_1 hypothesis is judged to be true.

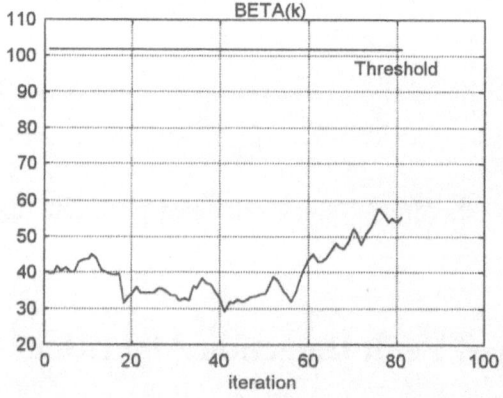

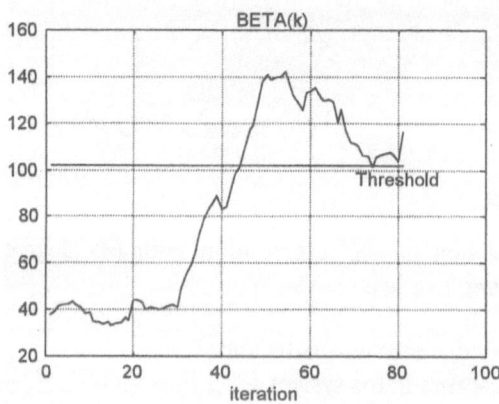

Fig 2. a. No fault
   b. Detection of sensor fault in the mean value.

b) Detection of faults changing the covariance matrix of the innovation sequence: In this case the mean value of the innovation sequence does not change, but the covariance matrix changes. The decision rule is:

$$H_0 : \lambda(k) \leq X^2_{\alpha, M\text{-}s+1} \qquad \forall k$$

$$H_1 : \lambda(k) > X^2_{\alpha, M\text{-}s+1} \qquad \exists k$$

$X^2_{\alpha, M\text{-}s+1} = 27.6$ for $\alpha=0.95$ and M-s+1=17 (degree of freedom).

The simulation results in this case are given in Fig. 3. The graphs of statistical values of $\lambda(k)$ are shown in Fig.3a when no sensor fault occurs, and in Fig.3b when a fault occurs in the pitch rate gyroscope at the step 30. This fault causes a change in the covariance matrix of the innovation sequence.

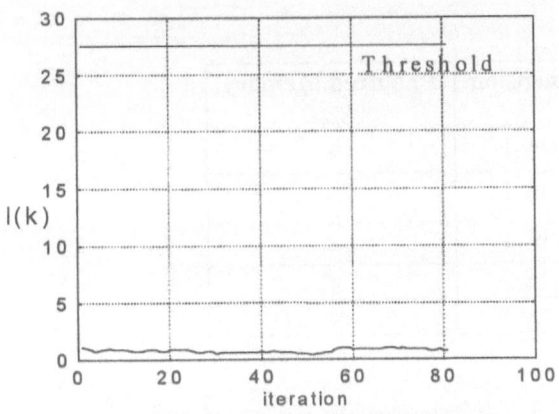

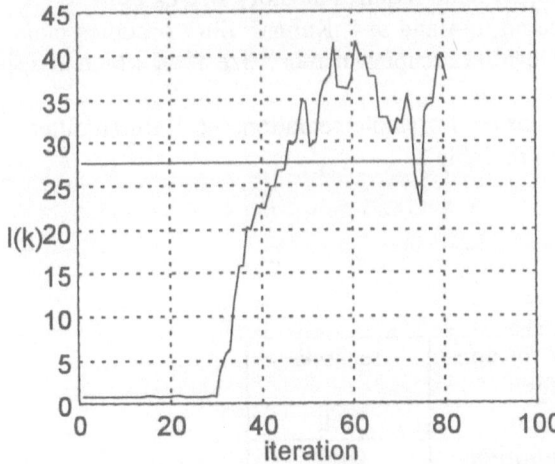

Fig 3. a. No fault
   b. Detection of sensor fault in the covariance.

As seen in Fig.3, when there is no sensor fault $\lambda(k)$ is lower than the threshold, and when a fault occurs in the pitch rate gyroscope $\lambda(k)$ grows rapidly and after 16 steps it exceeds the threshold. Hence H_1 hypothesis is judged to be true.

5 Computational features of the algorithms

In Table 1, the required memory size to implement the proposed algorithm is given. The size is closely related to the dimensions of the matrices and vectors used in the algorithms.

Table 1

Variable	Dimension	Required Memory Size
$\widetilde{\Delta}(k)$	s by 1	s
$\overline{\overline{\Delta}}(k)$	s by 1	s
$L_{opt}(k)$	s by 1	s
S	s by s	s^2
P_Δ	s by s	s^2

Kalman filter requires memory size of $5n^2 + s^2 + 2ns+2n+s$ where n is the order of the system, and s is the number of measurements. The monitoring algorithm, on the other hand requires memory size of $2s^2 + 3s$. As in the example used in the simulation n=4 and s=4, Kalman filter occupies memory size of 140, and monitoring algorithm occupies memory size of 44 which is respectively small.

Execution times for implementations of Kalman filter, and monitoring algorithm are given in Table 2.

Table 2

Execution time for one iteration for:	seconds
Kalman filter	0.08
Monitoring algorithm	0.05

6 Conclusion

In the paper, an approach based on the ratio of two quadratic forms of which matrices are theoretic and selected covariance matrices of Kalman filter innovation sequence for sensor fault detection, is presented. The optimal arguments of the quadratic forms are found to quickly detect the faults in sensors. The approach does not require a priori information about the faults and statistical characteristics of the system. Although the approach, like other fault detection approaches based on innovation sequence, cannot isolate the faults, it is quite useful to detect the faults considerably affecting the statistical characteristics of the innovation sequence, and will be augmented to isolate sensor faults.

References

1. Patton R., Frank P., Clark R. 1989 *Fault Diagnosis in Dynamic Systems, Theory and Applications*. Prentice Hall.

2. Beard R.D 1971 Failure accommodation in linear systems through self reorganization. Ph.D. Thesis, MTV 71-1, MIT.

3. Frank P.M 1990 Fault diagnosis in dynamic systems using analytical and knowledge based redundancy. A survey and some new results. *Automatica*, V.26, No.3, 459-474.

4. Gertler J. 1988 Survey of model based failure detection and isolation. *Automatica*, V.26, No.2, 3-11.

5. Isermann R 1982 Process fault detection based on modelling and estimation methods- A survey. *Automatica*, V.20, No.4, 387-404.

6. Willsky A.S. (1976). A survey of design methods for failure detection in dynamic systems. Automatica, V.12, No.6, 601-611.

7. Basseville M. and Benveniste A. (Eds.), 1986 *Detection of Abrupt Changes in Signals and Dynamics Systems*. LNCIS No.77, Springer, Berlin.

8. Hajiyev Ch. M. 1994 Fault detection in multidimensional dynamic systems based on statistical analysis of Kalman filter. *IFAC Symposium on Fault Detection, Supervision and Safety for Technical Processes, SAFEPROCESS'94*, Helsinki, Finland, V.1, 45-49.

9. Himmelblau D. M. 1978 *Fault Detection and Diagnosis in Chemical and Petrochemical Processes*. Elsevier Press, Amsterdam.

10. Sage E. and Mells J. 1976 *Estimation Theory and Its Application in Communication and Control*. (Russian Translation), Svyaz', Moscow.

11. Rao S.R. 1968 *Linear Statistical Methods and Their Applications*. (Russian Translation), Nauka, Moscow.

12. Horn R. and Johnson C. 1989 *Matrix Analysis*. (Russian Translation), Mir, Moscow.

13. Gantmacher F.R. 1959 *The Theory of Matrices*, Chelsea.

14. McLean D. 1990 *Automatic Flight Control Systems*. Prentice Hall International UK.

7

Analysis of Properties of Multitime-Scale Systems in 2D Approach

K. Galkowski, A. Gramacki and J. Gramacki

1 Introduction

The multitime scale systems are being systematically investigated by many authors since many years. The classical, that is one-dimensional, approach is well-documented in the survey papers by Kokotovic et al. [1] and Saksena et al. [2]. For systems with slow and fast dynamics the singular perturbation and multitime scale methods are attractive and being intensively developed. For large scale systems of very high order this approach is very interesting. Usually such systems involve interacting dynamic phenomena of widely different speeds (for example in a model of power system frequency and voltage transients at the user side range from intervals of seconds while generator voltage in a power plant ranges from several minutes). The underlying assumption is that during the fast transients the slow variables remain constant and that by the time their changes become noticeable, the fast transients have already reached their quasi steady state. The result is that we obtain the order reduction - the whole model is divided for two (or more in general) subsystems of smaller order which are easier to solve and probably more stable from numerical point of view.

On the other hand, many physical systems have a natural two-dimensional structure due to the presence of more than one independent variable. There is a rich literature dealing with 2D, or more generally, nD systems. The basic choice is for example Bose [3], Fornasini et al. [4], Kaczorek [5] and others.

Comparing the multitime scale systems with 2D systems one may observe the essential differences. First of all, multitime scale systems have, in fact, only one independent variable (usually time). However, there are two mutually-coinciding dynamics - the slow and the fast.

In what follows, performing one of a discretization method to such a system may yield a situation known as the multirate sampling (see for example Akari et al. [6],

Longhi [7], Berg *et al.* [8]) where, in fact, there are two independent variables (natural numbers) with the short and long sampling period. This resembles the aforementioned nD systems approach.

Some initial work has been undertaken in the authors' previous chapter (Gałkowski *et al.* [9]) where two scale systems have been approximated by linear discrete repetitive processes which are, in fact, the particular case of 2D systems.. For more information on repetitive processes see for example Rogers [10], Gałkowski [11] and others. The state-space model, when using the trapezoidal discretization method, has been derived there.

To deal with multitime scale systems from practical, not only theoretical, point of view there is a need to be in possession of a computer tool which could support a user in making some practical simulations. Such a tool (called MTSS Toolbox), based on MATLAB environment, is also presented in the chapter as well as some simulations results obtained. The current contents of the Toolbox is presented as well as definition of the multitime scale systems data structures implemented in MATLAB.

Also some initial steps in testing stability of 2D multitime scale systems are presented. For that purpose an equivalent 1D model was proposed.

2 Background

In practice, 2D systems are frequently characterised by a finite region of support in one of the two independent variables, e.g. so-called repetitive processes which are the subject of our investigations. The state-space model of a discrete linear repetitive process has the structure (Rogers *et al.* [10])

$$x(k+1, p+1) = \Phi x(k+1, p) + \Delta_0 y(k, p) + \Delta u(k+1, p) \tag{1}$$

$$y(k+1, p) = Cx(k+1, p) + D_1 y(k, p) + D_0 u(k+1, p). \tag{2}$$

Here, on pass k, $x(k, p)$ is the $n \times 1$ state vector, $y(k, p)$ is the $m \times 1$ vector pass profile and $u(k, p)$ denotes the $t \times 1$ vector of control inputs and Φ, Δ_0, Δ, C, D_1, D_0 are matrices of appropriate dimensions. This model has the so-called unit memory property, i.e. it is only the pass profile on pass k which (explicitly) contributes to its counterpart on pass $k+1$, $k \geq 0$. In the more general case, it is the previous $M > 1$ pass profiles which (explicitly) contribute to the current one. These so-called non-unit memory processes are not considered further here since the results developed generalise in a natural manner. The intrinsic feature of repetitive process is that all passes have the finite and fixed length α, i.e. $p = 0, 1, \ldots, \alpha - 1$.

Now following Sueur [12] the basic continuous multitime scale equations are defined as follows:

$$\dot{X}_1 = A_{11} X_1 + A_{12} X_2 + B_1 U \tag{3a}$$

$$\varepsilon \dot{X}_2 = A_{21} X_1 + A_{22} X_2 + B_2 U \tag{3b}$$

$$Y = C_1 X_1 + C_2 X_2 + DU \tag{4}$$

where

$$X_1 \in \mathfrak{R}^n, \quad X_2 \in \mathfrak{R}^m, \quad U \in \mathfrak{R}^p, \quad Y \in \mathfrak{R}^q$$
$$X_1(t_0) \triangleq X_{10}, \quad X_2(t_0) \triangleq X_{20}.$$

The intrinsic feature of these systems is presence of a small scalar ε in the equation of (3b). This is interpreted as that there are two time scales: the slow $t - t_0$ and the fast $\tau = (t - t_0) / \varepsilon$. Hence, the dynamics of the system represented by the variables X_2 and U have two parts: the slow and the fast, when, due to the lack of the small scalar, the variable X_1 has only the slow dynamics:

$$X_1 \cong X_{1s}$$
$$X_2 \cong X_{2s} + X_{2f}$$
$$U \cong U_s + U_f$$
$$Y \cong Y_s + Y_f \tag{5}$$

where the slow dynamics is obtained when substituting $\varepsilon = 0$ to the equation of (1b). It is a straightforward task to show that (Sueur [12]) both the slow and the fast dynamics are governed by the following equations

$$\dot{X}_{1s} = \left(A_{11} - A_{12} A_{22}^{-1} A_{21} \right) X_{1s} + \left(B_1 - A_{12} A_{22}^{-1} B_2 \right) U_s$$
$$X_{2s} = -A_{22}^{-1} A_{21} X_{1s} - A_{22}^{-1} B_2 U_s$$
$$Y_s = \left(C_1 - C_2 A_{22}^{-1} A_{21} \right) X_{1s} + \left(D - C_2 A_{22}^{-1} B_2 \right) U_s \tag{6}$$

and

$$\varepsilon \dot{X}_{2f} = A_{22} X_{2f} + B_2 U_f$$
$$X_{1f} = 0$$
$$Y_f = C_2 X_{2f} \tag{7}$$

where the initial conditions clearly satisfy

$$X_{1s}(t_0) = X_1(t_0)$$
$$X_{2f}(t_0) = X_2(t_0) - X_{2s}(t_0) = X_2(t_0) + A_{22}^{-1} A_{21} X_1(t_0). \tag{8}$$

3 Multitime scale systems in 2D (repetitive processes) framework

First representations of the multitime scale systems in terms of 2D systems have been developed by the authors [9]. The continuous-time model is discretized by one of standard discretization rule. In the presented Toolbox two discretization methods have been implemented: the well-known trapezoidal method

$$X_1(i+1) = X_1(i) + (H/2)\left[\dot{X}_1(i+1) + \dot{X}_1(i)\right] \tag{9}$$

$$X_2(j+1) = X_2(j) + (h/2)\left[\dot{X}_2(j+1) + \dot{X}_2(j)\right] \tag{10}$$

and much more simpler, but not so accurate as the trapezoidal one, forward method

$$X_1(i+1) = X_1(i) + H\dot{X}_1(i) \tag{11}$$

$$X_2(j+1) = X_2(j) + h\dot{X}_2(j). \tag{12}$$

Note, here, that due to the fact that the state sub-vector X_1 does not have the fast dynamics and is discretized 'slowly' when X_2 'quickly'. Hence, the index 'i' concerns the slow dynamics and 'j' the fast and 'H' denotes a 'long period' when 'h' denotes the short. Moreover, both of them satisfy

$$h = H\varepsilon. \tag{13}$$

Hence, multitime scale systems are discretized with different discretization periods. This results in that the 1D in nature signal "becomes" of the 2D type where there are two independent discrete variables - one is bounded and the second is unbounded. In the sequel, this yield the new discrete models which are very similar to (1)-(2). Now, the following notation is introduce

$$\varphi(i,j) \triangleq \varphi(iH + jh), \, i=0,1,\dots; j=0,1,\dots, N\text{-}1 \tag{14}$$

which gives us the 2D model of multitime scale systems. Also

$$N = \frac{1}{\varepsilon}. \tag{15}$$

The discrete multitime scale systems model depends on discretization method employed and step-wise assumptions taken under consideration. If a continuous-time model (3)-(4) is discretized by using the forward rule (11)-(12) one obtains the following discrete-time model:

$$X_1(i+1) = \tilde{A}_{11}X_1(i) + \tilde{A}_{12}X_2(i) + \tilde{B}_1U(i) \tag{16}$$

$$X_2(i,j+1) = \tilde{A}_{21}X_1(i) + \tilde{A}_{22}X_2(i,j) + \tilde{B}_2U(i,j).$$ (17)

where $i=0,1,...; j=0,1,..., N-1$, and

$$\begin{aligned} \tilde{A}_{11} &= I + HA_{11} & \tilde{A}_{12} &= HA_{12} & \tilde{B}_1 &= HB_1 \\ \tilde{A}_{21} &= HA_{21} & \tilde{A}_{22} &= I + HA_{22} & \tilde{B}_2 &= HB_2 \end{aligned}$$ (18)

This method leads to very simple discrete models but it meets here serious difficulties from the numerical point of view. First of all, some problems with stability of the resulted discrete model may occur. Thus, very short discretization periods have to be chosen so as to obtain accurate results.

Using trapezoidal rule (9)-(10) yields more complex model. The final one is presented below and details can be found in Gałkowski *et al.*[9]

$$X_1(i+1) = \left\{ \mathbf{A}_1 + \mathbf{A}_2 \left[\sum_{j=0}^{N-1} \mathbf{A}_4^j \mathbf{A}_3 \right] \right\} X_1(i) + \mathbf{A}_2 \left[I + \mathbf{A}_4^N \right] X_2(i)$$ (19)

$$+ \mathbf{A}_2 \left[\sum_{j=1}^{N-1} \mathbf{A}_4^{N-j-1} \mathbf{B}_2 U(i,j) \right] + \left[\mathbf{B}_1 + \mathbf{A}_2 \mathbf{A}_4^{N-1} \mathbf{B}_2 \right] U(i) + \mathbf{B}_1 U(i+1)$$

$$X_2(i,j+1) = \mathbf{A}_3 X_1(i) + \mathbf{A}_4 X_2(i,j) + \mathbf{B}_2 U(i,j)$$ (20)

$$Y(i,j) = C_1 Z_1(i) + C_2 X_2(i,j) + DU(i,j) + C_1 B_1 U(i)$$ (21)

$$i=0,1,...; \quad j=0,1,..., N-1$$

where matrices $\mathbf{A}_1, \mathbf{A}_2, \mathbf{A}_3, \mathbf{A}_4, \mathbf{B}_1, \mathbf{B}_2$ come from the trapezoidal discretization rule and are as follows:

$$\mathbf{A}_1 = \left[I - \frac{H}{2} A_{11} \right]^{-1} \left[I + \frac{H}{2} A_{11} \right] \qquad \mathbf{A}_2 = \left[I - \frac{H}{2} A_{11} \right]^{-1} \frac{H}{2} A_{12}$$

$$\mathbf{B}_1 = \left[I - \frac{H}{2} A_{11} \right]^{-1} \frac{H}{2} B_1 \qquad \mathbf{A}_3 = \left[I - \frac{H}{2} A_{22} \right]^{-1} HA_{21}$$

$$\mathbf{A}_4 = \left[I - \frac{H}{2} A_{22} \right]^{-1} \frac{H}{2} A_{22} \qquad \mathbf{B}_2 = \left[I - \frac{H}{2} A_{22} \right]^{-1} HB_2.$$ (22)

This model has been derived under the following assumptions. The inputs have been assumed to be step-wise along the fast variable, i.e. 'j'

$$U(i,j) = U(i,j+1), \quad j = 0,1,..., N-2.$$ (23)

Note also, that the subvector X_2 has simultaneously the slow and the fast dynamics, the previous assumption performed to inputs that the function is step wise, could not hold, i.e. one must assume that

$$X_2(i) \neq X_2(i+1), \quad U(i) \neq U(i+1). \tag{24}$$

In the sequel, the fact that the subvector X_1 has only the slow dynamics, i.e.

$$X_1(i,j) = X_1(i), \quad j = 0,1,\dots,N-1 \tag{25}$$

implies that this is constant along the fast 'direction', i.e., also,

$$X_1(i,j) = X_1(i,j+1), \quad j = 0,1,\dots,N-2. \tag{26}$$

In many applications, like repetitive processes global controllability and asymptotic stability investigation, it is necessary to build an equivalent 1D model describing the 2D, in fact, systems. In general such 1D global state-space models for 2D systems are infinite-dimensional but, here, due to the truncated dynamics along the one variable, this becomes finite, however possibly very large, dimension (Rogers *et al.* [11]). Start, first, with (16)-(17) and note that

$$X_2(i+1) \hat{=} X_2(i,N). \tag{27}$$

Hence, the value of $X_2(i+1)$ can be calculated from (17) as

$$X_2(i+1) = \left(\sum_{k=0}^{N-1} \tilde{A}_{22}^k \tilde{A}_{21}\right) X_1(i) + \tilde{A}_{22}^N X_2(i) + \sum_{k=0}^{N-1} \tilde{A}_{22}^k \tilde{B}_2 U(i,N-1-k). \tag{28}$$

Introduce now a global input vector

$$\mathbf{U(i)} = \begin{bmatrix} U(i,0) & U(i,1) & \cdots & U(i,N-1) \end{bmatrix}^T \tag{29}$$

which allows rewriting (16)-(17) in the below form

$$\begin{bmatrix} X_1(i+1) \\ X_2(i+1) \end{bmatrix} = \begin{bmatrix} \tilde{A}_{11} & \tilde{A}_{12} \\ \sum_{k=0}^{N-1} \tilde{A}_{22}^k \tilde{A}_{21} & \tilde{A}_{22}^N \end{bmatrix} \begin{bmatrix} X_1(i) \\ X_2(i) \end{bmatrix} + \begin{bmatrix} \tilde{B}_1 & 0 & 0 & 0 \\ \tilde{A}_{22}^{N-1} \tilde{B}_2 & \tilde{A}_{22}^{N-2} \tilde{B}_2 & \cdots & \tilde{B}_2 \end{bmatrix} \mathbf{U(i)} \tag{30}$$

which has the classical state-space structure.

Note that the same procedure of building 1D equivalent model can be introduced to the model described by (19)-(20). The final model has the same structure as (30) but the appropriate matrices are more complex.

It is also important to note that stability of the model as a whole is dependent of the both stabilities e.g. the slow and the fast ones when considered independently. Thus, in the particular case, both fractions are stable while the whole model remains unstable.

Such an 1D model may be very helpful in the process of system theoretic features investigation - and also in solving practical problems e.g. controllers design etc. These questions are the subject of on-going work and will be reported in due course.

4 Proposed data structure in Matlab

The presented MATLAB Toolbox can simulate MIMO systems of any order. The below picture describe it in more details:

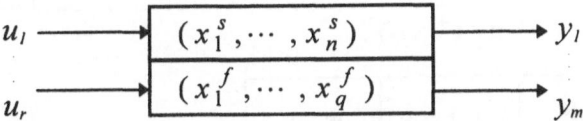

where
r - number of inputs,
m - number of outputs,
n - number of slow states,
q - number of fast states.
Model described by (16)-(17) or (19)-(20) requires the following inputs given by a user

$A_{11}, A_{12}, A_{21}, A_{22}, B_1, B_2, C_1, C_2, D$ - input vectors/matrices/scalars,
u - input vector/matrix,
x_{0s} - initial slow state vector/matrix,
x_{0f} - initial fast state vector/matrix,
K - number of slow states to be simulated,
H - long period,
h - short period.

The following outputs are calculated by the Toolbox

x_s - system slow states vector/matrix,
x_f - system fast states vector/matrix,
y - system outputs vector/matrix.

The u, x_{0s}, x_{0f}, K, H, h are represented in MATLAB in the following form

K - any positive integer number,
H - any positive number,
h - any positive number,
N - number of fast points within a slow point. This is calculated as H/h. There is also an additional requirement that remainder after division (H/h) equals zero (because N must be also an integer number).

The picture below shows the exact structure of the required inputs and outputs

5   Example

$K=3$, $r=3$, $m=2$, $n=2$, $q=1$, $H=4$, $h=1$

Input matrices (discrete model):

$$A_{11} = \begin{bmatrix} 0 & 1 \\ 1 & 0 \end{bmatrix} \quad A_{12} = \begin{bmatrix} 1 \\ 0 \end{bmatrix} \quad A_{21} = \begin{bmatrix} 1 & -1 \end{bmatrix} \quad A_{22} = \begin{bmatrix} 1 \end{bmatrix}$$

$$B_1 = \begin{bmatrix} 1 & 1 & 1 \\ -1 & -1 & -1 \end{bmatrix} \quad B_2 = \begin{bmatrix} 0 & 0 & 1 \end{bmatrix}$$

$$C_1 = \begin{bmatrix} 1 & 0 \\ 0 & 0 \end{bmatrix} \quad C_2 = \begin{bmatrix} 1 \\ 1 \end{bmatrix}$$

$$D = \begin{bmatrix} 1 & 0 & 0 \\ 1 & 0 & 0 \end{bmatrix}$$

Initial conditions:

$$x_{0s} = \begin{bmatrix} 0 \\ 0 \end{bmatrix} \quad x_{0f} = [0]$$

Inputs:

$$u = \begin{bmatrix} 1 & 1 & 1 & 1 & 1 & 1 & 1 & 1 & 1 & 1 & 1 & 1 \\ 1 & 1 & 1 & 1 & 1 & 1 & 1 & 1 & 1 & 1 & 1 & 1 \\ 1 & 1 & 1 & 1 & 1 & 1 & 1 & 1 & 1 & 1 & 1 & 1 \end{bmatrix}$$

System states (slow and fast):

$$x_s = \begin{bmatrix} 0 & 3 & 4 & 35 \\ 0 & -3 & 0 & 1 \end{bmatrix} \quad x_f = \begin{bmatrix} 0 & 1 & 2 & 3 & 4 & 11 & 18 & 25 & 32 & 37 & 42 & 47 \end{bmatrix}$$

Output matrices:

$$y = \begin{bmatrix} 1 & 2 & 3 & 4 & 8 & 15 & 22 & 29 & 37 & 42 & 47 & 52 \\ 1 & 2 & 3 & 4 & 5 & 12 & 19 & 26 & 33 & 38 & 43 & 48 \end{bmatrix}$$

6 Sample snapshots and simulation examples

To start the user-friendly graphical tool you need to type mstart at the MATLAB prompt. The main navigation window appears. In this window you specify all parameters of the simulated model. Figure 1 shows the snapshot of the main navigation window. The other graphical windows appear if needed and are used to enter input matrices, show plots and perform algebraic stability tests of the given model. To see the simulated results (that is x_s , x_f, and y matrices) one can display them in a dedicated window. Some examples are presented on the below pictures. All models were discretized by the trapezoidal method. Inputs and initial conditions are in all cases the same. They are: $u(i,j)=0$, $x_{0s}=1$, $x_{0f}=1$. Also the following matrices remain unchanged: $B_1 = 1$ $B_2 = 1$ $C_1 = 1$ $C_2 = 1$ $D = 2$.

Figures 2 and 3 present sample simulation results. On Figure 2 slow state is unstable while fast one is stable along slow periods (but asymptotically unstable). This result is obvious as A_{11} matrix has unstable value and A_{22} stable one. To stabilize the model we simply change A_{11} into negative (Figure 3).

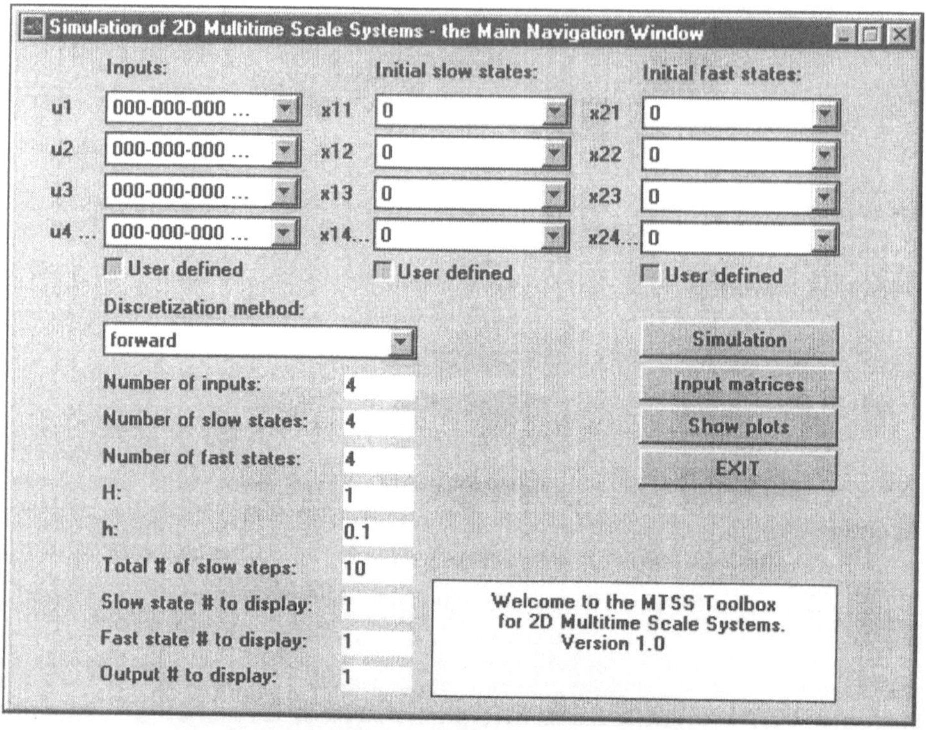

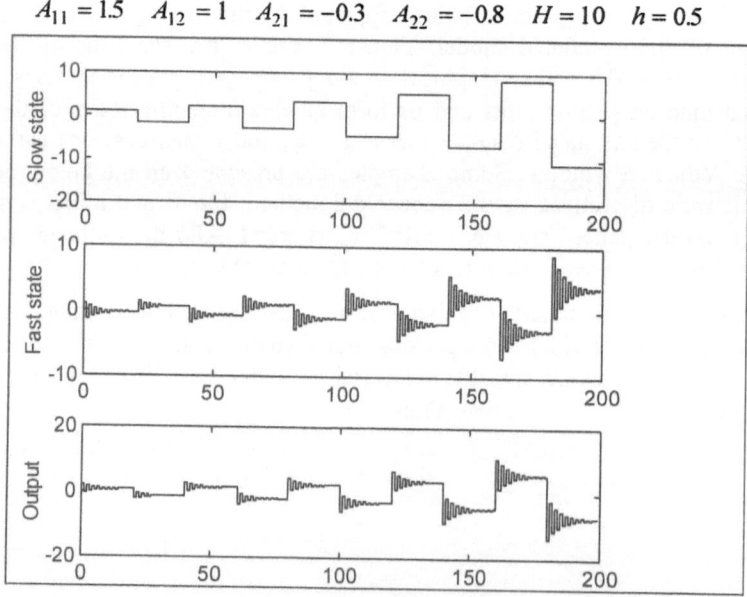

Figure 1. The main navigation window snapshot

$$A_{11} = 1.5 \quad A_{12} = 1 \quad A_{21} = -0.3 \quad A_{22} = -0.8 \quad H = 10 \quad h = 0.5$$

Figure 2. Discretization by trapezoidal rule. Fast variables are
stable while the slow and, hence, the overall are unstable.

$$A_{11} = -0.5 \quad A_{12} = 1 \quad A_{21} = -0.3 \quad A_{22} = -0.8 \quad H = 10 \quad h = 0.5$$

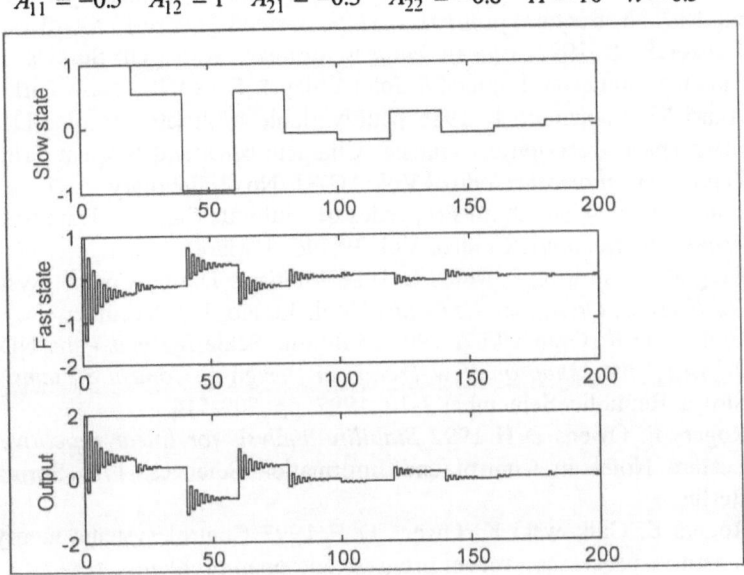

Figure 3. Discretization by trapezoidal rule. All states and outputs are stable.

7   Concluding remarks

The chapter presents a 2D approach of multitime scale systems as well as its 1D equivalent which is very useful while testing stability of 2D multitime models. Commonly known classical tools and methods can be used to make the tests. Also a brief introduction to a MATLAB based Toolbox for simulating multitime 2D scale systems was presented. This Toolbox is a part of computer tools developing by the authors and devoted to the 2D systems as a whole. The another Toolbox is presented on the Conference (LRP Toolbox - Linear Repetitive Processes Toolbox) and more information can be found in the proceeding materials.

References

1. Kokotovic P V, O'Malley R E, Sannuti P 1976 Singular Perturbations and Order Reduction in Control Theory -An Overview. *Automatica* Vol. 12:123-132
2. Saksena V A, O'Reilly J, Kokotovic P V 1984 Singular Perturbations and Time-scale Methods in Control Theory: Survey 1976-1983. *Automatica* Vol. 20, No. 3:273-293
3. Bose N K B 1985 *Multidimensional Systems Theory. Progress, Directions and Open Problems in Multidimensional System.* D Reidel, Dortrecht, Boston, Lancaster

4. Fornasini E, Marchesini G 1978 Doubly indexed dynamical systems: state-space models and structural properties. *Math. Systems Theory* 12:59-72
5. Kaczorek T 1992 *Linear control systems.* Research Studies Press LTD, Taunton, Somerset, England & John Wiley & Sons INC., New York
6. Akari M, Yamamoto K 1986 Multivariable Multirate Sampled-Data Systems: State_space Description, Transfer Characteristics and Nyquist Criterion. *IEEE Trans. On Automatic Control* Vol. AC-31, No. 2, February
7. Longhi S 1994 Structural Properties of Multirate Sampled-Data Systems. *IEEE Trans. On Automatic Control* Vol. 39, No. 3, March
8. Berg M C, Amit N, Powell J D 1988 Multirate Digital Control System Design. *IEEE Trans. On Automatic Control* Vol. 33, No. 12, December
9. Gałkowski K, Gramacki A 1997 Multitime Scale Systems - the ND Approach. *2nd IFAC Workshop on New Trends in Design of Control Systems, Smolenice, Slovak Republic*, September 7-10, 1997, pp. 509-514
10. Rogers E, Owens D H 1992 *Stability analysis for linear repetitive processes.* Lecture Notes in Control and Information Sciences, 175, Springer Verlag, Berlin
11. Rogers E, Gałkowski K, Owens D H 1997 Control systems theory for linear repetitive processes - recent progress and open problems. *Applied Mathematics and Computer Science, Zielona Góra, Poland* Vol. 7 No. 4:737-774
12. Sueur C, Dauphin-Tanguy G 1991 Bond graph approach to multi-time scale systems analysis. *J. of the Franklin Inst.*, Vol. 328, No. 5/6:1005-1026

PART II

CONTROL SYSTEM DESIGN

8

Bridging the Gap

C.H. Houpis

1   General Introduction

As an introduction to the main theme of this presentation, the author presents what he feels has occurred in some disciplines in academia. Since the early 50's, it is apparent that a shift of emphasis has occurred from a "true engineering curricula" to one that can be best described as an "engineering science curricula." This is more so in the graduate curriculum. Based upon the author's experience in the control system design area, a distinction is made between the scientific and engineering methods. This distinction is enhanced by the development of engineering rules, using the control area as the vehicle to demonstrate the concept of "bridging the gap" between these two methods. These engineering rules, based upon the control area, focus on achieving a successful "real-world" control system design. The design of control systems whose nonlinear characteristics are significant is addressed. A qualitative explanation of "structured plant parameter uncertainty" is presented. Some real-world control system designs will be highlighted. A design example is used to illustrate how the "real-world" knowledge of the plant to be controlled and the desired performance specifications can be utilized in trying to achieve a successful robust design for a nonlinear control problem. This presentation provides an overview of "using robust control system design to increase quality" in attempting to demonstrate the "bridging the gap" between control theory and the realities of a successful control system design. This "Bridging the Gap" must be addressed to better prepare the future engineers for the 21st century.

2   Scientific vs Engineering Method

A. *The Scientific Method* - Uses mathematical methods to gain insights into, to generalize, and to expand the state- of-the-art in many areas of science and technology.

B. *The Engineering Method* - Once the scientific method has successfully advanced the state-of-the-art, and when applicable, the engineer must take over and apply the new results to real-world problems.

C. *"Bridging The Gap"* between the two methods is best illustrated by the following anonymous saying:

> "In *THEORY* (scientist)
>> There is no difference between theory and practice.
> In *PRACTICE* (engineer)
>> There is a difference between practice and theory."

D. *The_IEEE_Trans._of_Automatic_Control* -- Stresses "Mathematical Control Theory" and emphasizes asymptotic results with what appears as down playing of the "engineering approach" to design. Consequently, "finite time," "small sample" real-world control system design and estimation are not properly addressed. Some engineering graduate schools have a tendency to do likewise in their approach to teaching control theory.

3 Structured Uncertainty (Plant Parameter Uncertainty) [1],[2],[4],[5]

A.. Most systems are nonlinear. Many of these systems can be described as systems containing structured plant parameter uncertainty.

B. What is "structured plant parameter uncertainty" (parametric uncertainty).

   1. *A_simple_illustration* of parametric uncertainty is described in Fig. 1.

   2. *A_simple_example_of_parametric_uncertainty_is_depicted_by_the following_simple second-order plant*:

$$P(s) = \frac{Ka}{s(s+a)}$$

(1)

where $K_{min} \leq K \leq K_{max}$ and $a_{min} \leq a \leq a_{max}$

    (a) Fig. 2 depicts the region of the structured plant parameter uncertainty for the min. and max. value of the gain K and the time constant a.

    (b) Fig. 3 shows the Bode plots of Lm $P(j\omega)$ vs ω and $\angle P(j\omega)$ vs ω for the 6 points shown in Fig. 2.

    (c) Templates, for various values of ω_i, representing the region of plant parameter uncertainty are shown in Fig. 4.

   3. *A_more_complex_plant_example* -- Templates representing the region of plant parameter uncertainty, for a given aircraft flight scenario, are shown in Fig. 5.

4 Robust Control System: definition

A control system (Fig. 6 or Fig.7) that has been designed to meet the desired system performance specifications despite structured plant parameter uncertainty, control effector failure(s), and plant disturbances is defined as a robust control system.

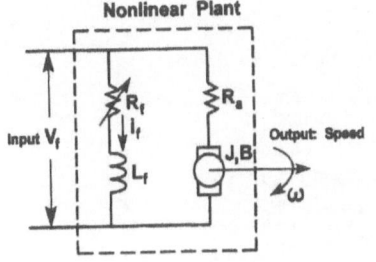

Figure 1: What is Parametric Uncertainty.

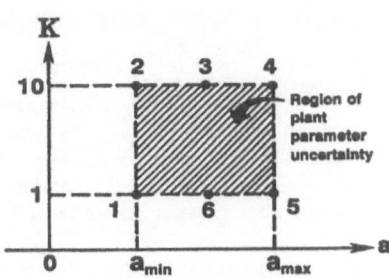

Figure 2: Region of plant uncertainty

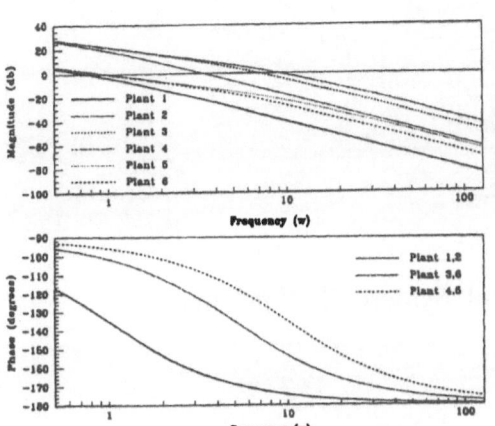

Figure 3: The Bode plot of six LTI plants that represent the range of the plant's parameter uncertainty.

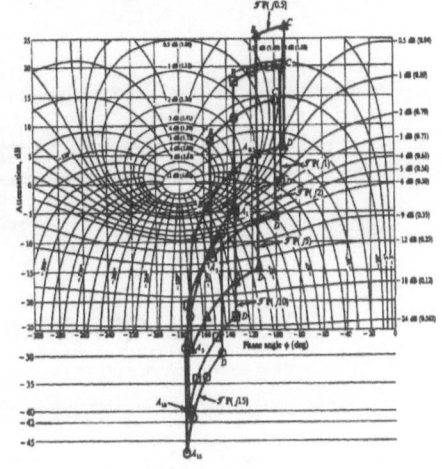

Figure 4: Templates drawn, for various values of ω_i from Fig. 6.

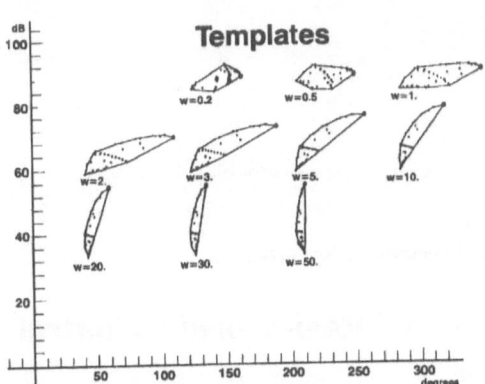

Figure 5: Templates for various values of ω_i, for a high order plant.

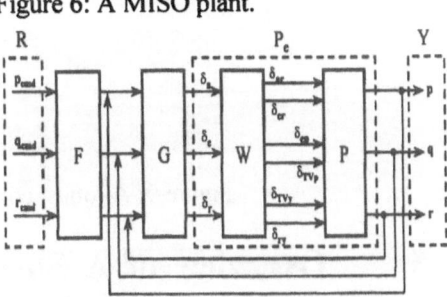

Figure 6: A MISO plant.

Figure 7: MIMO QFT control structure block diagram.

92

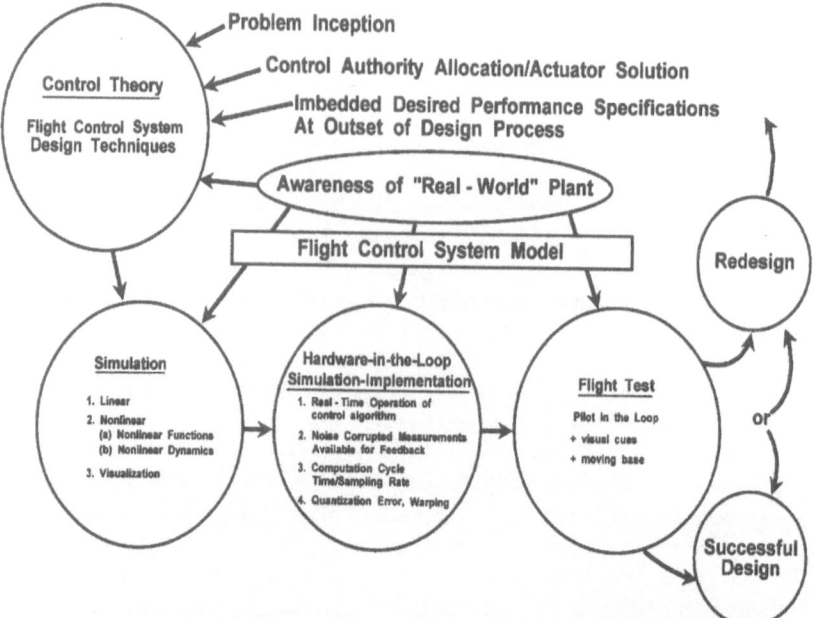

Fig. 8 The QFT Control System Design Process

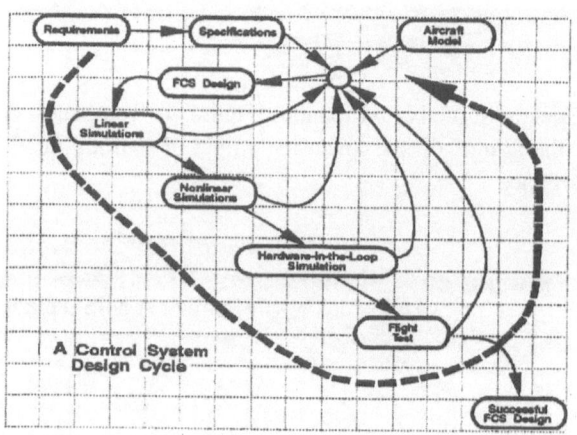

Figure 9: A robust control system design process.

5   Overview of A Successful "Real-World" Control System Design Process

A. *The_"QFT_Design_Process"_Flow_Chart_[8]*   There are many important factors that must be considered, both from the theoretical viewpoint and the real-world aspects of the control application, during and upon the completion of a successful control system design. The major factors that play a vital role in the design process are depicted in Fig. 8 for a flight control system. This figure shows four major aspects of the control design process: Control Theory (Design Techniques), Simulation, Implementation, and Flight Test.

B. *A_Robust_Control_System_Design_Process_Flow_Chart*[3] -- Figure 9 depicts the key elements in achieving a robust control system design process.

C. *"Bridging_The_Gap"* -- An engineer who has a firm understanding of the results of the "scientific method" and has a firm understanding of the nature and characteristics of the plant to be controlled must develop his or her own "engineering method" by developing appropriate "Engineering Rules" (see Sec. VIII) in order to "bridge the gap."

6 Items of Concern for a Successful Practical Control System Design Process[6],[7]

Examples_of_Engineering_Rules - Through the years of applying the QFT robust control design technique to many real world nonlinear control problems Engineering Rules have evolved. Many of these are applicable for other design techniques. Examples are:

E.R.-1 *Weighting_Matrix* -- When a weighting matrix $W = \{w_{ij}\}$ is required to achieve a square equivalent plant, i.e.,

$$P_e = PW \tag{2}$$

where

$$W = \{w_{ij}\}, P_e = \{p_{ij}\}, P_e^{-1} = \{p_{ij}^*\}, Q = \{q_{ij}\}, \text{ and } q_{ij} = 1/p_{ij}^* \tag{3}$$

It is desired to know at the onset if it is possible to achieve minimum-phase (m.p.) q_{ii}'s $(q_{ii} = 1/p_{ii}^{-1})$ by the proper selection of the w_{ij} elements. Now, m.p. q_{ii} plants are most desirable for they allow the full exploitation of the "benefits of feedback," i.e., high gain. It turns out that one can apply the Binet-Cauchy theorem (see Chap V in Reference 2) to determine if m.p. q_{ii}'s are possible. Also, it may be desirable to obtain complete decoupling for the nominal plant case, i.e.,

$$P_{e_{diag}} = \begin{bmatrix} p_{11} & 0 & \cdots & 0 \\ 0 & p_{22} & \cdots & 0 \\ \vdots & \vdots & \cdots & \vdots \\ 0 & 0 & \cdots & p_{mm} \end{bmatrix} = PW \tag{4}$$

Although for the non-nominal plants complete decoupling, in general, will not occur, the degree of decoupling will have been enhanced.

E.R.2 *n.m.p._q_{ii}'s* -- For q_{ii}'s that are non-minimum-phase (n.m.p.) one must determine if the location of the RHP zero(s) is in a region which will not present a problem for the real-world design problem being considered. For manual flight control systems,, if a RHP zero happens to be "close" to the origin (see Fig. 10), this is not necessarily deleterious since the pilot inputs a new command before its effect is noticeable; in other words, it is assumed that the unstable pole is outside the closed loop system's lower bandwidth. If this RHP zero is "far out" to the right, it is outside the bandwidth of concern in manual control and it does not present a problem. For these cases a satisfactory QFT design may be achievable.

E.R.3 *Templates* -- The adage "a picture is worth a thousand words" applies to the preliminary task of determining if a robust control solution exists, bearing in mind the need to satisfy tracking specifications, external disturbance and cross-coupling effects rejection, and satisfying the (robust) stability bounds. The theorems, corollaries, and/or lemmas pertaining to these bounds, obtained by the scientific method, may reveal that no loop shaping solution is possible. If so, then one must be attuned to stepping back and doing a "trade-off" in which some specifications are relaxed in order to achieve a solution, or one must be willing to live with a degree of gain scheduling. Thus, a graphical analysis of Fig. 5 can reveal the following:

(a) The maximum template height, in dB, is too large, indicating that the tracking specifications, at a given frequency ω_i, is not is not achievable. One can then decide if gain scheduling is required and is feasible in order to yield a tracking bound that satisfies the tracking specification at this frequency.

(b) The situation where the templates are too "wide" (the width reflects the magnitude of phase angle uncertainty at a given frequency) thus prohibiting a QFT solution or a solution by any other multivariable design technique. This is especially true for real-world control problems that involve control effector failures accommodation. In these design problems, generally, the worst failure case is the culprit in generating this large "angle width." Thus, in order to achieve a solution one needs to relax the requirement that the "worst failure case" be accommodated. Naturally, when this situation arises, it is necessary to stipulate for what failure cases a successful design is achievable. In determining "reasonable failure cases" that can be accommodated by robust (not adaptive) control one must consider if *10%, 25%, 50%, 80%* failure still permits enough control authority! This % of failure is a judgement call that can only be made by a person who is knowledgeable of the physical plant to be controlled. In general, knowledge of the plant (application) "is king" when it comes to the design of a feedback compensator or controller for the said plant.

(c) The effects of structured uncertainty on the template's geometry are now discussed. Thus, in flight control[3], linearized plants that represent different flight conditions in the flight envelope are extracted from a nonlinear truth model. An attempt is made to choose flight conditions in such a way as to fully cover the flight envelope with the templates. To do this, a nominal flight condition for an unmanned research vehicle was chosen to be *50 kts* forward velocity, *1000 ft* altitude, a weight of *205 pounds* and center of gravity at *29.9%* of the mean aerodynamic chord. From this nominal flight condition, each parameter was varied, in steps, through maximum and minimum values. These variations produced an initial set of templates.

It should be noted that if all the p_{ij}'s of P do not have the same value of λ (excess of poles over zeros) then as $\omega \to \infty$ the templates may not become straight lines. A possible method of reducing the size of the templates is given by E.R.7.

E.R.4 *Design_Techniques* -- No matter what design method one uses, performance specifications must be realistic and commensurate with the real world plant being controlled. Situations have occurred where the conclusion was reached that no acceptable designs were possible. For these situations when one "stepped back" and asked the pertinent question "was something demanded that this plant physically cannot deliver regardless of the control design technique?" it was determined that some or all of the prescribed performance specifications were unrealistic.

E.R.5 *Minimum_order_Compensator_(MOC)* -- In order to ensure the smallest possible order compensator/controller, one starts the loop shaping process by using the loop's nominal plant $L_{ol} = q_{11o}$, and then zeros and poles are successively added in order to obtain the required loop shape, resulting in:

$$L_o = \frac{L_{ol}(s - z_1)\ldots(s - z_w)}{(s - p_1)\ldots(s - p_v)} \tag{5}$$

Finally, the compensator is obtained from $g_l = L_o/q_{11o}$. Thus, the nominal plant's poles and zeros are being used to shape the loop. This insures that the ensuing compensator/controller is of the lowest order, which is highly desirable.

E.R.6 *Minimum_Compensator_Gain* -- To minimize the effects of noise, saturation, etc., it is desirable to minimize the amount of gain required in each loop *i*, while at the same time meet the performance specifications in the face of the given structured uncertainty. To achieve this goal, a control system designer, with a good understanding of the Nichols Chart and a good interactive QFT CAD package, can use his "engineering talent" to make use of the "dips" (troughs) in the optimal or composite $B_{oi}(j\omega_j)$ (see Fig. 11). The designer by shaping L_{oi} to pass through these dips, *where feasible*, can ensure achieving the minimum compensator gain that is realistically possible. To achieve this by an automatic loop shaping routine may be difficult.

E.R.7 *Basic_mxm_Plant_Preconditioning* -- When appropriate, utilize unity feedback loops for the *mxm* MIMO plant P which will yield an *mxm* preconditioned plant matrix P_P. The templates $\Im P_p(j\omega_i)$, in general can be smaller in size than the templates $\Im P(j\omega_i)$. This template reduction size is predicted by performing a sensitivity analysis (see Sec. 14.2 of Reference 4). The QFT design is performed utilizing the preconditioned matrix P_P. This concept has been used in a number of MIMO QFT designs (see References 56,77,78,121 in Ref. 2).

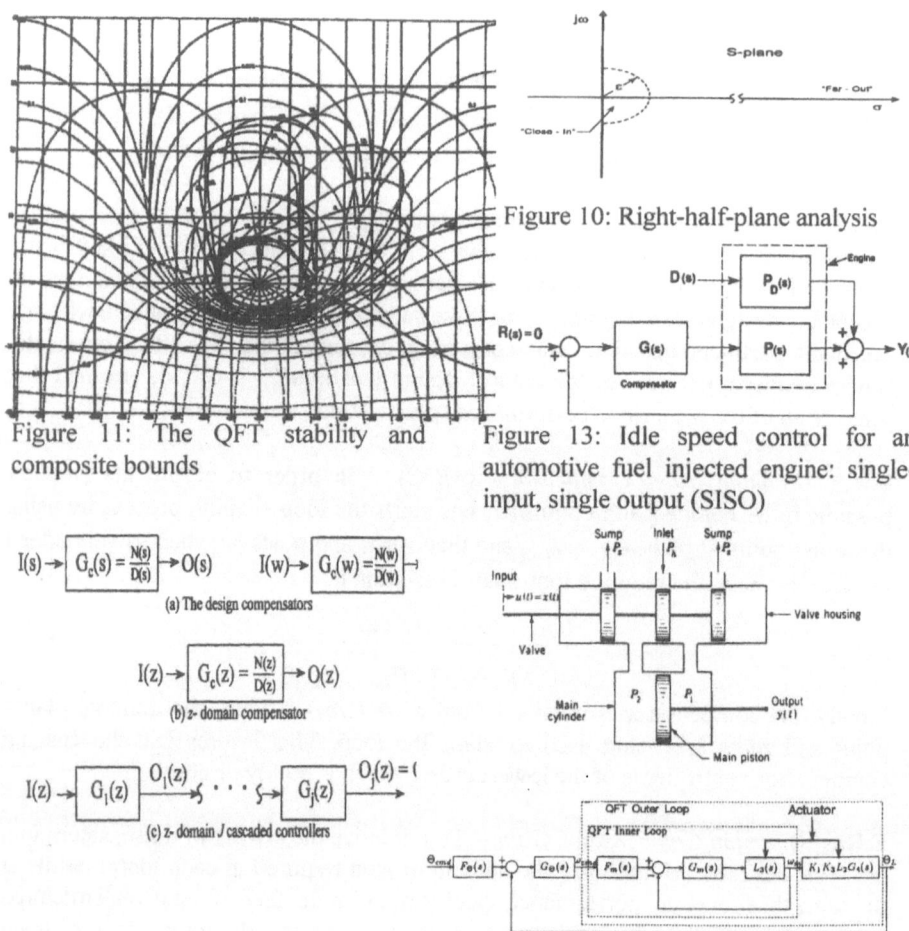

Figure 10: Right-half-plane analysis

Figure 11: The QFT stability and composite bounds

Figure 13: Idle speed control for an automotive fuel injected engine: single-input, single output (SISO)

Figure 12: s- or w- domain to z domain bilinear transformation: formulation for implementation of the G(z) controller.

Figure 14: A robust actuator control system: (a) hydraulic actuator [10], (b) Two feedback loops in a QFT two degrees of freedom structure

E.R.8 *Nominal_Plant_Determination* -- It is easy to determine the phase margin angle γ, the gain margin, and the phase margin frequency ω_ϕ of a feedback control system using the NC. Thus, QFT affords the robust establishment of these FOM's. Indeed, by choosing the nominal plant: (a) to correspond to the maximum dB plant on the $\omega_{\phi i} = BW(L_i)$ template *ensures_that* $\omega_{\phi i} \leq BW(L_i)$ for all plants; (b) for the achievement of *a_robustly_guaranteed_gain_margin* is easily accomplished provided the nominal plant is uniformly the maximum dB plant for all templates; and (c) which is the "left-most" plant on all templates ensures that the desired γ is robustly achieved.

E.R. 9 *Optimization_and_Simulation_Run_Time* -- In many real-world manual feedback linear or nonlinear control problems, the goodness of the design is judged on a pre-specified and limited planning time horizon beyond which the control performance is less important since the human operator will inject new inputs to the system. For example, in manual flight control the time horizon is determined by the aircraft's short period dynamics, e.g., *5 seconds* and there is no interest in the long time intervals commensurate with the slow phugoid dynamics.

E.R. 10 *Asymptotic_Results* -- Asymptotic results by mathematical analysis are not as useful as they seem to be. Consider the manual control disturbance rejection case where fast disturbance attenuation is more desirable than, say, total disturbance rejection which entails a very long "settling time."

E.R. 11 *Controller_Implementation* -- Tight performance specifications and a high degree of uncertainty require small sampling intervals *T*. Unfortunately, the smaller the value of *T*, the greater the degree of accuracy that is required to be maintained. The numerical accuracy is enhanced in using a factored representation of the controller and prefilter[5]. For example, by use of the bilinear transformation the controller $G_z(z)$ of Fig. 12(b) is obtained from the compensator $G_c(s)$ or $G_c(w)$ of Fig. 12(a). The equivalent cascaded transfer function representation (factored representation) of $G_z(z)$, shown in Fig. 12(c), is utilized to obtained the algorithm for the software implementation of $G_z(z)$.

E.R. 12 *Non-Ideal_Step_Forcing_Function* – For mathematical analysis of LTI systems ideal step functions do not exist. Thus, for a more realistic test in determining a control system's performance through simulation a *ramped-up step function* is used. Use of this type of a realistic forcing function has the tendency to minimize the saturating aspects of the system.

7 Deciding on a Control System Design Method To Use

What is the controlled variable?	(e.g., in flight control: pitch rate, normal acceleration, or a combination thereof)
Is the plant nonlinear?	Analog or a sampled-data control system
Order of compensator	Order of controller: *m-th* order m-time delays & controller implementation

Gain requirement for tracking performance, disturbance rejection, and robustness versus sensor noise and actuator saturation tradeoffs

8   Design Examples

8.1   Idle Speed Control for an Automotive Fuel Injected Engine [9]

The following is the abstract from the paper entitled "Robust Controller Design and Experimental Verification of I.C. Engine Speed Control" by Dr. M. A Franchek and R. Herrick, School of Mechanical Engineering, Purdue University:

Presented in this paper is the robust idle speed control ol of a Ford 4.6L V-8 fuel injected engine.   The   goal of this investigation is to design a robust feedback controller that maintains the idle speed within a 150 rpm tolerance about 600 rpm despite a 20 Nm step torque disturbance delivered by the power steering pump. The controlled input is the by-pass air valve which is subjected to an output saturation constraint. Issues complicating the controller design include the nonlinear nature of the engine dynamics, the induction-to-power delay of the manifold filling dynamics, and the saturation constraint of the by-pass air valve. An experimental verification of the proposed controller, utilizing the nonlinear plant, is included.

The authors' control system is shown in Fig. 13 that represents a SISO regulator control system. The authors show in their paper that they met all the design objectives and have achieved excellent results.

8.2   Control System for an Actuator Plant [10]

Actuators can be regarded as feedback control systems in their own right. In this work the mechanical feedback link commonly used in actuators is opened and replaced by a sensor and an electronic controller that drives the valve. The latter governs the piston, which is the power element. This control design problem entails the design of a robust controller for an actuator that: (1) takes into account the aging of some of the actuator's components over its expected life-time, and (2) the actuator's manufacturing tolerances such that when the actuator needs to be replaced, the overall control system's robustness is maintained by its replacement. Based upon these two factors, the structured parametric uncertainty of the hydraulic actuator[10] of Fig. 14(a), for various values of frequency is described by the templates of Fig. 15(a). The control system of Fig. 14(b) is designed to achieve the desired robust performance. As indicated in this figure, an inner loop QFT design is first accomplished to primarily "shrink" (E.R.7) the size of templates, as shown in Fig. 15(b), in order to enhance the eventual achievement of the desired degree of robustness when the outer loop is closed. The resultant QFT design of the actuator's control system shown in Fig. 14(b) yielded the desired degree of robustness. More complex examples exist in the literature.

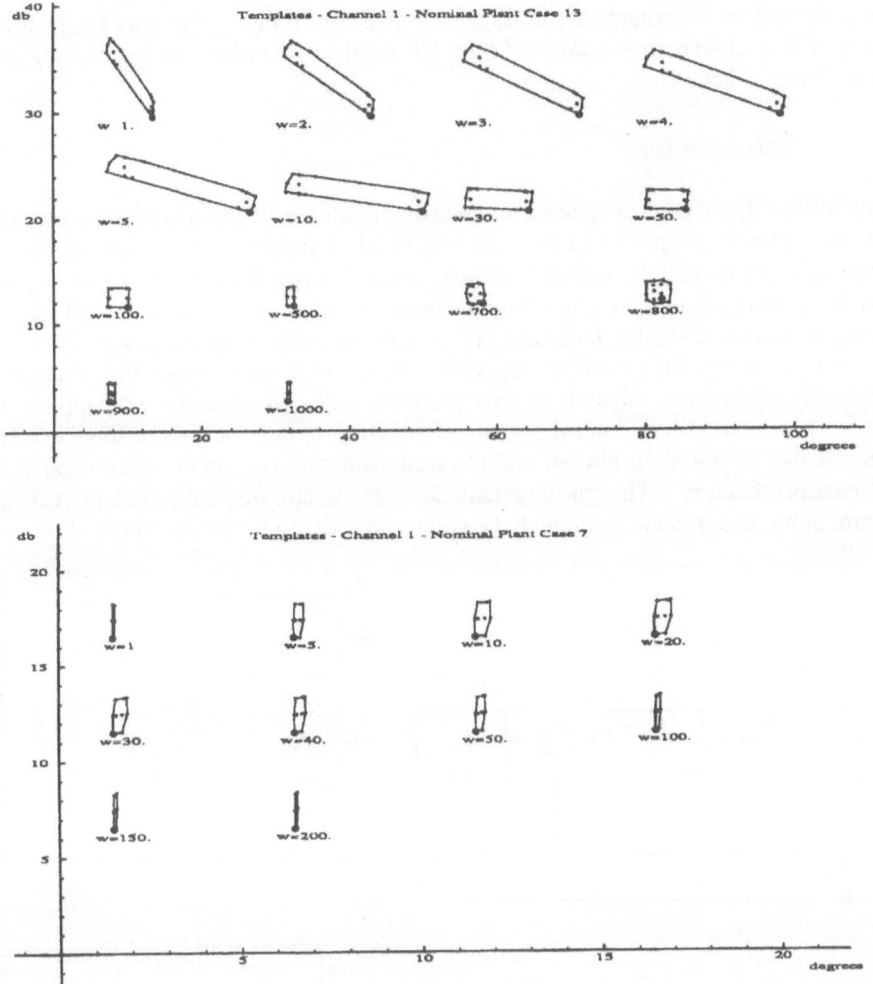

Figure 15: Templates of (a) Inner loop plant templates (b) Outer loop plant templates

9   An Illustrative Example: "A Quantitative Feedback Theory FCS Design for the Subsonic Envelope of the Vista F-16 Including Configuration Variation and Aerodynamic Control Effector Failures[13] "

It is the intent of this section to illustrate the "bridging of the gap" between the scientific and the engineering methods. To bridge this gap it is necessary for the control system designer to have a good understanding of the physical characteristics of the plant and of the control application. To assists in bridging this gap Engineering Rules (E.R.), similar to the ones of Sec. VI, need to be utilized in performing and analyzing a design.

9.1   Introduction

Fault tolerant flight control systems for combat aircraft are an alternative to excessively redundant aircraft designs or adaptive reconfigurable control laws. However, due to the range of flight conditions within a combat aircraft's operational flight envelope, the variety of its configurations, and the unavailability of an aerodynamic data-base for damaged aircraft, designing fault tolerant systems is a complicated endeavor. QFT is a robust control design technique especially well suited to manage the structured parametric uncertainty inherent in this problem, and consequently was applied in achieving a robust flight control system. Furthermore, realistic failure models were used for the VISTA F-16 aircraft and physical saturation constraints were applied to the control effectors. The ensuing fault tolerant design was subjected to realistic control inputs and was validated with the applicable MIL STD specifications.

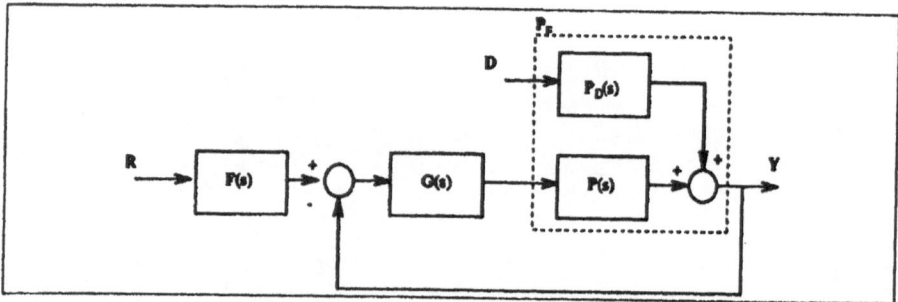

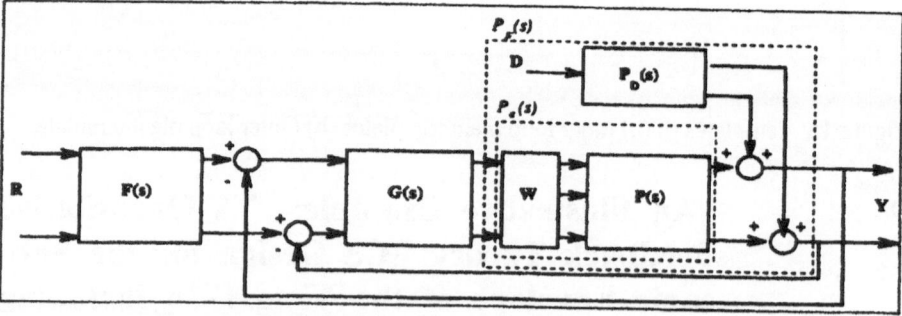

Figure 16: (a) Longitudal QFT feedback structure (b) General MIMO QFT system with an external disturbance included.

9.2   Scope

The robust design problem involved compensator and prefilter designs for both the longitudinal and lateral/directional aircraft channels. The longitudinal design is strictly a MISO structure [see Fig. 16(a)] with single control surface failures, while the lateral/directional FCS is a *2x2* MIMO control system [see Fig. 16(b)] allows for single, double, and triple control surface failures. In addition to the failure conditions imposed on the flight control design, rate and actuator saturation nonlinearities are included to simulate real world physical constraints. It should be noted that the external disturbance inputs shown in Fig. 16 are included to model the coupling of the lateral/directional and the longitudinal aircraft channels. Success of the design was determined by meeting the MIL STD 1797A Level 1 flying qualities for the healthy aircraft and Level 2 or 3 for the effector failure cases.

Table 1 Aircraft 's Healthy/Failure Scenario

	Longitudinal	Lateral/Directional		
Types of Failures	**Elevator**	**Differential/ Tail**	**Rudder**	**Ailerons**
No Failures	Healthy (H)	Healthy	Healthy	Healthy
Single Failure	Fail 15%	Fail 25%	H	H
	Fail 25%	Fail 45%	H	H
	H	H	Fail 25%	H
	H	H	Fail 45%	H
	H	H	H	Fail 25%
	H	H	H	Fail 45%
Multiple Failures	H	Fail 25%	Fail 25%	H
	H	Fail 25%	H	Fail 25%
	H	H	Fail 25%	Fail 25%
	H	Fail25%	Fail 25%	Fail 25%
	H	Fail 45%	Fail 45%	H
	H	Fail 45%	H	Fail 45%
	H	H	Fail 45%	Fail 45%
	H	Fail 45%	Fail 45%	Fail 45%

9.3 Healthy/Failure Scenario

The Table 1 lists the healthy(H)/failure test cases as the basis for this robust design. E.R.s 3(b) and 4 were kept in mind in determining the failure cases.

9.4 Weighting Matrix

A MIMO QFT design requirement is that the number of system inputs must equal the number of system outputs, in other words the plant must be square. To square the plant, a weighting matrix W is introduced (E.R.1). It is well established that there is a connection between the level of uncertainty and the weighting matrix. Weighting matrices were determined that reduced the level of the uncertainty (E.R.7) and thus the difficulty of achieving a robust QFT design.

9.5   Design Specifications

The QFT robust design technique involved designing the prefilters and the compensators of Fig. 16 so that the resultant closed-loop frequency responses for all flight scenarios lie between the specified upper T_{RU} and the lower T_{RL} bounds of Fig. 17. E.R.s 6 and 9 were invoked in order to achieve a minimum order compensator.

9.6   Longitudinal Channel Design

Template Analysis -- Template analysis, based on E.R.s 3 and 8, was used to select the nominal plant to be used for the design and to assist in the loop shaping (E.L. 6) process.

Stability Validation -- Since the QFT design process is based on the manipulation of one nominal plant, it is necessary to guarantee that *all 398* plants (*199* healthy plus

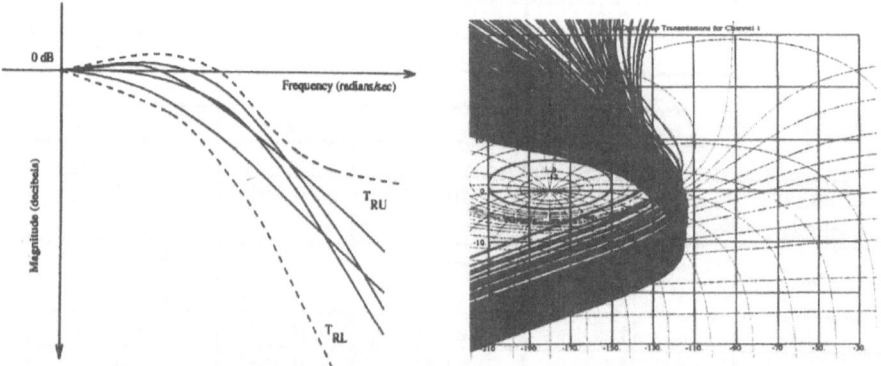

Figure 17: Compensated closed-loop frequency response with prefilter

Figure 18: QFT stability validation

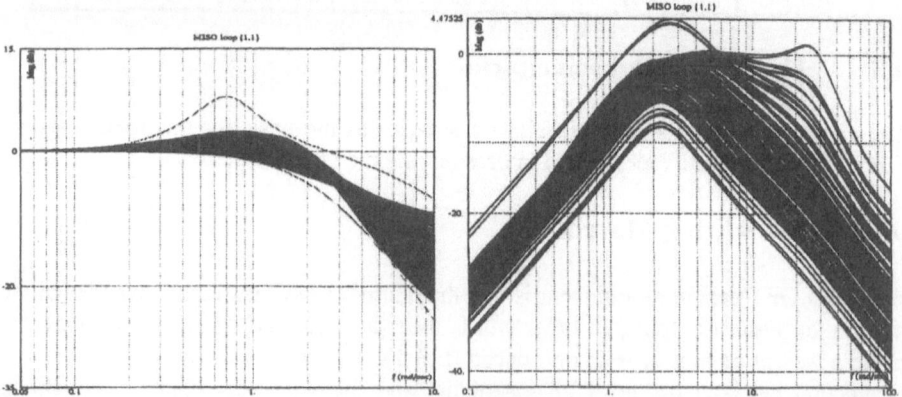

Fig. 20 QFT disturbance rejection validation.

Figure 19: QFTCAD tracking validation

Figure 20: QFT disturbance rejection validation

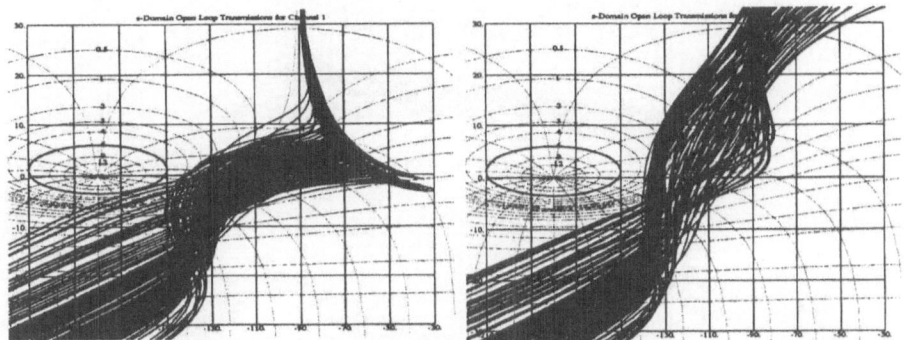

Figure 21:QFT stability validation for sideslip (*p*) channel.

Figure 22: QFT stability validation for roll (*p*) channel.

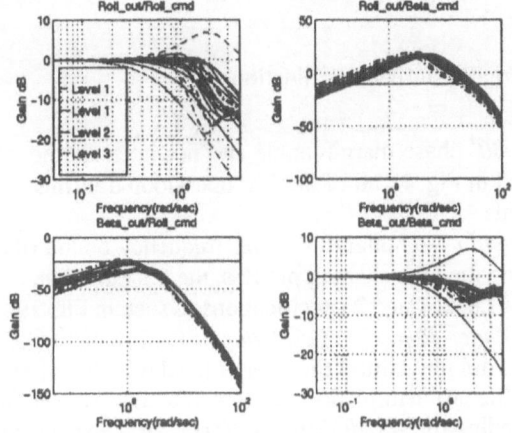

Figure 23: QFT tracking validation for the lateral / directional channel ($\bar{q}$ <150 lbs/ft^2)

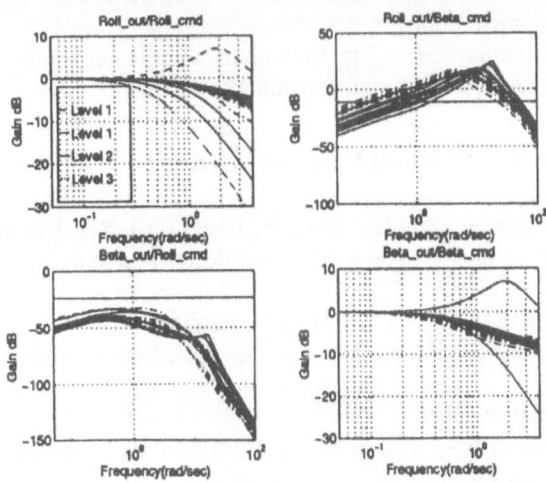

Figure 24:QFT tracking validation for the lateral / directional channel ($\bar{q}$ <150 lbs/ft^2)

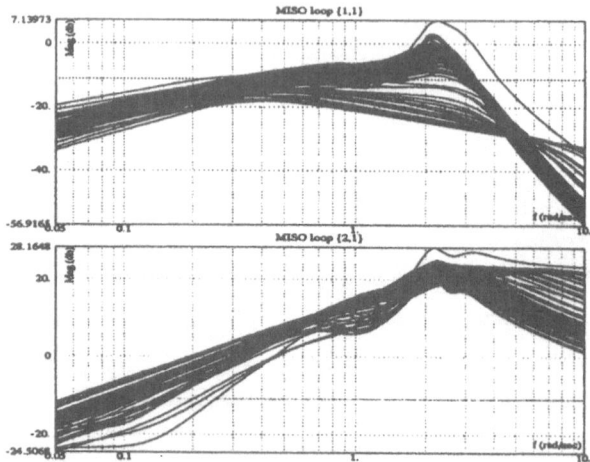

Figure 25: QFT external disturbance validation

199 failed) satisfy a *30°* phase margin angle and not intersect the M_L = *6 dB* stability contour. As indicated in Fig. 18 all of the *398* open-loop transmission functions satisfy these two requirements.

Tracking Validation -- Though the lower frequency bound (dashed curve) could not be achieved during the loop shaping process, the tracking time responses within the pilot's bandwidth met Level 1 and 2 specifications as seen in Fig. 19.

External Disturbance Rejection Validation -- Unfortunately, as shown in Fig. 20 illustrates, the system did not exhibit the overall level of external disturbance rejection initially mandated. The problem plants represented by those exceeding the *0 dB* line at *2 rps* and those exceeding the *-11 dB* limit at *30 rps*, are identified as the low dynamic pressure failure plants. Due to the conservatism inherent in the QFT robust design technique it is left up to the time simulation to determine if these trouble plants could be controlled by a human pilot.

Time Domain Validations -- The time domain nonlinear simulations (E.R.s 9 and 12) verify that all the required design benchmarks are met.

9.7   Lateral/Directional Channel Design

Template Analysis -- Template analysis, based on E.R.3, was used to select the nominal plant (E.R.s 5 and 8) to be used for the design and to assist in the loop shaping (E.R. 6) process.

Stability Validation -- None of the resultant open-loop sideslip channel transmission functions (see Fig. 21) and the open-loop roll channel transmission functions (see Fig. 22) intersected their M_L = *6 dB* contour thus the stability requirement has been achieved.

Tracking Validation -- The tracking validation plots are shown in Figs. 23 (low $\bar{q}$) and 24 (high $\bar{q}$). The tracking specifications were met except for the Roll$_{out}$/Beta$_{cmd}$ system that does not meet *the -11 dB* specification as seen in both figures. Thus, due to the conservatism of the design technique, it is left up to the nonlinear time domain simulation for the determination of the goodness of the design.

External Disturbance Rejection Validation -- The external disturbance validation plots in Fig. 25 are deceiving due to a subtlety of the external disturbance modeling process. Traditionally the external disturbance rejection plot represents how the system responds to a unit step external disturbance input. However, in this design, the external disturbance magnitude was enhanced by the longitudinal and lateral models. Thus, using a unit step external disturbance input in this design, represents the effect of applying a *5°* stabilator deflection. In other words, the external disturbance rejection plots in Fig. 25 are approximately *14 dB* greater then they should be. After reducing each of these figures by *14 dB*, the responses [MISO(1,1)] are within acceptable limits.. Finally, the external disturbance simulations are necessary to determine if the impaired plants can be controlled by a human pilot.

Actuator Saturation -- While actuator saturation is an important concern, as pointed out previously, for mathematical analysis of LTI systems, ideal step forcing functions are utilized. In the real world, ideal step forcing functions do not exist. Thus, a *Non-ideal Step Forcing_Function* (E.R. 12) is a more realistic nonlinear simulation test forcing function in determining a control system's performance; i.e., a *ramped-up step function* is used. Use of this type of realistic forcing function has the tendency to minimize the saturating aspects of a system.

Time Domain Validation (E.L.s 9 and 12) -- The time domain validation verifies that the FCS does meet the time domain specifications when realistically high amplitude inputs, and rate and deflection saturation limitations are applied. The FCS does indeed maintain stability, even though the extreme low $\bar{q}$ failed plants saturate the aileron deflection. The low $\bar{q}$ cases however do not meet the roll angle performance criteria since there is simply insufficient control authority available at low dynamic pressure to adhere to the MIL STD specifications. The external disturbance rejection simulations for the low dynamic pressure clearly satisfy the performance specifications. The high-dynamic pressure external disturbance time responses exhibit superior rejection in comparison to the low $\bar{q}$ plants. These high $\bar{q}$ plants easily satisfy the disturbance rejection settling, stability, and saturation requirements.

9.8   Design Example Summary

As much as time and space allowed, this example, as stated earlier, is intended to present an insight into the concept of bridging the gap between the scientific and engineering methods.

10   Summary

Engineers are applying the results of the scientific method to achieving solutions for real world problems. The QFT robust design technique was utilized in this paper to illustrate how the bridging of the gap between the scientific and engineering methods can be achieved. As an example, an aerospace firm has stated to Professor M. Grimble, University of Strathcldye, the following: "The QFT approach has the obvious advantage that it is close to engineers' existing experience in classical design methods. However, it provides facilities to deal with uncertainty which are not available in traditional methods. More recent tools such as H_∞ design also show promise but are very different to the existing procedures used in parts of the Aerospace industry. The QFT approach therefore appears to have the attractive features of providing a link with existing techniques while at the same time providing many of the advanced features needed for the 90's high performance systems. What might be needed are tools for the future which combine the attractive features of the QFT and H_∞ approaches." The guidelines, such as Engineering Rules, are presented to illustrate to the control engineer and to engineers in other disciplines on how to use Engineering Rules to assist them in interfacing between the scientific method and the engineering method – "Bridging the Gap." In conclusion, an attempt is made to bridge the often lamented gap between the scientific and engineering methods in order to prepare the future engineers for the complex engineering problems of the 21st century ."

References

1. Horowitz I M 1982 Quantitative Feedback Theory. *Proceedings IEE* 129D(6)
2. Houpis C H et al 1995 Quantitative Feedback Theory for the Engineer. *Wright Laboratory Technical Report, WL-TR-95-3061, Wright Laboratory, Wright-Patterson AFB, OH* (Available from National Technical Information Service, 5285 Port Royal Road, Springfield, VA 22151, document number AD-A297574.)
3. Rasmussen S J Houpis C H 1994 Development, Implementation, and Flight Test of a MIMO Digital Flight Control System for an Unmanned Vehicle. *The Winter Meeting of ASME Chicago, ILL., DSC-12C*
4. D'Azzo J J Houpis C H *1996 Linear Control System Analysis and Design; Conventional and Modern.* McGraw-Hill, Inc., 4th ed., New York
5. Houpis C H Lamont G B *1992 Digital Control Systems: Theory, Hardware, Software.* McGraw-Hill, Inc., Chapter 10, 2nd ed., New York
6. Houpis C H Pachter M 1995 Application of QFT to Control System Design - An Outline. *The Winter Meeting of ASME, San Francisco, CA.*
7. Houpis C.H Pachter M 1997 Application of QFT to Control System Design -- For Engineers. *Int. J. of Robust and Nonlinear Control* 17
8. Houpis C H 1992 Quantitative Feedback Theory Symposium *Proceedings WL-TR-92-3063, Air Force Wright Aeronautical Laboratories, Wright-Patterson AFB, OH,* (Available from Defense Technical Information Center, Cameron Station, Alexandria, VA 22314, document number AD-176883.)
9. Hamilton G K Franchek M A 1997 Robust Controller Design and Experimental Verification of I.C. Engine Speed Control, *Int. J. Robust and Nonlinear Control* 17
10. Kang K Pachter M Houpis C H 1995 Modeling and Control of an Electro-Hydrostatic Actuator. *National Aerospace and Electronics Conference (NAECON), Dayton, OH.*
11. Cacciatore V J 1995 *A Quantitative Feedback Theory FCS Design for the Subsonic Envelope of the VISTA F-16 Including Configuration Variation and Aerodynamic Control Effector Failures.* MS Thesis, AFIT/GE/ENG/95D-04, Graduate School of Engineering, Air Force Institute of Technology, Wright-Patterson AFB, OH

9

Control System Design Using Global Optimization Techniques

D. Famularo, P. Pugliese and Ya D. Sergeyev

1 Introduction

In the last decade, the development of new computational theories and tools has led to the reformulation of some classical problems in Control Theory, and to efficient algorithms for their solution. The best known example of these new paradigms is probably the Linear Matrix Inequalities approach to the analysis and synthesis of control systems [1].

Even more classical approaches, which translate the various control problems into optimization problems, have received renewed consideration as efficient optimization techniques have become available. Examples are the $\mathcal{H}_\infty$-Optimization approach to robust control system design [2] or the Complex Structured Singular Value (μ) for robustness/performance analysis [3]. Such examples let us say that computational system analysis is today an important chapter in Control Theory.

In this spirit, we present here the application of some recent results in the field of Global Optimization to the synthesis of linear control systems. The algorithm we consider has shown to be effective for the problem of stability and performance verification, and for the related problem of stability and performance margin computation [4].

All such problems were unconstrained, besides the obvious requirement of bounded decision variables. Here we apply the algorithm to the design of control systems, formulating this problem as a constrained optimization problem, and taking the constraints into account using penalty functions. The ability of the algorithm to deal with non-smooth functions makes the choice of the penalty functions and coefficients an easy task.

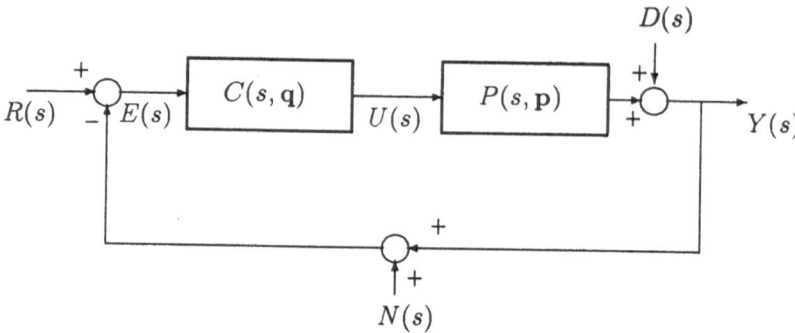

Figure 1: Standard Feedback System

2 Problem formulation

Let us consider the linear, time-invariant, SISO control system shown in Figure 1. It is well-known [5] that many control problems for such class of systems (e.g., multiobjective robust synthesis, parametric stability margin computation) can be formulated as the task of minimizing a nonlinear function, subject to some nonlinear constraints, i.e.:

$$\min_{\mathbf{q}} \qquad \mathcal{J}(\mathbf{q}, \mathbf{p}_0)$$

$$\text{subject to:} \quad \begin{cases} \mathcal{G}_i(\omega, \mathbf{q}, \mathbf{p}) < 0, \ i = 1, \dots, N \\ \mathbf{p} \in \mathbf{\Pi}, \mathbf{q} \in \mathbf{\Gamma}, \\ \omega \in [\omega_{\min}, \omega_{\max}]. \end{cases} \qquad (1)$$

In the above formulation, the vector $\mathbf{p}$ represents the uncertain plant parameters, $\mathbf{p}_0$ its nominal value, the vector $\mathbf{q}$ represents the controller parameters to be tuned, $\mathbf{\Pi}$ and $\mathbf{\Gamma}$ are hyperrectangles, and ω is the angular frequency.

The objective function and constraints in the above problem are derived from closed loop stability (or α-stability) requirements (Routh-Hurwitz criterion), from loop-shaping inequalities, or from specifications on the control energy, static and dynamic precision, sensitivity to disturbances, robustness against physical parameter variations, etc.

In particular, loop-shaping specifications take the form

$$f_{\min}(\omega) < |H(j\omega, \mathbf{p}, \mathbf{q})| < f_{\max}(\omega),$$

where, denoting by $L(s) = C(s, \mathbf{q})P(s, \mathbf{p})$ the loop-gain of the system, the function $H(s, \mathbf{p}, \mathbf{q})$ can be either the *Sensitivity* $S(s, \mathbf{p}, \mathbf{q}) = 1/(1 + L(s))$, or the *Complementary sensitivity* $T(s, \mathbf{p}, \mathbf{q}) = L(s)/(1 + L(s))$, or the *Control effort* $M(s, \mathbf{p}, \mathbf{q}) = C(s, \mathbf{q})/(1 + L(s))$.

For some classes of problems (e.g., parametric stability margin computation), no constraints are imposed, but the solution of the problem requires a bisection search over the dimensions of $\mathbf{\Pi}$ [6] and the solution of a global minimization problem at each step, therefore the efficiency of the algorithm being used is crucial.

From the computational point of view, difficulties arise because the objective function and the constraints are *multivariate*. It is important to recognize that only the *global* minimum of the problem is of interest [5], in particular when stability has to be imposed or checked, and that this problem belongs to the $\mathscr{NP}$-Hard complexity class [7].

3 Global Optimization

A number of numerical methods have been proposed to solve Global Optimization problems; see [8] for a comprehensive survey. Among them, some have been used to solve control problems: Branch and Bound techniques [5], Simulated Annealing [9], Genetic Algorithms [9, 10], just to name a few.

We remark that the suitability of such algorithms to solve control problems is seldom guaranteed by convergence results; more often the success of a method is given by successful application to practical problems [5].

A novel Global Optimization technique that seems suitable to compute the solution of problem (1) is presented here. This technique is based upon the algorithm proposed in [11] for unconstrained, unidimensional global optimization problems, that we describe briefly.

Such algorithm solves the problem

$$\min_{x \in [a,b]} f(x),$$

for a Lipschitz multiextremal function $f(\cdot)$. It belongs to the class of the so-called *information methods* [12], and it has been shown that it performs much better than many other methods from the literature on some typical test problems [11].

In the stochastic framework of the information approach, little is assumed to be known *a priori* about the objective function, beside of some technical assumptions as $f(\cdot)$ being Lipschitz. Rather, the function is regarded as a black-box subroutine, which returns a value $f(x)$ for an input value x, and it is postulated that we may gain *information* about $f(\cdot)$ each time we evaluate it at a new point.

The algorithms in this class compute, at the $(k+1)$-th step, a so-called *characteristic sequence* R_j, $j = 2, \ldots, k$, on the basis of the available data $(x_j, f(x_j))$, $j = 1, \ldots, k$, with $a = x_1 < x_2 < \cdots < x_k = b$ (the points x_j are reordered at each step in increasing order). Each value R_j is interpreted in this approach, after normalization, as the probability that the global minimum lies in the interval $[x_{j-1}, x_j]$. The algorithm then generates a new point x_{k+1}, which is interpreted as the maximum likelihood estimate of the global minimizer based on the available data.

At the next step, the objective function is evaluated at this new point and the R_j are recomputed. In the Bayesian sense, this accounts to transform the *a priori* probability R_j of the previous step into the *a posteriori* one, with respect to the new observation $(x_{k+1}, f(x_{k+1}))$. A global convergence result will be reported after the description of the algorithm.

3.1 The multidimensional case

Let now $\phi: \mathbf{R}^n \to \mathbf{R}$ a Lipschitz continuous function; we wish to find a solution of the multidimensional, multiextremal problem

$$\min_{y \in \mathbf{B}} \phi(y),$$

where $\mathbf{B}$ is an hyperrectangle in $\mathbf{R}^n$:

$$\mathbf{B} = \{y \in \mathbf{R}^n : a_j \le y_j \le b_j, j = 1, 2, \ldots, n\}.$$

The extension of the unidimensional algorithm to multidimensional problems is accomplished using space-filling curves of Peano-Hilbert type [13]. Such curves (see Figure 2) are fractals constructed on the principle of self-similarity, and represent a continuous mapping from the interval $[a, b]$ to $\mathbf{B}$. Let $y(\cdot)$ denote one of such curves; from the continuity of $\phi(\cdot)$ it follows:

$$\min_{y \in \mathbf{B}} \phi(y) = \min_{x \in [a,b]} \phi(y(x)),$$

and this reduces the original multidimensional problem to one dimension. Moreover, if $\phi(\cdot)$ satisfies Lipschitz condition with constant L over $\mathbf{B}$, then $\phi(y(\cdot))$ satisfies Hölder condition over $[a, b]$, i.e., for $x_1, x_2 \in [a, b]$,

$$|\phi(y(x_1)) - \phi(y(x_2))| \le K |x_1 - x_2|^{1/n},$$

where $K = 4\sqrt{n} L d$ is the Hölder constant and $d = \max\{b_j - a_j, j = 1, \ldots, n\}$.

Approximations to the Peano-Hilbert curve can be efficiently constructed, up to the desired level of accuracy, by subdivision of the initial hyperrectangle; see [14] for the details.

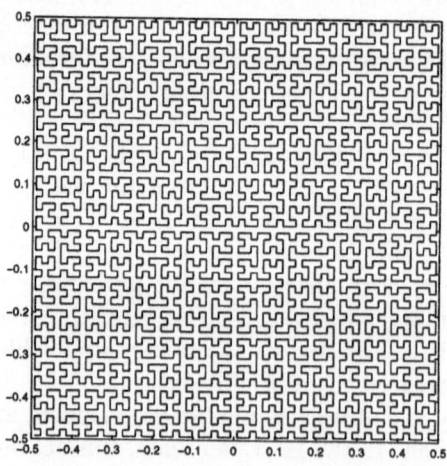

Figure 2: Approximation of level 6 to the Peano curve in two dimensions

The construction of the characteristic sequence $\{R_j\}$ needs an estimate of the Lipschitz constant of the function. A distinguishing feature of the algorithm is a procedure to estimate such constant *locally* during the search. Adaptive estimates of local constants (instead of the global one) can accelerate the search significantly, because of the tuning on the local behavior of the objective function. The following algorithm reflects these ideas.

Algorithm for global unconstrained multivariate optimization [11]

Initialization

1. Choose the points $x^1, \ldots, x^m$, $m \geq 2$, where $x^1 = 0, x^2 = 1, x^3, \ldots, x^m$ are arbitrary, and calculate $\phi(y(x^i))$, $i = 1, \ldots, m$;

2. Set the values of the parameters $r > 1$, $\varepsilon > 0$, and $\xi > 0$;

Repeat

1. Let $k + 1$ be the current step, $k \geq m$. Reorder the points $x^1, \ldots, x^k$ in increasing order, and denote them by $x_1, \ldots, x_k$, in such a way that $0 = x_1 < x_2 \cdots < x_k = 1$;

2. For $j = 2, \ldots, k$, calculate the quantities

$$X = \max_{i=2,\ldots,k} \{(x_i - x_{i-1})^{1/n}\},$$

$$\mu = \max_{i=2,\ldots,k} \{|z_i - z_{i-1}|/(x_i - x_{i-1})^{1/n}\},$$

$$\gamma_j = \mu(x_j - x_{j-1})^{1/n}/X, \quad j = 2, \ldots, k,$$

$$\lambda_j = \max_{i \in \mathcal{N}_j} \{|z_i - z_{i-1}|/(x_i - x_{i-1})^{1/n}\},$$

where $z_i = \phi(y(x_i))$ and $\mathcal{N}_2 = \{2, 3\}$, $\mathcal{N}_j = \{j-1, j, j+1\}$ for $2 < j < k$ and $\mathcal{N}_k = \{k-1, k\}$;

3. For $j = 2, \ldots, k$, estimate the local Lipschitz constants by

$$\widehat{L}_j = \max\{\lambda_j, \gamma_j, \xi\};$$

4. For $j = 2, \ldots, k$, calculate the characteristics

$$R_j = r\widehat{L}_j(x_j - x_{j-1})^{1/n} - 2(z_j + z_{j-1}) + (z_j - z_{j-1})^2/(r\widehat{L}_j(x_j - x_{j-1})^{1/n});$$

5. Let $t = \arg\max_{j=2,\ldots,k}\{R_j\}$; set

$$x^{k+1} = (x_t + x_{t-1})/2 - \text{sign}(z_t - z_{t-1})(|z_t - z_{t-1}|/\widehat{L}_t)^n/(2r);$$

6. Evaluate $\phi(y(x^{k+1}))$ and set $k = k + 1$;

until $(x_t - x_{t-1})^{1/n} \leq \varepsilon$.

The following proposition resumes the convergence results given in [11].

Proposition 1 *Denote by x^* the global minimizer of $\phi(y(x))$. If there exists an iteration number k^* such that, for all $k > k^*$, the condition*

$$r\widehat{L}_j \geq 2^{1-1/n}K_j + (4^{1-1/n}K_j^2 - M_j^2)^{1/2}$$

is met, where $j = j(k)$ is such that $x^* \in [x_{j(k)-1}, x_{j(k)}]$, then the sequence $\{x^k\}$ generated converges to the global minimizer x^*; in the above formula:

$$
\begin{aligned}
K_j &= \max\{\, (z_{j-1} - \phi(y(x^*)))/(x^* - x_{j-1})^{1/n}, \\
&\qquad (z_j - \phi(y(x^*)))/(x_j - x^*)^{1/n}\,\}, \\
M_j &= |z_{j-1} - z_j|/(x_j - x_{j-1})^{1/n}.
\end{aligned}
$$

It is easy to see that the convergence condition of the above result can always be fulfilled by choosing the parameter r large enough. However, we must be aware that large values of r lead to the generation of a large number of trial points. Reasonable values of r are in the range $[1.5, 10]$. The parameter ξ in step 3 of the algorithm ensures that $\widehat{L}_k$, the local estimates of Lipschitz's constant, are at least as large as ξ. This reflects the belief that the objective function is not quasi-constant over the j-th subinterval.

3.2 The constrained case

To take the constraints into account, we have used the simple approach of the penalty functions. To solve the problem

$$
\begin{aligned}
&\min_{y \in \mathbf{B}} && \phi(y) \\
&\text{subject to:} && \gamma_i(y) < 0, \; i = 1, \dots, N
\end{aligned}
\tag{2}
$$

we have defined the penalized function

$$\psi(y) = \phi(y) + \max\{0, c_1\gamma_1(y), \dots, c_N\gamma_N(y)\},$$

where the c_i, are large positive constants, and have used the reported algorithm to minimize it on $\mathbf{B}$. Though this is equivalent to solve Problem (2) only for $c_i \to \infty$ [15], we may expect that, for sufficiently large c_i, the solution thus obtained is a good approximation of the solution of Problem (2), being comforted in this belief by successful experience in a previous work [16].

It is important to note that the penalty function we use is not differentiable, but this is not a problem for the algorithm, which only require that the objective function is Lipschitz continuous.

4 Application to control problems

Here we test the proposed method on two typical control problems. In both we have used an approximation to the Peano curve of level 36: see [14] for the details on the approximation; ξ has been set to 10^{-4}. The algorithm has been coded in FORTRAN and run on a Sun UltraSparc, which has about the same floating-point performance as a Pentium 200 MHz PC.

Problem 1

Let us consider the following plant:

$$P(s) = \frac{2}{s\,(s-2)};$$

we wish to find the parameters of a (proper) PD regulator

$$C(s) = \frac{K_P(1 + K_D s)}{1 + 0.01\,s}$$

that stabilizes the closed-loop system and minimizes the ISE criterion

$$\mathcal{I}(K_P, K_D) = \int_0^\infty e^2(t)\,dt,$$

under a constraint on the control energy

$$\mathcal{U}(K_P, K_D) = \int_0^\infty u^2(t)\,dt < U_M.$$

Here, $e(\cdot)$ and $u(\cdot)$ are the input and the output of the regulator, respectively. Classical formulas for such indices (see e.g. [17, pg. 132]) allow to write:

$$\mathcal{I}(K_P, K_D) = \frac{196 + K_P\,(49.02 + 0.5\,K_D K_P)}{K_P\,(98\,(K_D\,K_P - 2) - K_P)},$$

$$\mathcal{U}(K_P, K_D) = \frac{5000\,K_P\,(0.0392 + K_P\,(K_D^2\,(K_D\,K_P - 2.04) - 0.01))}{98\,(K_D\,K_P - 2) - K_P}.$$

On denoting by $H_3(K_P, K_D)$ the Hurwitz determinant of the characteristic polynomial of the closed-loop system, we formalize the problem as:

$$\min_{(K_P, K_D) \in \Gamma} \mathcal{I}(K_P, K_D)$$

$$\text{subject to:} \quad \begin{cases} -H_3(K_P, K_D) < 0, \\ \mathcal{U}(K_P, K_D) < U_M. \end{cases}$$

The box is $\Gamma = [0.1, 10] \times [0, 2]$, the penalty coefficients have been set to $c_1 = 5 \cdot 10^7$, $c_2 = 5 \cdot 10^3$ and $U_M = 2000$.

The algorithm, using an accuracy $\varepsilon = 10^{-2}$ and $r = 1.15$, has found the minimum at $K_P^* = 9.9941$, $K_D^* = 0.62202$, to which a value $\mathcal{I}^* = 0.17791$ there correspond, with 235 evaluations of the objective function and a CPU-time of 0.043 seconds. The values of K_P^*, K_D^* and $\mathcal{I}^*$ did not change for smaller values of ε or larger values of r.

Figure 3 shows on the left the level curves of the penalized function $\psi(K_P, K_D)$ and the search points generated by the algorithm, and on the right a mesh of the logarithm of the same function.

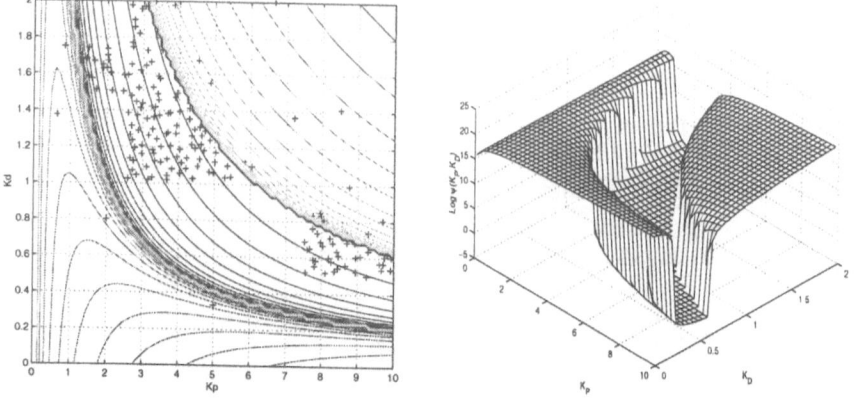

Figure 3: Contour levels and search points (left) and a mesh (right) of the penalized function of Problem 1.

Problem 2

The following plant has been considered in [18]:

$$P(s) = \frac{s-1}{(s+1)(s-2)};$$

note that this plant does not satisfy the Parity Interlace Property [19], i.e., an odd number of poles lie between two positive zeros (including the one at infinity), therefore an unstable compensator (with a pole $p > 1$) is needed.

We wish to find the parameters of the regulator

$$C(s) = \frac{K_P + K_D s}{s - p}$$

that locates the closed-loop poles to the left of the abscissa $\alpha = -1$, and minimizes the steady-state value of the control effort for a step input. This last is given, as long as the compensator stabilizes the overall system, by:

$$U_\infty = \lim_{t \to \infty} |u_s(t)| = \frac{2 K_p}{|K_p - 2p|}.$$

On denoting by $d(s)$ the closed-loop polynomial, the condition of α-stability corresponds to apply the Hurwitz test to:

$$d(s-1) = s^3 + (K_D - p - 4) s^2 + (K_P - 3 K_D + 3p + 3) s + 2 (K_D - K_P);$$

the parameter box is $\Gamma = [0.1, 200] \times [0.1, 25] \times [1.1, 15]$. A root-locus analysis shows that this problem is well-posed. The synthesis problem is then formulated as:

$$\min_{(K_P, K_D, p) \in \Gamma} \quad U_\infty(K_P, K_D, p)$$

$$\text{subject to:} \quad \begin{cases} p - K_D + 4 & < 0, \\ K_P - K_D & < 0, \\ -H_3(K_P, K_D, p) & < 0, \end{cases} \tag{3}$$

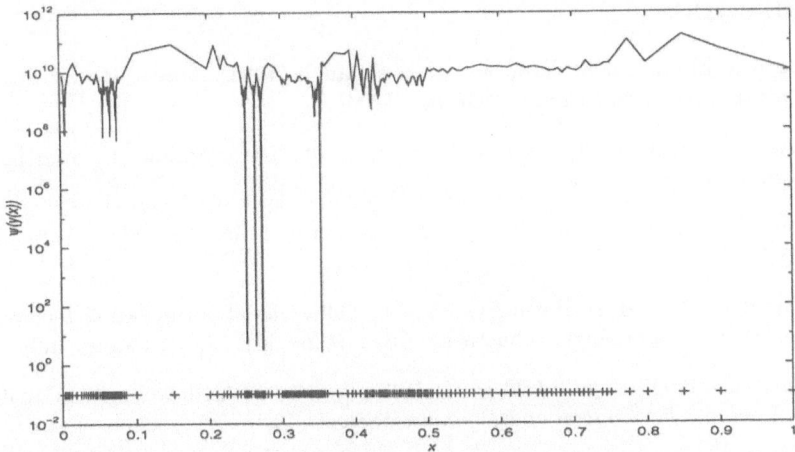

Figure 4: Reduced function $\psi(y(x))$ for the Problem 2.

where $-H_3(K_P, K_D, p) = (p - K_D + 4)(K_P - 3K_D + 3p + 3) + 2(K_D - K_P)$. We have defined the penalized function

$$\psi(K_P, K_D, p) = U_\infty(K_P, K_D, p) + C \max\{0, \gamma_1, \gamma_2, \gamma_3\},$$

where $\gamma_i = \gamma_i(K_P, K_D, p)$ are the constraints in Problem (3).

Using the values $C = 10^8$ and $r = 1.15$ we have found $U_\infty^* = 2.7792$ using $\varepsilon = 0.0002$ (1531 search points, 2.05 seconds) and $U_\infty^* = 2.7780$ using $\varepsilon = 0.0001$ (3045 search points, 7.50 seconds). No improvements in the value of the optimum were noticed for larger values of r. Figure 4 shows the reduced function $\psi(y(x))$, i.e., the objective function evaluated along the Peano curve, and the sequence of search points, for the case $\varepsilon = 0.001$.

5 Conclusions

The algorithm in [11] has shown to be effective in solving a number of optimization problems coming from Control Theory (two examples are reported).

One major feature is that we do not need any initial guess for the variables: the convergence to the global optimum, though it may be slow in some cases, is guaranteed. Even for the penalty coefficients, whose choice is often a hard task when joined to the choice of the initial guess, we have always found effective values in a very few trials.

One reason of the fast convergence of the method is the adaptive estimation of local Lipschitz constants. In fact, in the presence of penalty function, the function to be minimized can be very 'flat' where the constraints are satisfied but raises steep close to the frontier of the admissible region. This is precisely the situation where local estimates are most useful.

In conclusion, we believe that the algorithm may be a useful computational tool in the analysis and synthesis of control systems, and are working for making it more efficient and simple to use.

References

[1] Boyd S, El Ghaoui L, Feron E, Balakrishnan V (1994). *Linear Matrix Inequalities in System and Control Theory*, SIAM

[2] Kwakernaak H (1993). Robust control and $\mathcal{H}_\infty$-optimization. Tutorial paper, *Automatica* **29**:255-273

[3] Packard A, Doyle J (1993). The complex structured Singular Value, *Automatica* **29**:71-109

[4] Famularo D, Pugliese P, Sergeyev Ya D (1998). Global optimization techniques for checking parametric robustness, *Proc. of the MTNS '98*, Padue, Italy

[5] Balakrishnan V, Boyd S (1993). Global optimization in control system analysis and design. In: *Control and Dynamic Systems* **53**:1-55, Academic Press

[6] Malan S, Milanese M, Taragna M (1997). Robust analysis and design of control systems using interval arithmetic, *Automatica* **33**:1363-1372

[7] Murty K G, Kabady S N (1987). Some $\mathcal{NP}$-complete problems in quadratic and nonlinear programming, *Math. Programming* **39**:117-129

[8] Horst R, Pardalos P M, Eds. (1995). *Handbook of Global Optimization*, Kluwer

[9] Zhu X, Huang Y, Doyle J (1997). Genetic algorithms and simulated annealing for robustness analysis, *Proc. American Contr. Conf.*, Albuquerque

[10] Murdock T M, Schmittendorf W E, Forrest S (1991). Use of a genetic algorithm to analyze robust stability problems, *Proc. American Contr. Conf.*, Boston

[11] Sergeyev Ya D (1995). An information global optimization algorithm with local tuning, *SIAM J. Optimization* **5**:858-870

[12] Strongin R G (1989). The information approach to multiextremal optimization problems, *Stochastics and Stoch. Reports* **27**:65-82

[13] Sagan H (1994). *Space-Filling Curves*, Springer-Verlag

[14] Sergeyev Ya D, Strongin R G (1992). Global multidimensional optimization on parallel computers, *Parallel Computing* **18**:1259-1273

[15] Luenberger D G (1972). *Introduction to Linear and Nonlinear Programming*, Addison-Wesley

[16] Sergeyev Ya D, Markin D L (1995). An algorithm for solving global optimization problems with nonlinear constraints, *J. Global Optimization* **7**:407-419

[17] Izawa K (1963). *Introduction to Automatic Control*, Elsevier

[18] Keel L, Bhattacchryya S (1997). Robust, fragile, or optimal?, *IEEE Trans. Aut. Contr.* **42**:1098-1105

[19] Youla D C, Bongiorno J J, Lu C N (1974). Single-loop feedback stabilization of linear multivariable dynamical plants, *Automatica* **10**:159-173

10

An Optimal Model-Following Problem
for Linear Systems

C.Botan and A. Onea

1   Introduction

One of the most attractive ways to design an automatic system is to impose the dynamic behaviour. For instance, it is possible to impose that the system has a dynamic behaviour resembling, as much as possible, that of an imposed model. The model could either actually exist in the control system or could be an implicit one, whose behaviour is to be pursued. The paper tackles the latter variant.

A time invariant multivariable linear system, which is considered completely controllable and observable, is described by:

$$\dot{x}(t) = Ax(t) + Bu(t) + w(t)$$
$$x(t_0) = x^0 \tag{1}$$
$$y(t) = Cx(t) \tag{2}$$

where $x(t) \in R^n$, $u(t) \in R^m$, $y(t) \in R^p$, $w(t) \in R^n$ denote the state, control, output and disturbance vectors, respectively.

A main goal is to make the dynamic of the closed-loop system resemble, as closely as possible, that of the model. For this purpose one can introduce the term $\dot{y}(t) - Ly(t)$, $L \in R^{m \times m}$ in the criterion. This way was firstly adopted in [1] and was used in many other papers. As it is remarked in [2], a more convenient possibility is to use the term $\dot{x}(t) - Lx(t)$, $L \in R^{n \times n}$, in the criterion, because this way offers more possibilities in the choice of the dynamic behaviour of the model. Indeed, in the first variant, the order of the model is limited at the number of the outputs. However, the both indicated possibilities are properly only for the free response of the system, because the model do not includes the influence of the control $u(t)$. Moreover, for many applications, such as the electrical drive system, it is essential to consider the influence of the disturbance (e.g. the resistive torque) on the system and also on the model. Consequently, it is useful to have in the criterion a term

depending on the errors between the model and the system in the form
$\dot{x}(t) = Lx(t) + Bu(t) + w(t)$.

Taking into account the above considerations, the following quadratic criterion
was adopted:

$$J = \frac{1}{2}(x(t_f) - x_d)^T S(x(t_f) - x_d) +$$

$$+ \frac{1}{2}\int_{t_0}^{t_f}\left\{[\dot{x}(t) - Lx(t) - bu(t) - w(t)]^T Q_1[\dot{x}(t) - Lx(t) - bu(t) - w(t)] + \right. \tag{3}$$

$$\left. + [x(t) - x_d]^T Q_2[x(t) - x_d] + u^T(t)Pu(t)\right\}dt$$

where x_d denotes the desired value of the state and T denotes the transposition.
Note that $S = S_1^T \geq 0, Q_1 = Q_1^T \geq 0, P = P^T > 0$.

The significance of the terms of the criterion (3) will be discussed further for the
case of the electrical drive control system (chapter 3). It was adopted a finite final
time in the criterion (3) in order to ensure a small transient time.

The aim is to determine the closed-loop control *u(t)* so that the criterion (3) is
minimised with respect to the restriction (1) imposed by the system.

The solution of the problem is similar to that of the linear-quadratic (LQ)
optimisation problem with finite final time and the result is a time variant controller
[3].

The paper propose a different solution based on previous research of the authors
[4], [5] and leads to a more facile implementation.

2   Main results

For the above formulated optimisation problem, the Hamiltonian is:

$$H = \frac{1}{2}[\dot{x}(t) - Lx(t) - Bu(t) - w(t)]^T Q_1[\dot{x}(t) - Lx(t) - Bu(t) - w(t)]^T +$$

$$+ \frac{1}{2}[x(t) - x_d]^T Q_2[x(t) - x_d] + \frac{1}{2}u^T(t)Pu(t) \tag{4}$$

The canonical and transversality equations are:

$$u^*(t) = -P^{-1}B^T \lambda(t) \tag{5}$$

$$-\dot{\lambda}(t) = Qx(t) - Q_2 x_d + A^T \lambda(t) + \hat{A}^T Q_1 w(t) \tag{6}$$

with final condition:

$$\lambda(t_f) = S[x(t_f) - x_d] \tag{7}$$

where:

$$\hat{A}^T = A - L; \quad Q = \hat{A}^T Q_1 \hat{A} \tag{8}$$

The classical approach imposes a linear dependence between $\lambda(t)$ and $x(t)$ in the form:

$$\lambda(t) = \widetilde{R}(t)x(t), \qquad (9)$$

where $\widetilde{R}(t)$ is the time-variant solution to a Riccati differential equation.

This paper proposes instead of (9) the following relation:

$$\lambda(t) = \widetilde{R}(t)x(t) + v(t) \qquad (10)$$

with $v(t) \in R^n$ and $R = R^T$ a $n \times n$ constant matrix.

Equations (1), (5), (6), and (10) lead to:

$$(RNR - RA - A^T R - Q)x(t) - \dot{v}(t) + (RN - A^T)v(t) - Rw + Q_2 x_d = 0$$

This relation is true for any $x(t) \in R^n$ and $v(t) \in R^n$ if and only if:

$$RNR - RA - A^T R - Q = 0 \qquad (11)$$

and:

$$\dot{v}(t) = -F^T v(t) + Q_2 x - Rw(t) \qquad (12)$$

where:

$$N = BP^{-1}B^T \quad \text{and} \quad F = A - NR \qquad (13)$$

The final condition leads to:

$$v(t_f) = (S - R)x(t_f) - Sx_d . \qquad (14)$$

The optimal control variable $u^*(t)$ will be:

$$u^*(t) = u_f(t) + u_c(t) \qquad (15)$$

where

$$u_f(t) = -P^{-1}B^T Rx(t) \qquad (16)$$

and:

$$u_c(t) = -P^{-1}Bv(t) \qquad (17)$$

The optimal corrective vector $v(t)$ is the solution to the time-invariant linear differential equation (12), with the final condition (14), and R is the solution to the Riccati algebraic equation (11).

The component $u_f(t)$ of $u^*(t)$ is a feedback component and it is identical with the optimal control obtained in the similar optimal problem, but with infinite final time.

The component $u_c(t)$ given by (17) is a corrective component and it ensures the identity between the control $u^*(t)$ given by (15) and the control obtained if the constant vector $\lambda(t)$ is adopted in the form (9). In order to compute this corrective component within a real time controller, it is necessary to replace the final condition (14) by one depending on $x(t_0)$, which is the only known value at the beginning of the optimisation process. On this purpose the above differential equations for $x(t)$ and $v(t)$ are rewritten as:

$$\begin{bmatrix} \dot{x}(t) \\ \dot{v}(t) \end{bmatrix} = G\begin{bmatrix} x(t) \\ v(t) \end{bmatrix} + D\begin{bmatrix} w(t) \\ x_d \end{bmatrix}, \tag{18}$$

where:

$$G = \begin{bmatrix} F & -N \\ 0 & -F^T \end{bmatrix} \quad \text{and} \quad D = \begin{bmatrix} I & 0 \\ R & Q_2 \end{bmatrix}, \tag{19}$$

with I_n the identity matrix.

The solution of the equation (18) is:

$$\begin{bmatrix} x(t) \\ v(t) \end{bmatrix} = \Omega(t,t_f)\begin{bmatrix} x(t_f) \\ v(t_f) \end{bmatrix} + \int_{t_f}^{t} \Omega(t,\tau)D\begin{bmatrix} w(\tau) \\ x_d \end{bmatrix}d\tau, \tag{20}$$

where $\Omega(t,t_f)$ is the transition matrix for G and has the form[4], [5]:

$$\Omega(t,t_f) = \begin{bmatrix} \Psi(t,t_f) & \Omega_{12}(t,t_f) \\ 0 & \Phi(t,t_f) \end{bmatrix}; \tag{21}$$

$\Psi(t,t_f)$ and $\Phi(t,t_f)$ are the transition matrix for F and $-F^T$, respectively, and:

$$\Omega_{12}(t,t_f) = \int_{t}^{t_f} \Psi(t,\tau)N\Phi(\tau,t_f)d\tau \tag{22}$$

Using (14), (20), and (21), the state vector becomes:

$$x(t) = M(t,t_f)x(t_f) - \Omega_{12}(t,t_f)Sx_d +$$

$$+ \int_{t}^{t_f}[\Psi(t,t_f) - \Omega_{12}(t,t_f)R]w(\tau)d\tau + \int_{t}^{t_f}\Omega_{12}(t,t_f)Q_2 x_d d\tau \tag{23}$$

where:

$$M(t,t_f) = \Psi(t,t_f) + \Omega_{12}(t,t_f)(S-R), \tag{24}$$

and it is nonsingular matrix.

The vector $x(t_f)$ can be expressed in terms of $x(t_0)$ from (23) ,and consequently. from (14), $v(t_0)$ can written in terms of $x(t_0)$. Finally, the corrective vector $v(t)$ can be computed as:

$$v(t) = \Phi(t,t_0)v(t_0) - \int_{t_0}^{t} + \Phi(t,\tau)[Rw(\tau) - Q_2 x_d]d\tau \qquad (25)$$

where:

$$v(t_0) = V(t_0,t_f)[x(t_0) + \Omega_{12}(t_0,t_f)Sx_d + \int_{t_0}^{t_f}[\Psi(t_0,\tau) - \Omega_{12}(t,\tau)R]w(\tau)d\tau +$$

$$+ \int_{t_0}^{t_f}\Omega_{12}(t_0,t_f)Q_2 x_d d\tau] - \Phi(t_0,t_f)Sx_d + \int_{t_0}^{t_g} - \Phi(t_0,\tau)[Rw(\tau) + Q_2 x_d]d\tau \qquad (26)$$

and

$$V(t_0,t_f) = \Phi(t_0,t_f)(S-R)M^{-1}(t_0,t_f) \qquad (27)$$

The deduction of the initial value $v(t_0)$ is conditioned by the information on the disturbance $w(t)$ upon the entire interval $[t_0, t_f]$, or, at least, its variation form on the interval $[t_0, t_f]$ must be known. It means that its magnitude at the initial moment t_0 must be available (measured or estimated). For simplicity, in the sequel, the disturbance is considered constant (w_0) on the period $[t_0, t_f]$. In this case, the above relations can be written:

$$v(t_0) = V(t_0,t_f)x(t_0) + [V(t_0,t_f)H_{11} + H_{21}]w_0 + [V(t_0,t_f)H_{12} + H_{22}]x_d \quad (28)$$

where:

$$H_{11} = \int_{t_0}^{t_f}[\Psi(t_0,\tau) - \Omega_{12}(t,\tau)R]d\tau \qquad H_{12} = \int_{t_0}^{t_f}\Omega_{12}(t_0,t_f)d\tau Q_2 + \Omega_{12}(t_0,t_f)S$$

$$H_{21} = \int_{t_0}^{t_f}\Phi(t_0,\tau)d\tau R \qquad\qquad H_{22} = -\int_{t_0}^{t_f}\Phi(t_0,\tau)d\tau Q_2 - \Phi(t_0,t_f)S \qquad (29)$$

The computation of the control variable $u^*(t)$ given by (15) and especially of the corrective vector $v(t)$ is rather complicated, but most of them can be performed off-line: the solution to the Riccati algebraic equation, the computation of the matrix Ω_{12}, M, V, and the computation of the constant vectors from (26) or (28). The on-line control algorithm implies the following steps:

    (a) read the value of disturbance or compute it with an observer;

    (b) read the value of state variables at the sample instants t_i;

    (c) compute for each $t_i \in [0, t_f]$ the component $u_c(t)$ and $u_f(t)$;

    (d) repeat the steps (b) and (c) for all $t_i \in [t_0, t_f]$

After the first step, when $w(t)$ is estimated, the initial values $v(t_0)$ can be computed. For the next steps, only $v(t)$ given by (25) must be computed.

$\Phi(t,t_0)$ and $\int\limits_{t_0}^{t} \Phi(t,\tau)d\tau$ in (25) can be computed iteratively, by multiplying the obtained vector at the previous iteration with a constant matrix.

3   Electrical drive system application

The electrical drive system is a direct application of the model-following problem with disturbances. In this application, a DC motor drive system is considered, described by the equations (1) and (2), where [6]:

$$x(t) = \begin{bmatrix} \omega(t) \\ i(t) \end{bmatrix}, \quad A = \begin{bmatrix} -\rho/J_r & C_m/J_r \\ -C_e/L_a & -R_a/L_a \end{bmatrix}$$

$$B = \begin{bmatrix} 0 \\ 1/L_a \end{bmatrix}, \quad C = \begin{bmatrix} 1 & 0 \end{bmatrix}, \quad w(t) = \begin{bmatrix} (-1/J_r)m(t) \\ 0 \end{bmatrix} \tag{30}$$

ω- angular speed, u- armature voltage, i- rotor current, ρ- drive system parameter, Jr- inertia momentum, C_m, C_r - motor parameters, L_a, R_a - rotor circuit inductance and resistance, respectively, Mr- resistant torque.

The matrices S, Q_1, Q_2, P, L from criterion (3) are chosen based on the following considerations:

- $S = diag(s_1, 0)$ is chosen in order to select a desired weight of the final error $\omega(t) - \omega_d$. The desired value for current must not be imposed, because the values of current and speed are related through the equation (1) of the system;

- $Q_1 = diag(q'_1, q'_2) > 0$ and $Q_2 = diag(q_1, q_2) > 0$ are chosen in order to select the weight of the corresponding terms within the criterion;

- $L \in R^{2x2}$ is chosen in order to obtain a certain imposed transitory behaviour for $x(t)$. For instance, L can be chosen in the form $L = diag(l_1, l_2)$ or in the same form as the matrix A in (30), but with a lower value for J_r.

- $x_d = [\omega_d \; 0]^T$. This form is adopted because we are interested to penalise the big values of the current $i(t)$ and not of the difference $i_d - i(t)$.

The criterion has in this case the form:

$$J = \frac{1}{2} s_1 [\omega(t_f) - \omega_d]^2 + \frac{1}{2} \int\limits_{t_0}^{t_f} [q'_1 (\hat{a}_{11}\omega + \hat{a}_{12}i)^2 + q'_2 (\hat{a}_{21}\omega + \hat{a}_{22}i)^2 +$$

$$q_1(\omega(t_f) - \omega_d) + q_2 i^2(t) + pu^2(t)]dt$$

where $\hat{a}_{ij}, \quad i, j = 1, 2$ is the element of the matrix $\hat{A}$, or in the form:

$$J = \frac{1}{2} s_1 [\omega(t_f) - \omega_d]^2 + \frac{1}{2} \int_{t_0}^{t_f} [q'_1 (\dot{\omega}(t) + l_1 \omega(t))^2 + q'_2 (i(t) - l_2 i(t))^2 +$$

$$+ q_1 (\omega(t) - \omega_d)^2 + q_2 i^2(t) + pu^2(t)] dt$$

if the matrix L is adopted in the form $L = diag(l_1, l_2)$.

The criterion penalises the large value of the current (the dissipated energy on the rotor winding), armature voltage, angular speed deviations from the imposed dynamic behaviour.

The implementation of the optimal controller for this case is based on the relations indicated in paragraph 2.

4 Experimental results

The optimal control system was simulated with the MATLAB package. The matrices and vectors of the system in (1) are chosen:

$$x(t) = \begin{bmatrix} \omega(t) \\ i(t) \end{bmatrix}, \quad A = \begin{bmatrix} 0 & 20 \\ -3.5 & -19.4 \end{bmatrix}$$

$$B = \begin{bmatrix} 0 \\ 6.25 \end{bmatrix}, \quad C = \begin{bmatrix} 1 & 0 \end{bmatrix}, \quad w(t) = \begin{bmatrix} -35.7 \\ 0 \end{bmatrix} m$$

These numerical values are corresponding to an electrical drive system with a motor CI 12 made by IMEB Bucharest with the following parameters: Un=110V, In=3.3A, Ra=3.1Ω, La=0.16H, Ce=0.58Vs/rad, Cm=0.58Nm/A, Jr=0.028Nms^2/rad. The experimental results presented in this section are done for: ω_d=25rad/s, m=0.78Nm, S=diag(10, 1), Q_1=diag(0.8, 1), Q_2=diag(1, 3.1), P=p=1,

$$L = \begin{bmatrix} 0 & 31.11 \\ -3.5 & -19.4 \end{bmatrix}, t_0\text{=}0, t_f\text{=}0.3. \text{ The sampling period is 2 ms in all cases.}$$

The Fig. 1 presents the response of the optimal system for a step variation of ω_d from 0 to 25 rad/s in the presence of a disturbance torque m=0.78Nm.

A comparison between the variations of the angular speed is presented in Fig. 2: – curve a - variation of the output of the motor model; curve b – variation of the the motor speed of the optimal system; curve c – variation of the motor speed in open loop, for a step variation of the armature voltage u(t). There is a small difference between the response of the optimal system and the response of the reference model. A relatively large difference can be seen in the last part of the transient response because the optimal system is forced arise in a neighbourhood of the prescribed speed at $t=t_f$ when the matrix S is chosen with a big value of his first element.

For all the performed tests was computed the energy consumption on the optimisation period. The energy consumption is smaller than in the case of the classical cascade control system with 20% - 30%, depending on the weight matrices selected in the criterion.

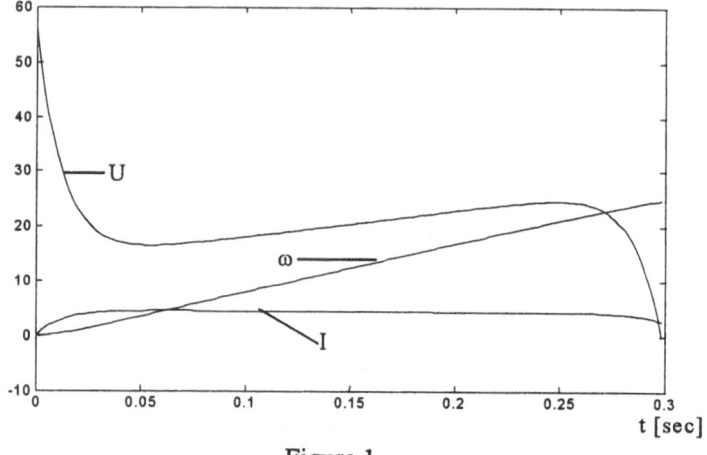

Figure 1

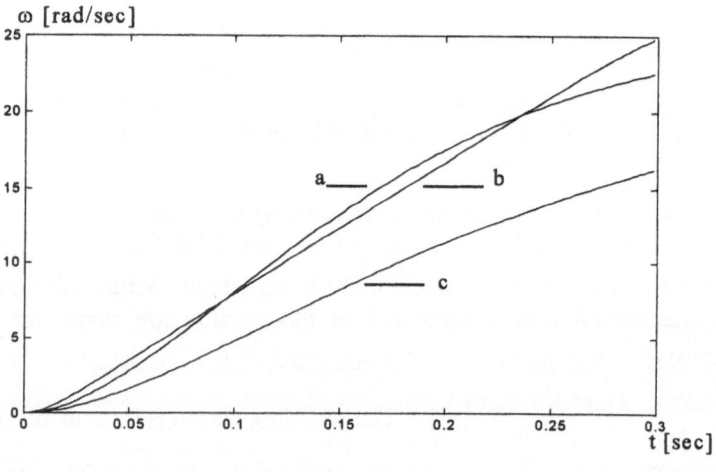

Figure 2

The presented algorithm shows that a big part of the computing time is used to calculate the corrective component $u_c(t)$. This time can be decreased by the increase of the sampling period for the corrective component. This possibility is illustrated in Fig. 3 and 4. Fig. 3 presents the behaviour of the optimal system in the case when the sampling period for the component $u_c(t)$ is increased 10 times. Fig. 4 presents the case when the sampling period for both components of the control variable, $u_c(t)$ and $u_f(t)$ are increased 10 times. Comparing the Fig. 1, 3 and 4, several remarks can be done:

- the increasing of the sampling period only for the corrective component $u_c(t)$ leads to a small difference from the optimal response,

- the increasing of the sampling period for both components leads to a significant difference.

The result is that it is possible to obtain a smaller computing real time amount without significant modification of the optimal behaviour increasing only the sampling period for $u_c(t)$. The computing amount will not exceed very much the necessary time for a usual state feedback control system.

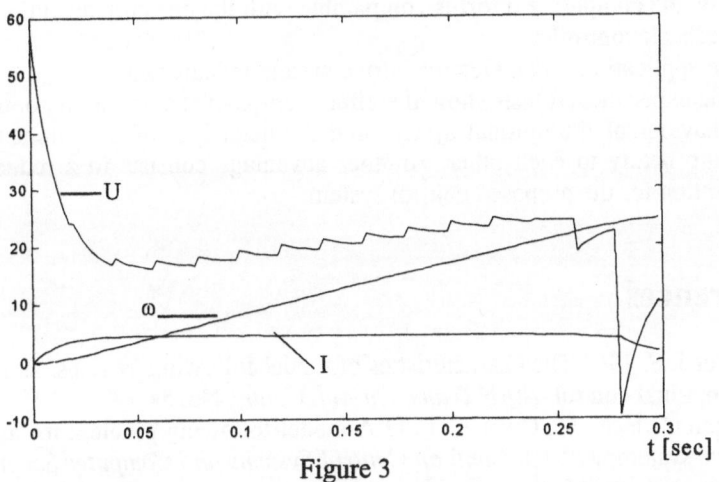

Figure 3

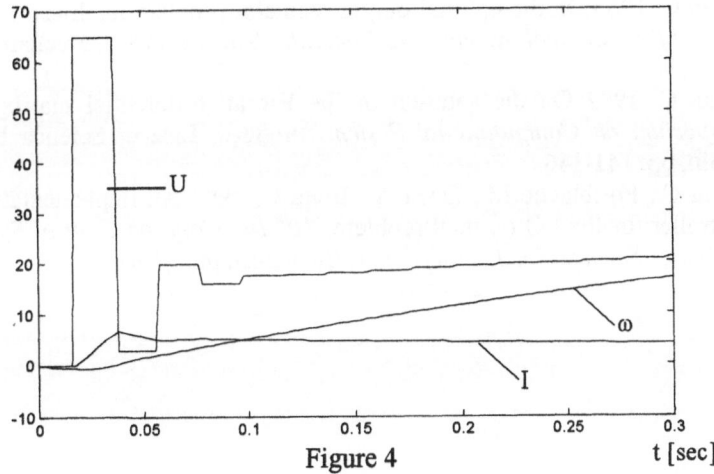

Figure 4

4   Conclusions

The paper studies a model following problem for a linear time-invariant multivariable system in the presence of disturbance. The problem can be solved if the disturbance or at least its form is beforehand known and its magnitude is estimated at the beginning of the optimisation interval.

The problem is studied as a linear-quadratic optimal problem with finite final time. A possibility to use a time-invariant controller is indicated. This controller computes a usual constant feedback component and a corrective component. The last one depends on the initial state, the desired value of the final state and the disturbance. The corrective component can be computed with a sampling period considerably greater than the period of the feedback component. This way, the computing effort is comparable with the one corresponding to a usual state feedback controller.

An application for an electrical drive system is indicated.

The experimental tests show the effectiveness of the proposed control system. The behaviour of the optimal system and the behaviour of the implicit reference model are nearly to each other. Another advantage consists in a reduced energy consumption for the proposed control system.

References

1. Tyler J. S. 1964 The characteristics of model-following systems, as synthesised by optimal control, *IEEE Trans. On Aut. Contr.*, No. 54.
2. Boţan C., Belţa C., Onea A. 1997 A model-following problem for an electrical drive system, *11th Int. Conf. on Control Systems and Computer Science, Vol. 1*, Bucharest, pp. 42-45.
3. Athans M., Falb P.L. 1966 *Optimal Control*, Mc Graw Hill, New York.
4. Boţan C. 1985 On the optimal output regulator problem for linear systems, *6th Int. Conf. on Control Systems and Computer Science, Vol. 1*, Bucharest, pp. 90-94.
5. Boţan C. 1992 On the solution of the Riccati differential matrix equation, *Symposium on Computational System Analyses*, Elsever Science Publishers, Berlin, pp. 141-146.
6. Boţan C., Postolache M., Onea A., Belţa C. 1995 An implementation of the controller for the LQ optimal problem, *10th Int. Conf. on Control Systems and Computer Science, Bucharest, Vol. 1*, Bucharest, pp. 83-87.

11

Optimal Control of Time-Varying Dynamic Systems

A.E. Kanarachos and K.T. Geramanis

1 Introduction

Nonlinear control laws often yield superior performance, compared to linear compensators, in both linear and nonlinear dynamic systems. While in linear dynamic systems different techniques can be used for optimum linear control design, in nonlinear dynamic systems optimum control design is based on heuristic rules and on standard control design procedures. Ostojic [1] presented a tracking controller based on numerical methods and corresponding recurrence relations. The structure of the controller is simply and easy to implement but wide application of the above methodology is limited, due to necessity of estimating system-order time derivatives of the control error. Yanushevsky [2] designed a controller for nonlinear systems by considering auxiliary sub-problems based on the original control system and applying Lyapunov theory for stable control of each sun-system. A special function is then formed based on Bellman equation for designing a optimal controller for the original system. Although Lyapunov theory provides a general approach, for designing control structures for nonlinear dynamic systems, often leads to complex control laws.

Additionally, Pandian *et. al.* [3], proposed a controller, for pneumatic robot manipulators including piston, pressure and valve dynamics, based on the sliding mode technique and incorporating differential pressure information instead of acceleration feedback. Model-based methodologies are also used for nonlinear control design. Chaney and Beaman [4] designed a model-based controller for a wind-tunnel type compressible flow process, with feed-forward and feed-back components. Comparison with gain-scheduling linear controller, showed that model-based controller yields improved performance and simpler tuning procedure. Healey [5] proposed a model following control design based on optimal control techniques, for the depth control of an autonomous underwater vehicle. The command parameters are generated off-line and selected by the vehicle's control system leading to improved control performance.

Time-varying dynamic systems with time-varying coefficients, is also a major topic of interest for several researchers. Sinha and Joseph [6] considered a linear dynamic system with periodically varying coefficients and designed a controller utilising the Lyapunov-Floquet transformations. The state transition matrix was based on Chebyshev polynomials, where the Liapunov-Floquet matrix was then used to transform the dynamic system into a form which control theory of time-invariant systems could be applied. Hu et. al. [7] investigated the problem of designing a linear state feedback controller to stabilise multi–input linear dynamic systems with time-varying bounded uncertain parameters. A Lyapunov function was constructed under the condition which the sign-invariant uncertainties in the system matrices are at least equal to the system order. A quadratically stabilisation of the system was then performed by a linear controller. Hong [8] investigated the asymptotic behaviour of a part of a solution of a linear dynamic system. The approach was based on a Lyapunov function, where the derivative is negative semidefinite, ensuring stability of the system, while further investigation was performed and showed that the partial state which remains in the derivative of the Lyapunov function converges to zero asymptotically, leading to convergence of state error to zero.

Additionally, Lee et. al. [9] proposed a receding horizon tracking control law based on H^∞ control concept, for time-varying discrete linear systems, where time-varying parameters and tracking commands needed, only for a finite future time. By considering new conditions on terminal weighting matrices, the controller guarantees closed loop stability but is only limited to linear dynamic systems. Minimum-variance controllers have also been considered for regulating linear time-varying dynamic systems. Extension of the above standard technique for the development of a d-step ahead minimum variance controller, was studied by Li and Evans [10], for linear dynamic systems under stochastic disturbances. The designed controller guarantees both closed-loop stability and minimum-variance control.

In nonlinear time-varying dynamic systems, the design procedure of an optimum controller becomes also a tedious task, where most efforts have been placed in nonlinear and/or adaptive controllers. Shin and Tsai [11] designed a neuro-fuzzy controller to control the position of a servohydraulic cylinder. The network structure was constructed by using bell-shaped membership functions and a look-up table, while optimum gains were determined by the back-propagation method. Bobrow and Lum [12] developed an adaptive controller for a hydraulic servovalve system, with full-state feedback for simultaneous parameter identification and tracking control. Experimental results demonstrated superior performance compared to a conventional fixed gain proportional controller. Sepheri et. al. [13] also considered the problem of position control of an industrial hydraulic manipulator. A nonlinear PI controller was designed with modifications in the integral portion of a conventional PI controller, while the position tracking accuracy was improved by a factor of five relative to the conventional PI controller.

This chapter proposes the design of a nonlinear controller for a nonlinear time-varying dynamic system common in textile industry. Position control of a transported flexible material with irregular contour, is considered, with nonlinear time-varying coefficients. The proposed controller is constructed using system-based information and actuator dynamics, with time-varying compensators and a dead zone nonlinearity. Numerical simulations show the satisfactory performance of the proposed approach, while the theoretical results are fully sustained by experiments, with conditions in a real industrial environment.

2 Optimal control design for a nonlinear time varying dynamic system

2.1 The dynamic system

In intelligent sewing environments, subsystem automation is a highly active filed, due to demands in increased productivity and improved quality. Patton *et. al.* [14] described and tested a controller for cloth handling using an adaptive force feedback, providing correct tension and straightened wrinkles on the fabric. Implementation of the controller on a PUMA 560 robot, demonstrated the performance of the designed controller. Berrett and Clapp [15] designed an electromagnetic actuator to eliminate adverse presser foot dynamics and a controller to maintain stable control of the fabric during sewing. Stylios *et. al.* [16] presented an extensive study on automatic systems for textile manufacture. Automatic measurement of fabric properties, seam appearance and sewing force penetration, sewing machines selecting automatically static and dynamic settings have been developed showing improved performance of the sewing process compared to conventional methods.

This chapter considers the problem of position control of a transported flexible material with irregular contour, common in sewing environments. The edge of the flexible material must pass form a fixed point, zero-point in Figure 1, while it is transported in x-direction and can also be turned around it. A model of the system dynamics is:

$$J(t) \cdot \ddot{\phi} = r_f \cdot F$$

$$T \cdot \dot{F} + F = P$$

$$P \begin{cases} = P_{\max} \\ = 0 \\ = -P_{\max} \end{cases} \tag{1}$$

where ϕ is the rotation angle of the flexible material, $J(t)$ the mass moment of inertia, $r_f \cdot F$ the time delayed applied torque, P the commanded force with values $\pm P_{\max}$ or 0, considering an on-off actuator and T the relevant time constant. Friction and spring coefficients are negligible so they are omitted from the differential equation of motion of the rigid body (eq. 1). The mass moment of inertia $J(t)$ for the flexible material with length L and mass m_o is equal to:

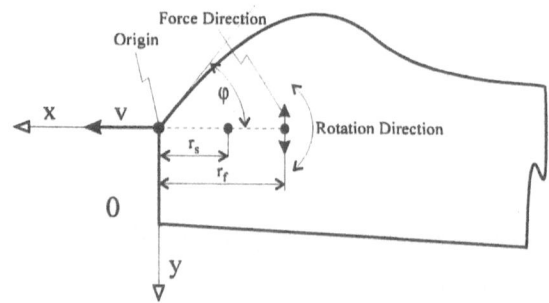

Figure 1. Geometrical characteristics of flexible material transportation.

$$J(t) = J_o \cdot (1 - \frac{tv}{L})^3 \qquad \text{with} \qquad J_o = \frac{m_o \cdot L^2}{4} \qquad (2)$$

and v denoting the transport velocity, L the length and m_0 the mass of the flexible material. Relation between the angle φ is necessary to perform a realistic evaluation of the results, since system requirements are expressed in terms of $y(t)$. If a sensor measuring y_s is positioned at a distance r_s from P, and φ is the rotation angle of the flexible material, then the sensor is measuring at $t=0$ the distance:

$$y_\phi = \phi \cdot r_s \qquad (3)$$

If the flexible material is transported with the velocity v, then the distance r_s is equal to:

$$r_s = v \cdot t \qquad (4)$$

then the sensor will be measuring the distance:

$$y_\phi = \phi \cdot (r_s + v \cdot t) \qquad (5)$$

In addition, if the flexible material has a contour $y_c(x)$, then for $t=0$:

$$y_c = y_c(r_s) \qquad (6)$$

while after a time period t the distance y_c will be equal to:

$$y_c = y_c(r_s + v \cdot t) \qquad (7)$$

Consequently, the sensor located at r_s will be measuring the distance:

$$y_s = y_\phi + y_c = \phi \cdot (r_s + v \cdot t) + y_c(r_s + v \cdot t) \qquad (8)$$

Substituting eq. 8 in eq. 1, where $y(t)=y_s(t)$, the dynamic of the system can be expressed in terms of $y(t)$. The objective of the designed controller is the edge position of the flexible material, within a predefined range ($|y(t)| \leq y_{max}$), while it moves in x direction with constant velocity v.

2.2 Nonlinear control design

Besides the existence of nonlinear time-varying parameters (eq. 1), system requirements denote that the applied force must be close to the edge, thus making the system response more difficult to control. Additionally the designed controller must cope with irregular edge curvature and non-uniform flexible material transportation. The proposed control law is a function of position, velocity and acceleration error of the edge, thus forming a proportional-derivative controller:

$$u = u(e_y, \dot{e}_y, \ddot{e}_y) \tag{9}$$

where $P=P(u)$, $e_y=-y(t)$, $\dot{e}_y = -\dot{y}(t)$ and $\ddot{e}_y = -\ddot{y}(t)$, as the reference state is zero.

In order to compensate with time-varying parameters of the dynamic system, a time-varying control law is proposed:

$$u = h(t) \cdot (k_1 \cdot e_y + k_2 \cdot \dot{e}_y + k_3 \cdot \ddot{e}_y) \tag{10}$$

where $h(t) = \left(1 - \dfrac{t \cdot v}{x_L}\right)^3$ is the time-varying function of control gains, x_L is a scale coefficient and k_i ($i=1,2,3$) are the control coefficients.

In order to avoid undesirable command signals which cause unacceptable system performance due to time delay of the actuator system, and thus designing an optimum controller with minimum energy control, a "window" of zero control is also proposed. If the calculated command force (eq. 10) is less than the "window" value w, the control command is zero, thus allowing a wider range for the zero state and making less sensitive the control law:

$$P \begin{cases} = \quad 0 & \text{if } |u| \leq w \\ = sign(u) \cdot P_{max} & \text{if } |u| > w \end{cases} \tag{11}$$

Equations (1)-(11) correspond to the structure of the designed controller, while selection of optimum control coefficients (k_1, k_2, k_3, w, x_L), is performed by heuristic procedure.

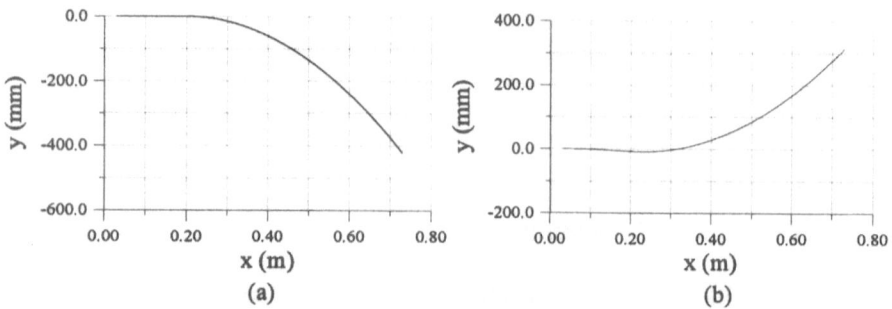

Figure 2. Edge geometry with high concave curvature (a) and convex-concave curvature (b).

3 Numerical example

A typical process in textile industry is considered, where a flexible material of mass $m_0 \in [0.03,...,0.08]$ Kgr and length $L=0.72$ m, is transported with $v=0.2$ m/sec, having irregular edge geometry with both convex and concave curvatures. The sensor is placed at a distance $r_s=0.03$ m (eq. 8) form zero point (Figure 1), and the actuator at a distance $r_f=0.05$ m (eq. 1). The maximum allowed movement of the edge near the zero point is $y_{max}=2$mm, the time constant of the actuator is $T=0.1$ and the maximum applied force is $P_{max}=0.8$ Nt.

Control coefficients have been determined with heuristic procedure, corresponding to:

$$(k_1, k_2, k_3, w, x_L) = (20.0, 8.0, 0.1, 0.005, 0.95) \qquad (12)$$

Figure 2 (a)-(b) shows the edge geometry of the flexible material, which must be placed within the desired range, during the transportation in x-axis. Both convex and concave curvatures have been considered, to ensure robustness of the controller.

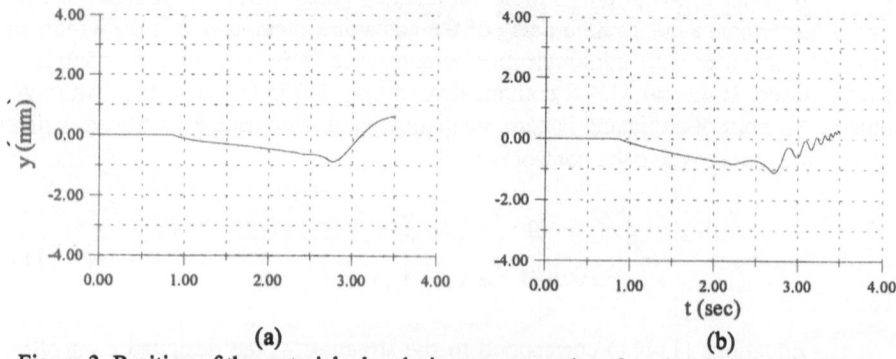

Figure 3. Position of the material edge during transportation, for edge curvature of Figure 2(a) and material mass $m_0=0.08$ Kgr (a), $m_0=0.03$ Kgr (b).

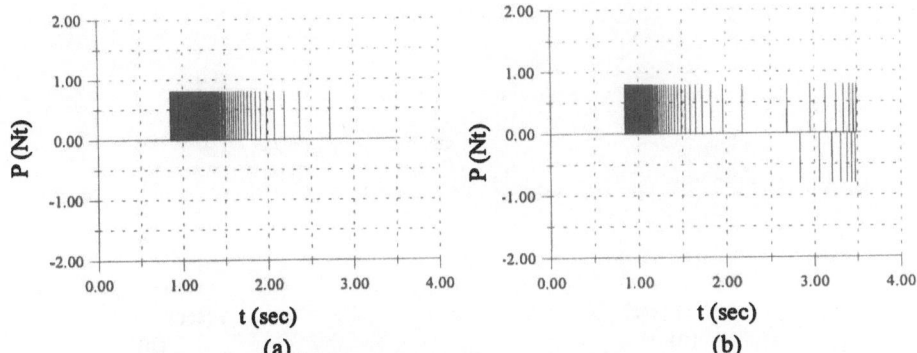

Figure 4. Commanded force, for edge curvature of Figure 2(a) and material mass
m_0=0.08 Kgr (a), m_0=0.03 Kgr (b).

Figure 3 shows the position $y(t)$ of the material edge for the edge curvature
of Figure 2(a) and material mass of m_0=0.08 Kgr and m_0=0.03 Kgr, respectively. As
the material mass decreases, the natural frequency of the system increases, leading
to faster dynamic response. Actuator response consider to have possitive, negative
and zero values, as it is shown on Figure 4.

Considering the edge curvature in Figure 2(b), the material edge position
during transportation does not differ significant from edge curvature of Figure 2(a).
Figure 5 (a) and (b) shows the material edge position for the second edge geometry
and material mass of m_0=0.08 Kgr and m_0=0.03 Kgr, respectively. Control
requirements are fulfilled, while the maximum peak in edge position is almost 1 mm
for both cases. Actuator responses are shown on Figure 6 (a) and (b), yielding
positive, negative and zero values due to existence of both convex and concave edge
curvature.

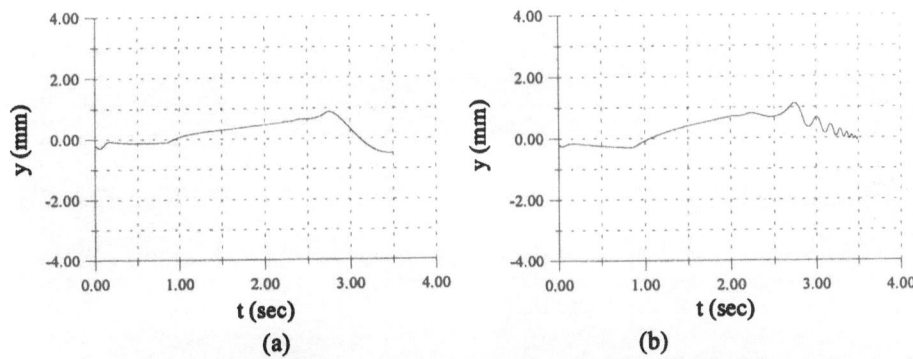

Figure 5. Position of the material edge during transportation, for edge curvature of
Figure 2(b) and material mass m_0=0.08 Kgr (a), m_0=0.03 Kgr (b).

134

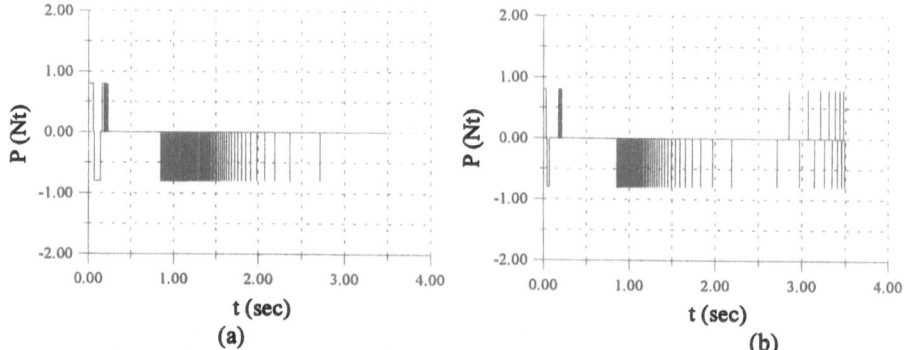

Figure 6. Commanded force, for edge curvature of Figure 2(b) and material mass
m_0=0.08 Kgr (a), m_0=0.03 Kgr (b).

4 Experimental results

The layout of the experimental system is shown in Figure 7 (a), while
Figure 7 (b) shows a close view of the actuator and the sensor. A non-contact
actuator was considered, thus avoiding kinematics interaction of the acting force P
(eq. 11) and the transportation of the flexible material in x direction. The actuator
consists of two one-way digital valves, where the air acts the necessary force via
properly designed air nozzles in a metallic plate (Figure 7(b)). An analogue sensor
with 5 mm working range is selected and a look-up table provides the current
position $y(t)$ of the material edge. A computer controls the whole process via a 12-
bit A/D converter for the sensor input and 2 Solid State Relays for the valves output.

A typical flexible material for a sewing process is selected, corresponding
to a trouser. Figure 8 shows the response of material edge during the transportation,
which is constantly less than 2 mm. A common optical sensor has been considered,
leading to small oscillations of the material edge, while smoother response can be
achieved utilising a more advanced and less sensitive sensor.

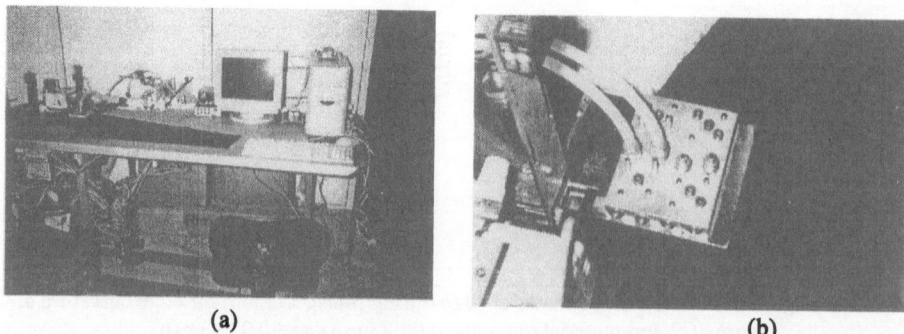

Figure 7. Layout of the experimental system (a) and close view of the actuator and the sensor

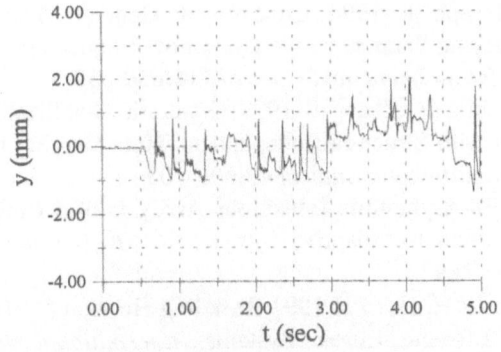

Figure 8. Position of edge curvature with nonlinear control, for a convex edge
geometry.

5 Conclusions

A nonlinear controller is proposed in this chapter for nonlinear time-
varying dynamic systems. Position control of a transported flexible material with
irregular contour is considered with nonlinear time-varying parameters, while the
proposed controller achieves fulfilment of system requirements and minimum
energy concept, leading to optimum performance.

Numerical simulations demonstrate the robustness of the controller,
considering different types of flexible materials, different weights and different edge
curvature. The theoretical results are fully sustained by experiments, with real
conditions that prove also the non-sensitivity of the controller design in real
industrial environment.

References

1. Ostojic M 1996 Numerical Approach to Nonlinear Control Design. *Journal of
   Dynamics, Measurement, and Control* 118:332-337
2. Yanushevsky R T 1992 A Controller Design for a Class of Nonlinear Systems
   Using the Lyapunov-Bellman Approach'. *Journal of Dynamics, Measurement,
   and Control* 114:390-393
3. Pandian S. R, Hayakawa Y, Kanazawa Y, Kamoyama Y, Kawamura S 1997
   Practical Design of a Sliding Mode Controller for Pneumatic Actuators. *Journal
   of Dynamics, Measurement, and Control* 119:666-674
4. Chaney M J, Beaman J J 1992 Comparison of Nonlinear Tracking Controllers
   for a Compressible Flow Process. *Journal of Dynamics, Measurement, and
   Control* 114:493-499
5. Healey A J 1992 Model-based Maneuvering Controls for Autonomous
   Underwater Vehicles. *Journal of Dynamics, Measurement, and Control* 114:
   614-622

6. Sinha S C, Joseph P 1994 Control, of General Dynamic Systems with Periodically Varying Parameters via Lyapunov-Floquet Transformation. *Journal of Dynamics, Measurement, and Control* 116:650-658
7. Hu S., Dai Q, Jing Y, Zhang S 1997 Quadratic Stabilizability of Multi-Input Linear Systems with Structural Independent Time-Varying Uncertainties. *IEEE Transactions on Automatic Control* 42:699-703.
8. Hong K S 1997 Asymptotic Behaviour Analysis of a Coupled Time-Varying System: Application to Adaptive Systems. *IEEE Transactions on Automatic Control* 42:1693-1697
9. Lee J W, Kwon W H, Lee J H 1997 Receding Horizon H^{∞} Tracking Control for Time-Varying Discrete Linear Systems. *International Journal of Control* 68:385-399
10. Li Z, Evans R J 1997 Minimum-Variance Control of Linear Time-Varying Systems. *Automatica* 8:1531-1537
11. Shih M C, Tsai C P 1995 Servohydraulic Cylinder Position Control Using a Neuro-fuzzy Controller. *Mechatronics* 5:497-512
12. Bobrow J E, Lum K 1996 Adaptive, High Bandwidth Control of a Hydraulic Actuator. *Journal of Dynamics, Measurement, and Control* 118:714-720
13. Sepehri N, Khayyat A A, Heinrichs B 1997 Development of a Nonlinear PI Controller for Accurate Positioning of an Industrial Hydraulic Manipulator. *Mechatronics* 7:683-700
14. Patton R, Swen F, Tricamo S, van der Veen A 1992 Automated Cloth Handling Using Adaptive Force Feedback. *Journal of Dynamics, Measurement, and Control* 114:731-735
15. Berrett G R, Clapp T G 1995 Coprime Factorization Design of a Novel Maglev Presser Foot Controller. *Mechatronics* 5:279-294
16. Stylios G, Sotomi O J, Zhu R, Xu Y M, Deacon R 1995 The Mechatronics Principles for Intelligent Sewing Environments. *Mechatronics* 5:309-319

12

Response Optimization of a Discrete-Time Bang-Bang Optimal Control Problem

A.G. Petridis, G.N. Charalampopoulos and A.E. Kanarachos

1 Introduction

The optimal control problem of an initial-value ordinary differential equation, with Bolza objectives and mixed constraints has the following form:

$$\min_{x,u} \int_0^T L\big(x(t),u(t),t\big)\cdot dt + \phi_f\big(x(T)\big), \tag{1.a}$$

$$\text{s.t.} \quad \dot{x}(t) = f\big(x(t),u(t),t\big), \qquad x(0) = x_{init}, \tag{1.b}$$

$$g\big(x(t),u(t),t\big) \le 0, \qquad t \in [0,T], \qquad g_f\big(x(T)\big) \le 0. \tag{1.c}$$

Here, $x:[0,T] \to \mathbf{R}^{n_x}$, $u:[0,T] \to \mathbf{R}^{n_c}$, $L:\mathbf{R}^{n_x} \times \mathbf{R}^{n_c} \times [0,T] \to \mathbf{R}$, $\phi_f:\mathbf{R}^{n_x} \to \mathbf{R}$, $g:\mathbf{R}^{n_x} \times \mathbf{R}^{n_c} \times [0,T] \to \mathbf{R}^{n_s}$, $g_f:\mathbf{R}^{n_x} \to \mathbf{R}^{n_f}$.

A discrete-time counterpart is the problem:

$$\min_{x_i,u_i} \sum_{i=1}^N L_i\big(x_i,u_i\big) + \phi_N\big(x_{N+1}\big), \tag{2.a}$$

$$\text{s.t.} \quad x_{i+1} = f_i\big(x_i,u_i\big), \qquad i = 1,\ldots,N, \qquad x_1 \text{ fixed}, \tag{2.b}$$

$$g\big(x_i,u_i\big) \le 0, \qquad i = 1,\ldots,N, \qquad g_f\big(x_{N=1}\big) \le 0 \tag{2.c}$$

Many computational algorithms have been proposed for various special classes of the above problems. In the unconstrained case (when the terms g, g_i and g_f do not exist), Newton-like and conjugade gradient methods for class (1) are discussed by Polak (Ref. 1) and for case (2) Newton's methods and its efficient implementation are discussed by Dunn and Bertsekas (Ref. 2). A variety of quasi-Newton methods have also been applied for the unconstained version of case (1) problems, by Edge and Powers (Ref. 3) and Kelley and Sachs (Ref. 4). In the control-constrained case (when g_f does not exist and the states x and x_i do not appear in terms g and g_i), the optimal control problem is traditionally treated as a constrained optimization problem in u or u_i. Because of the pointwise or separable nature of the constraints, methods of the gradient projection class are easily implementable; see for example, Demyanov and Rubinov (Ref. 5) and Dunn (Refs. 6,7). In the finite-dimensional problem, these methods have the advantage that the set of currently active constraints can change extensively at each iteration, whereas active set methods only allow a single change to the active constraints set. Newtonian scaling has been added to gradient projection algorithms by Gafni and Bertsekas (Ref. 8) and Dunn (Ref. 9) to improve their asymptotic rate of convergence and the resulting methods have proven to be efficient for the control-constrained version of class (2) by Wright (Ref. 0) Pantoja and Mayne (Ref. 11) described a stage-wise algorithm for the control-constrained case that, in a neighborhood of a solution of (2), produces iterations that are identical to sequential quadratic programming iterations. Instead of utilizing the inherent structure in (2) at the level of the linear algebra computations, as Wright do (Refs. 12,13), Pantoja and Mayne exploit the structure at a somewhat higher level.

The most general cases of (1) and (2) classes, in which the functions g, g_i and g_f are nontrivial in both states and controls, are significantly more difficult to solve than the special cases described above. Algorithms based on nonlinear programming techniques appear to be the most promising. In these algorithms, both states and controls are treated as unknown and the state equations and auxiliary constraints are treated as equality and inequality constraints, respectively. Polak, Yang and Mayne (Ref. 14) propose a first-order algorithm which utilizes barrier functions for the inequality constraints. Evtushenko (Ref. 15) describes a variety of augmented Langrangian penalty function methods, in which a class (2) problems is reduced to an unconstrained problem. Also, Di Pillo, Grippo and Lampariello (Ref. 16) describe a structured quasi-Newton method for a particular augmented Lagrangian.

Several efficient control parametrization algorithms have already been developed for solving optimal control problems involving inequality state constraints by many authors (Refs. 17-27). In this paper, an optimal control problems with state and mixed control/state inequality constraints is solved, using a parameter optimization Newton methhodology proper for optimal control problems (Ref. 28). In order to face the inequality constraints, the technique of their incorporation into a pseudo-objective function in the form of penalties, is adopted. During each iteration of the proposed optimization algorithm, an internal iterative Newton procedure is performed, for the minimization of the Smoothed Quadratic Approximation (S.Q.A) surrogate function. The solution of this optimization sub-problem, provides the new estimation of the original problem optimum. The surrogate function is formed utilizing the total or a suitably selected set of the zeroth order information (values of the objective function and the constraints), that have been resulted from the analyses of the previous estimations of the original optimum and combine many Smoothed Quadratic Approximations of the objective or pseudo-objective function. After each main iteration has been completed, the surrogate function is altered -at least in the neighborhood of the last optimum estimation- modifying the result of the next internal optimization procedure.

2 Problem Formation

We turn to an optimal control problem with state and mixed control/state inequality constraints and a bang bang solution. Here, $n_s = 4$, $n_c = 1$, $T = 4.2$.

$$\min \; \sum_{i=1}^{4} x_i^2(T) \tag{3.a}$$

$$\text{s.t.} \quad \dot{x}_1 = -0.5 \cdot x_1 + 5 \cdot x_2, \tag{3.b}$$
$$\dot{x}_2 = -5 \cdot x_1 - 0.5 \cdot x_2 + u,$$
$$\dot{x}_3 = -0.6 \cdot x_3 + 10 \cdot x_4,$$
$$\dot{x}_4 = -10 \cdot x_3 - 0.6 \cdot x_4 + u,$$

$$\left| u(t) \right| \le 1, \tag{3.c}$$

$$x_i(0) = 10, \qquad\qquad i = 1,...,4, \tag{3.d}$$

$$x_i(T) \le 1, \qquad\qquad i = 1,...,4. \tag{3.e}$$

Versions of this problem have been discussed by a number of authors, including Jacobson and Mayne (Ref. 29), who exclude the terminal inequality constraints, by Longsdon (Ref. 30) and Wright (Ref. 10). The problem has a bang bang solution with eight switching times. It is solved in Ref. 30 utilizing a discrete nonlinear programming formulation and in Ref. 29 utilizing a second-order method for problems with control bounds and bang bang solutions. In Ref. 10 the problem is discetized by approximating the ODEs by a midpoint rule.

Let t_i ($i = 0,1,...,7$) be the consequent switching times, when the control signal is inverted between the two extreme values $u = \pm 1$ and $u(0) = 1$. The initial switching time is predetermined ($t_0 = 0$) and the remains are corresponding to the seven control variables of the current optimal control problem. Each combination of the seven undetermined switching times has a considerable number of other equivalent combinations. In order to avoid the existence of a great number of equivalent global optima ($5040 = 7 \cdot 6 \cdot 5 \cdot ... \cdot 2$), an additional set of inequality constraints has to be adopted.

$$t_0 < t_1 < ... < t_7 < T. \tag{4}$$

The sequence of the switching times can be easily succeeded by the next anadromous relation.

$$t_i = t_{i-1} + y_i \cdot (T - t_{i-1}) \Leftrightarrow y_i = \frac{t_i - t_{i-1}}{T - t_{i-1}}, \qquad\qquad i = 1,...,7. \tag{5}$$

Thus, an alternative set of decision variables y_i is introduced. Each one of the new variables represents the percentage of the available time interval between the end time and the last switching time, that is covered by the time interval between the corresponding switching time and the previous one. The proposed variables enable the replacement of the inequality constraints (4), by a more convenient (according to the structure of S.Q.A algorithm) set of side constraints. The inequality constraints (3.e) are incorporated into the objective function in the form of extended quadratic penalties. After these modifications, a parameter optimization problem (6) is formulated, to be solved, instead of the original optimal control problem (3).

$$\min \quad L(y_1, y_2, ..., y_7) = \sum_{i=1}^{4} x_i^2(T) + K \cdot \sum_{i=1}^{4} \max\{x_i(T) - 1, 0\}^2 \qquad (6.a)$$

s.t. $0 < y_i < 1,$ $\qquad i = 1,...,7.$ $\qquad\qquad\qquad$ (6.b)

$t_i = t_{i-1} + y_i \cdot (T - t_{i-1})$ $\qquad i = 1,...,7,$ $\qquad t_0 = 0,$ $\qquad$ (6.c)

$u(t) = (-1)^j,$ $\qquad\qquad j = \max\{i = 1,...,7, \quad t_i < t\},$ $\qquad$ (6.d)

$\dot{x}_1 = -0.5 \cdot x_1 + 5 \cdot x_2,$ $\qquad\qquad\qquad\qquad\qquad\qquad\qquad$ (6.e)

$\dot{x}_2 = -5 \cdot x_1 - 0.5 \cdot x_2 + u,$

$\dot{x}_3 = -0.6 \cdot x_3 + 10 \cdot x_4,$

$\dot{x}_4 = -10 \cdot x_3 - 0.6 \cdot x_4 + u,$

$x_i(0) = 10,$ $\qquad\qquad\qquad i = 1,...,4.$ $\qquad\qquad\qquad$ (6.f)

3 Optimization Procedure

The proposed parameter optimization algorithm, named S.Q.A, has been designed and developed according to the demands of the parametric response optimization of linear or nonlinear dynamic systems. The relevant problems have singular objective function, difficult to manipulate equality and inequality constraints, many local optima and high computational cost, since a time integration of a system of differential equations is required for each analysis. Moreover, the derivatives of the objective function and constraints are not directly available, due to high computational cost. The peculiarity of the optimized function is intensified when a sequence of pseudo-objective functions, including penalties, is optimized instead of the original function. The conventional iterative algorithms are inefficient for the above mentioned problems because of: a) the insufficient amount of information that a typical approximation function can contain, b) the direct discarding of the available data form previous minimum estimations. This discarding is usually performed after one or limited number of iterations, leading to aimless reexaminations of acceptable sub-regions, even if the available data could be sufficient for the rejection of the possibility for a minimum existence in the specific areas, c) the local nature of the evaluated data (value, gradient vector, Hessian matrix, etc.). Therefore, the global optimum cannot be located, when the algorithm has been trapped around a local minimum. There is also a wide tendency of using semi-stochastic and stochastic algorithms for the solution of problems with many local optima. These algorithms avoid being trapped by using a cloud of analyzed points and a respectable amount of randomness at the generation of the next point or cloud of points. Meanwhile, semi-stochastic and stochastic algorithms converge slowly, locate the optimum with a low accuracy level and usually require a great number of iterations and multiple number of analyses to be performed.

For the solution of an optimization problem, a surrogate problem is usually formulated and solved. The surrogate problem has to: a) approximate the original one with a sufficient accuracy, at least at the critical sub-regions of the feasible domain b) have a simple (typical) mathematical form c) have optima that correspond with the original ones, and these only, at a sufficient accuracy level. When the objective function is differentiable, the surrogate problem is usually modeled using first or higher order Taylor approximations. On the other hand, at some mathematical and technical problems the global minimum of non-classical functions have to be explored. Such problems appear in the approximation theory, nonlinear dynamic systems control, variable structure control systems applications, control of discrete events systems, cost optimization, etc. The results presented by numerous authors encourage the exploitation of the second order derivatives for piece-wise differentiable functions, in accordance with the evolution of normal functions numerical optimization methods, where Newton type methods tend to displace Gradient type methods.

The exploitation of information that is provided from previous iterations with simultaneous use of surrogate functions has considerable advantages, such as a) more accurate objective function approximation, due to higher order provided data b) the capability of modeling the original function through many simple local approximations, with diversified starting points c) the appropriate formation of the surrogate problem so as to be easily solvable by a predetermined classical optimization algorithm, through indirect approximate calculation of the necessary data.

In each S.Q.A iteration, an internal iterative optimization procedure is performed. The consequent internal optimization result is a surrogate function minimum that approximates a minimum of the original objective function. The internal procedure consists of direction and step size selection loops, that minimize the surrogate function, according to a typical Newton methodology.

The surrogate function is defined through a series of smoothed quadratic approximations of the n variables objective function through the definition of the relevant $N = (n+1) \cdot (n+2)/2$ approximation coefficients. The classical quadratic approximation evaluates a considerable amount of N information. In order to avoid the entire existence of objective function local formations and to augment the available information exploitation, the S.Q.A technique is adopted, which permits the exploitation of as many and whichever available zeroth order data. This wider capability is not succeeded thanks to the S.Q.A formation, but the procedure of data exploitation.

When the objective function has many different local optima, discontinuities and/or non defined or discontinuous derivatives, no single quadratic function can be found with adequate global approximation accuracy. Besides, the required global accuracy level can be succeeded by a set of quadratic functions with local influence. The surrogate function is defined for any feasible point via a smoothed quadratic approximation starting from the specific point. After starting point location, the stored information are partially rearranged according to their corresponding reduced Euclidean distances. This way, M of the K available points are selected ($M \geq N$) around the starting point. After the quadratic equation is asked to be fulfilled for each member of the set of selected information an $M \times N$ system of equations is provided. This over-defined system cannot be generally solved when $M > N$. Therefore, an optimization sub-problem of minimizing the second power of the norm of the corresponding error vector, is solved. The current solution defines the local S.Q.A and the specific surrogate function's data (value and first and second order derivatives). The above surrogate function is well defined. In other words, each point of the explored area corresponds to one and only one surrogate function value. This correspondence is changed between two or more sequential external iterations due to the participation of the last analyzed points. This change is imperative for the internal optimization results differentiation.

The surrogate function formation and the S.Q.A optimization methodology activation presupposes the availability of efficient number of information, that result from the analysis of a minimum number of preliminary points analyses. The corresponding feasible combinations of the decision variables should be generated, so as to facilitate the subsequent essential optimization procedure. From the examination of S.Q.A algorithm performance arise that it is expedient to augment the number of preliminary points over the minimum one (N). Two alternative methods of preliminary generations have been tested, the random and the evolutionary ones. **a)** *Random Generation.* Each time one point is generated with equal probability for each point located in the n-dimensional orthogonal parallelepiped feasible domain. **b)** *Evolutionary Generation.* A group of points is produced, corresponding to the offspring created by means of recombination and mutation of a multimembered evolution strategy (ES).

The internal optimization initialization rule affects the internal minimization results, the available data distribution and consequently influences the S.Q.A algorithm performance. Two alternative rules have been applied suitable for different classes of problems. **a)** *Best Point Initialization.* From the set of the available points, the one that results the best (minimum) objective function value is selected. This rule is preferable for problems without many local optima or singular local formations. **b)** *Random Initialization.* The starting point of the internal optimization is randomly located in the feasible domain. This way the S.Q.A algorithm is never trapped in a local optimum neighborhood. In opposite, all the optima are gradually approached by different iterations results. This rule is preferable when the location of the global optimum sub-region has not been confirmed. In order to combine the advantages of the two initialization rules, the first one can be applied in the primary stage of the optimization procedure and afterwards the second one, or both of them can be cyclically activated.

4 Results

In the following chart (chart1) we examine the quick convergence of the algorithm (iteration 350 finally but it had converge from equation 102 and in next equations tried to improve the accuracy. Also in chart 2 we see the initial and final values of optimization variables in optimization procedure.

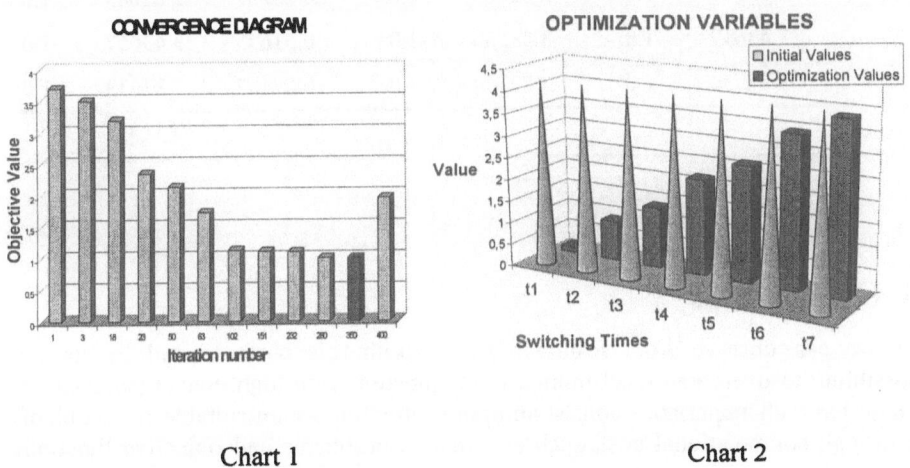

Chart 1   Chart 2

According to the terminal state vector coordinates values for the optimal switching times all the terminal inequality constraints are found to be satisfied.

$$x_1 = 0.10044, x_2 = 0.74775, x_3 = 0.62367, x_4 = 0.2138$$

Table 1. Comparison of Control Profiles for Computed Bang Bang Solutions

	Longsdon (Ref. 30) $f=1.00349$		JACOBSON and MAYNE (Ref. 29) $f=1.00357$		S.Q.A Algorithm $f=1.00346$		
	Switching Time	u	Switching Time	u	y_i	Switching Time	U
0.	0.00000	-1.0	0.00000	-1.0	0.00000	0.00000	-1.0
1.	0.11198	1.0	0.11000	1.0	0.02643	0.11101	1.0
2.	0.89979	-1.0	0.89800	-1.0	0.19274	0.89712	-1.0
3.	1.36428	1.0	1.36500	1.0	0.14138	1.36580	1.0
4.	2.16960	-1.0	2.16800	-1.0	0.28312	2.16822	-1.0
5.	2.62063	1.0	2.62000	1.0	0.22257	2.62043	1.0
6.	3.43619	-1.0	3.43560	-1.0	0.51675	3.43667	-1.0
7.	3.87530	1.0	3.87740	1.0	0.57467	3.87533	1.0
N.O.I						60 - 80	

5 Conclusions

As we can observe from results SQA Algorithm is characterized by its big possibility to overcome local minina in conjunction with high convergence speed. These basic characteristics consist an optimization algorithm suitable for problems with high computational cost, such as dynamic problems, and objective functions with many peculiarities.

References

1. POLAK, E., *Computational Methods in Optimization*, Academic Press, New York, New York, 1970.

2. DUNN, J. C., and BERTSEKAS, D. P., *Efficient Dynamic Programming Implementations of Newton's Method for Unconstrained Optimal Control Problems*, Journal of Optimization Theory and Applications, Vol.63, pp. 23-38, 1989.

3. EDGE, E. R., and POWER, W., F., *Functions Space Quasi-Newton Algorithms for Optimal Control Problems with Bounded Controls and Singular Arcs*, Journal of Optimization Theory and Applications, Vol. 20, pp. 455-479, 1976.

4. KELLEY, C. T. and SACHS, E. W., *A Pointwise Quasi-Newton Method for Unconstrained Optimal Control Problems*, Numerische Mathematik, Vol. 55, pp. 159-176, 1989.

5. DEMYANOV, V. F., and RUBINOV, A. M., *Approximate Methods in Optimization Problems*, American Elsevier, New York, New York, 1970.

6. DUNN, J. C., *Global and Asymptotic Convergence Rate Estimates for a Class of Projected Gradient Processes*, SIAM Journal on Control and Optimization, Vol. 19, pp. 368-400, 1981.

7. DUNN, J. C., *On the Convergence of Projected Gradient Proccesses to Singular Critical Points*, Journal of Optimization Theory and Applications, Vol. 55, pp. 203-216, 1987.

8. GAFNI, E. M., and BERTSEKAS, D. P., *Two-Metric Projection Methods for Constrained Optimization*, SIAM Journal on Control and Optimization, Vol. 22, pp. 936-964, 1984.

9. DUNN, J. C., *A Projected Newton Method for Minimization Problems with Nonlinear Inequality Constraints*, Numerische Mathematik, Vol. 53, pp. 377-409, 1988.

10. WRIGHT, S. J., *Interior Point Methods for Optimal Control of Discrete Time Systems*, Journal of Optimization Theory and Applications, Vol. 77, pp. 161-187, 1993.

11. PANTOJA, J. F. A. D., and MAYNE, D. Q., *Sequential Quadratic Programming Algorithm for Discrete Optimal Control Problems with Control Inequality Constraints*, International Journal of Control, Vol. 53, pp. 823-836, 1991.

12. WRIGHT, S. J., *Solution of Discrete-Time Optimal Control Problems on Parallel Computers*, Parallel Computing, Vol. 16, pp. 221-238, 1990.

13. WRIGHT, S. J., *Partitioned Dynamic Programming for Optimal Control*, SIAM Journal on Optimization, Vol. 1, pp. 620-642, 1991.

14. YANG, T. H., POLAK, E., and MAYNE, D. Q., *A Method of Centers Based on Barrier Functions for Solving Optimal Control Problems with Continuum Constraints*, Proceedings of the 29th Conference on Decision and Control, pp. 2327-2332, 1990.

15. EVTUSHENKO, Y. G., *Numerical Optimization Techniques*, Optimization Software, New York, New York, 1985.

16. DI PILLO, G., GRIPPO, L., and LAMPARIELLO, F., *A Class of Structured Quasi-Newton Algorithms for Optimal Control Problems*, Proceedings of the IFAC Conference on Applications of Nonlinear Programming to Optimization and Control, pp. 101-107, 1983.

17. TEO, K. L., and JENNINGS, L. S., *Nonlinear Optimal Control Problems with Continuous State Inequality Constraints*, Journal of Optimization Theory and Applications, Vol. 63, pp. 1-22, 1989.

18. GOH, C. J., and TEO, K. L., *Control Parametrization: A Unified Approach to Optimal Control Problems with General Constraints*, Automatica, Vol. 24, pp. 3-18, 1988.

19. TEO, K. L., and GOH, C. J., *A Computational Method for Combined Optimal Parameter Selection and Optimal Control Problems with General Constraints*, Journal of the Australian Mathematical Society, Series B, Vol. 30, pp. 350-364, 1989.

20. TEO, K. L., and GOH, C. J., *A Simple Computational Procedure for Optimization Problems with Functional Inequality Constraints*, IEEE Transactions on Automatic Control, Vol. 32, pp. 940-941, 1987.

21. WONG, K. H., KLEMENTS, D. J., and TEO, K. L., *Optimal Control Computation for Nonlinear Time-Lag Systems*, Journal of Optimization Theory and Applications, Vol. 47, pp. 91-107, 1985.

22. SIRISENA, H. R., *Computation of Optimal Controls Using a Piecewise Polynomial Parametrization*, IEEE Transactions on Automatic Control, Vol. 18, pp. 409-411, 1973.

23. SIRISENA, H. R., and CHOU, F. S., *Convergence of Control Parametrization Ritz Method for Nonlinear Optimal Control Problems*, Journal of Optimization Theory and Applications, Vol. 29, pp. 369-382, 1979.

24. SISIRENA, H. R. and TAN, K. S., *Computational of Constrained Optimal Controls Using Parametrization Techniques*, IEEE Transactions on Automatic Control, Vol. 19, pp. 431-433, 1974.

25. TEO, K. L., and WOMERSLEY, R. S., *A Control Parametrization Algorithm for Optimal Control Problems Involving Linear Systems and Linear Inequality Constraints*, Numerical Functional Analysis and Optimization, Vol. 6, pp. 291-313, 1983.

26. TEO, K. L., WONG, K. H., and CLEMENTS, D. J., *Optimal Control Computation for Linear Time-Lag Systems with Linear Terminal Constraints*, Journal of Optimization Theory and Applications, Vol. 44, pp. 509-529, 1984.

27. KAJI, K., and WONG, K. H., *Nonlinearly Constrained Time-Delayed Optimal Control Problems*, Journal of Optimization Theory and Applications, Vol. 82, pp. 295-313, 1994.

28. PETRIDIS, A. G., HARALABOPOULOS, G. N., and KANARACHOS, A. E., *A New Global Optimization Algorithm Combining the Natural Evolution Model and the Deterministic Newton Methodology*, EURISCON '98, 1998.

29. JACOBSON, D. H., and MAYNE, D. Q., *Differential Dynamic Programming*, American Elsevier, New York, New York, 1970.

30. LONGSDON, J. S., *Efficient Determination of Optimal Control Profiles for Differential Algebraic Systems*, PhD Thesis, Carnegie-Mellon University, 1990.

13

Adaptive LQ Optimal Autopilots for Tankers Based on Two-Point Multirate Controllers

P.N. Paraskevopoulos, K.G. Arvanitis and A.A. Vernardos

1   Introduction

Changing circumstances, interwoven with variations in the steering characteristics of a ship, as well as changes of the required performance due to varying traffic situations, often require automatic re-adjustment of a series of settings of course keeping ship autopilots. In this respect, adaptive autopilots, based on identification of ship steering dynamics, that can be used to apply a course feedback system, in those situations where varying conditions give rise to difficulties, are of great importance and have long been the focus of interest by many marine engineers and specialized control designers [1]-[7].

Ship steering dynamics with a constant forward speed, can usually be modeled by Nomotos's 2nd order model [2]-[5], [7]-[9], which represents the sway-yaw dynamics of a ship. It is essentially a third order linear state-pace dynamical system, whose state variables are the sway velocity, the ship's angular velocity and the deviation in heading angle. For many ships, this sway-yaw dynamics is unstable [2], [7], [8]. LQ optimal regulators can be used to stabilize this dynamics. Moreover, LQ control theory can be applied to ships, in order to obtain increased performance when sailing in restricted waters and reduced fuel consumption when sailing in open sea. In particular, the trade-off between accurate and economical steering of tankers can be easily related to a quadratic criterion [7], [9].

The adaptive LQ regulation problem has extensively been studied in the past [10]-[14]. In particular, adaptive LQ regulators for tankers have been studied in [3], [4], [7]. A common feature of most of the control strategies proposed in the past, is that they are based on the design of full order adaptive state observers, providing suboptimal control schemes, since state observers offer only an approximate estimation of the state vector, needed to implement these control strategies. Moreover, high order exogenous dynamics is introduced in the control loop. Finally, most of the above control strategies are largely affected upon several disadvantages of estimator based controllers, related to increased on-line computational effort as well as reduced robustness.

To overcome these difficulties, a new technique to the adaptive LQ regulation problem has recently been presented in [13]. It is based on Two-Point-Multirate Controller (TPMRC), a control strategy, that is a combination of the control strategies reported in [14] and [15]. The suggested adaptive control strategy relies on solving the continuous-time LQ regulation problem. Based on this strategy, the original problem is reduced to an associate discrete-time LQ regulation problem, for the performance index with cross product terms, for which a fictitious static state feedback controller is needed to be computed. Thus, the present technique essentially resort to the computation of simple gain controllers rather than to the computation of state observers, as compared to known techniques. On the basis of TPMRCs, a useful indirect adaptive control scheme is obtained, which estimates the unknown plant parameters (and hence the parameters of the desired multirate controller) on line, from sequential data of the inputs and the outputs of the plant, which are recursively updated within the time limit imposed by a fundamental sampling period T_0. The designed TPMRCs based adaptive LQ optimal regulators can possess any prescribed degree of stability, since there is the possibility to choose the transition matrices of the controllers arbitrarily. Note also that, using the proposed technique, the exogenous dynamics introduced in the control loop is of low order, while a satisfactory robustness level is guaranteed. On the other hand, the proposed approach incorporates all the advantages of multirate control strategies over relevant single rate ones (for a detailed analysis of these advantages, see [15]). It is worth noticing that, the a priori knowledge, needed to implement the proposed adaptive algorithm, is controllability and observability of the continuous and the discretized system under control and known order. Finally, persistency of excitation of the controlled system is assured without making any assumption on the existence of special convex sets in which the estimated system parameters belong, or on the coprimeness of the polynomials describing the ARMA model, as in known techniques. In the present paper, the proposed adaptive regulator is applied to the unstable sway-yaw dynamics of a minesweeper as well as of a large tanker. Simulation results clearly illustrate the effectiveness of the proposed adaptive controller which has a satisfactory performance.

2 Adaptive LQ Regulation Using TPMRCs

Consider the linear single-input, single output system of the form

$$\dot{x}(t) = \mathbf{A}x(t) + \mathbf{b}u(t) \ , \ y(t) = \mathbf{c}^T x(t) \tag{1}$$

where, $\mathbf{x}(t) \in \mathbf{R}^n$, $u(t) \in \mathbf{R}$, $y(t) \in \mathbf{R}$, and $\mathbf{A}$, $\mathbf{b}$, $\mathbf{c}^T$ are real matrices having appropriate dimensions. The matrix triplet $(\mathbf{A}, \mathbf{b}, \mathbf{c}^T)$ is arbitrary and unknown, except for the information summarized in the following Assumption:

Assumption: (a) System (1) is controllable and observable and of known order n. (b) There is a sampling period $T_0 \in \mathbf{R}^+$ and an integer L > n, such that the systems with matrix triplets $(\Phi, \mathbf{B}_N, \mathbf{c}^T)$, $(\Phi, \hat{\mathbf{b}}, \mathbf{c}^T)$ and $(\Phi_\tau, \mathbf{b}_\tau, \mathbf{c}^T)$, are minimal, where $\tau = \dfrac{T_0}{(2n+1)L}$ and

$$\Phi = \exp(\mathbf{A}T_0), \quad \Phi_\tau = \exp(\mathbf{A}\tau), \quad \mathbf{b}_\tau = \int_0^\tau \exp(\mathbf{A}\lambda)\mathbf{b}d\lambda, \quad \hat{\mathbf{b}} = \int_0^{T_0} \exp(\mathbf{A}\lambda)\mathbf{b}d\lambda$$

and where $\mathbf{B}_N \in \mathbf{R}^{n \times p_N}$, is the full rank matrix defined by

$$\mathbf{B}_N \mathbf{B}_N^T = \mathbf{W}_N(T_0, 0) \geq 0, \quad \mathbf{W}_N(T_0, 0) = T_N^{-1} \sum_{\mu=0}^{N-1} \delta_\mu \delta_\mu^T, \quad p_N = \text{rank } \mathbf{W}_N(T_0, 0)$$

$$T_N = \frac{T_0}{N}, \quad \delta_\mu = \hat{\mathbf{A}}_c^{(N-\mu-1)} \hat{\mathbf{b}}_{T_N}, \quad \hat{\mathbf{A}}_c = \exp(\mathbf{A}T_N), \quad \hat{\mathbf{b}}_{T_N} = \int_0^{T_N} \exp(\mathbf{A}\lambda)\mathbf{b}d\lambda$$

To system (1) apply the TPMRC based feedback strategy. More precisely, let the input of the plant be constrained to the following piecewise constant control

$$u(kT_0 + \mu T_N + \zeta) = \Pi_\mu \hat{\mathbf{u}}(kT_0) \equiv T_N^{-1} \delta_\mu^T \mathbf{B}_N^r \hat{\mathbf{u}}(kT_0), \quad \hat{\mathbf{u}}(kT_0) \in \mathbf{R}^{p_N} \tag{2}$$

for $t = kT_0 + \mu T_N + \zeta$, $\mu = 0, \ldots, n-1$, $\zeta \in [0, T_N)$, where $\mathbf{B}_N^r = \mathbf{B}_N (\mathbf{B}_N^T \mathbf{B}_N)^{-1}$.

Also, let the plant output be detected at every $T_M = \dfrac{T_0}{M}$, where, $M \in \mathbf{Z}^+$ is the output multiplicity of the sampling. Note that, in general, $M \neq N$. The sampled values of the plant output obtained over $[kT_0, (k+1)T_0)$, are stored in the M-dimensional column vector $\hat{\gamma}(kT_0)$ of the form

$$\hat{\gamma}(kT_0) = [y(kT_0) \quad y(kT_0 + T_M) \quad \cdots \quad y(kT_0 + (M-1)T_M)]^T$$

The vector $\hat{\gamma}(kT_0)$ is used in the control law of the form

$$\hat{\mathbf{u}}[(k+1)T_0] = \mathbf{L}_u \hat{\mathbf{u}}(kT_0) - \mathbf{K}\hat{\gamma}(kT_0), \quad \mathbf{L}_u \in \mathbf{R}^{p_N \times p_N}, \quad \mathbf{K} \in \mathbf{R}^{p_N \times M} \tag{3}$$

The adaptive LQ regulation problem treated in this paper is as follows: Find a TPMRC, which when applied to system (1), minimizes the following cost criterion

$$J = \frac{1}{2} \int_0^\infty (qy^2(t) + ru^2(t))dt \tag{4}$$

where $q \geq 0$, $r > 0$, and where $(\mathbf{A}, q\mathbf{c}\mathbf{c}^T)$ is an observable pair.

For the case of known systems, according to the results in [15], the LQ regulation problem considered here is equivalent to the problem of designing a control law of the form (2), (3), in order to minimize the following performance index

$$J = \frac{1}{2}\sum_{k=0}^{\infty}\left[\mathbf{x}^T\!\left(kT_0\right) \quad \hat{\mathbf{u}}^T\!\left(kT_0\right)\right]\begin{bmatrix}\tilde{\mathbf{Q}}_N & \tilde{\mathbf{G}}_N \\ \tilde{\mathbf{G}}_N^T & \tilde{\mathbf{R}}_N\end{bmatrix}\begin{bmatrix}\mathbf{x}\!\left(kT_0\right) \\ \hat{\mathbf{u}}\!\left(kT_0\right)\end{bmatrix}$$

for the system

$$\mathbf{x}\!\left[(k+1)T_0\right] = \Phi\mathbf{x}\!\left(kT_0\right) + \mathbf{B}_N\hat{\mathbf{u}}\!\left(kT_0\right)$$

where

$$\tilde{\mathbf{Q}}_N = q\!\int_0^{T_0}\exp(\mathbf{A}^T\lambda)\mathbf{c}\mathbf{c}^T\exp(\mathbf{A}\lambda)d\lambda \equiv \sum_{\mu=0}^{N-1}\left(\hat{\mathbf{A}}_c^T\right)^{\mu}\Xi\!\left(T_N\right)\hat{\mathbf{A}}_c^{\mu}$$

$$\tilde{\mathbf{G}}_N = T_N^{-1}\left\{\sum_{\mu=0}^{N-1}\left(\hat{\mathbf{A}}_c^T\right)^{\mu}\left[\Xi\!\left(T_N\right)\mathbf{V}_{\mu}^T + \Lambda\!\left(T_N\right)\delta_{\mu}^T\right]\right\}\mathbf{B}_N^r$$

$$\tilde{\mathbf{R}}_N = T_N^{-2}\left(\mathbf{B}_N^r\right)^T\left\{\sum_{\mu=0}^{N-1}\left(\left[\mathbf{V}_{\mu} \quad \delta_{\mu}\right]\begin{bmatrix}\Xi\!\left(T_N\right) & \Lambda\!\left(T_N\right) \\ \Lambda^T\!\left(T_N\right) & N\!\left(T_N\right)+rT_N\end{bmatrix}\begin{bmatrix}\mathbf{V}_{\mu}^T \\ \delta_{\mu}^T\end{bmatrix}\right)\right\}\mathbf{B}_N^r$$

$$\Xi\!\left(T_N\right) = q\!\int_0^{T_N}\exp(\mathbf{A}^T\lambda)\mathbf{c}\mathbf{c}^T\exp(\mathbf{A}\lambda)d\lambda \;,\; \Lambda\!\left(T_N\right) = q\!\int_0^{T_N}\exp(\mathbf{A}^T\lambda)\mathbf{c}\mathbf{c}^T\hat{\mathbf{b}}_{\lambda}\,d\lambda$$

$$N\!\left(T_N\right) = q\!\int_0^{T_N}\hat{\mathbf{b}}_{\lambda}^T\mathbf{c}\mathbf{c}^T\hat{\mathbf{b}}_{\lambda}\,d\lambda \quad,\quad \mathbf{V}_{\mu} = \hat{\mathbf{A}}_c^{N-\mu}\Theta_{\mu}\!\left(T_N\right)\Theta_{\mu}^T\!\left(T_N\right)$$

$$\hat{\mathbf{b}}_{\zeta} = \int_0^{\zeta}\exp(\mathbf{A}\lambda)\mathbf{b}d\lambda \quad,\quad \Theta_{\mu}\!\left(T_N\right) = \left[\hat{\mathbf{b}}_{T_N} \quad \hat{\mathbf{A}}_c\hat{\mathbf{b}}_{T_N} \quad \cdots \quad \hat{\mathbf{A}}_c^{\mu-1}\hat{\mathbf{b}}_{T_N}\right]$$

Now let

$$\hat{\mathbf{A}} = \exp(\mathbf{A}T_M) \;,\; \mathbf{B}_M^*\!\left(\rho\right) = T_N^{-1}E_M\!\left(\rho\right)\mathbf{B}_N^r$$

$$E_M\!\left(\rho\right) = \sum_{j=0}^{a(\rho)-1}\left(\int_{jT_N}^{(j+1)T_N}\exp\left\{\mathbf{A}\left(\rho\frac{N}{M}T_N - \lambda\right)\right\}\mathbf{b}d\lambda\right)\delta_j^T$$

$$+ \left(\int_{a(\rho)T_N}^{\rho\frac{N}{M}T_N}\exp\left\{\mathbf{A}\left(\rho\frac{N}{M}T_N - \lambda\right)\right\}\mathbf{b}d\lambda\right)\delta_{a(\rho)}^T \quad,\quad a(\rho) = \mathrm{INT}_s\!\left(\rho\frac{N}{M}\right)$$

$$\mathbf{H} = \begin{bmatrix}\mathbf{c}^T\left(\hat{\mathbf{A}}^M\right)^{-1} \\ \vdots \\ \mathbf{c}^T\hat{\mathbf{A}}^{-1}\end{bmatrix} \;,\; \mathbf{D} = \begin{bmatrix}\mathbf{c}^T\hat{\mathbf{B}}_0 \\ \vdots \\ \mathbf{c}^T\hat{\mathbf{B}}_{M-1}\end{bmatrix} \;,\; \hat{\mathbf{B}}_{\rho} = \mathbf{B}_M^*\!\left(\rho\right) - \hat{\mathbf{A}}^{\rho-M}\mathbf{B}_N \;,\; \rho = 0,1,\ldots,M-1$$

Also, let $\mathrm{INT}_s\!\left(v\right)$ be the greatest integer that is less than or equal to $v \in \mathbf{R}^+$
Then, the following relation holds

$$\mathbf{H}\mathbf{x}\!\left[(k+1)T_0\right] = \hat{\gamma}\!\left(kT_0\right) - \mathbf{D}\hat{\mathbf{u}}\!\left(kT_0\right) \;,\; k \geq 0$$

Moreover, if $M \geq n$, then, matrix $\mathbf{H}$ has full column rank. In this case, for almost every T_0, we can make the control law (3) equivalent to the control law

$$\hat{\mathbf{u}}(kT_0) = -\mathbf{Fx}(kT_0) \text{, for } k \geq 1 \tag{5}$$

by choosing properly the controller pair $(\mathbf{K}, \mathbf{L}_u)$, such that

$$\mathbf{KH} = \mathbf{F} \text{ , } \mathbf{L}_u = \mathbf{KD}$$

Finally, assuming that for some $M = n^* \geq n + p_N$, matrix $[\mathbf{H} \ \mathbf{D}]$ has full column rank, then, for almost every sampling period T_0, there is a matrix $\mathbf{K}$ such that

$$\mathbf{K}[\mathbf{H} \ \mathbf{D}] = [\mathbf{F} \ \mathbf{L}_u]$$

where $\mathbf{F}$ and $\mathbf{L}_u$ are arbitrarily specified matrices.

Hence, we can realize any state feedback matrix $\mathbf{F}$ by a TPMRC of the form (2), (3), possessing any prescribed degree of stability, since we can choose the matrix $\mathbf{L}_u$ arbitrarily. The choice $\mathbf{L}_u = 0$ is of course permissible, leading to a static TPMRC. A *fictitious* state feedback law of the form (5), that minimizes (4) is

$$\mathbf{F} = \left(\widetilde{\mathbf{R}}_N + \mathbf{B}_N^T \mathbf{PB}_N\right)^{-1}\left(\widetilde{\mathbf{G}}_N + \mathbf{B}_N^T \mathbf{P\Phi}\right)$$

$$\mathbf{P} = \mathbf{\Phi}^T \mathbf{P\Phi} + \widetilde{\mathbf{Q}}_N - \left(\widetilde{\mathbf{G}}_N + \mathbf{\Phi}^T \mathbf{PB}_N\right)\left(\widetilde{\mathbf{R}}_N + \mathbf{B}_N^T \mathbf{PB}_N\right)^{-1}\left(\widetilde{\mathbf{G}}_N^T + \mathbf{B}_N^T \mathbf{P\Phi}\right)$$

When $\mathbf{L}_u$ is not prespecified, the TPMRC gains $\mathbf{K}$ and $\mathbf{L}_u$ are given by

$$\mathbf{K} = \left(\widetilde{\mathbf{R}}_N + \mathbf{B}_N^T \mathbf{PB}_N\right)^{-1}\left(\widetilde{\mathbf{G}}_N + \mathbf{B}_N^T \mathbf{P\Phi}\right)\mathbf{H}^1 \tag{6a}$$

$$\mathbf{L}_u = \left(\widetilde{\mathbf{R}}_N + \mathbf{B}_N^T \mathbf{PB}_N\right)^{-1}\left(\widetilde{\mathbf{G}}_N + \mathbf{B}_N^T \mathbf{P\Phi}\right)\mathbf{H}^1 \mathbf{D} \tag{6b}$$

where $\mathbf{H}^1 \mathbf{H} = \mathbf{I}$. When $\mathbf{L}_u$ is prespecified, the gain $\mathbf{K}$ is given by

$$\mathbf{K} = \left[\left(\widetilde{\mathbf{R}}_N + \mathbf{B}_N^T \mathbf{PB}_N\right)^{-1}\left(\widetilde{\mathbf{G}}_N + \mathbf{B}_N^T \mathbf{P\Phi}\right) \ \mathbf{L}_u\right]\hat{\mathbf{H}}^1 \text{ , } \hat{\mathbf{H}}^1[\mathbf{H} \ \ \mathbf{D}] = \mathbf{I} \tag{7}$$

In the case of unknown systems, we introduce in the control loop the persistent excitation signal

$$\mathbf{v}(t) = \mathbf{q}^T(t)\mathbf{v} \text{ , } \mathbf{q}^T(t) = [q_0(t), \cdots, q_{N-1}(t)]$$

Here, $\mathbf{q}(t)$ is the T_N-periodic vector function with elements having the form

$$q_i(t) = q_{i,\mu} \text{ , } \text{ for } t \in [\mu T_N, (\mu+1)T_N) \text{ , } i = 0,1,...,N-1 \text{ , } \mu = 0,1,...,N-1$$

where $q_{i,\mu}$ is constant taking the following values

$$q_{i,\mu} = \begin{cases} 1 \text{, for } \mu = i \\ 0 \text{, for } \mu \neq i \end{cases}$$

Note that, if $(\mathbf{A}, \mathbf{b})$ is controllable, then for almost all $T_0 \in \mathbf{R}^+$ and $N > n$, such that controllability is preserved with sampling, matrix $\mathbf{B}_N$ is nonsingular and the set of the zeros of the analytic function $\psi(T_N) = \det\left[\hat{\mathbf{b}}_{T_N} \ \ \hat{\mathbf{A}}_c\hat{\mathbf{b}}_{T_N} \ \ \cdots \ \ \hat{\mathbf{A}}_c^{N-1}\hat{\mathbf{b}}_{T_N}\right]$, does not have any limiting points except infinity. Then [13], $\mathbf{v}^T = [\rho_1 \ \ \cdots \ \ \rho_{N-n}]\mathbf{B}_0^{*T}$, where $\mathbf{B}_0^*$ is the $N \times (N-n)$ matrix whose columns are the linearly independent

N-dimensional vectors which are orthogonal to the rows of

$$\mathbf{B}^* = \left[\hat{\mathbf{A}}_c^{N-1} \hat{\mathbf{b}}_{T_N} \quad \hat{\mathbf{A}}_c^{N-2} \hat{\mathbf{b}}_{T_N} \quad \cdots \quad \hat{\mathbf{b}}_{T_N} \right]$$

and where ρ_j, $j = 1, 2, \ldots, N - n$ are arbitrary real parameters.

Furthermore, in the unknown systems case, the computation of the controller parameters relies on estimates of the plant parameters. By discretizing system (1) with sampling period $\tau = \dfrac{T_0}{(2n+1)L}$, $L = \text{lcm}\{N, M\}$, we obtain

$$\mathbf{x}\left[(\nu + 1)\tau\right] = \Phi_\tau \mathbf{x}(\nu\tau) + \mathbf{b}_\tau u(\nu\tau) \ , \quad y(\nu\tau) = \mathbf{c}^T \mathbf{x}(\nu\tau) \ , \quad \nu \geq 0 \tag{8}$$

It is easy to see that all matrices and vectors involved in the forms of the TPMRC parameters and the vectors $\mathbf{v}$ can be expressed and/or can be calculated on the basis of the matrix triplet $\left(\Phi_\tau, \mathbf{b}_\tau, \mathbf{c}^T \right)$ (see [13], for details). Moreover, since

$$\begin{bmatrix} \mathbf{A} & \mathbf{b} \\ 0 & 0 \end{bmatrix} = \frac{1}{\tau} \ln\left(\begin{bmatrix} \Phi_\tau & \mathbf{b}_\tau \\ 0 & 1 \end{bmatrix} \right)$$

the pair $(\mathbf{A}, \mathbf{b})$ can be computed on the basis of the pair $(\Phi_\tau, \mathbf{b}_\tau)$. Then, the integral forms $\widetilde{\mathbf{Q}}_N$, $\widetilde{\mathbf{G}}_N$ and $\widetilde{\mathbf{R}}_N$, can be computed using Van Loan's algorithm [16], on the basis of $(\Phi_\tau, \mathbf{b}_\tau, \mathbf{c}^T)$. Now, fixing the coordinate system such that

$$\Phi_\tau = \begin{bmatrix} 0 & 0 & \cdots & 0 & -\alpha_n \\ 1 & 0 & \cdots & 0 & -\alpha_{n-1} \\ \vdots & \vdots & \ddots & \vdots & \vdots \\ 0 & 0 & \cdots & 1 & -\alpha_1 \end{bmatrix}, \quad \mathbf{b}_\tau = \begin{bmatrix} \beta_n \\ \beta_{n-1} \\ \vdots \\ \beta_1 \end{bmatrix}, \quad \mathbf{c}^T = \begin{bmatrix} 0 & \cdots & 0 & 1 \end{bmatrix} \tag{9}$$

only α_i and β_i, $i = 1, 2, \ldots, n$ are considered as unknown parameters. Note that relations (8) and (9), are equivalent to the following difference equation

$$y(\nu\tau) + \sum_{\rho=1}^{n} \alpha_i y(\nu\tau - \rho\tau) = \sum_{\rho=1}^{n} \beta_i u(\nu\tau - \rho\tau) \ , \quad \nu \geq 0 \tag{10}$$

which can be used for the identification of the parameters of the unknown system. To this end, relation (10) can be written in the following linear regression form

$$y(\nu\tau) = \varphi^T(\nu\tau)\theta \ , \quad \theta = \begin{bmatrix} \alpha_1 & \cdots & \alpha_n & \beta_1 & \cdots & \beta_n \end{bmatrix}$$

$$\varphi(\nu\tau) = \begin{bmatrix} -y(\nu\tau - \tau) & \cdots & -y(\nu\tau - n\tau) & u(\nu\tau - \tau) & \cdots & u(\nu\tau - n\tau) \end{bmatrix}^T$$

Next, define

$$\mathbf{Y}(kT_0) = \begin{bmatrix} y(kT_0) & y(kT_0 - \tau) & \cdots & y\left[(k-1)T_0\right] \end{bmatrix}$$

$$\mathbf{Z}(kT_0) = \begin{bmatrix} \varphi(kT_0) & \varphi(kT_0 - \tau) & \cdots & \varphi\left[(k-1)T_0\right] \end{bmatrix}$$

$$\hat{\theta}_k = \begin{bmatrix} \hat{\alpha}_1(kT_0) & \cdots & \hat{\alpha}_n(kT_0) & \hat{\beta}_1(kT_0) & \cdots & \hat{\beta}_n(kT_0) \end{bmatrix}$$

Clearly, we have the relation $\mathbf{Y}(kT_0) = \mathbf{Z}^T(kT_0)\theta$. We now choose the recursive algorithm for the estimation of $\hat{\theta}_k$ as

$$\hat{\theta}_{k+1} = \hat{\theta}_k - \left[a\mathbf{I} + \mathbf{Z}^T(kT_0)\mathbf{Z}(kT_0) \right]^{-1} \mathbf{Z}(kT_0) \left[\mathbf{Z}^T(kT_0)\hat{\theta}_k - \mathbf{Y}(kT_0) \right]$$

where $a \in \mathbf{R}^+$ and $\hat{\theta}_0$ are arbitrarily specified.

Details regarding boundedness, convergence, persistent excitation and global stability of the above adaptive scheme can be found in [13].

3   Ship-Steering Dynamics

The equations describing the motion of a ship can be easily obtained from conservation of momentum and angular momentum [1]-[9]. It is customary to write these equations using a coordinate frame fixed to the ship. Considering the ship as a rigid body with six degrees of freedom, it is worth noticing that, in a ship like a large tanker there is little coupling between the surge, sway, heave, roll, pitch and yaw motions. Therefore, one can describe the steering dynamics of the ship by considering the surge, sway and yaw motions separately. The nonlinear equations of motion of the ship are then as follows

$$m(\dot{u} - \upsilon r - x_G r^2) = X \quad , \quad m(\dot{\upsilon} + ur + x_G \dot{r}) = Y \quad , \quad I_z \dot{r} + mx_G(\dot{\upsilon} + ru) = N \quad (11)$$

where u and v are the x- and y-coordinates of the ship's velocity V, $r = \dfrac{d\psi}{dt}$ is the component of the angular velocity on the z-axis, ψ is the heading angle and x_G denotes the x-coordinate of the center of mass, which is assumed that is located in the x-z plane. The mass of the ship is m and its moment of inertia with respect to the z-axis is I_z. Moreover, X and Y are the components of the hydrodynamic forces on the x- and y-axis respectively, while N is the z-component of the torque due to the hydrodynamic forces. These hydrodynamic forces are complicated functions of the motion which are usually expressed as functions of acceleration, velocity and helm angle (rudder deflection) δ, i.e.

$$Y = Y\left(\upsilon, r, \delta, \dot{\upsilon}, \dot{r} \right) \quad , \quad N = N\left(\upsilon, r, \delta, \dot{\upsilon}, \dot{r} \right)$$

The projection of ship's velocity on the x-axis is assumed to be constant with value $u = u_0$. With this assumption, the first of equations (11) can be neglected when analyzing steering, and the last two of equations (11) can be rewritten in the form

$$m(\dot{\upsilon} + u_0 r + x_G \dot{r}) = Y \quad , \quad I_z \dot{r} + mx_G(\dot{\upsilon} + ru_0) = N \quad (12)$$

It is also customary to normalize (12) by the use of the so-called "*prime*" system, in which the length unit is the length of the ship L, the time unit is L/V and the mass unit is $\rho_w L^3 / 2$, where ρ_w is the mass density of the water. Furthermore, in order to linearize the normalized equations, one is forced to introduce the partial derivatives of the hydrodynamic force Y and of the induced torque N. The partial derivative $N_\upsilon = \partial N\left(\upsilon, r, \delta, \dot{\upsilon}, \dot{r} \right) / \partial \upsilon$, where the right-hand side is evaluated at arguments zero, is called a *"hydrodynamic derivative"*. The derivatives Y_υ, $Y_{\dot\upsilon}$, Y_δ, Y_r, $Y_{\dot r}$, $N_{\dot\upsilon}$, N_δ, N_r and $N_{\dot r}$ have analogous definitions. Linearization of (12) around the stationary solution $\upsilon = r = 0$ and normalization yields

$$\begin{bmatrix} m'-Y_{\dot\upsilon}' & m'x_G'-Y_{\dot r}' \\ m'x_G'-N_{\dot\upsilon}' & I_z'-N_{\dot r}' \end{bmatrix} \frac{d}{dt'}\begin{bmatrix} \upsilon' \\ r' \end{bmatrix} = \begin{bmatrix} Y_\upsilon' & Y_r'-m' \\ N_\upsilon' & N_r'-m'x_G' \end{bmatrix}\begin{bmatrix} \upsilon' \\ r' \end{bmatrix} + \begin{bmatrix} Y_\delta' \\ N_\delta' \end{bmatrix}\delta \quad (13)$$

All parameters and variables in (13) are dimension free. In deriving (13) it has been assumed that $u_0 / V = 1$. Solving the normalized equations of motion (13) for the derivatives $\frac{d\upsilon'}{dt'}$ and $\frac{dr'}{dt'}$, and introducing the heading angle ψ, defined by $r' = d\psi / dt'$ and viewed as the ship's output, as an extra variable, we can easily convert (13) to the standard state space form (1), with

$$\mathbf{x} = \begin{bmatrix} \upsilon' \\ r' \\ \psi \end{bmatrix}, \mathbf{u} = \delta, \mathbf{y} = \psi, \mathbf{A} = \begin{bmatrix} \alpha_{11} & \alpha_{12} & 0 \\ \alpha_{21} & \alpha_{22} & 0 \\ 0 & 1 & 0 \end{bmatrix}, \mathbf{b} = \begin{bmatrix} b_1 \\ b_2 \\ 0 \end{bmatrix}, \mathbf{c}^T = \begin{bmatrix} 0 & 0 & 1 \end{bmatrix}$$

$$\begin{bmatrix} \alpha_{11} & \alpha_{12} \\ \alpha_{21} & \alpha_{22} \end{bmatrix} = \begin{bmatrix} m'-Y_{\dot\upsilon}' & m'x_G'-Y_{\dot r}' \\ m'x_G'-N_{\dot\upsilon}' & I_z'-N_{\dot r}' \end{bmatrix}^{-1}\begin{bmatrix} Y_\upsilon' & Y_r'-m' \\ N_\upsilon' & N_r'-m'x_G' \end{bmatrix}$$

$$\begin{bmatrix} b_1 \\ b_2 \end{bmatrix} = \begin{bmatrix} m'-Y_{\dot\upsilon}' & m'x_G'-Y_{\dot r}' \\ m'x_G'-N_{\dot\upsilon}' & I_z'-N_{\dot r}' \end{bmatrix}^{-1}\begin{bmatrix} Y_\delta' \\ N_\delta' \end{bmatrix}$$

The above model for ship steering dynamics is usually called "the 2nd order Nomoto's model" [7]. It is worth noticing, at this point, that the values of the parameters α_{ij} and b_i, change with the trim and the draught of the ship.

The criteria used in the evaluation and design of autopilots for ship steering depend on many factors, like safety, propulsion economy and accuracy in pathkeeping. LQ optimal control theory can be applied to ships, in order to obtain increased performance and accurate control when sailing in restricted waters and reduced fuel consumption, when sailing in the open sea. The trade-off between accurate and economical steering can be related to a discrete quadratic criterion of the form (4). In particular, it has been shown in [9], that when designing and evaluating autopilots for steering in open sea, it seems natural to use the following values for the weighting factors q and r of the cost J

$$q=0.014 \,, \quad \text{(calm sea)} \quad 0.1q \le r \le 10q \quad \text{(rough sea)}$$

4 Simulation of the Proposed Adaptive LQ Regulator

In this Section, the proposed adaptive LQ regulator is tested through two simulation examples. As a first example, we address the sway-yaw motion of a minesweeper, with a length of 55 m and a velocity of 4 m/sec. The data of the system parameters have been taken from [2], [8]. Their nominal values are

$$\alpha_{11} = -0.863, \ \alpha_{12} = -0.482, \ \alpha_{21} = -5.25, \ \alpha_{22} = -2.45, \ b_1 = 0.175, \ b_2 = -1.38$$

The nominal sway-yaw motion of the minesweeper is unstable. Our aim here is to find a TPMRC based adaptive optimal LQ regulator such that (4), with q=0.014 and r=0.042, to be minimized. By choosing $T_0 = 0.5 \sec$, N=8 and M=4, the sampling period $\tau = 8.9 msec$. In the case where matrix L_u is not prespecified, the parameters of the admissible TPMRC, as computed by (6a) and (6b), are

$$K = \begin{bmatrix} 4.8885 & -7.6640 & -4.9401 & 7.9005 \\ -2.2968 & 3.5726 & 2.3220 & -3.6438 \\ 0.0708 & -0.1073 & -0.0718 & 0.1053 \end{bmatrix}, L_u = \begin{bmatrix} 0.2768 & 0.1876 & 0.1410 \\ -0.1320 & -0.0837 & -0.0670 \\ 0.0043 & 0.0022 & 0.0021 \end{bmatrix}$$

Note that the eigenvalues of L_u lies inside the unit circle. In the case where, matrix L_u is desired to have a prespecified value, for example the value $L_u = 0_{3 \times 3}$, we select N=M=6. Then, the admissible TPMRC gain matrix K, as computed by (7), has the value

$$K = 10^3 \times \begin{bmatrix} 0.2181 & -1.0708 & 2.0522 & -1.9165 & 0.8620 & -0.1447 \\ -0.0951 & 0.4675 & -0.8968 & 0.8390 & -0.3786 & 0.0641 \\ 0.0054 & -0.0264 & 0.0507 & -0.0476 & 0.0217 & -0.0037 \end{bmatrix}$$

In the unknown system case the simulation has been performed using the proposed modified recursive least square algorithm. The nominal parameter vector θ, in the case of the unconstrained TPMRC, has the value

$$\theta = \begin{bmatrix} -2.9709 & 2.9417 & -0.9709 & -5.4717 \times 10^{-5} & -4.5201 \times 10^{-7} & 5.3689 \times 10^{-5} \end{bmatrix}$$

while in the case of the static TPMRC has the value

$$\theta = \begin{bmatrix} -2.9464 & 2.8927 & -0.9463 & -1.8979 \times 10^{-4} & -2.9026 \times 10^{-6} & 1.8320 \times 10^{-4} \end{bmatrix}$$

The identification algorithm has been initialized with a sequence of random numbers having normal distribution, with zero mean and covariance 1. Random noise, having normal distribution with zero mean and covariance 0.03, has been introduced in the system. The forgetting factor is a=100. Simulations are given in Figures 4.1 and 4.2.

As a second example, we next address the sway-yaw motion of a large tanker of 190000 dwt. The tanker has a length of 305 m and a velocity of 7.7 m/sec. The numerical data of the system parameters have been taken from [2]. Their nominal values are

$$\alpha_{11} = -0.597, \ \alpha_{12} = -0.372, \ \alpha_{21} = -3.66, \ \alpha_{22} = -1.87, \ b_1 = 0.103, \ b_2 = -0.80$$

Note that the nominal state space model of the tanker is unstable. Our aim here is to find a TPMRC based adaptive optimal LQ regulator such that the performance index of the form (4), with q=0.014 and r=0.07, to be minimized. By choosing $T_0 = 0.6 \sec$, N=8 and M=4, the sampling period $\tau = 10.7 msec$. When matrix L_u is not prespecified, the parameters of the admissible TPMRC, are given by

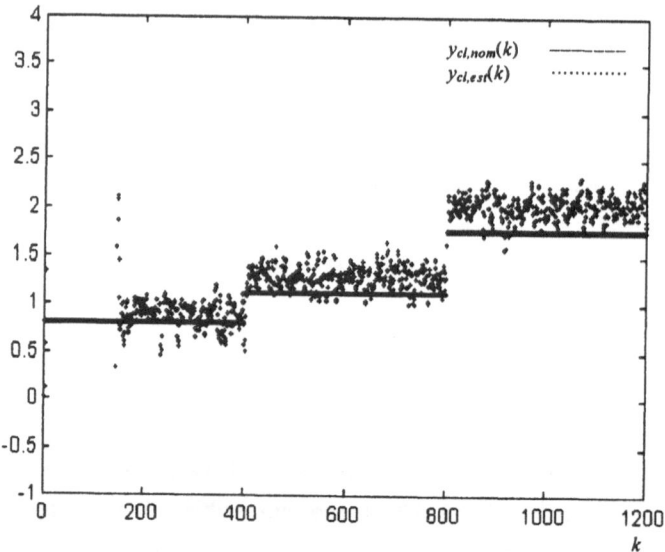

Figure 4.1. Closed loop heading angle under identification process versus nominal closed loop heading angle of the minesweeper, in the case of non-prespecified $\mathbf{L}_u$ and for step changes in the set-point.

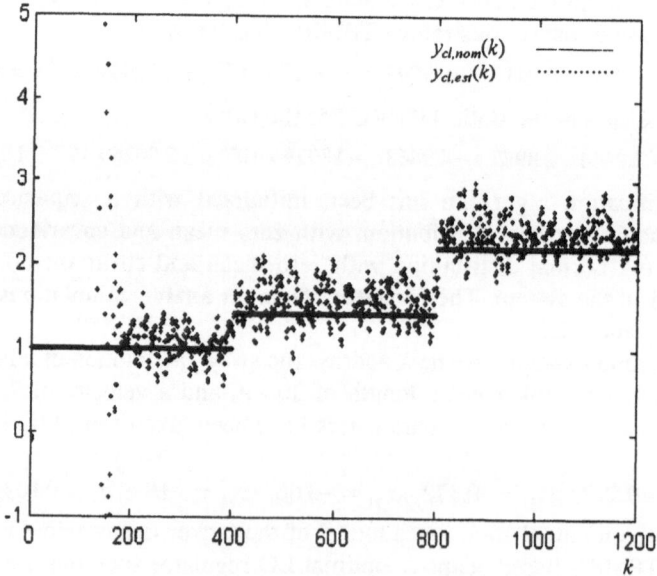

Figure 4.2. Closed loop heading angle under identification process versus nominal closed loop heading angle of the minesweeper, in the case of $\mathbf{L}_u = 0$ and for step changes in the set-point.

$$K = \begin{bmatrix} 5.8703 & -8.7650 & -5.9388 & 8.9679 \\ -2.8917 & 4.2893 & 2.9245 & -4.3525 \\ 0.0823 & -0.1197 & -0.0832 & 0.1182 \end{bmatrix}, L_u = \begin{bmatrix} 0.2639 & 0.1408 & 0.1339 \\ -0.1315 & -0.0659 & -0.0664 \\ 0.0039 & 0.0016 & 0.0019 \end{bmatrix}$$

The eigenvalues of L_u lies inside the unit circle. Analogous results can be obtained in the case where matrix L_u is desired to have a prespecified value.

The nominal parameter vector θ, in the case of unconstrained TPMRC, is

$$\theta = \begin{bmatrix} -2.9739 & 2.9479 & -0.9739 & -4.5691 \times 10^{-5} & -2.9228 \times 10^{-7} & 4.4946 \times 10^{-5} \end{bmatrix}$$

In the unknown system, the identification algorithm has been initialized with a sequence of random numbers having normal distribution with zero mean and covariance 1. Random noise having normal distribution with zero mean and covariance 0.03 has been introduced. The forgetting factor is once again a=100. Simulation results are given in Figure 4.3.

To obtain the simulation results, both nominal and estimated plant are subject to step changes in the set point. From these results, it is easy to see that the proposed TPMRC based adaptive LQ regulator has an acceptable performance.

5   Conclusions

The feasibility of using TPMRC based adaptive LQ regulators for tankers has been explored. In TPMRCs for tankers, feedback control relies on measurements obtained by multirate sampling of the heading angle, while the helm angle is constrained to a certain piecewise constant signal. The proposed control strategy is very simple, since it reduces the original problems to an associate discrete problem for which a fictitious state feedback controller is needed to be computed. The proposed TPMRC based adaptive LQ regulator can possess any prescribed degree of stability, since there is the possibility of choosing the transition matrices of the controller arbitrarily. Persistency of excitation of the controlled system is assured without making assumptions on the plant under control, other than minimality and known order. Finally, the proposed techniques can be implemented easily by the use of digital computers. Simulation results show that the proposed autopilot has a satisfactory performance.

References

[1]. Van Amerongen J., Udink Ten Cate A.J. 1975 Model reference adaptive autopiots for ships. *Automatica* 11, 441-449.

[2]. Astrom K.J., Kallstrom C.G. 1976 Identification of ship steering dynamics. *Automatica* 12, 9-22.

[3]. Kallstrom C.G., Astrom K.J., Thorell N.E., Erikson J., Sten L. 1979. Adaptive autopilots for tankers. *Automatica* 15, 241-254.

[4]. Astrom K.J. 1980 Why use adaptive techniques for steering large tankers. *Int. J. Control* 32, 689-708.

158

[5]. Kallstrom C.G., Astrom K.J. 1981 Experiences of system identification applied to ship steering. *Automatica* 17, 187-198.

[6]. Van Amerongen J. 1984 Adaptive steering of ships-A model reference approach. *Automatica* 20, 3-14.

[7]. Fossen T.I. 1994 *Guidance and Control of Ocean Vehicles*, John Wiley & Sons Inc., New York.

[8]. Goclowski J., Gelb A. 1966 Dynamics of an automatic ship steering system. *IEEE Trans. Autom. Control* AC-11, 513-524.

[9]. Norrbin N.H. 1972 On the added resistance due to steering on a straight course, *Proc. 13th ITTC*, Hamburg, Germany.

[10]. Samson C. 1982 An adaptive LQ control for nonminimum phase systems. *Int. J. Control* 35, 1-28.

[11]. Chen H.-F., Zhang J.-F. 1990 Identification and adaptive control for systems with unknown orders, delay and coefficients, *IEEE Trans. Autom. Control* AC-35, 866-877.

[12]. Sun J., Ioannou P. 1992 Robust adaptive LQ control schemes. *IEEE Trans. Autom. Control* AC-37, 100-106.

[13]. Arvanitis K.G. 1998 A new adaptive optimal LQ regulator for linear systems based on two-point multirate controllers. *Circuits Syst. Sign. Proces.*, accepted for publication.

[14]. Arvanitis K.G. 1996 Design of adaptive LQ regulators for MIMO systems based on multirate sampling of the plant output. *J. Opt. Th. Appl.* 91, 35-60.

[15]. Al-Rahmani H.M., Franklin G.F. 1990 A new optimal multirate control of linear periodic and time-invariant systems. *IEEE Trans. Autom. Control* AC-35, 406-415.

[16]. Van Loan C.F. 1978 Computing integrals involving the matrix exponential. *IEEE Trans. Autom. Control* AC-23, 395-404.

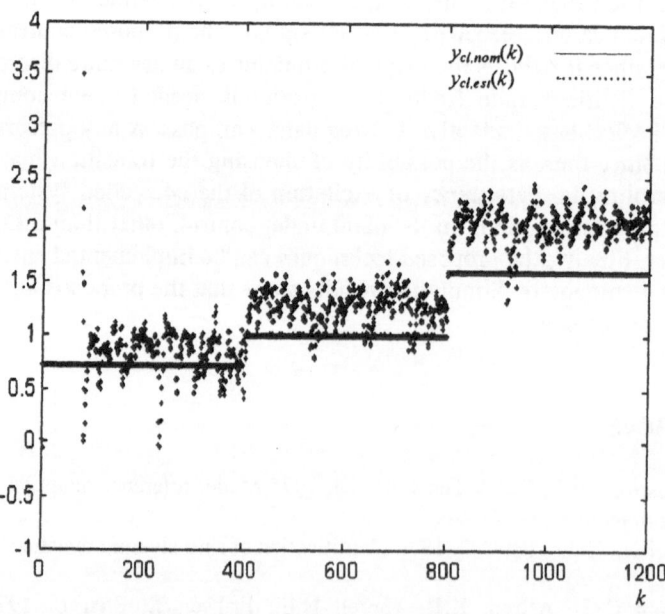

Figure 4.3. Closed loop heading angle under identification process versus nominal closed loop heading angle of the 190000 dwt tanker, in the case of non-prespecified $\mathbf{L}_u$ and for step changes in the set-point.

14

Design of Adaptive LQ Optimal Trackers Based on Multirate Sampling of the Plant Output

K.G. Arvanitis and G. Kalogeropoulos

1   Introduction

Periodically varying and multirate feedback strategies for linear systems have long been the focus of interest by many control designers (see, for example, [1]-[10]). This increased interest is warranted by the new dimensions of flexibility of the design procedure, offered by these control schemes, which also provide a series of remarkable advantages over ordinary time-invariant feedback schemes, such as classical state feedback, dynamic compensation or state observers (see [3], [7] for an overview). Among the most interesting control strategies of this type, is feedback control based on multirate-output controllers (MROCs). MROCs have been proposed first in [3], and subsequently have successfully been applied in solving a variety of important control problems (see e.g. [3], [4], [6]-[8]).

In this respect, the use of a MROCs, in order to achieve adaptive LQ optimal tracking in linear systems with unknown parameters, is investigated, in the present paper. In particular, we use the certainty equivalence principle to combine a MROC based LQ control stucture, which could be used to meet the control objective in the case of plants with known parameters, with the identification procedure. Adaptive LQ control has been studied in the past, in [11]-[16]. In [11], the adaptive tracking of a reference signal by the output of any deterministic discrete SISO linear system with unknown parameters is investigated. In [12], the implementation aspects of an adaptive LQ control scheme for discrete-time plants is studied intuitively. In [13], the continuous-time adaptive LQ control problem with average cost per unit time is studied and its solution is obtained via the separation principle. In [14], optimal adaptive controls are designed for both LQ optimal tracking and LQ regulation of both stochastic and deterministic discrete-time systems, when the system parameters, including time-delay, orders and coefficients, are unknown. In [15], the design, analysis and robustness of continuous-time and hybrid adaptive LQ control schemes for continuous-time plants have been presented. Finally, in [16], some practical issues of the design procedure for adaptive LQ regulation is investigated. Up to now, the feedback strategies proposed to solve both adaptive LQ regulation and adaptive LQ optimal tracking problems rely on the design of full order adaptive state observers. As a consequence of this fact, suboptimal adaptive control schemes are obtained, since state observers provide only an approximate estimation of the state vector needed to implement known adaptive LQ control schemes. Moreover, the exogenous dynamics introduced in the control loop is of high order. Finally, most of these control strategies are largely affected upon several disadvantages of estimator based controllers, such as increased on-line computational effort and reduced robustness.

In the present paper, the adaptive LQ optimal tracking problem is solved on the basis of an alternative approach, which relies on an indirect adaptive control scheme. The proposed technique makes use of a modified version of the fundamental results of [3], concerning the equivalence between MROCs and static state feedback, of some new ideas, regarding persistent excitation of a continuous-time plant and of some well known results concerning the static state feedback LQ optimal tracking problem. On the basis of this technique, the designed MROC based adaptive trackers are optimal, since MROCs reconstruct the state vector exactly and no approximate estimation of the state vector through state observers is made here, as in known techniques. Furthermore, using the proposed technique, the exogenous dynamics introduced in the control loop is of low order. The a priori knowledge needed in order to implement the proposed adaptive LQ optimal trackers, is minimality of the continuous and the discretized plant under control, and known order. Finally, persistent excitation of the controlled plant is assured without making assumptions on the existence of special convex sets in which the estimated parameters belong, or on the coprimeness of the polynomials describing the ARMA model, as in known techniques.

2 Preliminaries and Problem Formulation

Consider the linear single-input, single-output system of the form

$$\dot{x}(t) = Ax(t) + bu(t) \ , \ y(t) = c^T x(t) \tag{1}$$

where, $x(t) \in R^n$ is the state, $u(t) \in R$ is the input and $y(t) \in R$ is the output of the system. With regard to system (1), the following two assumptions are made:

Assumption 2.1. System (1) is controllable, observable and of known order n.

Assumption 2.2. There is a sampling period $T_0 \in R^+$, such that the system

$$\left(\Phi \equiv \exp(AT_0) \ , \ \hat{b} \equiv \int_0^{T_0} \exp[A(T_0 - \lambda)]b d\lambda \ , \ c^T \right)$$

is controllable and observable.

Except for this a priori information, the matrix triplet (A, b, c^T) is arbitrary and unknown.

The MROCs sampling mechanism involves detecting the plant output at every $T^* = T_0 / N$, where T_0 is the so-called frame sampling period and $N \in Z^+$ is known as the output multiplicity of the sampling. The sampled values of the plant output obtained over $[kT_0, (k+1)T_0)$, are stored in the vector $\hat{\gamma}(kT_0)$ of the form

$$\hat{\gamma}(kT_0) = \left[y(kT_0), y(kT_0 + T^*), \ldots, y\left[kT_0 + (N-1)T^*\right] \right].$$

The vector $\hat{\gamma}(kT_0)$ is used in the control law of the form

$$u\left[(k+1)T_0\right] = l_u u(kT_0) - k^T \hat{\gamma}(kT_0) \tag{2}$$

It is pointed out that the following basic multirate relation holds [3]

$$Hx\left[(k+1)T_0\right] = \hat{\gamma}(kT_0) - du(kT_0), \quad \text{for } k \geq 0$$

where

$$H = \begin{bmatrix} c^T(\hat{A}^N)^{-1} \\ \vdots \\ c^T \hat{A}^{-1} \end{bmatrix}, \quad d = \begin{bmatrix} c^T \hat{b}_N \\ \vdots \\ c^T \hat{b}_1 \end{bmatrix} \tag{3}$$

$$\hat{A} = \exp(AT^*), \quad \hat{b}_j = \int_0^{-jT^*} \exp(A\lambda)d\lambda \triangleq -\int_{-jT^*}^0 \exp(A\lambda)d\lambda, \quad \text{for } j = 1,\ldots,N \tag{4}$$

Note also that if $N > n$, then matrix H has full column rank. In this case, for almost every T_0, the control law (2) is equivalent to the control law

$$u(kT_0) = -f^T x(kT_0), \quad \text{for } k \geq 1 \tag{5}$$

provided that the MROC pair (k^T, l_u), is chosen such that

$$k^T H = f^T, \quad l_u = k^T d \tag{6}$$

Moreover, matrix $[H\,d]$ has full column rank, for almost every T_0 if $N \geq n+1$, and

$$\text{rank} \begin{bmatrix} A & b \\ c^T & 0 \end{bmatrix} = n+1$$

In this case, for almost every period T_0, there is a vector k^T such that

$$k^T[\,H\,d\,] = \left[f^T\,l_u \right] \tag{7}$$

where f^T and l_u are arbitrarily specified.

Therefore, we can equivalently realize any state feedback vectror f^T by a MROC possesing any prescribed degree of stability, since we can choose the gain l_u arbitrarily. The choice $l_u = 0$ is permisssible, leading to the static MROC

$$u\left[(k+1)T_0\right] = -k^T \hat{\gamma}(kT_0)$$

The adaptive LQ tracking problem treated here consists in finding an appropriate sequence $r(kT_0)$ and a MROC of the form (2), such that the output $y(kT_0)$ of the closed-loop system to track-out the output $y_r(kT_0)$ of a reference model, according to the following quadratic cost related to k_f sampling steps

$$J_{k_f} = k_f^{-1}\left\{\left(y(k_f T_0) - y_r(k_f T_0)\right)^2 + \sum_{k=0}^{k_f-1}\left[\left(y(kT_0) - y_r(kT_0)\right)^2 + \lambda u^2(kT_0)\right]\right\} \qquad (8)$$

with λ being any strictly positive scalar.

To solve the above problem, we next present a solution to the LQ tracking problem via MROCs, for known systems. This is done in Section 3. Next, using this result, the tracking problem is solved for a slightly different configuration, in which a persistent excitation signal is introduced in the control loop for future identification purposes. This is accomplished in Section 4. Finally, in Section 5, the proposed adaptive control scheme is derived and its global stability is studied.

3   Solution of the Problem for Known Systems

The procedure for LQ optimal tracking using MROCs, consists in finding a control law of the form (5), which minimizes the performance index (8), and then either determining the MROC pair $\left(k^T, 1_u\right)$ by (6) or the vector k^T by (7). A state feedback law of the form (5), which minimizes the performance index (8), is well known to be [11]

$$u(k, k_f) = -f^T(k, k_f)x(k, k_f) + r(k, k_f)$$

$$f^T(k, k_f) = \left(\hat{b}^T P(k, k_f)\hat{b} + \lambda\right)^{-1}\hat{b}^T P(k, k_f)\Phi$$

$$r(k, k_f) = -\left(\hat{b}^T P(k, k_f)\hat{b} + \lambda\right)^{-1}\hat{b}^T w(k, k_f)$$

where $P(k, k_f)$ satisfies the following difference Riccati equation

$$P(k-1, k_f) = \Phi^T P(k, k_f)\Phi - \Phi^T P(k, k_f)\hat{b}\left(\hat{b}^T P(k, k_f)\hat{b} + \lambda\right)^{-1}\hat{b}^T P(k, k_f)\Phi + cc^T$$

and $w(k, k_f)$ is the solution of the following matrix difference equation

$$w(k-1, k_f) = \left(\Phi - \hat{b}f^T(k, k_f)\right)^T w(k, k_f) - cy_r(kT_0)$$

with the final conditions $P(k_f - 1, k_f) = cc^T$, $w(k_f - 1, k_f) = -cy_r(k_f T_0)$. Note that matrix $P(k, k_f)$ and vector $w(k, k_f)$ are computed recursively in "reverse time". Thus, the performance index (8) can be minimized only if the reference signal $y_r(kT_0)$ is known in advance.

For practical reasons, we are interested in calculating a control on $[0, +\infty)$. If $k_f \to \infty$, then $P(k, k_f) \to P$, where P is the unique positive definite solution of the discrete algebraic Riccati equation

$$\mathbf{P} = \mathbf{\Phi}^T\mathbf{P}\mathbf{\Phi} - \mathbf{\Phi}^T\mathbf{P}\hat{\mathbf{b}}(\hat{\mathbf{b}}^T\mathbf{P}\hat{\mathbf{b}} + \lambda)^{-1}\hat{\mathbf{b}}^T\mathbf{P}\mathbf{\Phi} + \mathbf{c}\mathbf{c}^T \tag{9a}$$

for which there is a unique $\mathbf{P}$, if $(\mathbf{\Phi}, \hat{\mathbf{b}}, \mathbf{c}^T)$ is controllable and observable. In this case $\mathbf{\Phi} - \hat{\mathbf{b}}\mathbf{f}^T$ is exponentially stable, and

$$\mathbf{f}^T = (\hat{\mathbf{b}}^T\mathbf{P}\hat{\mathbf{b}} + \lambda)^{-1}\hat{\mathbf{b}}^T\mathbf{P}\mathbf{\Phi} \tag{9b}$$

However, it is not possible, in general, to compute $\lim\limits_{k_f \to \infty} \mathbf{w}(k, k_f)$ and hence it is not possible to calculate the control $u(k) = \lim\limits_{k_f \to \infty} u(k, k_f)$. To remove this difficulty, we assume that $y_r(kT_0)$ is bounded and we propose the suboptimal solution

$$u(k, M) = -\mathbf{f}^T\mathbf{x}(kT_0) + r(k, M) \tag{10a}$$

where $\mathbf{f}^T$ is given by (9b), M is an appropriate positive integer and

$$r(k, M) = (\hat{\mathbf{b}}^T\mathbf{P}\hat{\mathbf{b}} + \lambda)^{-1}\hat{\mathbf{b}}^T\mathbf{w}(k, M) \tag{10b}$$

with $\mathbf{w}(k, M)$ being calculated from the M reference signals $\{y_r(k+1), \cdots, y_r(k+M)\}$ as follows (see [11] for details)

$$\begin{aligned}
\mathbf{w}(k, 1) &= -\mathbf{c}y_r(k+M) \\
\mathbf{w}(k, 2) &= (\mathbf{\Phi} - \hat{\mathbf{b}}\mathbf{f}^T)\mathbf{w}(k, 1) - \mathbf{c}y_r(k+M-1) \\
&\vdots \\
\mathbf{w}(k, M) &= (\mathbf{\Phi} - \hat{\mathbf{b}}\mathbf{f}^T)\mathbf{w}(k, M-1) - \mathbf{c}y_r(k+1)
\end{aligned} \tag{10c}$$

The control (10a)-(10c) approximates the optimal control, can be calculated on $[0, +\infty)$ and stabilizes any system (1) for all M. Note also that M can be determined by the designer with regard to his knowledge of the reference signal following time k and his desire for simplicity in the calculation of the vector $\mathbf{w}(k, M)$, since by (10c), the vector $\mathbf{w}(k, M)$ is obtained after M iterations. Obviously, the greater M is, the better the control. Finally, note that a good value for M depends on the number of equivalent delays of the system with matrix triplet $(\mathbf{\Phi}, \hat{\mathbf{b}}, \mathbf{c}^T)$ (see [11], for details).

In the case where, $y_r(kT_0)$ is a constant y_r (or slowly varying), then we can approximate the steady-state solution of $\mathbf{w}(k, k_f)$ as

$$\mathbf{w}_\infty = -[\mathbf{I} - (\mathbf{\Phi} - \hat{\mathbf{b}}\mathbf{f}^T)]^{-1}\mathbf{c}y_r \tag{11a}$$

In this case

$$r_\infty = -(\hat{\mathbf{b}}^T\mathbf{P}\hat{\mathbf{b}} + \lambda)^{-1}\hat{\mathbf{b}}^T[\mathbf{I} - (\mathbf{\Phi} - \hat{\mathbf{b}}\mathbf{f}^T)]^{-1}\mathbf{c}y_r \tag{11b}$$

$$u_\infty(kT_0) = -(\hat{\mathbf{b}}^T\mathbf{P}\hat{\mathbf{b}} + \lambda)^{-1}\hat{\mathbf{b}}^T\mathbf{P}\mathbf{\Phi}\mathbf{x}(kT_0) - (\hat{\mathbf{b}}^T\mathbf{P}\hat{\mathbf{b}} + \lambda)^{-1}\hat{\mathbf{b}}^T[\mathbf{I} - (\mathbf{\Phi} - \hat{\mathbf{b}}\mathbf{f}^T)]^{-1}\mathbf{c}y_r$$

We are now able to compute the MROC gains. In the case where l_u is not prespecified, the MROC pair $(\mathbf{k}^T, l_u)$ is given by.

$$\mathbf{k}^T = (\hat{\mathbf{b}}^T\mathbf{P}\hat{\mathbf{b}} + \lambda)^{-1}\hat{\mathbf{b}}^T\mathbf{P}\mathbf{\Phi}\mathbf{H}^l \quad , \quad l_u = (\hat{\mathbf{b}}^T\mathbf{P}\hat{\mathbf{b}} + \lambda)^{-1}\hat{\mathbf{b}}^T\mathbf{P}\mathbf{\Phi}\mathbf{H}^l\mathbf{d} \tag{12}$$

where $\mathbf{H}^{1}\mathbf{H} = \mathbf{I}$. In the case where l_u is prespecified, the MROC gain $\mathbf{k}^T$ is

$$\mathbf{k}^T = \left[\left(\hat{\mathbf{b}}^T\mathbf{P}\hat{\mathbf{b}} + \lambda\right)^{-1}\hat{\mathbf{b}}^T\mathbf{P}\Phi \quad l_u\right]\hat{\mathbf{H}}^1 \quad , \quad \hat{\mathbf{H}}^1\left[\mathbf{H} \quad \mathbf{d}\right] = \mathbf{I} \qquad (13)$$

4   A Solution Appropriate for the Adaptive Case

To obtain a solution of the problem which will be more appropriate for application in the case of unknown systems, we next slightly modify the above control stategy by introducing in the control loop the persistent excitation signal

$$v(t) = \mathbf{q}^T(t)\mathbf{v} \quad , \quad \mathbf{q}^T(t) = \left[q_0(t), \cdots, q_{N-1}(t)\right]$$

Here, $\mathbf{q}(t)$ the T^*-periodic vector function with elements having the form

$$q_i(t) = q_{i,\mu} \quad , \quad \text{for } t \in \left[\mu T^*, (\mu+1)T^*\right] \quad , \quad i = 0,1,\dots,N-1 \quad , \quad \mu = 0,1,\dots,N-1$$

where $q_{i,\mu}$ is constant taking the following values

$$q_{i,\mu} = 1, \text{ for } \mu = i \text{ and } q_{i,\mu} = 0, \text{ for } \mu \neq i$$

Note that $\mathbf{v}$ is as yet unknown. The additive term $v(t) = \mathbf{q}^T(t)\mathbf{v}$, in the input of the continuous-time system, is used only for identification purposes and it is selected such as it will not influence the LQ tracking problem.

The following Theorem can now be established (for proof, see [7]).

Theorem 4.1. If $N > n$, then, the closed-loop system has the form

$$\xi\left[(k+1)T_0\right] = \left(\Phi - \hat{\mathbf{b}}\mathbf{f}^T\right)\xi(kT_0) + \mathbf{B}^*\mathbf{v} \quad , \quad y(kT_0) = \mathbf{c}^T\xi(kT_0) \quad , \quad k > 0$$

where

$$\mathbf{B}^* = \left[\hat{\mathbf{A}}^{N-1}\hat{\mathbf{b}}^* \quad \hat{\mathbf{A}}^{N-2}\hat{\mathbf{b}}^* \quad \cdots \quad \hat{\mathbf{b}}^*\right] \qquad (14)$$

$$\hat{\mathbf{b}}^* = \int_0^{T^*} \exp\left[\mathbf{A}(T^* - \lambda)\right]\mathbf{b}\,d\lambda$$

Note that, if $N > n$, then matrix $\mathbf{B}^*$ has full row rank for almost every sampling period T_0 [5].

Since the MROC gains $\mathbf{k}^T$ and l_u can be computed as in the previous Section, it only remains to determine the appropriate vector $\mathbf{v}$ which does not influence the LQ tracking problem. That is $\mathbf{v} \in \ker \mathbf{B}^*$, or $\mathbf{B}^*\mathbf{v} = 0$. Hence, $\mathbf{v}$ can be selected as

$$\mathbf{v}^T = \left[\rho_1 \quad \rho_2 \quad \cdots \quad \rho_{N-n}\right]\mathbf{B}_0^{*T} \qquad (15)$$

where $\mathbf{B}_0^*$ is a basis for $\ker \mathbf{B}^*$ and $\rho_j, j = 1,\dots,N-n$ are arbitrary real parameters.

The introduction of the reference signal $v(t)$ in the control loop, greatly facilitates the estimation of the unknown plant parameters. For these reason, the modified control strategy presented above is more appropriate than the control strategy of Section 3, for the development of the indirect adaptive control scheme, presented in the following Section.

5 Control Strategy for the Adaptive Case

The control scheme presented in the previous Sections has a corresponding scheme in the case of unknown systems. In this case, the control strategy relies on the computation of the MROC gains k^T and l_u and of the vector v from suitable estimates of the parameters of the plant with update taking place every kT_0, $k \geq 0$ and results to a globally stable closed-loop system.

5.1. Identification of the System

System (1), descretized with sampling period $\tau = \dfrac{T^*}{2n+1}$, takes the form

$$x\big[(v+1)\tau\big] = \Phi_\tau x(v\tau) + \hat{b}_\tau u(v\tau) \ , \ y(v\tau) = c^T x(v\tau) \ , \ v \geq 0 \qquad (16)$$

where

$$\Phi_\tau = \exp(A\tau) \ , \ \hat{b}_\tau = \int_0^\tau \exp\big[A(\tau - \lambda)\big] b \, d\lambda$$

Iterating equation (16), $2n+1$ times and observing that $u(v\tau)$ is constant, yields

$$x\big[(m+1)T^*\big] = \Phi_{T^*} x(mT^*) + \hat{b}_{T^*} u(mT^*) \ , \ y(mT^*) = c^T x(mT^*) \ , \ m \geq 0$$

where

$$\Phi_{T^*} \equiv \hat{A} = \Phi_\tau^{2n+1} \ \text{and} \ \hat{b}_{T^*} \equiv \hat{b}^* = \sum_{\rho=0}^{2n} \Phi_\tau^\rho \hat{b}_\tau \qquad (17)$$

We also note that matrix Φ and vector $\hat{b}$ can be written as

$$\Phi \equiv \hat{A}^N = \Phi_\tau^{(2n+1)N} \ , \ \hat{b} = \sum_{\rho=0}^{N-1} \hat{A}^\rho \hat{b}^* \equiv \sum_{\rho=0}^{(2n+1)N-1} \Phi_\tau^\rho \hat{b}_\tau \qquad (18)$$

Furthermore, the vectors $\hat{b}_j$, $j = 1,2,\ldots,N$ may be expressed as

$$\hat{b}_j = -\big(\Phi_\tau^{(2n+1)j}\big)^{-1} \left(\sum_{\rho=0}^{(2n+1)j-1} \Phi_\tau^\rho \hat{b}_\tau \right) \qquad (19)$$

From the above analysis it becomes clear that the matrices Φ, $\hat{b}$, $\hat{A}$ and $\hat{b}_j$ can be computed on the basis of the pair $\big(\Phi_\tau , \hat{b}_\tau\big)$. Moreover, fixing the coordinate system such that

$$\Phi_\tau = \begin{bmatrix} 0 & 0 & \cdots & 0 & -\alpha_n \\ 1 & 0 & \cdots & 0 & -\alpha_{n-1} \\ \vdots & \vdots & \ddots & \vdots & \vdots \\ 0 & 0 & \cdots & 1 & -\alpha_1 \end{bmatrix} \ , \ \hat{b}_\tau = \begin{bmatrix} \beta_n \\ \beta_{n-1} \\ \vdots \\ \beta_1 \end{bmatrix} \ , \ c^T = \begin{bmatrix} 0 & \cdots & 0 & 1 \end{bmatrix} \qquad (20)$$

only α_i and β_i, $i = 1,2,\ldots,n$ are considered as unknown parameters. Note that relations (16) and (20), are equivalent to the following difference equation

$$y(v\tau) + \sum_{\rho=1}^{n} \alpha_i y(v\tau - \rho\tau) = \sum_{\rho=1}^{n} \beta_i u(v\tau - \rho\tau) \ , \quad v \geq 0 \qquad (21)$$

Relation (21), can now be used for the identification of the unknown plant parameters. To this end, (21) can be written in the linear regression form

$$y(v\tau) = \varphi^T(v\tau)\theta$$

where

$$\varphi(v\tau) = \left[-y(v\tau - \tau) \quad \cdots \quad -y(v\tau - n\tau) \quad u(v\tau - \tau) \quad \cdots \quad u(v\tau - n\tau) \right]^T$$

$$\theta = \left[\alpha_1 \quad \cdots \quad \alpha_n \quad \beta_1 \quad \cdots \quad \beta_n \right]$$

Next, define

$$\mathbf{Y}(kT_0) = \left[y(kT_0) \quad y(kT_0 - \tau) \quad \cdots \quad y\left[(k-1)T_0\right] \right]$$

$$\mathbf{Z}(kT_0) = \left[\varphi(kT_0) \quad \varphi(kT_0 - \tau) \quad \cdots \quad \varphi\left[(k-1)T_0\right] \right]$$

$$\hat{\theta}_k = \left[\hat{\alpha}_1(kT_0) \quad \cdots \quad \hat{\alpha}_n(kT_0) \quad \hat{\beta}_1(kT_0) \quad \cdots \quad \hat{\beta}_n(kT_0) \right]$$

Clearly, we have the relation

$$\mathbf{Y}(kT_0) = \mathbf{Z}^T(kT_0)\theta$$

We now choose the recursive algorithm for the estimation of $\hat{\theta}_k$ as

$$\hat{\theta}_{k+1} = \hat{\theta}_k - \left[a\mathbf{I} + \mathbf{Z}^T(kT_0)\mathbf{Z}(kT_0) \right]^{-1} \mathbf{Z}(kT_0) \left[\mathbf{Z}^T(kT_0)\hat{\theta}_k - \mathbf{Y}(kT_0) \right] \ , \quad a \in \mathbf{R}^+ \qquad (22)$$

The convergence and the boundedness properties of the proposed identification procedure are summarized in the following Proposition (for proof see [17]).

Proposition 5.1. Let $\tilde{\theta}_k$ be the parameter estimation error, defined as $\tilde{\theta}_k = \hat{\theta}_k - \theta$. Then, denoting by $\lambda_{\min}(\mathbf{M})$ the minimum eigenvalue of $\mathbf{M}$, we have

(I) $\left\| \hat{\theta}_k \right\| \leq \xi$, for some finite $\xi \in \mathbf{R}^+$.

(II) If $\lim_{k \to \infty} \sum_{\rho=0}^{k} \lambda_{\min}\left(\mathbf{Z}(\rho T_0)\mathbf{Z}^T(\rho T_0) \right) = \infty$ then $\lim_{k \to \infty} \hat{\theta}_k = \theta$

5.2. Adaptive Controller Synthesis Algorithm

On the basis of the estimated parameter vector $\hat{\theta}_k$ obtained from (22), as well as on the basis of the relations (17)-(20), one can take the estimates, which are needed for the computation of matrices $\hat{\mathbf{A}} \equiv \hat{\mathbf{A}}(\hat{\theta}_k)$, $\Phi \equiv \Phi(\hat{\theta}_k)$, $\hat{\mathbf{b}} \equiv \hat{\mathbf{b}}(\hat{\theta}_k)$ and $\hat{\mathbf{b}}_j \equiv \hat{\mathbf{b}}_j(\hat{\theta}_k)$, which are involved in the algorithms presented in the previous Sections. Moreover, since matrices $\mathbf{P}$, $\mathbf{H}$ and vectors $\mathbf{w}(k,M)$ (resp. $\mathbf{w}_\infty$), $\mathbf{d}$ can be constructed on the basis of the matrices $\hat{\mathbf{A}}(\hat{\theta}_k)$, $\Phi(\hat{\theta}_k)$, $\hat{\mathbf{b}}(\hat{\theta}_k)$ and $\hat{\mathbf{b}}_j(\hat{\theta}_k)$, then provided that the matrix triplet $\left(\Phi_{T^*}(\hat{\theta}_k), \hat{\mathbf{b}}_{T^*}(\hat{\theta}_k), \mathbf{c}^T \right)$ is minimal, for all possible values of $\hat{\theta}_k$, we obtain

$$\mathbf{k}^T \equiv \mathbf{k}^T(\hat{\theta}_k) \ , \ \mathbf{l}_u \equiv \mathbf{l}_u(\hat{\theta}_k), \ r(k,M) = r(k,M,\hat{\theta}_k) \ (\text{resp. } r_\infty(k) = r_\infty(k,\hat{\theta}_k)$$

Overall, the procedure for the synthesis of an adaptive MROC based LQ tracker, consists on the main steps given bellow:

1. Case of nonprespecified l_u.

Step 1. Choose the sampling period τ such that $\tau = \dfrac{T_0}{(2n+1)N} = \dfrac{T^*}{2n+1}$.

Step 2. Update the estimates using (22).

Step 3. Use (20) to compute the matrices Φ_τ, $\hat{b}_\tau$ and c^T.

Step 4. Use (18) to compute the matrices Φ, $\hat{b}$ and (17) to compute matrix $\hat{A}$.

Step 5. Choose M as suggested in [11], and compute the signal w(k,M) using (10c) (resp. compute $\mathbf{w}_\infty$ using (11a)).

Step 6. Compute r(k,M) using (10b) (resp. compute r_∞ using (11b)).

Step 7. Use (9a), (9b) to implement $\mathbf{f}^T$.

Step 8. Find matrix $\mathbf{B}^*$ using relation (14).

Step 9. Form the matrix $\mathbf{H}$ and the vector $\mathbf{d}$, using relations (3), (4).

Step 10. Implement the MROC tracker sought using (12) and (15).

2. Case of prespecified l_u.

In this Case repeat steps 1-9 of Case 1 and furthermore:

Step 10. Use (19) to compute the vectors $\hat{\mathbf{b}}_j$.

Step 11. Implement the MROC tracker sought using (13) and (15).

5.3 Stability Analysis of the Adaptive Control Scheme

The following Theorem can now be established.

Theorem 5.1. The regressor sequence $\varphi(v\tau)$ is persistently exciting, i.e. there is a $\delta > 0$, such that

$$\mathbf{Z}(kT_0)\mathbf{Z}^T(kT_0) = \sum_{v=0}^{(2n+1)N} \varphi(kT_0 - v\tau)\varphi^T(kT_0 - v\tau) \geq \delta\mathbf{I} \tag{23}$$

Proof: Let $u(t) = \mathbf{q}^T(t)\mathbf{v}$. Introducing the pseudovariable $\zeta(v\tau)$, (21) yields

$$\zeta(v\tau) + \sum_{i=1}^{n} \alpha_i \zeta(v\tau - i\tau) = u(v\tau) \ , \ y(v\tau) = \sum_{i=1}^{n} \beta_i \zeta(v\tau - i\tau) \ , \ v \geq 1 \tag{24}$$

Defining the following vectors

$$\hat{\varphi}(v\tau) = \left[u(v\tau) \ \cdots \ u(v\tau - n\tau) \ y(v\tau - \tau) \ \cdots \ y(v\tau - n\tau) \right]^T$$

$$\hat{\zeta}(v\tau) = \left[\zeta(v\tau) \ \cdots \ \zeta(v\tau - n\tau) \right]^T$$

it is easy to see that

$$\hat{\varphi}(v\tau) = \mathbf{R}\hat{\zeta}(v\tau) \tag{25}$$

where $\mathbf{R}$ is the nonsingular Sylvester-matrix of the form

$$\mathbf{R} = \begin{bmatrix} 1 & \alpha_1 & \alpha_2 & \alpha_3 & \cdots & \alpha_n & 0 & 0 & 0 & \cdots & 0 \\ 0 & 1 & \alpha_1 & \alpha_2 & \cdots & \alpha_{n-1} & \alpha_n & 0 & 0 & \cdots & 0 \\ \vdots & \vdots & \vdots & \vdots & \cdots & \vdots & \vdots & \vdots & \vdots & \cdots & \vdots \\ 0 & 0 & 0 & 0 & \cdots & 0 & 1 & \alpha_1 & \alpha_2 & \cdots & \alpha_n \\ 0 & 0 & \beta_1 & \beta_2 & \cdots & \beta_{n-1} & \beta_n & 0 & 0 & \cdots & 0 \\ 0 & 0 & 0 & \beta_1 & \cdots & \beta_{n-2} & \beta_{n-1} & \beta_n & 0 & \cdots & 0 \\ \vdots & \vdots & \vdots & \vdots & \cdots & \vdots & \vdots & \vdots & \vdots & \ddots & \vdots \\ 0 & 0 & 0 & 0 & \cdots & 0 & \beta_1 & \beta_2 & \beta_3 & \cdots & \beta_n \end{bmatrix}$$

Observe now that vectors $\varphi(v\tau)$ and $\hat{\varphi}(v\tau)$ are interrelated as

$$\varphi(v\tau) = \mathbf{T}\hat{\varphi}(v\tau) \tag{26}$$

where $\mathbf{T} \in \mathbf{R}^{2n \times (2n+1)}$ is the full row rank matrix of the form

$$\mathbf{T} = \begin{bmatrix} \mathbf{0}_{2n \times 1} & \begin{matrix} \mathbf{0}_{n \times n} & -\mathbf{I}_{n \times n} \\ \mathbf{I}_{n \times n} & \mathbf{0}_{n \times n} \end{matrix} \end{bmatrix}$$

Obviously, excitation of $\hat{\zeta}(v\tau)$ implies excitation of $\varphi(v\tau)$. Therefore, we next investigate excitation of $\hat{\zeta}(v\tau)$. To this end, from relation (24), we can write

$$\gamma^T \hat{\zeta}(v\tau) = u(v\tau) \tag{27}$$

where $\gamma^T \in \mathbf{R}^{2n+1}$ is the following vector

$$\gamma^T = \begin{bmatrix} 1 & \alpha_1 & \alpha_2 & \cdots & \alpha_n & 0 & \cdots & 0 \end{bmatrix}$$

Now, let $\mathbf{X}(v\tau) \in \mathbf{R}^{2n \times 2n}$ be the following symmetric matrix

$$\mathbf{X}(v\tau) = \begin{bmatrix} \hat{\zeta}(v\tau) & \hat{\zeta}(v\tau - \tau) & \cdots & \hat{\zeta}(v\tau - 2n\tau) \end{bmatrix} \tag{28}$$

and $\hat{\mathbf{u}}(v\tau) \in \mathbf{R}^{2n}$ be the following vector

$$\hat{\mathbf{u}}(v\tau) = \begin{bmatrix} u(v\tau) & u(v\tau - \tau) & \cdots & u(v\tau - 2n\tau) \end{bmatrix}^T \tag{29}$$

Combining relations (27)-(29), we obtain

$$\gamma^T \mathbf{X}(v\tau) = \hat{\mathbf{u}}^T(v\tau)$$

Therefore, for every column vector η, with norm equal to unity, we have

$$\left| \eta^T \hat{\mathbf{u}}(v\tau) \right|^2 = \left| \eta^T \mathbf{X}^T(v\tau)\gamma \right|^2 = \left| \gamma^T \mathbf{X}(v\tau)\eta \right|^2 \leq \|\gamma\|^2 \|\mathbf{X}^T(v\tau)\eta\|^2$$

Summing over the interval $\left[kT_0 + (2n+1)\tau, \ kT_0 + (4n+1)\tau \right]$ and observing that the following relation holds

$$\left[\hat{\mathbf{u}}\left(kT_0 + (2n+1)\tau\right) \ \hat{\mathbf{u}}\left(kT_0 + (2n+2)\tau\right) \ \cdots \ \hat{\mathbf{u}}\left(kT_0 + (4n+1)\tau\right) \right] = \hat{\mathbf{U}}(kT_0)$$

where $\hat{\mathbf{U}}(kT_0)$ is the $(2n+1) \times (2n+1)$ upper triangular matrix whose non-zero elements are equal to 1, we obtain

$$\sum_{v=2n+1}^{4n+1} \left| \eta^T \hat{\mathbf{u}}(kT_0 + v\tau) \right|^2 = \left\| \hat{\mathbf{U}}(kT_0)\eta \right\|^2 \le \left\| \gamma \right\|^2 \sum_{v=2n+1}^{4n+1} \left\| \mathbf{X}^T(v\tau)\eta \right\|^2$$

$$\le \left\| \gamma \right\|^2 (2n+1) \sum_{v=1}^{4n+1} \left[\hat{\zeta}^T(kT_0 + v\tau)\eta \right]^2$$

Hence,

$$\sum_{v=1}^{4n+1} \left[\hat{\zeta}^T(kT_0 + v\tau)\eta \right]^2 \ge \left[\left\| \gamma \right\|^2 (2n+1) \right]^{-1} \left\| \hat{\mathbf{U}}(kT_0)\eta \right\|^2$$

Since the smallest singular value of $\hat{\mathbf{U}}(kT_0)$ is greater than a constant, there is a constant $\delta > 0$ such that

$$\sum_{v=1}^{4n+1} \hat{\zeta}(kT_0 + v\tau)\hat{\zeta}^T(kT_0 + v\tau) \ge \delta$$

Hence, the vector $\hat{\zeta}(v\tau)$ is persistently exciting. According to relations (25) and (26), the regressor sequence $\varphi(v\tau)$, is also persistently exciting. This completes the proof. □

Since $\varphi(v\tau)$ is persistently exciting, the difference $\hat{\theta}_k - \theta$, where θ is the true value of the parameters, converges to zero. This guarantees convergence of the controller parameter estimates to their true values, uniform boudedness of $\xi(kT_0)$, $y(kT_0)$, $\forall k \ge 0$ and $y(t)$ and assymptotic LQ optimal tracking. Moreover, the ada-ptive scheme ensures exponential convergence of the estimated parameters, since

$$\hat{\theta}_{k+1} - \theta = \left[1 + a^{-1} \mathbf{Z}(kT_0)\mathbf{Z}^T(kT_0) \right] \left(\hat{\theta}_k - \theta \right) \qquad (30)$$

Relation (30) together with relation (23), ensures that $\hat{\theta}_k \to \theta$ exponentially as $k \to \infty$.

6 Conclusions

A new indirect adaptive scheme has been derived for adaptive LQ optimal tracking of continuous-time linear time-invariant single-input, single-output systems using multirate-output controllers. Using the proposed technique, the adaptive LQ optimal tracking problem is reduced to the determination of a fictitious static state feedbak controller, due to the merits of multirate-output controllers. Known tehniques do not have this flexibility and they resort to the direct computation of full order state observers. Moreover, the exogenous dynamics introduced in the control loop by MROCs, is of low order. Finally, persistency of excitation of the plant under control and hence parameter convergence, is provided, without making any assumption on the existence of special convex sets in which the estimated parameters belong or on the coprimeness of the polynomials describing the ARMA model, as compared to known adaptive LQ optimal control schemes.

References

[1]. Chammas A.B., Leondes C.T. 1978 On the design of linear time invariant systems by periodic output feedback: Parts I and II. *Int.J.Control* 27, 885-903.

[2]. Paraskevopoulos P.N., Arvanitis K.G. 1994 Exact model matching of linear systems using generalized sampled-data hold functions. *Automatica* 30, 503-506.

[3]. Hagiwara T., Araki M. 1988 Design of a stable state feedback controller based on the multirate sampling of the plant output. *IEEE Trans. Autom. Control* AC-33, 812-819.

[4]. Hagiwara T., Fujimura T., Araki M. 1990 Generalized multirate-output controllers. *Int.J.Control* 52, 597-612.

[5]. Araki M., Hagiwara T. 1986 Pole assignment by multirate sampled data output feedback. *Int.J.Control* 44, 1661-1673.

[6]. Arvanitis K.G., Paraskevopoulos P.N. 1995 Sampled-data minimum H^{∞}-norm regulation of continuous-time linear systems using multirate-output controllers. *J. Optim. Theory Appl.* 87, 235-267.

[7]. Arvanitis K.G. 1996 An indirect model reference adaptive controller based on the multirate sampling of the plant output. *Int. J.Adapt. Contr. Signal Process.* 10, 673-705.

[8]. Er M.-J., Anderson B.D.O. 1991 Practical issues in multirate-output controllers. *Int.J.Control* 53, 1005-1020.

[9]. Qiu L., Chen T. 1994 H_2-optimal design of multirate sampled-data systems. *IEEE Trans.Autom. Control* AC-39, 2506-2511.

[10]. Chen T., Qiu L. 1994 H_{∞} design of general multirate sampled-data control systems. *Automatica* 30, 1139-1152.

[11]. Samson C. 1982 An adaptive LQ control for nonminimum phase systems. *Int. J. Control* 35, 1-28.

[12]. Clarke D.W., Kanjilal P.P., Mohtadir C. 1985 A generalized LQG approach to self-tuning control - Part I: Aspects of design. *Int.J.Control* 41, 1509-1523.

[13]. Morimoto H. 1990 Adaptive LQG regulator via the separation principle. *IEEE Trans. Autom. Control* 35, 85-88..

[14]. Chen H.-F., Zhang J.-F. 1990 Identification and adaptive control for systems with unknown orders, delay and coefficients, *IEEE Trans. Autom. Control* AC-35, 866-877.

[15]. Sun J., Ioannou P. 1992 Robust adaptive LQ control schemes. *IEEE Trans. Autom. Control* AC-37, 100-106.

[16]. Polderman J.W. 1986 On the necessity of identifying the true system in adaptive LQ control. *Syst. Control Lett.* 8, 87-91.

[17]. Arvanitis K.G. 1998 An adaptive decoupling compensator for linear systems based on periodic multirate-input controllers. *J. Math. Syst. Estimation Contr.* 8, 373-376.

15
Control of an Automaton Using Uncertain Information

G.Tsirigotis and M. Naranjo

---

1 Introduction

The understanding and utilisation of the Speech Recognition technology allow auguring, in the next future, a wide establishment of vocal control systems in the production units. One can consider that the decisive progress originates, on the one hand in time/frequency algorithms for the speech signal pre-processing, and on the other in the simplification of discriminators using neuromimetic algorithms. The communication between a human being and a machine contains a technical device that emits, or is submitted to important noise perturbations. The received speech signal is degraded and only a low information quantity is perceived.

The control of well-modelled continuous or discrete systems has been greatly studied in the past and many algorithms are now available to solve various real - time control problems. In the case of complex systems, it is very difficult to obtain by identification methods, any models accurate enough to be used for efficient control. For some kinds of control objectives, it is sometimes easier to consider these systems automata [1]. In spite of this conceptual simplification, it is often necessary to deal with uncertain information on their functioning and this infers problems of modelling and control of automata. In Robotics for example, it is absolutely imperative to take a final decision that leads to an unambiguous result for the location of autonomous mobile robots or for location of pieces that must be picked up by manipulator robots.

The vocal interface introduces a new problem of the control of a deterministic automaton using uncertain information. We have now two predictive systems. One model takes information about the past and present state of the system and makes a prediction of its state in the near future. The human - machine interface provides orders taken about information from the past and present state of the surrounding environment.

We have the essential elements of an Anticipatory System [2] (fig. 1) which forms an expectation of future events and renders a decision accordingly.

The whole Speech Processing Procedure has for input the temporal speech signal of a word, and for output a recognised word labelled with an intelligibility index given by the recognition quality. We present an automatic calculation procedure of the loudness and the architecture of the Time Delay Neural Network (TDNN). We give the organisation of the network when the input is constituted by the output of the critical bands obtained after calculation of the loudness. The output of the network is under the form of a vector where the components have values comprised among 0 and 1. While assimilating this vector to a fuzzy vector, the Information Theory allows us to extract a recognised word. A large part of this paper is devoted to the problem control of a deterministic automaton using, on the one hand, stochastic information of its Pattern Recognition Module, and on the other, fuzzy information provide by the Speech Processing Module.

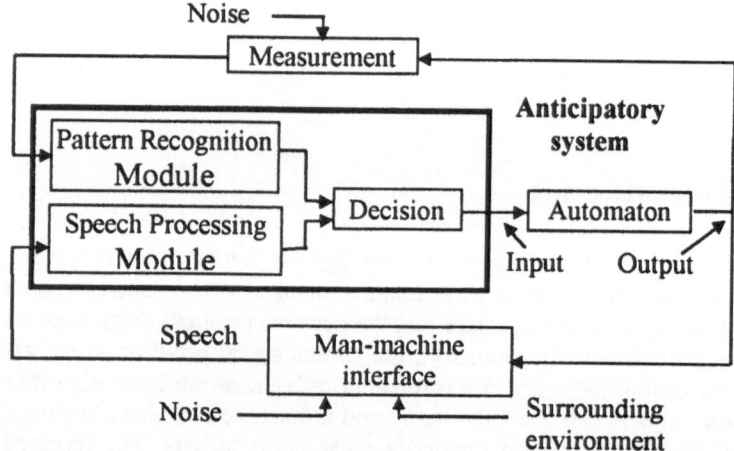

Figure 1: Place of the anticipatory system in the control problem

2 Loudness Calculation

2.1 Psychoacoustic Definitions

A sound becomes a stimulus when it attains the sensorial organ. The stimulus components are the values of excitation, physically measurable. When stimulus corresponds to an audible sound, it becomes a sensation, which can be only qualitative. The loudness (or relative loudness) is the associated value of sensation magnitude to a loudness level. One attributes arbitrarily a relative loudness $N = 1$ sone, for a sound of 1 kHz with an acoustic level of $L=40$ dB and a duration of 1 second.

Different types of interactions between sounds having close frequencies put the resolution frequency limit in evidence [3]. The frequency range, bellow these interactions are produced, is called critical band. The loudness is independent of the bandwidth until this one is inferior to a certain value. Such this analysis, the whole audible frequency range is divided in 24 adjacent critical bands. The passage of a critical band to the superior critical band corresponds, by convention, to a growth of 1 *Bark*. The subjective intensity of a sound to a given level is diminished when another sound is simultaneously present. This is the phenomenon of masking in frequency. A Temporal masking is also observed. It is due to the response time that necessitates the ear when it is submit to a stimulus.

2.2   Loudness Automatic Calculation

The schema for the automatic calculation of the loudness is given in fig. 2. The heart loudness (or main loudness) is the maximal specific loudness widened to an interval of 1 bark. Thank to Dr Stephane Veste, Engineer in the society « Teamlog », we can use a realisation based on the « LabWiev » software. The sampling frequency S varies from 150HZ for the lowest band, to 25 kHz for the highest band. FIR filters having an attenuation of approximately 50 dB per octave, first filter the signal. The different filter width bands are given by critical band. The coefficients of difference equations are to the number of 20. The intensity is obtained in calculating the absolute value logarithm of the signal for each critical band. The F/D factor takes into account the fact that sometimes, instead of an ideal plan acoustic field (F=Free), the sound simultaneously arrives, to the point of measure, of all the directions (D=Diffus).

The outputs are on the form a time/frequency representation giving a psychoacoustic model of the hearing. These are the (temporal) output impulses of the different frequency bands that are used to be computed by the neuromimetic recognisers [4]. The calculation of the loudness concerned words find in an ordinary vocabulary of control (industrial or militarily). The complete basis has 32 words taken from the French data "GRECO-PRC COMMUNICATION HOMME MACHINE. The length of a corresponding frame is 84 impulses. To test if the

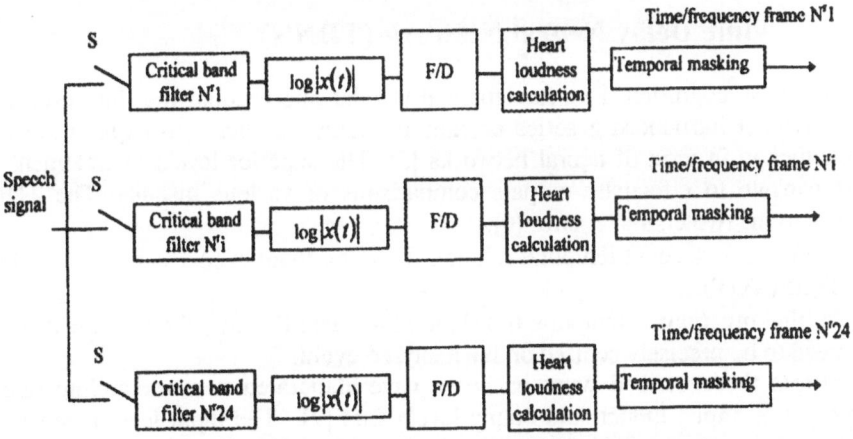

Figure 2: Loudness Automatic Calculation

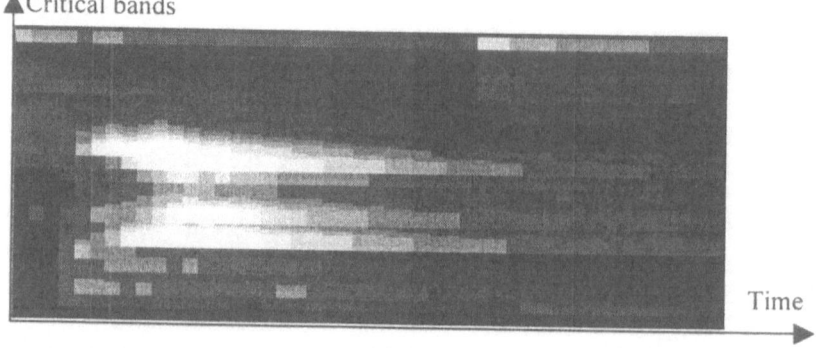

Figure 3: 2D representation of the loudness (back view) for the word "quatre"

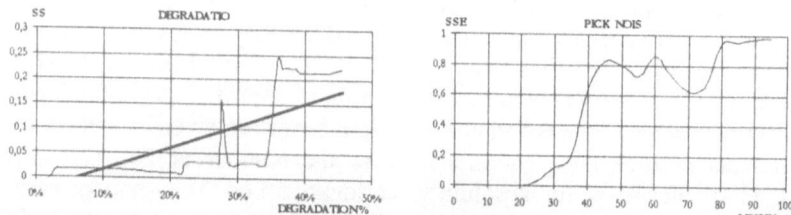

Figure 4: Noise influence in Sum Square Error (SSE)

recognition rate can be right, the recording of words was effectuated in an ideal acoustics environment. "LabView" software was excellent for the realisation of the device. A 2D time/critical bands representation can be extracted (fig. 3) where the values of loudness is represented with black and white colours for 1 and 0 respectively.

3 Recognition

3.1 Time Delay Neural Network (TDNN)

In Speech Recognition, it is fundamental to take into account the time evolution. Alex Waibel introduced a series of captors highly paralleled to explain the time phenomenon in term of neural networks [5]. The superior levels of treatment are thus capable to effectuate outputs comparisons of various instants. The TDNN replies to the two characteristics [6]:

- Taking into account temporal relations between input events as the Voice Onset Time (VOT).
- Putting invariant in temporal translation this identification: the window does not need to be precisely centred on the analysed event.

From the described word basis we have elaborated network architecture of three levels: input, hidden and output levels data [7]. The input level receives the output of the 23 obtained frequency bands in the loudness calculation. The 24th band is not taken to a count because is not significant for the recognition. Therefore the number of feature units is 23, the dimension of the matrix who describes the loudness of each word is 23x84. This matrix furnishes the TDNN. The output level possesses 5 units, that we can code in binary $2^5 = 32$ words. An important delay length, therefore a strong interconnection, allows to diminish the number of hidden level neurons (of about 300). We have used the excellent SNNS software (Stuttgart Neural Network Simulator) developed, under the direction of Andreas Zell at the « Institute for Parallel and Distributed High Performance Systems ». The obtained results are satisfactory: the average quadratic error is inferior to 0.05 and the recognition rates very close of 100% on the learning.

3.2   Influence of the noise in recognition

As we say in the introduction the system may be works in a noisy environment where the received speech signal is degraded and only a low information quantity is perceived. We have examined the influence of different kind of noise in the recognition process and we approved (as we can see in the following figure 4) that the system has an important resistance to the noise coming from an arbitrary degradation and from cutting the picks of the signal. This resistance in the noise of the psychoacoustic model of the ear that we took approves the good choice. Contrary the performance is not satisfactory for the gaussian noise but we can pre-process the sound signal with one of good commercial software reducing the noise.

4   Control of a Deterministic Automaton using uncertain information

4.1   Formulation of the control problem

Let $A = (Q, E, F, \Box, \Box, q^0, q^*)$ be an automaton.
- Q is the finite set of t states,
- E the finite set of u inputs,
- F the finite set of v outputs, also called forms (state-classes, patterns),
- $\Box$ is a mapping : $QxE \rightarrow Q$,
- $\Box$ is a mapping : $Q \rightarrow F$,
- $q^0 \in Q$ is the initial state of the automaton
- $q^* \in Q$ a pre-specified final state.

Example 1 $Q = \{q_1, q_2, q_3, q_4\}, E = \{a, b\}, F = \{f_1, f_2\}, q_0 = unknown; \quad q^* = q_4$

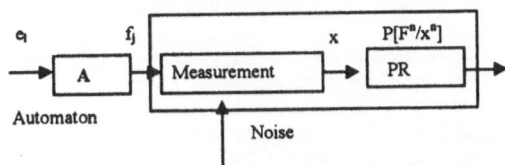

Figure 5 : The automaton and information System

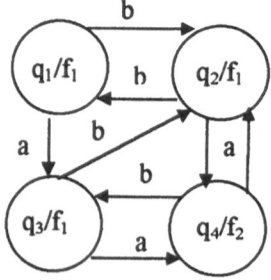

Fig. 6: Transition graph

The knowledge of the output of the automata is only accessible through a noisy measurement subsystem that produces a vector $x \in R^p$ followed by a Pattern Recognition subsystem that gives, at the step n of the procedure, a probability vector $P[F^n / x^n]$ after the complete application to the automaton of a command $d^n = d_l = (e_{l_1}, e_{l_2}, \ldots, e_{l_r}); d_l \in D$. D is the set of all command, d^n connotes the command decided and applied at the nth step. Similarly (q^n, f^n) will be respectively the state and the output of the automaton after the application of d^n. The control problem is to find a sequence of commands $(d^1 d^2 \ldots d^m)$ such as $P[q^m = q*] = 1$ and the total number of inputs has to be minimum, $q*$ being a final state ordered by human voice.

The automaton is controllable but not fully observable. It is obvious that the complexity of the problem is due on the one hand to the presence of the mapping γ in the definition of the automaton, and on the other, to the Pattern Recognition subsystem provides an output probability vector $P[F^n / x^n]$. It is supposed that all the variables whose values are needed to calculate this vector, have been previously obtained by learning.

4.2   Modelling the Anticipatory System

In the automaton A, two automata A_Q and A_F can be distinguished, respectively called state and form automaton.

$$A_Q = (Q, E, \delta, q^0, q*)$$

$$A_F = (F, E, \alpha, f^0, f*) \quad with \quad \alpha = \gamma \delta \gamma^{-1}$$

$$FxE \to F; f^0 = \gamma(q^0); f* = \gamma(q*)$$

The nature of the automaton A_F depends on the mapping

- if γ is bijective, it is obvious that A_Q and A_F are equivalent ; then A_F is a deterministic automaton ;
- if γ is surjective, A_F may be a non-deterministic automaton. For this it is sufficient that :

$$\exists (e_i \in E)(f_j \in F): \gamma^{-1}(f_j) = \left(q_{k_1}, q_{k_2} \right);$$

$$\delta \left(q_{k_1}, e_i \right) = q_{l_1}; \delta \left(q_{k_2}, e_i \right) = q_{l_2}; \gamma \left(q_{l_1} \right) = f_{j_1}; \gamma \left(q_{l_2} \right) = f_{j_2}$$

$$\alpha \left(f_j, e_i \right) = \gamma \delta \gamma^{-1} \left(f_j, e_i \right) = \left(f_{j_1}, f_{j_2} \right)$$

The knowledge of the state of A_Q and the forms of A_F is probabilistic since the system of Pattern Recognition provides a probability vector. The Anticipatory System must take a decision d^n in function of the last calculated state probability vector and a fuzzy ordered final state. Then after the application of each command d^n followed by a measure x^n, it is necessary to calculate the state and form probability vectors.

Notations : Let

- $\Delta(e_i)$ be the binary transition matrix associated with the input e_i,
- $\delta_D : QxD \to Q$ be the command mapping,
- $\Delta_D(e_l) = \Delta\left(e_{l_r}\right)\Delta\left(e_{l_{r-1}}\right)...\Delta\left(e_{l_1}\right)$ be the transition matrix associated with d_l
- Γ be the binary matrix of the mapping γ.

  An element Γ_{ij} of Γ is such as $\begin{cases} \Gamma_{ij} = 1 & \text{if } \gamma(q_i) = f_j \\ \Gamma_{ij} = 0 & \text{othewise} \end{cases}$

- $P\left[Q^n / d^n\right]$ and $P\left[F^n / d^n\right]$ be respectively the state and form probability vectors knowing the command d^n

- $P\left[Q^n / d^n, x^n\right]$ and $P\left[F^n / d^n x^n\right]$ be respectively the state and form probability vectors knowing the measure x^n after the command d^n.

Anticipatory scheme: For a command d^n, the states of A_Q and the forms of A_F can be predicted as follows :

$$P\left[Q^n / d^n\right] = \Delta_D\left(d^n\right).P\left[Q^{n-1} / x^{n-1}\right]$$

$$P\left[F^n / d^n\right] = \Gamma^t.P\left[Q^n / d^n\right]$$

Updating scheme: After obtaining the measure x^n, the knowledge on the states and the forms can be updated.

Let $P\left[f_j^n / d^n x^n\right]$ be the jth component of the vector $P\left[F^n / d^n x^n\right]$. We have used a non linear scheme :

$$P\left[f_j^n \mid d^n x^n\right] = \frac{P\left[f_j^n \mid d^n\right] \cdot P\left[f_j^n \mid x^n\right]}{\sum_{k=1}^{v} P\left[f_k^n \mid d^n\right] \cdot P\left[f_k^n \mid x^n\right]}$$. Noting that $\Gamma_{ji} = P\left[f_i \mid q_j\right]$,

Bayes theorem enables to calculate :

$$P\left[q_j^n \mid d^n f_i^n\right] = \frac{P\left[q_j^n \mid d^n\right]\Gamma_{ji}}{\sum_{k=1}^{t} P\left[q_k^n \mid d^n\right]\Gamma_{ki}} = \left(\Gamma^{-1}\right)_{ji}^n$$ Then

$$P\left[Q^n \mid d^n x^n\right] = \left(\Gamma^{-1}\right)^n \cdot P\left[F^n \mid d^n x^n\right]$$

Initialisation : For $n=0$, $\left(\Gamma^{-1}\right)^0$ must be defined since

$$P\left[Q^0 \mid x^0\right] = \left(\Gamma^{-1}\right)^0 \cdot P\left[F^0 \mid x^0\right].$$ In the case of lack of initial data, arbitrary

values may be chosen for $P\left[q_j^0 \mid f_i^0\right]$. For example :

$$P\left[q_{j_h}^0 \mid f_i^0\right] = \frac{1}{r}, \forall h = 1,\dots,r \quad \text{if} \quad \gamma^{-1}(f_i) = \left(q_{j_1}, q_{j_2},\dots,q_{j_r}\right)$$

Convergence : Let f_a^n be the most probable form at step n, before knowing x^n, we get : $\lim_{n\to\infty} P\left[f_a^n \mid d^n x^n\right] = 1$. In this case, the successive concordance of the

predicted values and the information brought by the observation x^n, for the most probable form, are strongly taken into account.

Example 2: From the automaton of the example 1 :

$$\Delta(a) = \begin{pmatrix} 0 & 0 & 0 & 0 \\ 0 & 0 & 0 & 1 \\ 1 & 0 & 0 & 0 \\ 0 & 1 & 1 & 0 \end{pmatrix}; \Delta(b) = \begin{pmatrix} 0 & 1 & 0 & 0 \\ 1 & 0 & 1 & 0 \\ 0 & 0 & 0 & 1 \\ 0 & 0 & 0 & 0 \end{pmatrix}; \Delta(d' = ba) = \begin{pmatrix} 0 & 0 & 0 & 0 \\ 0 & 0 & 0 & 0 \\ 0 & 1 & 0 & 0 \\ 1 & 0 & 1 & 1 \end{pmatrix};$$

$$\Gamma = \begin{pmatrix} 1 & 0 \\ 1 & 0 \\ 1 & 0 \\ 0 & 1 \end{pmatrix};$$

$$P[Q^0 / x^0] = \begin{vmatrix} 0.25 \\ 0.25 \\ 0.25 \\ 0.25 \end{vmatrix}; \quad P[Q^1 / d^1] = \Delta(ba).P[Q^0 / x^0] = \begin{vmatrix} 0 \\ 0 \\ 0.25 \\ 0.75 \end{vmatrix};$$

$$P[F^1 / d^1] = \begin{vmatrix} 0.25 \\ 0.75 \end{vmatrix} \quad and \quad after \quad PR: \quad P[F^1 / x^1] = \begin{vmatrix} 0.4 \\ 0.6 \end{vmatrix};$$

The results are : $P[Q^1/d^1x^1] = (0 \quad 0 \quad 0.18 \quad 0.82)^T$

4.3 Control Strategy

Every control strategy must satisfy the two objectives previously defined :

- $P[q_*^m / x^m] = 1$, q_* is ordered by human voice,
- Minimum number of inputs applied to the automaton.

- The first objective is related to the knowledge that has been obtained on the state of the automaton. A classical measure of this is the value of an information function defined from the state probability vector. The well-known Shannon information function will be used here for its simplicity. When the real state of the automaton is not precisely known, it is absolutely necessary to choose a command d_l after the application of which the measure x will reduce this uncertainty as much as possible. A first criterion L_1 for this choice using information function is then useful.

- The second objective is clearly related to the displacements in the state transition graph. A second criterion L_2 may be associated with them that is a function of the length of the shortest path from any state to the final state q_*.

This decision problem is then a multi-criteria's one and a classical way of solving it consists on defining a global criterion that is a weighted sum of L_1 and L_2. Finally, it must be noted that the nature of this control problem imposes a heuristic search of the solution, which is characterised by the determination of the best successive commands using a Bayesian strategy, at each step.

Bayesian control strategy

At the step $n+1$, a command $d^{n+1} = d_a$ is chosen according to a Bayesian strategy if :

$$E\left[L(q_i^n, d^{n+1} = d_a\right] \le E\left[L(q_i^n, d^{n+1} = d_l\right] \forall d_l \in D$$

$$with \quad L\left(q_i^n, d^{n+1}\right) = L_1\left(q_i^n, d^{n+1}\right) + \alpha L_2\left(q_i^n, d^{n+1}\right)$$

$$E\left[L(q_i^n, d^{n+1} = d_l)\right] = \sum_{i=1}^{t} L(q_i^n, d^{n+1} = d_l) . P\left[q_i^n / x^n\right]$$

$\Rightarrow$ **Criterion** $\left[L_1(q_i^n, d^{n+1})\right]$:

Supposing that the set x of all the measures x is finite, the a priori probability of q_i when the state of the automaton is q_j is defined by :

$$P\left[q_j / q_i\right] = \sum_x P\left[q_j / x\right] . P\left[x / q_i\right] \text{ where :}$$

$P\left[q_j / x\right] = P\left[q_j / F\right] . P\left[F / x\right]$. As seen before, the value of

- $P\left[q_j / F\right]$ changes during the control process. It is set here to its initial value $P\left[q_j^0 / F^0\right]$.

- $P\left[x / q_i\right] = P\left[x / F\right] . P\left[F / q_i\right]$; $P\left[F / q_i\right]^t$ is the ith row of the binary matrix Γ.

- $P\left[F / x\right]$ and $P\left[x / F\right]$ are learning data.

The confusion index associated with the state q_i is defined by the Shannon's information function :

$$H(q_i) = -\sum_{j=1}^{v} P\left[q_j / q_i\right] . \ln P\left[q_j / q_i\right]$$

It is known that $0 \le H(q_i) \le \ln t$ and it can be proved that :

i) $H(q_i) = 0$ if and only if the form f_i of q_i contains only the state q_i and the subset x_i of measures associated with f_i is such as $x_i \cap x_j = \phi, \forall j \ne i$ (x_j is the subset of measures associated with f_j). The state q_i is then absolutely discriminated by a measure x.

ii) When $H(q_i) = \ln t$ which implies that $P\left[q_j / q_i\right] = \frac{1}{t}, \forall j$, then the state q_i is totally non-discriminated.

Let $q_k^{n+1} = \delta\left(q_i^n, d^{n+1}\right)$ and $L_1\left(q_i^n, d^{n+1}\right) = H\left(q_k^{n+1}\right) - H\left(q_i^n\right)$;

$L_1 < 0$ indicates that the resulting state q_k^{n+1} after the application of the command d^n is better discriminated on an average than the starting state q_i^n.

Example 3 : From the transition graph of the example 1 and a given confusion matrix

$$P\left[q_j / q_i\right] = \begin{pmatrix} 0.3 & 0.3 & 0.3 & 0.1 \\ 0.3 & 0.3 & 0.3 & 0.1 \\ 0.3 & 0.3 & 0.3 & 0.1 \\ 0.1 & 0.1 & 0.1 & 0.7 \end{pmatrix};$$

state	input d^n		
	a	ba	b
q_1	0	-0.36	0
q_2	-0.36	0	0
q_3	-0.36	-0.36	0
q_4	+0.36	0	+0.36
L_1	+0.36	-0.72	+0.36

$$H(q_1) = H(q_2) = H(q_3) = 1.3$$
$$H(q_4) = 0.94$$
with $D = \{a, ba, b\}$

In regard of L_1, the choice is $d^n = ba$.

$\Rightarrow$ **Criterion $\left[L_2(q_i^n, d^{n+1}) \right]$** : A length c_{ij} is defined for each arc $\left(q_i, q_j \right)$ of the transition graph such as :

$$c_{ij} = \begin{vmatrix} 1 & if & \exists\, e_h \in E : \delta\left(q_i, e_h \right) = q_j \; (j \ne i) \\ 0 & if & i = j \\ \infty & otherwise \end{vmatrix}$$

The Bellmann-Kalaba's algorithm enables to calculate a matrix V, the value of its element v_{ij} being the length of the shortest path which leads from q_i to q_* through q_j. For a command d^{n+1} such as $q_j^{n+1} = \delta(q_i^n, d^{n+1})$, the criterion L_2 is then defined by : $L_2(q_i^n, d^{n+1}) = v_{ii} - v_{ij}$

Example 4 : From the transition graph of the example 1 and $q_4 = q_*$.

$$V = \begin{pmatrix} 2 & 2 & 2 & 2 \\ 3 & 1 & 3 & 1 \\ 4 & 2 & 1 & 1 \\ 4 & 2 & 2 & 0 \end{pmatrix} ; \text{ with } \alpha = \frac{1}{5}$$

state	input d^n		
	a	ba	b
q_1	0	0	0
q_2	0	2	2
q_3	0	1	1
q_4	2	0	2
$\square\, L_2$	0.4	0.6	1
$L_1 + \square L_2$	+0.04	-0.12	+1.36

The best choice in regard of the Bayesian strategy is $d^n = ba$. The weight α can be modified during the control process in order to favour either the discrimination of the states if the knowledge of the situation of the automaton is not sufficient, or the shortest displacement in the transition graph.

5 Conclusion

We have presented a method which allows to introduce a vocal control system in the production units.

From a speech signal of a word, the time/frequency procedure using the loudness calculation allows to take the psychoacoustic characteristics of hearing into account and make possible to recognise the right control order. The LabView software has revealed itself excellent for the realisation of the time/frequency device. We have used the very complete and documented Stuttgart Neural Network Simulator where the Time Delay Neural Network has permitted in the one hand to solve two important problems in Speech Recognition : taking into account the temporal relations between input events and giving them invariant in temporal translations, and on the other to minimize the noise effect in the recognition procedure.

The speech-processing module is put in the feedback loop of the machine. The human - machine interface provides orders taken about information from the past and present state of the surrounding environment. We have the essential elements of an Anticipatory System and we have provided a decision module control which chooses the right order for the convergence of the machine. This problem has been solved here by defining a criteria that depends on an information function which measure the probabilistic knowledge of the state of the automaton, and on shortest path length function, and by using a Bayesian decision strategy.

In a very noisy acoustic environment, the speech intelligibility considerably decreases. At some stage and due to the noise of transmission system that throws back a different perceived order from the vocal order, several compatible orders with the present state of the machine controlled are plausible. For this reason we will continue to elaborate a new automaton control problem who will take in to count one intelligibility index for each recognized order.

References

[1] Naranjo M, Richetin M, Rives G 1980 Study and realisation of a computer aided diagnosis and control system for a rolling mill. DGRST Contract Report 79-7-0299, edited by LASMEA, Blaise Pascal University of Clermont Ferrand (in French).

[2] Rosen R 1985 *Anticipatory Systems*. Pergamon Press.

[3] Zwicker E, Fast H 1991 *Psychoacoustics Facts and Models*. Springer Verlag.

[4] Naranjo M, Tsirigotis G 1997 Speech intelligibility measure for vocal control of an automaton. In: *Proceedings First International Conference on Computing Anticipatory Systems (CASY'97), Liege (Belgium), August 11-15.*

[5] Waibel A, Hanazawa T, Hinton G E, Shikano K, Lang K 1987 Phoneme Recognition Using Time Delay Neural Networks, Technical Report TR-1-0006, edited by Advanced Telecommunication Research Institute, Japan.

[6] Lang K, Hinton GE 1988. A time Delay Neural Network Architecture for Speech Recognition, TR CMU-CS n° 88-152, Carnegie Mellon University.

[7] Naranjo M, Tsirigotis G 1997 Time Delay Neural Network in a Psychoacoustic Model of Hearing, In: *World Multiconference on Systemics, Cybernetics and Informatics (SCI'97), Caracas (Venezuela), July 7-11.*

16

Learning Processes and Logic-Algebraic Method in Knowledge-Based Control Systems

Z. Bubnicki

1 Introduction

A great variety of definitions and approaches in the field of knowledge-based and learning control systems have been described. This chapter concerns a specific class of knowledge-based systems with a static plant described by a knowledge representation in the form of a set of facts given by an expert (logic knowledge representation). For the static plant described by a function $y = \phi(x)$, the control problem may consist in finding the decision $\bar{x}$ such that $\bar{y} = \phi(\bar{x})$ is the required output value. For the plant described by the knowledge representation presented in this chapter, the control problem consists in finding the proper **input property** which implies the required **output property**. In the case with the logic knowledge representation the facts, input property and output property are logic formulas concerning x, y and some additional variables. For this class of knowledge-based systems the **logic-algebraic method** has been developed [1, 2, 3, 4, 5, 6]. The main idea of the logic-algebraic method consists in replacing the individual reasoning concepts based on inference rules by unified algebraic procedures based on the rules in two-value logic algebra. The results may be considered as a unification and generalisation of the different individual reasoning algorithms for the class of systems determined by the form of the knowledge representation in Sec. 2.

The purpose of this chapter is to show how the logic-algebraic method may be used for the determination of learning algorithms in control systems with unknown parameters in the knowledge representation. Sec. 3 presents a formulation and solution of the control problem for the known parameters. The main idea of the learning process consists in a modification of the control decisions based on a current estimation of the unknown parameters. The

This work was supported by the State Committe for Scientific Research under Grant No 8 T11A 022 14.

different versions of the control algorithm are described in Sec. 4 and 5. The presented approach may be considered as an extension of the known ideas of pattern recognition (or identification) and adaptation for the special case of the knowledge representation and the specific logic-algebraic method for the generation of the control decisions based on the logic knowledge representation.

2 Logic Knowledge Representation

In many cases the knowledge representation given by an expert has a form of a set of relations

$$R_i(x, y, w, c), \qquad i = 1, 2, \ldots, k$$

where $x \in X$, $y \in Y$ are input and output vectors, respectively, $w \in W$ is a vector of additional variables, and $c \in C$ is a vector of parameters. Then the plant under consideration is described by

$$R(x, y, c) = \left\{ (x, y) \in X \times Y : \bigvee_{w \in W} \left[(x, y, w) \in \bigcap_{i=1}^{k} R_i \right] \right\}.$$

Now we shall consider the knowledge representation in which the relations R_i have the form of logic formulas concerning x, y, w, c. Let us introduce the following notation:

1. $\alpha_{xi}(x, c)$ — simple formula (i.e. simple property) concerning x and c, $i = 1, 2, \ldots, n_1$, e.g. $\alpha_{x1}(x, c) = $ "$x^T x \leq c^T c$".

2. $\alpha_{wr}(x, y, w, c)$ — simple formula concerning x, y, w and c, $r = 1, 2, \ldots, n_2$.

3. $\alpha_{ys}(y, c)$ — simple formula concerning y and c, $s = 1, 2, \ldots, n_3$.

4. $\alpha_x = (\alpha_{x1}, \ldots, \alpha_{xn_1})$, $\alpha_w = (\alpha_{w1}, \ldots, \alpha_{wn_2})$, $\alpha_y = (\alpha_{y1}, \ldots, \alpha_{yn_3})$.

5. $\alpha = (\alpha_x, \alpha_w, \alpha_y)$ — sequence of all simple formulas in the knowledge representation.

6. $F_j(\alpha)$ — j-th fact given by an expert. It is a logic formula composed with the subsequence of α and the logic operations: $\vee$ – **or**, $\wedge$ – **and**, $\neg$ – **not**, $\rightarrow$ – **if ... then**, $j = 1, 2, \ldots, k$. E.g. $F_1 = \alpha_1 \wedge \alpha_2 \rightarrow \alpha_4$, $F_2 = \alpha_3 \vee \alpha_2$ where $\alpha_1 = $ "$x^T x < c^T c$", $\alpha_2 = $ "the temperature is small or $y^T y \leq 3$", $\alpha_3 = $ "$y^T y = w^T x + c^T c$", $\alpha_4 = $ "$y^T y \geq 2c^T c$".

7. $F(\alpha) = F_1(\alpha) \wedge F_2(\alpha) \wedge \ldots \wedge F_k(\alpha)$.

8. $F_x(\alpha_x)$ — input property, i.e. the logic formula using α_x.

9. $F_y(\alpha_y)$ — output property.

10. $a_m \in \{0, 1\}$ — logic value of α_m, $m = 1, 2, \ldots, n$, $n = n_1 + n_2 + n_3$.

11. $a = (a_1, a_2, \ldots, a_n)$.

12. $F(a)$ — the logic value of $F(\alpha)$. All facts given by an expert are assumed to be true, i.e. $F(a) = 1$.

13. $\langle \alpha, F(\alpha) \rangle = KR$ (knowledge representation).

In this version of KR

$$R_i(x, y, w, c) = \left\{ (x, y, w) \in X \times Y \times W : F_i(a) = 1 \right\}, \qquad i \in \overline{1, k}.$$

The input and output properties may be expressed as follows:

$$x \in D_x(c), \qquad y \in D_y(c)$$

where

$$D_x(c) = \{x \in X : F_x(a_x) = 1\}, \qquad D_y(c) = \{y \in Y : F_y(a_y) = 1\}. \tag{1}$$

3 Control Problem

For the fixed value c the decision making or control problem may be formulated in the following way: given $F(\alpha), \alpha_x, \alpha_y$ and the required output property $F_y(\alpha_y)$ — find the best input property $F_x(\alpha_x)$ such that the implication $F_x \rightarrow F_y$ is satisfied. If it is satisfied for F_{x1} and F_{x2} and $F_{x2} \rightarrow F_{x1}$ then F_{x1} is better than F_{x2}.

It may be shown [4, 6] that finding F_x is reduced to solving two sets of equations:

$$\left. \begin{array}{r} F(a_x, a_w, a_y) = 1 \\ F_y(a_y) = 1 \end{array} \right\} \quad \text{and} \quad \left. \begin{array}{r} F(a_x, a_w, a_y) = 1 \\ F_y(a_y) = 0 \end{array} \right\} \tag{2}$$

with respect to a_x. If S_{x1} denotes the set of all solutions of the first equation (i.e. the set of all a_x for which there exist a_w, a_y such that $F = 1$ and $F_y = 1$), and S_{x2} denotes the set of all solutions of the second equation — then F_x is determined by $S_x = S_{x1} - S_{x2}$. Given S_x, the formula F_x is determined in the known way, such that $a_x \in S_x \leftrightarrow F_x(a_x) = 1$. E.g. if $\alpha_x = (\alpha_{x1}, \alpha_{x2}, \alpha_{x3})$ and $S_x = \{(1, 0, 1), (0, 1, 1)\}$ then

$$F_x(\alpha_x) = (\alpha_{x1} \wedge \neg \alpha_{x2} \wedge \alpha_{x3}) \vee (\neg \alpha_{x1} \wedge \alpha_{x2} \wedge \alpha_{x3}).$$

It is worth to note that the solution of our problem is reduced to solving the algebraic equations (2) where F, F_y are the algebraic expressions in two-value logic algebra. It is just the main idea of the **logic-algebraic method**. In general, solving the equations (2) requires to test all 0–1 sequences a. The algorithm generating the solution may be more effective when the following **decomposition** of the set of facts is applied:

$$F(a) = \overline{F}_1(\overline{a}_0, \overline{a}_1) \wedge \overline{F}_2(\overline{a}_1, \overline{a}_2) \wedge \ldots \wedge \overline{F}_p(\overline{a}_{p-1}, \overline{a}_p) \wedge \ldots \wedge \overline{F}_N(\overline{a}_{N-1}, \overline{a}_N)$$

where $\bar{a}_0 = (a_x, a_y)$, $\overline{F}_p$ is the conjunction of all facts containing the variables from the sequence $\bar{a}_{p-1}$ and $\bar{a}_p$ is the sequence of all other variables in $\overline{F}_p$ $(p = 1, 2, \ldots, N)$. Then the set $\overline{S}_0$ of all solutions (a_x, a_y) of the equation $F(a_x, a_w, a_y) = 1$ may be determined by the following recursive procedure

$$\overline{S}_{p-1} = \left\{ \bar{a}_{p-1} \in S_{p-1} : \bigvee_{\bar{a}_p \in \overline{S}_p} \left[\overline{F}_p(\bar{a}_{p-1}, \bar{a}_p) = 1 \right] \right\} \tag{3}$$

where S_p is the set of all $\bar{a}_p$, $p = N, N-1, \ldots, 1$, $\overline{S}_N = S_N$. Then, for the given $\overline{S}_p$ (starting from $p = N$) the next set $\overline{S}_{p-1}$ is generated by testing $\overline{F}_p$, i.e. by finding all $\bar{a}_{p-1}$ for which there exists $\bar{a}_p \in \overline{S}_p$ such that $\overline{F}_p = 1$. Having $\overline{S}_0$ we can determine S_x (and consequently F_x):

$$S_x = \left\{ a_x : \bigvee_{a_y \in S_y} (a_x, a_y) \in \overline{S}_0 \right\} - \left\{ a_x : \bigvee_{a_y \notin S_y} (a_x, a_y) \in \overline{S}_0 \right\}$$

where

$$S_y = \{ a_y : F_y(a_y) = 1 \}.$$

Thus, for the given requirement $D_y(c)$ one can obtain the result $D_x(c)$ such that

$$\left. \begin{array}{l} x \in D_x(c) \rightarrow y \in D_y(c) \\ x \notin D_x(c) \rightarrow y \notin D_y(c) \end{array} \right\} \tag{4}$$

where the sets D_x, D_y are determined by (1).

4 Pattern Recognition in Learning Control System

Assume for the futher consideration that the parameter c does not occur in α_y, i.e. in the formulation of $D_y(c)$ and that $D_x(c)$ is a continuous and closed domain in X. The parameter c in the knowledge representation has the value $c = \bar{c}$ and $\bar{c}$ is unknown. Consequently, for the fixed value x it is not known if x is a correct control decision, i.e. if $x \in D_x(\bar{c})$ which implies $y \in D_y$. Our problem may be considered as a classification or pattern recognition problem with two classes. The point x should be classified to class $j = 1$ if $x \in D_x(\bar{c})$ and to class $j = 2$ if $x \notin D_x(\bar{c})$. Assume that we can use the learning sequence

$$(x_1, j_1), (x_2, j_2), \ldots, (x_n, j_n) \triangleq S_n \tag{5}$$

where $j_i \in \{1, 2\}$ are the results of the correct classification given by a trainer (a teacher) for a sequence of points $x_1, x_2, \ldots, x_n$. The learning sequence (5) may be used to the current estimation of $\bar{c}$ and consequently to the current updating of the recognition algorithm determined by the form of $D_x(c)$.

A. Off-line pattern recognition

According to this approach two time intervals are introduced. In the first interval the learning sequence is obtained for the sequence of inputs x_i chosen randomly from X. In the second interval the control is based on the result of learning, i.e. x_i are chosen randomly from $D_x(c_n)$ where c_n is the estimated value of $\bar{c}$ obtained at the end of learning (**control based on the result of learning**). Let us dénote by $\bar{x}_i$ the subsequence for which $j_i = 1$, i.e. $\bar{x}_i \in D_x(\bar{c})$ and by $\hat{x}_i$ the subsequence for which $j_i = 2$, and introduce the following sets in C:

$$\overline{D}_c(n) = \left\{ c \in C : \bar{x}_i \in D_x(c) \text{ for each } \bar{x}_i \text{ in } S_n \right\}, \tag{6}$$

$$\hat{D}_c(n) = \left\{ c \in C : \hat{x}_i \in X - D_x(c) \text{ for each } \hat{x}_i \text{ in } S_n \right\}. \tag{7}$$

The set

$$\overline{D}_c(n) \cap \hat{D}_c(n) \triangleq \overline{\Delta}_c(n)$$

is proposed here as the estimation of $\bar{c}$. Assume that x_i is chosen randomly from X with probability density $f(x)$.

Theorem *If $f(x) > 0$ for each $x \in X$, and for each $c \neq \bar{c}$ $D_x(c) \neq D_x(\bar{c})$ then $\overline{\Delta}_c(n)$ converges to $\{\bar{c}\}$ with probability 1.*

Proof: From (6), if $x_{n+1} = \hat{x}_{n+1}$ then $\overline{D}_c(n+1) = \overline{D}_c(n)$, if $x_{n+1} = \bar{x}_{n+1}$ then

$$\overline{D}_c(n + 1) = \left\{ c \in C : \left[\bar{x}_i \in D_x(c) \text{ for each } \bar{x}_i \text{ in } S_n \right] \wedge \left[\bar{x}_{n+1} \in D_x(c) \right] \right\}.$$

Consequently, $\overline{D}_c(n + 1) \subseteq \overline{D}_c(n)$ which means that $\overline{D}_c(n)$ is a convergent sequence of sets. We shall show that

$$\lim_{n \to \infty} \overline{D}_c(n) = \overline{D}_c \tag{8}$$

with probability 1, where

$$\overline{D}_c = \left\{ c \in C : D_x(\bar{c}) \subseteq D_x(c) \right\} \tag{9}$$

and

$$\lim_{n \to \infty} \overline{D}_c(n) = \left\{ c \in C : \bar{x}_i \in D_x(c) \text{ for each } \bar{x}_i \text{ in } S_\infty \right\} \triangleq \tilde{D}_c.$$

Assume that (8) is not satisfied, i.e. there exists $\hat{c} \in \tilde{D}_c$ such that $D_x(\bar{c}) \not\subseteq D_x(\hat{c})$. There exists then the subset of $D_x(\bar{c})$

$$D_x(\bar{c}) - D_x(\hat{c}) \triangleq D_R$$

such that $\bar{x}_i \notin D_R$ for each $\bar{x}_i$ in S_∞. But according to the assumption about $f(x)$ — the probability of such a case is equal to 0. Then $\tilde{D}_c = \overline{D}_c$ with probability 1. In the same way it may be proved that

$$\lim_{n \to \infty} \hat{D}_c(n) = \hat{D}_c \tag{10}$$

with probability one, where

$$\hat{D}_c = \left\{ c \in C : X - D_x(\bar{c}) \subseteq X - D_x(c) \right\}, \text{ i.e.}$$

$$\hat{D}_c = \left\{ c \in C : D_x(c) \subset D_x(\bar{c}) \right\}.$$

From (8) and (10) one can derive that $\overline{\Delta}_c(n)$ converges with probability 1 to $\overline{D}_c \cap \hat{D}_c \triangleq \overline{\Delta}_c$ (the boundary of $\overline{D}_c$). Using the assumption about D_x it is easy to note that $\overline{\Delta}_c = \{\bar{c}\}$ □

For example, let $D_x(c)$ is described by inequality

$$(x^{(1)})^2 + (x^{(2)})^2 + \ldots + (x^{(k)})^2 \le c^2$$

where $x = [x^{(1)} \ldots x^{(k)}]^T$. Then

$$\overline{D}_c(n) = [c_{min}, \infty), \quad \hat{D}_c(n) = [0, c_{max}), \quad \Delta_c(n) = [c_{min}, c_{max})$$

and $\Delta_c(n) \to c$ w.p.1.

The recursive **learning algorithm** for $n > 1$ is then the following:
If $j_n = 1 \quad (x_n = \bar{x}_n)$
Prove if

$$\bigwedge_{c \in \overline{D}_c(n-1)} \left[x_n \in D_x(c) \right]$$

If yes then $\overline{D}_c(n) = \overline{D}_c(n-1)$. If not — determine new $\overline{D}_c(n)$

$$\overline{D}_c(n) = \left\{ c \in \overline{D}_c(n-1) : x_n \in D_x(c) \right\}.$$

Put $\hat{D}_c(n) = \hat{D}_c(n-1)$.
If $j_n = 2 \quad (x_n = \hat{x}_n)$
Prove if

$$\bigwedge_{c \in \hat{D}_c(n-1)} \left[x_n \in X - D_x(c) \right].$$

If yes then $\hat{D}_c(n) = \hat{D}_c(n-1)$. If not — determine new $\hat{D}_c(n)$

$$\hat{D}_c(n) = \left\{ c \in \hat{D}_c(n-1) : x_n \in X - D_x(c) \right\}.$$

Put $\overline{D}_c(n) = \overline{D}_c(n-1), \overline{\Delta}_c(n) = \overline{D}_c(n) \cap \hat{D}_c(n)$.
For $n = 1$, if $x_1 = \bar{x}_1$ determine

$$\overline{D}_c(1) = \left\{ c \in C : x_1 \in D_x(c) \right\},$$

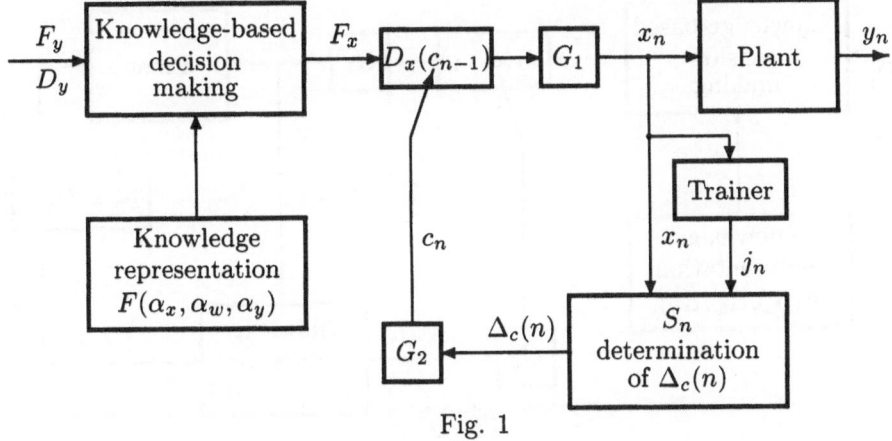

Fig. 1

if $x_1 = \hat{x}_1$ determine

$$\hat{D}_c(1) = \left\{ c \in C : x_1 \in X - D_x(c) \right\}.$$

If for all $i \le p$ $x_i = \overline{x}_i$ $(x_i = \hat{x}_i)$ — put $\overline{D}_p = X$ $(\hat{D}_p = \emptyset)$.

B. Learning process in the control system

The control decisions may be determined and put at the input of the plant currently during the learning process. In such a case the determination of the successive decision x_{n+1} is based on the estimation of $\overline{c}$ at the moment n. In the control algorithm proposed here the value c_n is chosen randomly from $\overline{\Delta}_c(n)$ and the value x_{n+1} is chosen randomly from $D_x(c_n)$ with the fixed probability distribution determined for $\overline{\Delta}_c(n)$ and $D_x(c_n)$, respectively. If $D_x(c_n) = \emptyset$ (the controllability condition is not satisfied for $\overline{c} = c_n$) then x_n is chosen randomly from X without the restriction to $D_x(c_n)$. Finally, the **control algorithm** based on the pattern recognition with learning is the following:

1. Determine F_x and $D_x(c)$ using logic-algebraic method presented in Sec. 3.

2. Using the learning sequence S_n (5) determine $\Delta_c(n)$ as it was described in the former algorithm.

3. Choose randomly c_n from $\Delta_c(n)$.

4. Put $c = c_n$ in $D_x(c)$.

5. Choose randomly x_{n+1} from $D_x(c_n)$

6. If $D_x(c_n) = \emptyset$, choose randomly x_{n+1} from X.

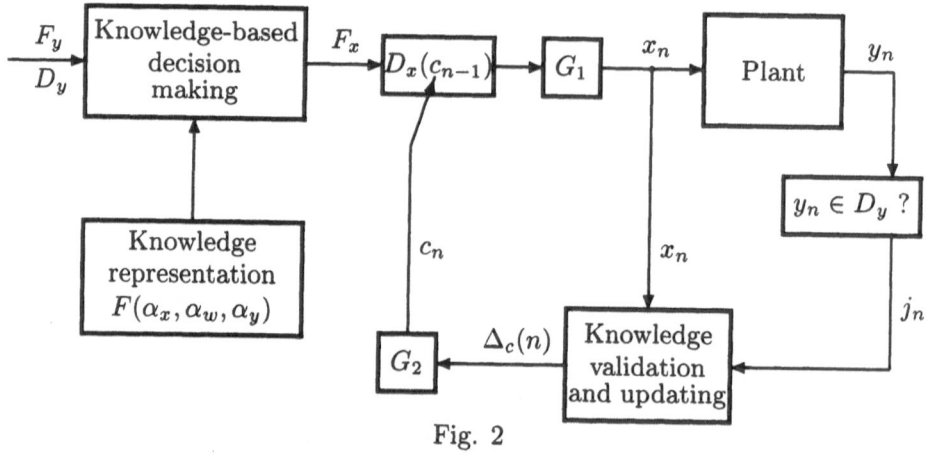

Fig. 2

The block scheme of the learning control system is presented in Fig. 1. For the random choice of c_n and x_n the generators G_1 and G_2 of the random numbers are required. The probability distributions are precised currently for $\Delta_c(n)$ and $D_x(c_n)$.

5   Learning Process in the Closed-loop Control System

The approach is a similar to that described in the former section and may be applied if it is possible to test if the requirement $y_i \in D_y$ is satisfied. To know if x_i is a correct control decision, i.e. if $x_i \in D_x(\overline{c})$, it is necessary to put x_i at the input, to measure the output y_i and to prove if $y_i \in D_y$. The plant generating y_i together with the verification of the relation $y_i \in D_y$ may be considered as a trainer giving $j_i \in \{1, 2\}$ for the successive values x_i in the learning sequence (5). The algorithms in the versions A and B are then the same as in the former section with $y_n \in D_y$ and $y_n \notin D_y$ in the places of $j_n = 1$ and $j_n = 2$, respectively. In version B the successive x_n is the control decision based on the current result of $\overline{c}$ estimation, i.e. x_n is chosen randomly from $D_x(c_n)$ where c_n is chosen randomly from $\overline{\Delta}_c(n)$. The control algorithm is the same as in version A with the difference that x_n is chosen from $D_x(c_n)$ and not from X. The block scheme of the learning control system is presented in Fig. 2.

6   Example

Let us consider a simple example with y, $c \in R^1$, $x = (x^{(1)}, x^{(2)})$. The simple formulas:

$$\begin{aligned}
\alpha_{x1} &= \alpha_1 = \text{``}(x^{(1)})^2 + (x^{(2)})^2 \leq c^{2}\text{''}, \\
\alpha_{x2} &= \alpha_2 = \text{``}x^{(1)} \leq 3c\text{''}, \\
\alpha_{w1} &= \alpha_3 = \text{``pressure is high''}, \\
\alpha_{w2} &= \alpha_4 = \text{``humidity is small''}, \\
\alpha_{w3} &= \alpha_5 = \text{``temperature is less than } x^{(2)} + y + c\text{''} \\
\alpha_{y1} &= \alpha_6 = \text{``}y \geq 5\text{''}, \\
\alpha_{y2} &= \alpha_7 = \text{``}2 \leq y \leq 3\text{''}.
\end{aligned}$$

The knowledge representation:

$$\begin{aligned}
F_1 &= \alpha_1 \wedge (\alpha_4 \vee \neg\alpha_6) \\
F_2 &= \alpha_2 \wedge \alpha_4 \to \alpha_6 \\
F_3 &= \neg\alpha_4 \vee \neg\alpha_3 \vee \alpha_5 \\
F_4 &= \alpha_4 \wedge (\alpha_3 \vee \neg\alpha_5) \\
F_5 &= (\alpha_4 \wedge \neg\alpha_2) \to \alpha_7
\end{aligned}$$

E.g. F_2 means "if $x^{(1)} \leq 3c$ and humidity is small then $y \geq 5$", F_3 means: "humidity is not small or pressure is not high or temperature is less than $x^{(2)} + y + c$". Substituting a_i in place of α_i we obtain the expression $F_i(a)$. The conjunction F of our facts can be easily simplified but it is not the purpose of our example. In our case $\bar{a}_0 = (a_x, a_y) = (a_1, a_2, a_6, a_7)$ and as a result of the decomposition we have

$$\overline{F}_1 = F_1 \wedge F_2 \wedge F_5, \quad \overline{F}_2 = F_3 \wedge F_4, \quad \bar{a}_1 = a_4, \quad \bar{a}_2 = (a_3, a_5)$$

Using the recursive procedure (3) (two steps for $p = 2, 1$) we obtain $\overline{S}_0$:

a_1	1	1	1	1
a_2	1	1	0	0
a_6	1	1	1	0
a_7	1	0	1	1

where the columns are the elements of $\overline{S}_0$. We can consider the different cases of $F_y(a_6, a_7)$. It is easy to see that for $F_y = a_6 \vee a_7$ the set S_x is

a_1	1	1
a_2	1	0

and $F_x = (a_1 \wedge a_2) \vee (a_1 \wedge \neg a_2) = a_1$. The result is then as follows: If $(x^{(1)})^2 + (x^{(2)})^2 \leq c^2$ then $2 \leq y \leq 3$ or $y \geq 5$. In our example $D_x(\bar{c})$ is determined by the inequality

$$(x^{(1)})^2 + (x^{(2)})^2 \leq \bar{c}^2,$$

$x_n = (x_n^{(1)}, x_n^{(2)}) = \bar{x}_n$ if x_n satisfies this inequality,

$$\overline{D}_c(n) = [c_{min,n}, \infty), \qquad \widehat{D}_c(n) = [0, c_{max,n})$$

where

$$c^2_{min,n} = \max_i \left[(\bar{x}_i^{(1)})^2 + (\bar{x}_i^{(2)})^2 \right], \qquad c^2_{max,n} = \min_i \left[(\widehat{x}_i^{(1)})^2 + (\widehat{x}_i^{(2)})^2 \right]$$

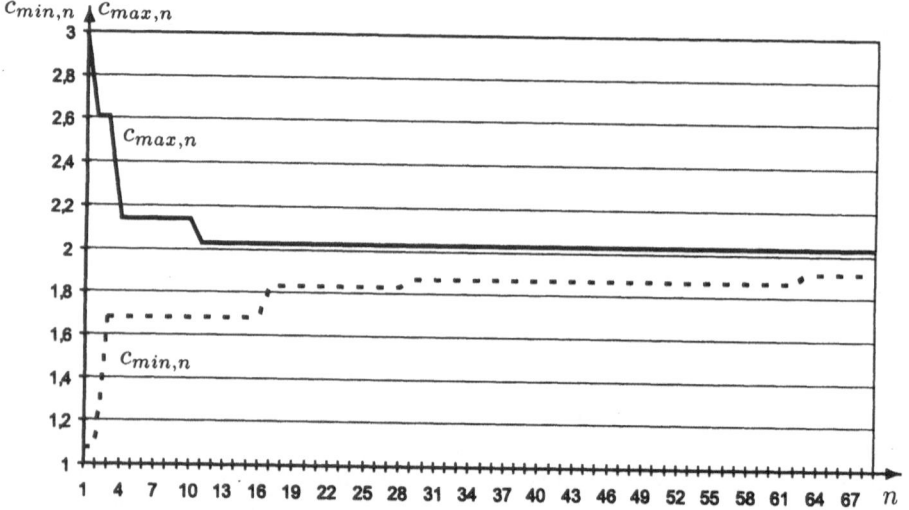

Fig. 3

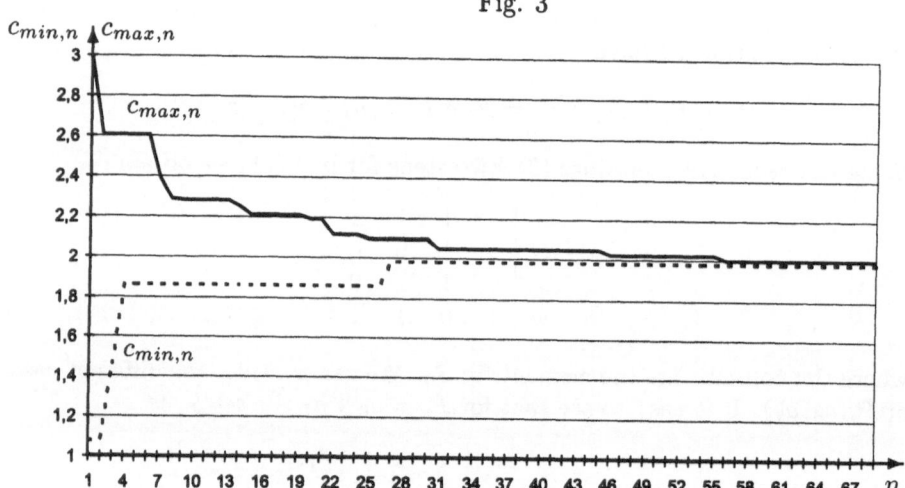

Fig. 4

$$\overline{\Delta}_c(n) = [c_{min,n}, c_{max,n}).$$

The control and learning process has been simulated for the different probability distributions $f(x)$ and the different numerical values of x_1 and c. The results of the simulations for $c = 2$ are presented in Fig. 3 for version A and in Fig. 4 for version B. In version A we assumed that X is a circle with the radius 3. In both versions $f(x)$ was constant in the corresponding circle and equal to 0 outside this circle. In version B c_n is chosen randomly with the rectangular distribution (from $c_{min,n}$ to $c_{max,n}$). In version A the convergence is slower and the control quality is smaller than in version B.

7   Final Remarks and Conclusions

For the plant described by a set of relation $R_i(x, y, w, \bar{c})$ and consequently by $R(x, y, \bar{c})$ the control problem may be formulated as follows: for the given R and the set $D_y \subset Y$ (user's requirement) to find the largest set $D_x(\bar{c}) \subset X$ such that the implication $x \in D_x(\bar{c}) \rightarrow y \in D_y$ is satisfied [7, 8]. It is easy to see that

$$D_x(\bar{c}) = \left\{ x \in X : D_y(x) \subseteq D_y \right\} \tag{11}$$

where

$$D_y(x) = \left\{ y \in Y : (x, y) \in R(x, y, \bar{c}) \right\}.$$

It is important to note that in the control problem presented in Sec. 3, the possibilities of the input and output properties formulation (i.e. the formulations of D_x and D_y) are restricted. They must be presented in the form of the logic formulas $F_x(\alpha_x)$ and $F_y(\alpha_y)$ with the simple formulas α_x, α_y which are used in the facts $F(\alpha)$. On the other hand, in this case the unified logic-algebraic method may be applied and the generation of control decision F_x is much easier than the determination of $D_x(\bar{c})$ (11) which requires individual approaches for different forms of R. For $R(x, y, \bar{c})$ the method of estimation and learning analogous to that presented in this chapter for $D_x(\bar{c})$ may be applied. If $R(x, y, \bar{c})$ is a continuous and closed domain in $X \times Y$ an we have the sequence of observations

$$(x_1, y_1), (x_2, y_2), \dots, (x_n, y_n), \qquad \bigwedge_i \left[(x_i, y_i) \in R(x, y, \bar{c}) \right]$$

then as the estimation of $\bar{c}$ we can use the boundary $\Delta_c(n)$ of the set

$$D_c(n) = \left\{ c \in C : \bigwedge_i \left[(x_i, y_i) \in R(x, y, c) \right] \right\}$$

which was described in [8]. The idea of the knowledge updating based on the input-output observations may be considered as the known concept of identification in the traditional case [3, 9].

The convergence of the algorithm presented in the chapter has been proved for the case when x_n are chosen from X_n. Numerous simulations with different numerical data and forms of probability distributions for the version B when x_n are chosen from $D_x(c_n)$ in learning control systems showed the convergence of $\overline{\Delta}_c(n)$ to $\bar{c}$, but till now it has not been proved.

The control problems presented in this chapter has been developed for the logic knowledge representation with fuzzy and uncertain parameters [10, 11, 12] — without the knowledge updating and learning. The possibility of the extension of the concept with learning process for the situation with *a priori* knowledge of the parameters in the knowledge representation in the form of membership functions, probability distributions or certainty distributions presented in [11, 12, 13] — seems to be interesting and promising.

References

[1] Bubnicki Z 1992 Algebraic approach to some class of reasoning problems. In: Shi Z (ed) 1992 *Proc. of IFIP Workshop on Automated Reasoning.* Beijing, pp 12–16

[2] Bubnicki Z 1992 Decomposition of a system described by logical model. In: Trappl R (ed) 1992 *Cybernetics and Systems Research, Vol 1.* World Scientific, Singapore, pp 121–128

[3] Bubnicki Z 1993 New problems and directions in the area of control and identification for knowledge based systems. In: Hamza M H (ed) 1993 *Proc. of 12th IASTED International Conference. Modelling, Identification and Control.* Acta Press, Zurich, pp 1–4

[4] Bubnicki Z 1996 Logic-algebraic foundations of a class of knowledge based control systems. In: Jamshidi M, Yuh J, Dauchez P (eds) 1986 *Proc. of the World Automation Congress, Vol 4.* TSI Press, Albuquerque, pp 89–94

[5] Bubnicki Z 1997 Logic-algebraic method for a class of dynamical knowledge-based systems. In: Sydow A (ed) 1997 *15th IMACS World Congress of Scientific Computation, Modelling and Applied Mathematics, Vol 4.* Wissenschaft und Technik Verlag, Berlin, pp 101–106

[6] Bubnicki Z 1997 Logic-algebraic method for a class of knowledge based systems. In: Pichler F, Moreno-Diaz R (eds) 1997 *Proc. of Eurocast 97,* Lecture Notes in Computer Science, Springer Verlag, pp 420–427

[7] Bubnicki Z 1998 Logic-algebraic method for knowledge-based relation systems. *Systems Analysis, Modelling and Simulation* (to be published)

[8] Bubnicki Z 1997 Knowledge updating in a class of knowledge-based learning control systems. *Systems Science,* 23:19–36

[9] Bubnicki Z 1980 *Identification of Control Plants.* Elsevier, Oxford, Amsterdam, New York

[10] Bubnicki Z 1997 Logic-algebraic approach to a class of knowledge based fuzzy control systems. In: 1997 *Proc. of European Control Conference ECC 97, Vol 1.* Brussels

[11] Bubnicki Z 1998 Uncertain variables and learning algorithms in knowledge-based control systems. In: Sugisaka M (ed) 1998 *Proc. of the Third International Symposium on Artificial Life and Robotics, Vol 2.* Oita, pp 490–494

[12] Bubnicki Z 1998 Uncertain variables and learning algorithms in knowledge-based control systems. *Artificial Life and Robotics,* 3: (to be published)

[13] Bubnicki Z 1998 Uncertain logics, variables and systems. In: 1998 *Proc. of The Third Workshop of The International Institute for General Systems Studies.* Tianjin People's Publishing House, Beidaihe, Qinhuangdao, pp 7–14

17

A Practical Approach to Motion Control for Varying Inertia Systems

T. Kaipio, L. Smelov, C. Morgan and N. Leighton

1   Introduction

In this paragraph we will remind some basic principals.

1.1   Position and Velocity Control Review

Constant load inertia is usually obtained in the mechanical configurations where a motor is connected to its load by a gearbox, leadscrew or equivalent. However, there are occasions where a linkage, such as a crank-slider or four-bar-linkage couples the motor to the load. Principal reasons for such arrangements are for example, higher mechanical advantage and/or higher velocity ratio at certain positions; reduction of friction forces and reduction of backlash. Such configurations are however subjected to varying inertia loads due to cyclical changes in the ratio produced. Typically, these mechanisms are used in sewing machines, packaging operations and materials handling.

1.2   The Work-Energy Principle

(1) shows that the system possesses position dependent variables. This means that every cycle (cyclically varying inertia) they will have same values no matter what is the velocity. Therefore when system is driven through complete cycles, each having a different velocities, the position dependent variables will be unchanged.

$$W_m = K(\theta,\omega)_x - K(\theta,\omega)_{x-1} + U(\theta)_x - U(\theta)_{x-1} - W(\theta)_f - W(\omega)_d \quad (1)$$

In the following methods, knowledge of work done by motor (W_m) has been used to define amount of kinetic energy in the system.

To validate the proposed methods a Four-Bar-Linkage mechanism has been used. The following calculations are meant to create an overall inertia table of the system.

2   The Inertia Estimation - Simulation

When the system's total moment of inertia values are known and at least one angular velocity value, any other velocity value can be calculated. Therefore positions x and x-1 can be situated at any point on the revolution. So if n is the number of points then $0<x\leq n$.

2.1   Applying Torque (Wm)

If there is a loss of energy, or energy is added to the system, then the system's total energy value at point x (E_x) and energy value at point x-1 (E_{x-1}) are no longer equal. However, if the energy added or removed (E_{change}) from the system is known, we could say at any point of revolution (2). It is assumed in this case that there is no potential energy change in this system.

$$I_x = \frac{2 * E_{change} + I_{x-1} * \omega_{x-1}^2}{\omega_x^2} \quad (2)$$

2.2   Removing Position Dependent Energy Transform (U and Wf) Effect

Friction (θ) has been applied to the simulation model. If the total inertia at position (x-1) is known, then one can calculate the total inertia value at position x, they even the position dependent energy transform would have an effect on the system. In the initial position energy input is zero. If the term (x-1) would always indicate the initial position (sub index 0) then motor energy input may be calculated from the initial position onwards for the whole cycle. In some case difference between ω_{x-1} and ω_x might increase, and therefore the velocity measurement would be less sensitive to measurement errors. Sub index 0 indicates the initial position. W_m indicates work done by motor during the cycle before point x as illustrated below.

$$I_x = \frac{2 * (W_{ma} - W_{mb}) + I_0 * (\omega_{a0}^2 - \omega_{b0}^2)}{\omega_{ax}^2 - \omega_{bx}^2} \qquad (3)$$

Assuming that inertia at the initial position is known, then a new set of inertia data may be obtained by using (3). It can be seen that the effects caused by position dependent energy transform were cancelled. The calculated inertia is now similar to the actual system inertia.

2.3 Effect of Velocity Dependent Energy Transform (Wd)

Velocity dependent energy transform (W_d), like the energy transform caused by damping, would cause errors in the results of (3). Force, which is changing based on velocity, would be seen in that case as an inertia change.

Applying damping to the simulation model is used to simulate this type of energy transform.

It can be easily proven that the method, which removes friction, does not remove damping.

To understand this difference consider the motor energy input/output graphs. Figure 1 shows motor energy graphs for both cases. On the left-hand side we can see that the overall extra energy feed is the same with both angular velocities (slow is dash dot and fast is solid line). On the right-hand side it can be seen that the demands (fast and slow) use different quantities of the motor energy. This explains the inability of (3) to calculate the inertia data accurately.

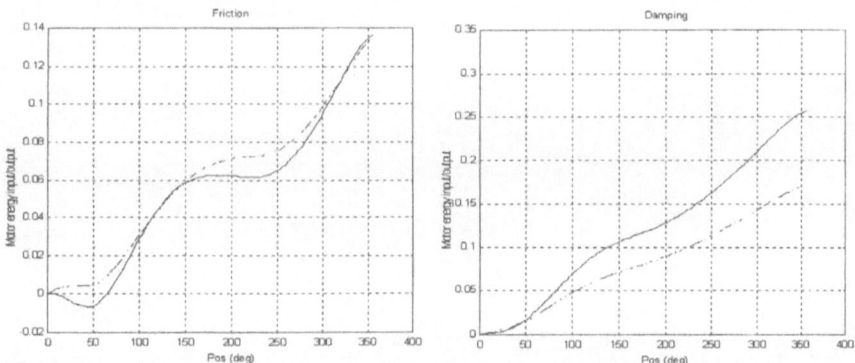

Figure 1

Let us assume that the last measured position (end of cycle) has a sub index n then (4) could be derived. By calculating the average angular velocity, the damping constant (Dc) can be found. The average angular velocity is not based on time but on position. The reason is that energy changes are based on position change.

$$W_{man} - Dc * \omega_{a(avg\theta)} * (\theta_n - \theta_0) = W_{mbn} - Dc * \omega_{b(avg\theta)} * (\theta_n - \theta_0)$$

$$Dc = \frac{W_{man} - W_{mbn}}{(\omega_{a(avg)} - \omega_{b(avg)}) * (\theta_n - \theta_0)} \qquad (4)$$

It is obvious that only damping is removed from work done by the motor in (5). New work done by the motor without damping is denoted W_{m_wd}.

$$W_{m\_wda(x)} = W_{ma(x)} - Dc * \omega_{a(avg\theta)(x)} * (\theta_x - \theta_0) \qquad (5)$$

Figure 2 shows how work done by the motor is effected when calculated by (5). The left side of the graph illustrates the case where only friction affects the system. The top graph (Friction A) does not use (5) and the lower graph (Friction B) includes this equation. It can be seen that removing the damping has no effect. The behaviour of the damped system is presented on the right side. The top graph (Damping A) shows how the motor work graphs are before removing the damping and the lower graph (Damping B) shows how it is after removing the damping.

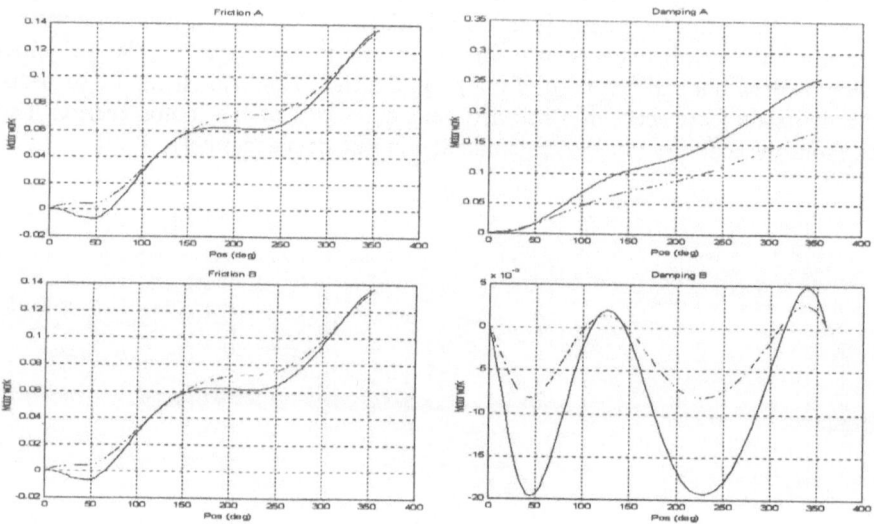

Figure 2

This is the method that is used to remove the damping effect from the inertia calculations (6).

$$I_x = \frac{2 * (W_{m\_wdax} - W_{m\_wdbx}) + I_0 * (\omega_{a0}^2 - \omega_{b0}^2)}{\omega_{ax}^2 - \omega_{bx}^2} \tag{6}$$

2.4 Defining the Moment of Inertia Value at the Initial Position

If there are 2 different sets of data representing two different velocity profiles denoted a, b, c and d. The direction of rotation is the same in a and b and in c and d. For example a and c may have different directions of rotation from c and d.

By using (7), we could find the value of I_0 by using velocity profiles a, b, c and d. However almost every single measured point would give a different value for I_0. This method is illustrated by (7) and the results given by Figure 3 as a dotted line. Every measurement has calculated its own I_0 value.

$$\frac{2 * (W_{m\_wdax} - W_{m\_wdbx}) + I_0 * (\omega_{a0}^2 - \omega_{b0}^2)}{\omega_{ax}^2 - \omega_{bx}^2} = \frac{2 * (W_{m\_wdcx} - W_{m\_wddx}) + I_0 * (\omega_{c0}^2 - \omega_{d0}^2)}{\omega_{cx}^2 - \omega_{dx}^2}$$

$$I_0 = \frac{2 * ((W_{m\_wdax} - W_{m\_wdbx}) * (\omega_{cx}^2 - \omega_{dx}^2) - (W_{m\_wdcx} - W_{m\_wddx}) * (\omega_{ax}^2 - \omega_{bx}^2))}{(\omega_{c0}^2 - \omega_{d0}^2) * (\omega_{ax}^2 - \omega_{bx}^2) - (\omega_{a0}^2 - \omega_{b0}^2) * (\omega_{cx}^2 - \omega_{dx}^2)} \tag{7}$$

Therefore, a different method of finding value of I_0 has to be considered. A different method could be one, which creates minimum error between two measurements. (8) and (9) show absolute error at the x^{Th} measurement.

$$Error_x = \left| \frac{2 * (W_{m\_wdax} - W_{m\_wdbx}) + I_0 * (\omega_{a0}^2 - \omega_{b0}^2)}{\omega_{ax}^2 - \omega_{bx}^2} - \frac{2 * (W_{m\_wdcx} - W_{m\_wddx}) + I_0 * (\omega_{c0}^2 - \omega_{d0}^2)}{\omega_{cx}^2 - \omega_{dx}^2} \right| \tag{8}$$

$$R_x = 2 * \left(\frac{W_{m\_wdax} - W_{m\_wdbx}}{\omega_{ax}^2 - \omega_{bx}^2} - \frac{W_{m\_wdcx} - W_{m\_wddx}}{\omega_{cx}^2 - \omega_{dx}^2} \right)$$

$$S_x = \frac{\omega_{a0}^2 - \omega_{b0}^2}{\omega_{ax}^2 - \omega_{bx}^2} - \frac{\omega_{c0}^2 - \omega_{d0}^2}{\omega_{cx}^2 - \omega_{dx}^2} \tag{9}$$

$$Error_x = \left| R_x + I_0 * S_x \right|$$

(10) represents the overall error sum.

$$\sum \text{Error}_n = \left| R_1 + I_0 * S_1 \right| + \left| R_2 + I_0 * S_2 \right| + \ldots + \left| R_n + I_0 * S_n \right| \qquad (10)$$

(10) can be solved into the second power, see (11):

$$\sum \text{Error}_n^2 = (S_1^2 I_0^2 + 2S_1 R_1 I_0 + R_1^2) + (S_2^2 I_0^2 + 2S_2 R_2 I_0 + R_2^2) + \ldots + (S_n^2 I_0^2 + 2S_n R_n I_0 + R_n^2) \qquad (11)$$

This is combined in (12).

$$\sum_{0..n} \text{Error}^2 = \sum_{0..n} (S^2) * I_0^2 + \sum_{0..n} (2 * S * R) * I_0 + \sum_{0..n} R^2 \qquad (12)$$

(12) gives in the form of normal parabolic curve. So I_0 value with a minimum second power error can be calculated by using (13).

$$I_0 = \frac{-\sum_{0..n} (2 * S * R)}{2 * \sum_{0..n} (S^2)} \qquad (13)$$

(13) has been used to calculate I_0 value in such a way that at point x it has x data points available and so on. The result of this is shown in Figure 3 as a solid line. The horizontal straight dash dot line is the true I_0 value in the simulation model. The left side graph simulation model does not consider friction or damping.

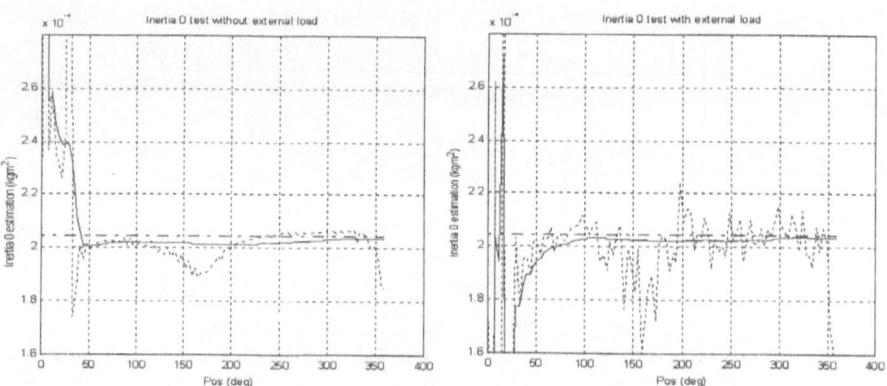

Figure 3

The value from the second method is used as the I_0 value in all the following graphs.

3   Test Rig Data

As the practical test rig is driven data is collected. This set of data is then imported into the Matlab program, which does the calculations, based on the formulas described earlier. Turbo C++ program has also been coded to do all this on-line. When total moment of inertia values for graphs are calculated a set of data has not been filtered.

3.1   Overview

Different kinds of resisting devices have been used when collecting data from the test rig. Both test linkages, light and heavy, have been driven without external load, with the spring, with the extra friction and with the spring and extra friction. As the inertia (θ) always remains the same the cases can be compared. Extra loads are so high that they change both shape and size of velocity and work done by the motor graphs. Velocity graphs can be seen on the left side and work done by motor graphs on the right side of Figure 4. The measurements correspond to the positive direction. Velocity demand has not been the same in each case. Linkage without load can be seen as a solid line. Linkage with the spring can be seen as x marks. Linkage with the friction can be seen as * marks. Linkage with spring and friction can be seen as o marks.

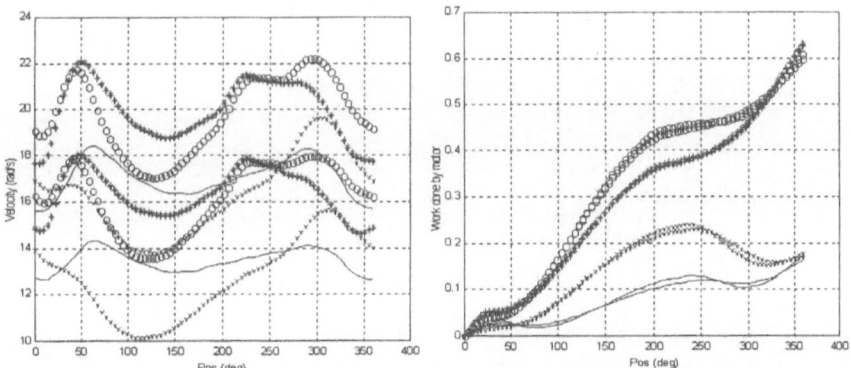

Figure 4

Conclusions based on those graphs
- Damping has a very small effect on these graphs. This can be seen from the fact that work done by the motor is ending to the same place even though the velocity is different.

- The spring does not add any extra friction or damping to the system as in both cases where spring is added work done by the motor has the same final value as in the case that did not have spring.

3.2 Practical Results

Similar effects can be seen in the practical results as are seen in simulation results. Graphs will be deformed if some of the energy transform properties have been ignored. The following example has been using linkage with the spring and extra friction. Figure 5 shows the importance of removing different disturbing properties, such as friction. In that figure:

- The top left inertia graph has been calculated by using (2) and assumes that I_0 value is known
- The top right inertia graph has been calculated by using (3) and assumes that I_0 value is known
- The bottom left inertia graph has been calculated by using (6) and estimated I0 value
- The bottom right I_0 estimation graph shows how I_0 value has reached the final estimated value

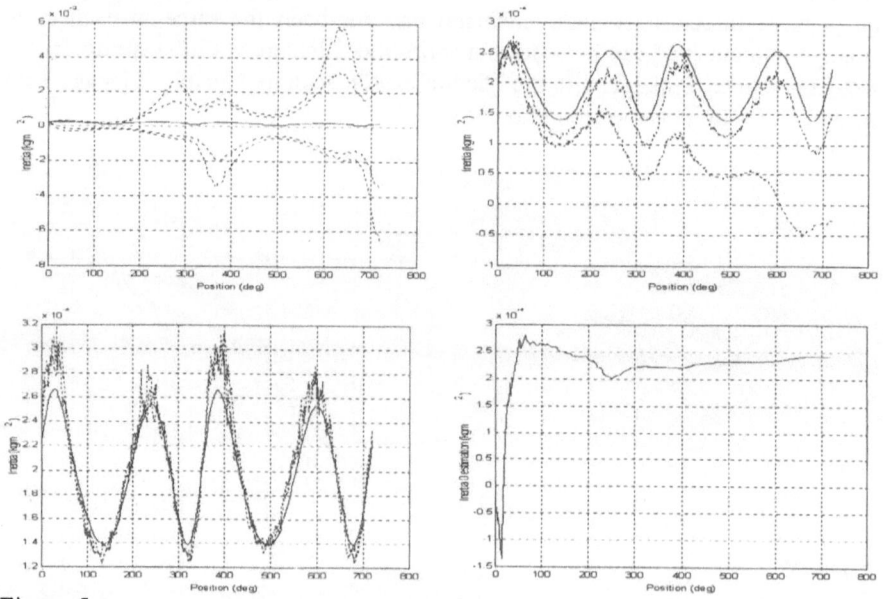

Figure 5

Figure 6 shows that in all these different load cases similar results can be achieved. In this figure:

- The top left graph has I_0 estimations in each different load case when the light linkage is used.
- The top right graph shows the final inertia graphs with light linkage (solid line is real total inertia graph).
- The bottom left graph has I_0 estimations in each different load case when the heavy linkage is used.
- The bottom right graph shows the final inertia graphs with the heavy linkage (solid line is the real total inertia graph).

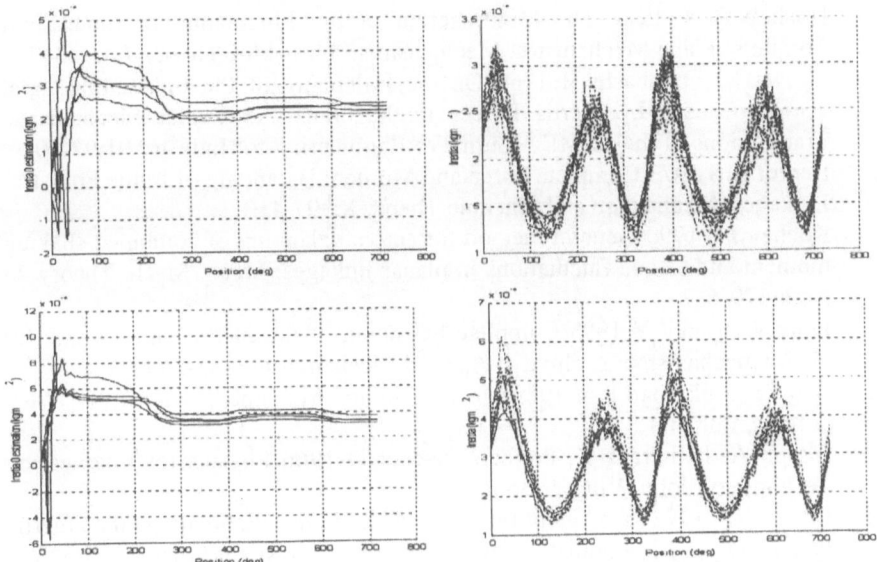

Figure 6

4   Conclusions

The methods described in this work to identify varying inertia mechanisms have been validated in the simulation environment as well as on the practical test rig. The proposed approach uses the identification results as a basis and aims to create an implementation of a PC based system, which changes the control parameters in response to inertia changes. Final identification method and control parameters will be implemented to the microchip-controlled platform. The final product will be completely a stand-alone.

5 References

1. Jones B, Leighton N J 1984 Adaptive feed system for assembly and packing. ISATA '84, International Symposium on Automotive Technology
2. Jones B, Leighton N J 1988 Digitally controlled drives for flexibility in high speed machinery. ISATA '88, International Symposium on Automotive Technology and Automation
3. Berkof R S 1979 The Input Torque in Linkages. Mechanism and Machine Theory 14:61-73
4. Hockey B A 1972 The Minimization of the Fluctuation of Input-Shaft Torque in Plane Mechanisms. Mechanism and Machine Theory 7:335-347
5. Ogawa K, Funabashi H 1969 On the Balancing of the Fluctuating input Torque Caused by Inertia Forces in the crank-and-Rocker Mechanisms. Transactions of the ASME / Journal of Engineering for Industry 91:97-102
6. Berkof R S 1973 Complete Force and Moment Balancing of Inline Four-Bar Linkages. Mechanism and Machine Theory 8:397-410
7. Kochev I S 1990 General method for active balancing of combined shaking moment and torque fluctuations in planar linkages. Mech. Mach. Theory 25 No.6:679-687
8. Funk W, Han J Y 1993 Complete balancing of the inertia input torque for planar mechanisms. Archive of Applied Mechanics 63:353-360
9. Rao J S, Dukkipati R V 1989 Mechanism and Machine Theory. John Wiley & Sons, England
10. Joseph A, Pomeranz K, Prince J, Sacher D 1966 Physics for Engineering Technology. John Wiley & Sons, London
11. Lin S K 1992 An Identification Method for Estimating the Inertia Parameters of a Manipulator. Journal of Robotic Systems 9, Iss.4:505-528
12. Karanadasa J P, Renfrew A C Nov.1991 Design and implementation of microprocessor based sliding mode controller for brushless servomotor. IEE Proceedings-B 138, No.6:345-363
13. Karanadasa J P, Renfrew A C 1992 Robust microprocessor control of a brushless D.C. motor driving variable inertia loads. Mechatronics 2, No.4: 347-361
14. Awaya I, Kato Y, Miyake I, Ito M 1992 New Motion Control with Inertia Identification Function using Disturbance Observer. Conf.Rec. IEEE Ind.Electro.Soc.Ann.Meeting:77-81
15. Kim N J, Hyun D S 1994 Very Low Speed Control of Induction Machine by Instantaneous Speed and Inertia Estimation. Conf.Rec. IEEE Ind.Electro.Soc.Ann.Meeting:605-610
16. Kim N J, Moon H S, Hyun D S Nov-Dec.1996, Inertia Identification for the Speed Observer of the Low Speed Control of Induction Machines. IEEE Transactions on Industry Applications 32, No.6:1371-1379

18

Iterative Model Based H_2/H_∞ Synthesis for Active Suspension System

P. Gáspar and J. Bokor

1 Introduction

The basis of the model-based controller design scheme is an input-output model that usually derives from an identification process. Since the identified model tends to the actual plant more or less because of the modeling error and uncertainties, therefore the difference between the model and the plant has to be considered in the controller design process. On the other hand, the model has to be identified as accurately as possible in frequency domains that are important in the sense of the controlled system, and in other frequency domains accuracy is less important, so the control law has to be considered in the identification process. Consequently, one should identify the model for controller design and at the same time, one should consider the controller in the identification process. However, this is an impossible task because controller design depends on the identified model while model identification depends on the controller. The identification and the controller design are not independent during the design process, but they pertain to each other, so the design procedure should be performed in an iterative way. Recently, the iterative identification and controller design has come to the forefront in the field of control research. Its aim is to enhance the performance of the controller based on the identified model, which serves as a controller design step.

The idea of the iterative model-based controller design as a way of adaptive control theory has been formulated. In the adaptive scheme, the idea of updating the controller comes naturally, since the parameters of the model are revised in each time step. In the iterative scheme, updating of the controller is performed in an off-line way using the measured data, i.e. the performance of the controller is enhanced until it is possible [2,17]. The main paths of iterative approaches are summarized in several important survey papers [9,23]. The iterative method proposed in this paper is based on the mixed H_2/H_∞ control design. Two main directions are connected with this scheme, namely the Zang scheme and the three-stage scheme by Van den Hof and Schrama. The Zang scheme is based on the Linear Quadratic and Gaussian (LQG) optimization criterion with Least Squares (LS) identification [8,27]. The three-stage scheme is based on the H_∞ norm controller design, closed loop identification method and robustness analysis [25].

This paper presents an iterative model-based controller synthesis to design a mixed H_2/H_∞ controller. The mixed norm optimization results in a quadratic performance index that is close to the LQG performance index, and an H_∞ norm that is close to the H_∞ optimal solution. For practical application of the method, the iterative scheme will be demonstrated on an active suspension problem based on the so-called quarter-car model. The aim of the design is to find a stabilizing compensator that minimizes the harmful vibration of the vehicle body caused by road irregularities. This paper is organized as follows. In Chapter 2, the motivation background of this method for vehicle suspension design is summarized. In Chapter 3, the principle of the iterative scheme based on the mixed H_2/H_∞ design is introduced. In Chapter 4, the steps of the iterative algorithm, i.e. the mixed H_2/H_∞ controller design step, the LS identification step, and the verification step, will be presented with the whole algorithm. Finally, in Chapter 5, the applicability of the proposed algorithm for the active suspension design will be demonstrated.

2. Quarter-car model for suspension design

The well-known quarter-car vehicle model, which is shown in Figure 2.1, is widely used for active suspension design, because of its simplicity, low number of state variables and parameters, and because its easy to investigate the performance properties [11,19,20]. Let the vehicle body mass and axle mass be denoted by m_a and m_b, the suspension stiffness and tire stiffness be denoted by c_s and c_t. The quarter-car model contains two vertical degrees-of-freedom: let the displacement of the body mass and the axle mass be denoted by y_b and y_a. In the quarter-car model, the disturbance, d, is caused by road irregularities. The input signal, u, is generated by the actuator controlled.

The motion equations of the quarter-car model can be formalized as follows:

$$m_b\ddot{y}_b + c_s y_b - c_s y_a + u = 0,$$
$$m_a\ddot{y}_a + c_s y_a + c_t y_a - c_s y_b - u - c_t d = 0. \tag{2.1}$$

The road input is chosen to be Gaussian white-noise with autocorrelation function

$$E[d(t)d(t+\tau)] = S_v \delta(\tau), \tag{2.2}$$

where $\delta(\tau)$ is the unit impulse and S_v is a constant power spectral density, which depends on both road roughness and velocity displacement.

One of the most important difficulties in the suspension design is that the model contains a large number of components the behavior and properties of which are unpredictable, unknown or uncertain. Therefore, the nominal model of the plant cannot be determined accurately, it can only be approximated in an identification process. The vehicle consists of many components the properties of which are uncertain for different reasons. The imperfections in the manufacture of tires, wheels, etc. may result in different types of variation. The tire stiffness is a function of the internal tire pressure, which can change under the process. The vehicle mass varies with the cargo, the fuel, or the number of passengers. The most significant excitation source is the road irregularities, which can be divided into two groups in the probabilistic sense. The random road excitations are characterized by stochastic properties with long duration, while the deterministic road disturbances are usually a single excitation with short duration: e.g. kerbs, potholes, etc. [14,16].

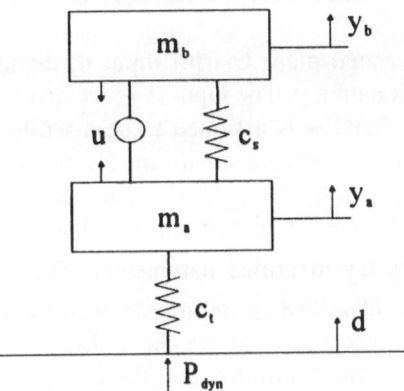

Figure 2.1: Quarter-car model for design

Suspension performance criteria include wheel-load variation, static and dynamic attitude control, working space, discomfort, and steering behavior. Using the simple quarter-car model wheel-load variation, suspension working space and body mass acceleration can be calculated. Since the displacement of the body mass can be linked with discomfort, the criteria above can be considered applying a quarter-car model. The investigation of the steering behavior can be solved using a very detailed model, such as a full-vehicle model [1,4,12]. The aim of this paper is to demonstrate an iterative model-based control design method for active suspension. In order to achieve this using a quarter-car model will be sufficient.

In the traditional quadratic performance criteria, the performance outputs to be controlled are the dynamic tire-load variation, the body mass acceleration, and the suspension working space. Therefore, the following performance index is used:

$$J = \lim_{T \to \infty} \frac{1}{T} \int_0^T \left[q_1 \{c_t(d - y_a)\}^2 + q_2 \{\ddot{y}_b\}^2 + q_3 \{y_a - y_b\}^2 \right] dt. \qquad (2.3)$$

It shows, J is a weighted quadratic sum of dynamic tire-load variation, the body mass acceleration and suspension working space with weighting factors q_1, q_2 and q_3 [15,19].

An iterative LQG controller design scheme for the active suspension problem was suggested by Michelberger, *et al.* [13]. During the iteration steps, the identified model approximates to the actual plant better and better and the design controller tends toward the optimal LQG solution. In this paper, the mixed H_2/H_∞ control design scheme is proposed. The H_∞ control law assumes that the disturbances are deterministic, and it assumes the worst case disturbance rather than one with a power spectrum. The main advantage of the H_∞ control is that it provides maximum robustness to the most destabilizing uncertainty, which is modeled as disturbance input. The H_2 performance criterion introduced above is extended with an H_∞ criterion for the body mass acceleration. This idea leads to an attempt of the mixed H_2/H_∞ control design scheme.

3. The principle of iterative scheme

Consider the linear actual plant G with input u, disturbance w, performance outputs z_2, z_∞, feedback output y. The input is generated by output feedback using the controller K. The signal w is assumed to be a white noise disturbance. The signal z_∞ is the performance associated with the H_∞ constraint, the signal z_2 is the performance associated with the H_2 criterion. The output signals y, z_2 and z_∞ are measured. The controller design in this paper is based on the identified model $\hat{G}(\theta)$, where θ denotes the identified parameters. The input, performance, and feedback outputs of the identified model are denoted by u^c, z_2^c, z_∞^c, and y^c. Two systems are distinguished according to the above, i.e. the actual controlled system with the actual plant and the controller, and the designed controlled system with the identified model and the controller. The actual and the designed controlled systems are illustrated in Figure 3.1.

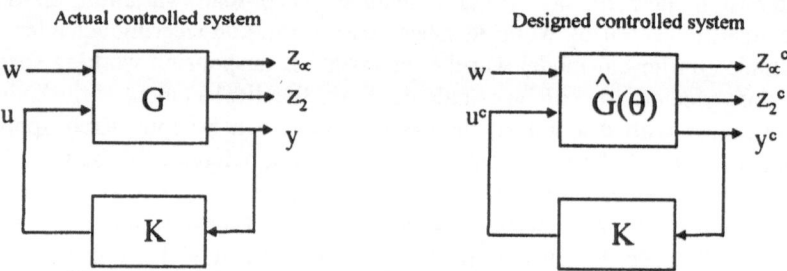

Figure 3.1: The actual and the designed controlled system

The controller design is performed in the designed controlled system, using the identified model, therefore, only the performance outputs of the designed system are able to be investigated instead of the performance outputs of the actual system. The controller has to be verified in the actual controlled system under the real circumstances. Consequently, the controller has to be designed in such way that it is suitable not only to the identified model but to the actual plant as well. The model identification is performed in the actual closed loop system, using the measured input-output signals. Since the final aim of the iteration is the controller design, the model identification process has to serve the controller design. Moreover, the model has to be identified more accurately in the frequency domains which are important in the performance of the closed loop. Therefore, the identification has to be performed by an indirect method, which takes the controller into account.

The actual controlled system in the sense of the mixed H_2/H_∞ controller is shown in Figure 3.2. The state space representation (SSR) of the plant, using the state vector x, can be written as follows:

$$\dot{x} = Ax + B_1 w + B_2 u,$$
$$z_\infty = C_\infty x + D_{\infty 2} u,$$
$$z_2 = C_2 x + D_{22} u,$$
$$y = C_y x + D_{y1} w.$$

(3.1)

The plant is given by the state space triple (A, B_2, C_y). The pairs (A, B_1) and (A, B_2) are assumed to be stabilizable, and the pairs (A, C_∞), (A, C_2) and (A, C_y) are assumed to be detectable. The dynamic controllers K considered in this paper are assumed to be strictly proper and have SSR in the following form:

$$\dot{x}_c = A_c x_c + B_c y,$$
$$u = C_c x_c.$$

(3.2)

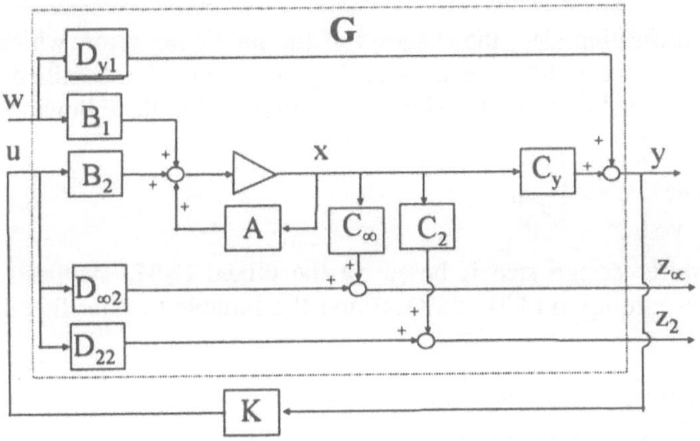

Figure 3.2: Actual controlled system for the mixed H_2/H_∞ controller synthesis

The mixed H_2/H_∞ controller synthesis problem is formulated as follows. For plant with SSR given by (3.1), find a dynamic controller with realization (3.2) such way that

- the closed loop system constructed from (3.1) and (3.2) is asymptotically stable,
- the H_∞ norm of the closed loop transfer function from w to z_∞ satisfies

$$\left\| T_{z_\infty w}(s) \right\| < \gamma,$$

(3.3)

for a given real positive γ, and

- the H_2 norm of the closed loop transfer function from w to z_2 satisfies

$$\min_K \left\| T_{z_2 w}(s) \right\|_2^2.$$ (3.4)

So, the task is to parametrize all suboptimal H_∞ dynamic controllers, which stabilize the closed-loop system and satisfy H_∞ norm constraint $\left\| T_{z_\infty w}(s) \right\| < \gamma$, and to find among them a controller which minimizes the standard Euclidean norm $\left\| T_{z_2 w}(s) \right\|_2^2$.

The controller design criteria formalized by (3.3) and (3.4) assume the knowledge of the actual plant (3.1). Since the transfer function of the actual plant is unknown, the applied criteria have to be based on the identified model $\hat{G}(\theta)$, where θ denotes the identified parameters. The SSR of the $\hat{G}(\theta)$ can be written as follows:

$$\begin{aligned}
\dot{x}^c &= \hat{A}x^c + \hat{B}_1 w + \hat{B}_2 u^c, \\
z_\infty{}^c &= \hat{C}_\infty x^c + \hat{D}_{\infty 2} u^c, \\
z_2{}^c &= \hat{C}_2 x^c + \hat{D}_{22} u^c, \\
y^c &= \hat{C}_y x^c + \hat{D}_{y1} w.
\end{aligned}$$ (3.5)

In the identification step, the criterion of the prediction error, which describes the difference between the predicted signals generated by the identified model and the measured actual input-output signals, is minimized in the following way:

$$\varepsilon = \begin{bmatrix} z_\infty \\ z_2 \\ y \end{bmatrix} - \hat{G}(\theta) \begin{bmatrix} w \\ u \end{bmatrix}.$$ (3.6)

The controller design step is based on the mixed H_2/H_∞ synthesis, which is formulated according to (3.3) and (3.4) and the suitable transfer functions are as follow:

$$\begin{aligned}
T_{z_\infty w} &= \hat{G}_{11} + \hat{G}_{12} K \left(I - \hat{G}_{32} K \right)^{-1} \hat{G}_{31}, \\
T_{z_2 w} &= \hat{G}_{21} + \hat{G}_{22} K \left(I - \hat{G}_{32} K \right)^{-1} \hat{G}_{31},
\end{aligned}$$ (3.7)

where matrices $\hat{G}_{ij}(\theta)$ come from the factorization of the identified model $\hat{G}(\theta)$.

In the next step of the iterative design, the examination of the stability and performance criteria is performed in the actual controlled system, and the iterative steps are continued until a proper control objective is achieved.

4. Mixed H₂/H_inf control design and the iterative algorithm

4.1 Mixed H₂/H_inf control design

In the classical control design, the well-known linear quadratic Gaussian (LQG) optimal synthesis has arbitrarily poor stability margins against uncertainties [5]. Thus, the LTR method is applied, where a μ parameter is used to achieve a desired regulator loop shape and to reach robustness to unmodelled dynamics. The H_{inf} optimal control law method provides maximum robustness to the most destabilizing uncertainty assuming the worst-case disturbance [7]. In this case, frequency-weighting functions are applied to shape the loop transfer function and to provide maximum robustness. The mixed H_2/H_{inf} controller synthesis method is applied, which minimizes an LQG performance index while a slightly suboptimal H_{inf} controller is preferred to an optimal one. The freedom of selecting a suboptimal H_{inf} controller can be used to minimize the upper bound of an LQG performance index. The theory for mixed H_2/H_{inf} method is well-developed [3,21,28], and several methods are proposed that produce control laws with specified loop shape within the mixed H_2/H_{inf} norm optimization framework [6,18,22].

In the mixed H_2/H_{inf} controller design, the desired loop shape can be reached by tuning two parameters, i.e. the γ parameter in the H_2/H_{inf} synthesis and the μ parameter in the LTR approach based on the identified model. Applying the H_{inf} versus H_2 curve, a γ value has to be selected, which reduces the H_{inf} norm significantly with little increase in the H_2 norm, therefore the selected γ gives a balance between stability robustness and quadratic performance. The advantage of the mixed H_2/H_{inf} controller synthesis is that the solution of the H_2 synthesis results in excellent disturbance rejection and the solution of the H_{inf} synthesis guarantees good performance and stability margins. The algorithm of the mixed H_2/H_{inf} synthesis covers both the pure H_2 and the pure H_{inf} synthesis.

The solution of the mixed H_2/H_{inf} synthesis is given by the following three Riccati equations:

$$X_M\left[\hat{A}+\gamma^{-2}(Y_M+P)\hat{C}_2^T\hat{C}_2\right]+\left[\hat{A}+\gamma^{-2}(Y_M+P)\hat{C}_2^T\hat{C}_2\right]^T X_M \\ -X_M\hat{B}_2\left(\hat{D}_{22}^T\hat{D}_{22}\right)\hat{B}_2^{\ T}X_M+\hat{C}_\infty^T\hat{C}_\infty=0, \tag{4.1}$$

$$\hat{A}Y_M+Y_M\hat{A}^T+Y_M\left[\gamma^{-2}\hat{C}_2^T\hat{C}_2-\hat{C}_y^T\hat{C}_y\right]Y_M+\hat{B}_1\hat{B}_1^T=0, \tag{4.2}$$

$$\left[\hat{A}+\gamma^{-2}Y_M\hat{C}_2^T\hat{C}_2-\hat{B}_2\left(\hat{D}_{22}^T\hat{D}_{22}\right)\hat{B}_2^{\ T}X_M\right]P \\ +P\left[\hat{A}+\gamma^{-2}Y_M\hat{C}_2^T\hat{C}_2-\hat{B}_2\left(\hat{D}_{22}^T\hat{D}_{22}\right)\hat{B}_2^{\ T}X_M\right]^T \\ +\gamma^{-2}P\hat{C}_2^T\hat{C}_2P+Y_M^T\hat{C}_y^T\hat{C}_yY_M=0, \tag{4.3}$$

where X_M, Y_M and P are symmetric and positive definite solutions of the above equations. It follows that Y_M can be determined by (4.2), however X_M and P can be estimated by (4.1) and (4.3) in an iterative way until their solutions converge to a desired accuracy. The Kalman filter and the regulator gains are given by the following equations:

$$L_M = -Y_M \hat{C}_y, \tag{4.4}$$

$$F_M = -\left(\hat{D}_{22}^T \hat{D}_{22}\right) \hat{B}_2^T X_M, \tag{4.5}$$

and the coefficients of the designed controller are as follows:

$$A_c = \hat{A} + \gamma^{-2} Y_M \hat{C}_2^T \hat{C}_2 + L_M \hat{C}_y + \hat{B}_2 F_M,$$

$$B_c = L_M, \tag{4.6}$$

$$C_c = -F_M.$$

The recovery parameter μ is applied through D_{22}, where $D_{22} = \mu I$. It is reduced in order that the loop shape GK approaches the desired loop shape. The solution of the mixed H_2/H_{inf} synthesis is a complicated process, it contains inner loops itself such as the three Riccati equations, the reduction of the recovery parameter μ, and the γ parameter in the H_2/H_{inf} synthesis.

4.2 LS identification step

The purpose of the identification is to estimate the transfer function $\hat{G}(\theta)$ based on prediction error criterion. The identification step can be performed by an indirect method that solves the closed loop identification problem by applying a classical direct open-loop identification technique. Several methods have to be applied for this task, such as the two-stage method, the coprime factorization method, or the dual Youla parametrization method [10,23]. In the iterative method, the so-called two-stage method is applied. In the first stage, the sensitivity transfer function is estimated possibly with a high-order model. Then a noise-free input signal is reconstructed with this model. In the second stage, the model is identified between the reconstructed input signal and the measured output signals [24].

4.3 Iterative algorithm

The steps of the iterative model-based mixed H_2/H_{inf} controller design are as follows:

Step 1: Initialization of the iterative scheme.

Step 2: Model identification.

    Step 2.1: The input signal is generated applying the controller K. The input and the output signals are measured.

    Step 2.2: Closed loop identification is performed, applying the two-stage method.

Step 3: Controller design.

Step 3.1: A large initial γ is chosen, and the Kalman filter gain is designed by solving (4.2) and (4.4). A parameter μ is chosen.

Step 3.2: The coupled Riccati equations (4.1) and (4.3) are solved in an iterative way, and the regulator gain is selected by (4.5).

Step 3.3: The controller K is formalized, then the loop shape GK is plotted together with the filter loop shape.

Step 3.4: The recovery parameter μ is reduced and from Step 3.2 the steps are repeated until the loop shape tends to the filter loop shape.

Step 3.5: The value γ is reduced and the steps are repeated from Step 3.1 as long as a solution is acceptable.

Step 4: Verification of the actual closed loop system. The stability and performance property of the actual closed loop system is verified. The iterative process is continued from Step 2 until a proper control objective is achieved.

5. Demonstration example

In this chapter the active suspension design is attempted by applying the mixed H_2/H_{inf} norm. In the simulation example, let the parameters of a conventional passive suspension system with parameters based on a large saloon car be as follows: m_b=200 kg, m_a=33 kg, c_s=9 kN/m, c_t=200 kN/m. Let the output signals be as follows: let the y signal be the velocity of the body mass, the z_2 performance output be the dynamic tire-load variation, and the z_∞ performance output be the acceleration of the body mass.

The identified model is assumed to differ from the actual plant due to the presence of uncertain components. The sets of maximal singular values (MaxSV) of the identified model and the plant are shown in the left-upper side of the Figure 5.1. The solid line corresponds to the model, and the dashed line to the actual plant. Both of the two peaks shift in the frequency domain due to the modeling error. Then the mixed H_2/H_{inf} controller based on the identified model is designed. Since the actual plant is known, the controller design can be performed accurately and the desired loop shape can be determined. Iteration is expected to yield the desired loop shape. It can be verified that the closed loop system on the design approximates the actual closed loop better and better during the iteration steps.

The sets of MaxSV of the mixed H_2/H_{inf} performance loop shape are shown in the right-upper side of the Figure 5.1. The solid line corresponds to the performance of the designed system, and the dotted line to the performance of the actual controlled system, i.e. to the desired performance. It shows that the two curves are different in the amplitude peaks. The MaxSV of the sensitivity functions and the complementary sensitivity functions are plotted in the lower side of the Figure 5.1.

Using the mixed H_2/H_{inf} controller, the model is identified again by applying the indirect two-stage method. The MaxSV functions of the identified model with the actual plant, the MaxSV functions of the performance loop shape and the MaxSV functions of the sensitivity functions are shown in Figure 5.2. It shows that the designed loop tends toward the desired loop in the sense of the frequency domain.

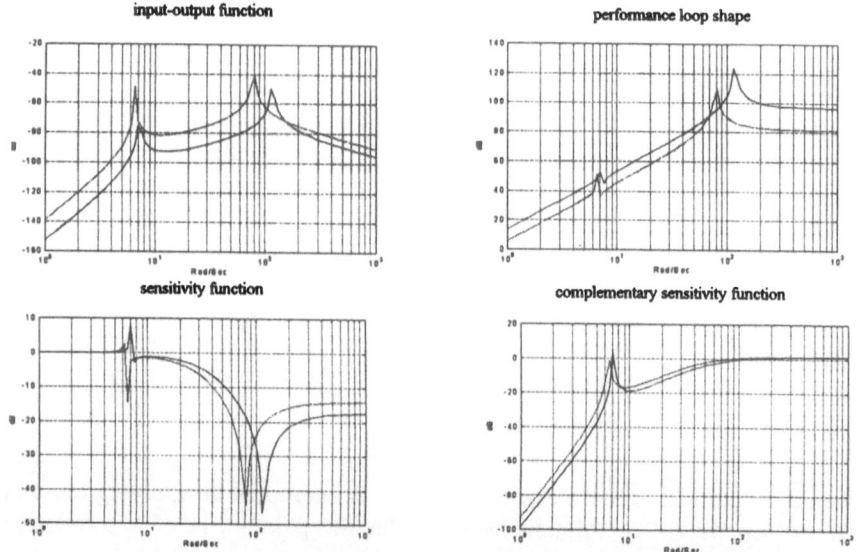

Figure 5.1: The MaxSV of the input-output function, the performance loop shape, and the sensitivity functions

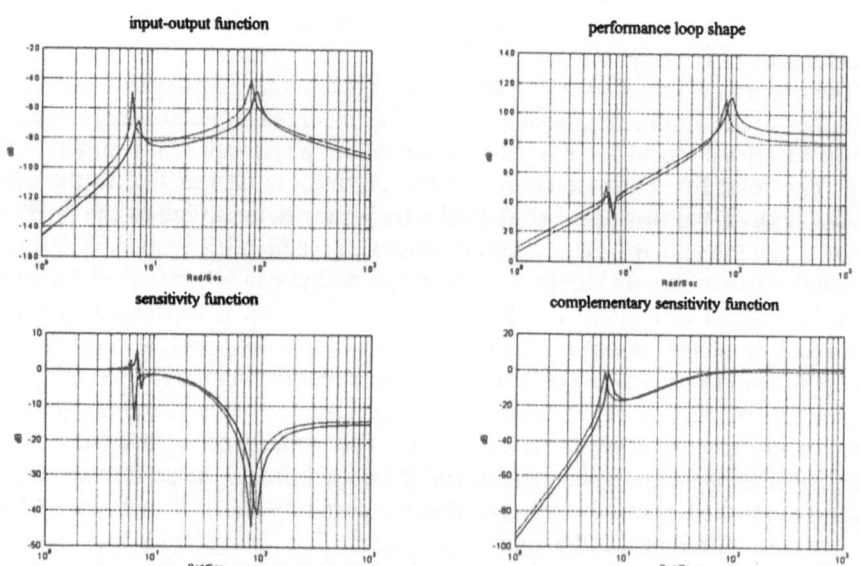

Figure 5.2: The MaxSV of the input-output function, the performance loop shape, and the sensitivity functions after the iteration #2

6. Conclusion

This paper has developed an iterative model-based mixed H_2/H_∞ controller design method. This method creates excellent disturbance rejection because of the H_2 criterion, and a good performance and stability margin because of the H_∞ criterion. In this paper, the design of the active suspension system has been attempted, where the main objective is to minimize the harmful vibration of the vehicle body caused by road irregularities. The purpose of the mixed performance criteria is to combine the traditional quadratic form with the H_∞ norm of the body acceleration. The aim of this example has been to illustrate the fact that, if the design is performed on a nominal model in which the parameters are not exactly known, the iterative design is an important tool for practicing engineers. The research project is to be continued. In the future, several problems have yet to be investigated such as the convergence property of the iterative scheme, the balance between the H_2 and the H_∞, the tune of the parameters γ and the μ.

Acknowledgement

This project was supported by the Hungarian National Science Foundation (OTKA) under grant no. T-016418 which is gratefully acknowledged.

7. References

1. Abdel Hady, M.B.A., D.A Crolla 1992 Active suspension control algorithms for a four-wheel vehicle model, *International Journal of Vehicle Design*, 13:144-158.
2. Anderson, B.D.O., R.L. Kosut 1991 Adaptive robust control: On-line learning, *Proc. of the Conf. on Decision and Control*, Brighton, 297-298
3. Bernstein, D.S., W.M. Haddad 1989 LQG Control with an H_∞ performance bound, *IEEE Trans. on Automatic Control*, 34:293-305
4. Crolla, D.A., M.B.A. Abdel Hady 1991 Active Suspension Control; Performance Comparisons Using Control Laws Applied to a Full Vehicle Model, *Vehicle System Dynamics*, 20:107-120
5. Doyle, J.C. 1978 Guaranteed margins for LQG regulators, *IEEE Trans. on Automatic Control*, 23:756-757
6. Doyle, J., Zhou, K., K. Glover, B. Bodenheimer 1994 Mixed H_2 and H_∞ performance objectives. II. Optimal control, *IEEE Trans. on Automatic Control*, 39:1575-1587
7. Doyle, J., K. Glover, P.P. Khargonekar, B. A. Francis 1989 State space solutions to standard H_2 and H_∞ control problems, *IEEE Trans. on Automatic Control*, 34:831-846
8. Gáspár, P., J. Bokor 1997 Multivariable weighted LQG control design in the iterative Zang scheme, *Proc. of the European Control Conf.*, Brussels, 11.4
9. Gevers, M. 1993 Towards a joint design identification and control? In: H.L. Trentelman, J.C. Willems (eds) 1993 *Essays on Control: Perspectives in the Theory and Its Applications*, Birkhauser, Boston, 111-151
10. Gevers, M., L. Ljung, P. Van den Hof 1997 Asymptotic variance expressions for closed-loop identification and their relevance in identification for control, *Proc. of the 11th IFAC Symp. on System Identification*, Kitakyushu, 3:1449-1454

11. Hrovat, D. 1990 Optimal active suspension structures for quarter car vehicle models, *Automatica*, 26:845-860

12. Krtolica, R., D. Hrovat 1990 Optimal active suspesion control based on a half-car model, *Proc. American Control Conf.*, Honolulu, 2238-2243

13. Michelberger, P., J. Bokor, L. Palkovics, E. Nándori, P. Gáspár 1997 Iterative identification and control design for uncertain parameter suspension system, *8th IFAC Symp. on Transportation Systems*, 2:464-469

14. Palkovics, L., P. Gáspár, J. Bokor 1993 Design of active suspension system in the presence of physical parameter uncertainties, *Proc. American Control Conf.*, San Francisco, 1:696-700

15. Sharp, R.S., D.A. Crolla 1987 Road Vehicle Suspension System Design - a review, *Vehicle System Dynamics*, 16:167-192

16. Sharp, R.S., S.A. Hassan 1986 An Evaluation of Passive Automotive Suspension Systems with Variable Stiffness and Damping Parameters, *Vehicle System Dynamics*, 15:335-350

17. Skelton, R.E. 1989 Model error concepts in control design, *International Journal of Control*, 49:1725-1753

18. Steinbuch, M., O.H. Bosgra 1991 Necessary conditions for static and fixed order dynamic mixed H_2/H_∞ optimal control, *Proc. of the American Control Conf.*, 1137-1142

19. Thompson, A.G. 1984 Optimal and suboptimal linear active suspensions for road vehicles, *Vehicle System Dynamics*, 13:61-72

20. Thompson, A.G., B.R. Davis 1988 Optimal Linear Active Suspensions with Derivative Constraints and Output Feedback Control, *Vehicle System Dynamics*, 17:179-192

21. Yeh, H.H., S.S. Banda, B.C. Chang 1992 Necessary and sufficient conditions for mixed H_2 and H_∞ optimal control, *IEEE Trans. on Automatic Control*, 37:355-358

22. Yeh, H.H., S.S. Banda, A.G. Sparks, D.B. Ridgely 1992 Loop shaping in mixed H_2 and H_∞ control, *Int. J. Control*, 56:1059-1078

23. Van den Hof, P.M.J., R.J.P. Schrama 1994 Identification and control - closed loop issues, *Proc. of the 10th IFAC Symp. on System Identification*, Copenhagen, 2:1-13

24. Van den Hof, P.M.J., R.J.P. Schrama 1992 An indirect method for transfer function estimation from closed loop data, *Proc. of the 31st Conf. on Decision and Control*, Tucson, 1702-1706

25. Van den Hof, P.M.J., R.J.P. Schrama, O.H. Bosgra, R.A. Callafon 1993 Identification of normalized coprime plant factors for iterative model and controller enhancement, *Proc. of the 32nd Conf. on Decision and Control*, San Antonio, 2839-2844

26. Yamashita, M., K. Fujimori, K. Hayakawa, H. Kimura 1994 Application of H_∞ control to active suspension system *Automatica*, 30:1717-1729

27. Zang, Z., R.R. Bitmead, M. Gevers 1995 Iterative weighted least-squares identification and weighted LQG control design, *Automatica*, 31:1577-1594

28. Zhou, K., K. Glover, B. Bodenheimer, J. Doyle 1994 Mixed H_2 and H_∞ performance objectives. I. Robust performance analysis, *IEEE Trans. on Automatic Control*, 39:1564-1574

19
A Matlab-Based User-Friendly Graphical Environment for Control System Analysis and Design (COSAD)

S.G. Tzafestas and D.L. Kostis

1   Introduction

A user-friendly and comprehensive control system design tool called Control *System Analysis and Design (COSAD)* is presented. Actually it is a typical graphical environment built on the platform of MATLAB 4.2.c (the system editing, analysis and design functions are carried out via pull-down menus, push buttons, radio buttons, check and dialogue boxes etc.), which uses the Graphical User Utilities Libraries of Matlab [1-2]. Our motivation for designing this tool is that there are many benefits to be gained by having available software for system analysis and control system design, which does not require any computer code writing. The objective is that users of COSAD system can design and simulate controllers, even if they don't have much experience in control system design and without having to spend most of their time writing, debugging and perfecting their own programs, in fields such as *education*, *research* and *practical* engineering applications.

In *education*, where the aim is to teach undergraduate classical and modern control theory, it is inappropriate to let the students focus their attention and waste time on peripheral matters. A poor user interface can be frustrating and may result in students not only not gaining the educational benefits they should from the software, but also adopting a negative attitude to the course.

On the other hand an appropriate tool which is appreciated by the students has many benefits such as :
1. the ability to focus on analysis and control design issues that simply cannot be taught in the real sense in a traditional lecture format.
2. the understanding which can be obtained by displaying both the system graphs (Root-locus, Bode and Nyquist diagrams) and the system response to the different kinds of inputs.

3. the opportunity to simulate high-order control systems by using special routines, to ensure the stability of the numerical integration in the case of ill-positioned systems.

4. the possibility of self-paced learning, (since the users may check their own calculations) and understanding by entering appropriate system descriptions into the software tool.

In *research* it is convenient to be able to simulate in short time and easily the control schemes under investigation for different sets of control parameters and various kinds of desired trajectories.

In *engineering* applications the objective is to make control system design methods available to more engineers, in order to carry out control system design and obtain satisfactory control in short time, even if they don't have high experience or scientific knowledge in Control System design and without any knowledge of programming.

This philosophy has been the drawing line in the development of our tool, which runs with a windows version of Matlab 4.2.c. and makes use of the Control System Toolbox and the Graphical User Interface Utilities library of Matlab [1-2].

2   The COSAD Environment

As already mentioned the COSAD environment enables one to study any feedback control system that can be modeled by the classic block diagram form (control schemes of SISO systems), which is embedded in the central window of COSAD , as shown in Figure 1.

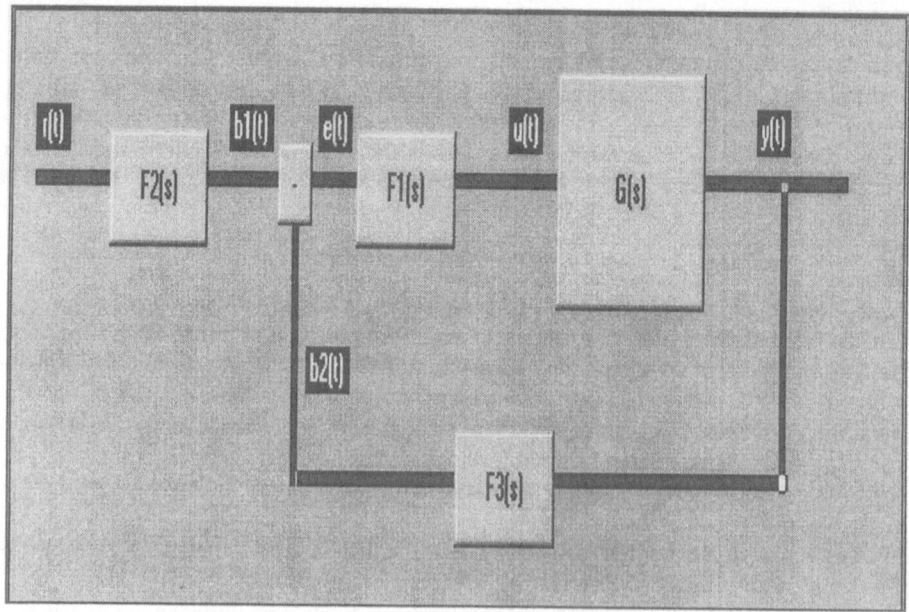

Figure 1

Since the case of linear systems is being considered, the software uses Matlab Script files (using also Control Toolbox Functions) in order to calculate the transfer function of the close-loop system, currently determined by the user. The user defines the transfer function of the open-loop system as G(s), the control structure by determining the sign of the feedback (±) and the transfer functions $F_1(s)$, $F_2(s)$ and $F_3(s)$ of the controller.

When the COSAD environment is invoked, the basic block diagram of Figure 1 appears in the main window as shown in Figure 2.

This screen stays alive during the use of the software and consists of a combination of *pull-down menus, radio buttons and push buttons*. The radio buttons are used to toggle between selection modes. The push buttons (Figure 1) are used to edit the description of the system under investigation and to perform various plots and animation options of open-loop and close-loop system response. Finally, the pull-down menus perform a variety of actions.

In general, the environment has four main modes which are the *pull-down menus*, the *system block*, the *display windows* and *a group of buttons* for plotting and animating the system response, which are described in the following .

2.1 Editing the Model

Editing a model which will consist of either entering a model from scratch or modifying an existing one, is carried out by pressing (clicking with the mouse) the corresponding element which opens up a pop-up dialogue box on the block diagram of the system (Control Scheme of SISO System), as shown in Figure 2.

The user is able to determine the transfer function of the system G(s) or the transfer functions of the controller $F_1(s)$, $F_2(s)$ and $F_3(s)$ (by simply clicking these elements on the block diagram), using the pop-up dialogue box as shown in Figure 3.

The user is asked to enter the numerator, Q(s), and denominator ,P(s), polynomials' coefficients in descending order. For example if he (she) wants to enter the transfer function H(s)= Q(s)/P(s) = $2/(s^2+2 s+1)$ the user must enter the vector [2] for the numerator and the vector [1 2 1] for the denominator polynomial.

The user is also able to alter the sign of the feedback loop by simply clicking on the cross symbol.

Once all parameters have been entered into a dialogue box, they can be stored by pressing *OK* or canceled by pressing *Cancel* while the previous set of parameters will remain active in the system.

220

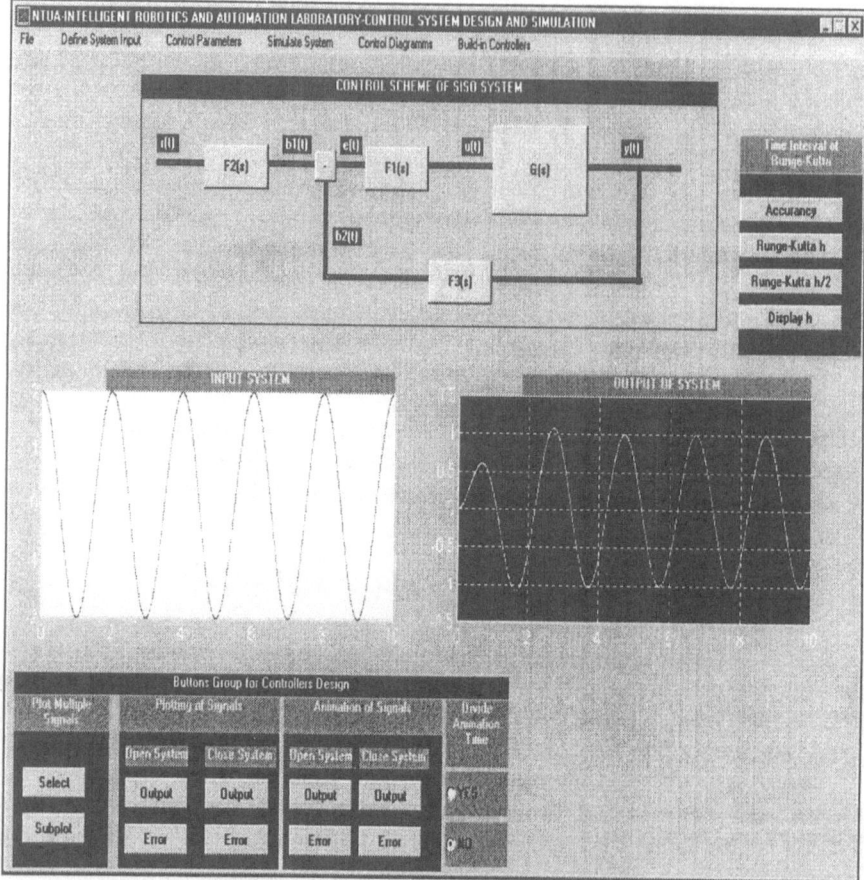

Figure 2

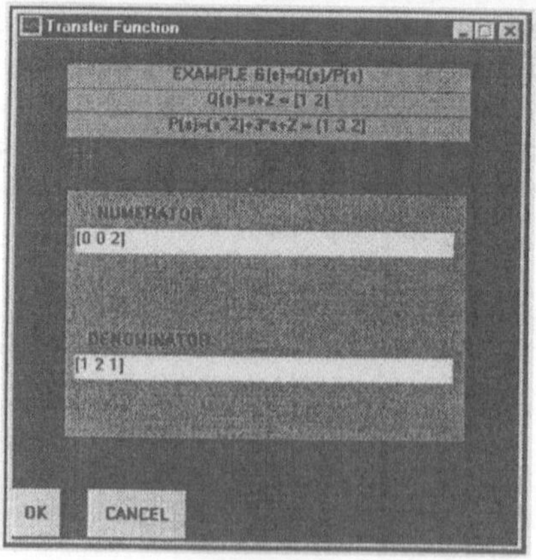

Figure 3

2.2   Pull-down Menus

There are six pull-down Menus: *File, Define System Input, Control Parameters, Simulate System, Control Diagrams, Build-in Controllers.*
The *File menu* is shown in Figure 4, and the

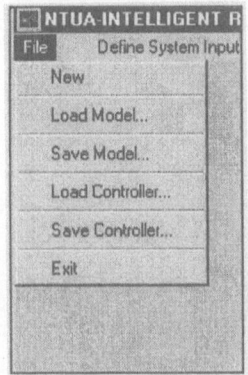

Figure 4

options perform the following functions :

- *New* resets all parameters to the default values. If the user decides to enter a system from scratch, all transfer functions ($G(s)$, $F_1(s)$, $F_2(s)$, $F_3(s)$) are defaulted to unity and the sign of feedback becomes negative.

- *Load Model* is used to load an existing model of the open-loop system $G(s)$ from the disk.

- *Save Model* is used to save a model into the disk.

- *Load Controller* is used to load an existing controller ($F_1(s)$, $F_2(s)$, $F_3(s)$ and the sign of feedback) from the disk.

- *Save Controller* is used to save a controller into the disk.

- *Exit* is used to quit the environment, which closes all windows and moves the user back to the Matlab workspace. It should be noted however, that throughout the session, the Matlab Command window remains active and the user may access it in the normal way if he (she) wishes to do so.

The *Define System Input Menu* is shown in Figure 5. Using the above Menu the analysis of the designed control system can be carried out with different kinds of inputs such as step, pulse, polynomial functions, etc.

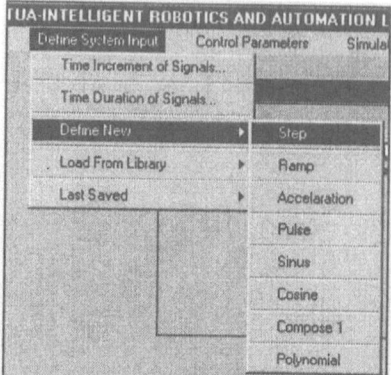

Figure 5

- *Time Increment of Signals and Time Duration of Signals* allows the user to define the number of the time instances and the duration of system input using a dialogue box.

- *Define New* is used to create a new input of the system choosing one from the pop up menu shown in Figure 5. A dialogue box presented in Figure 6 allows the user to determine his (her) preferences (input parameters, time duration of the signal).

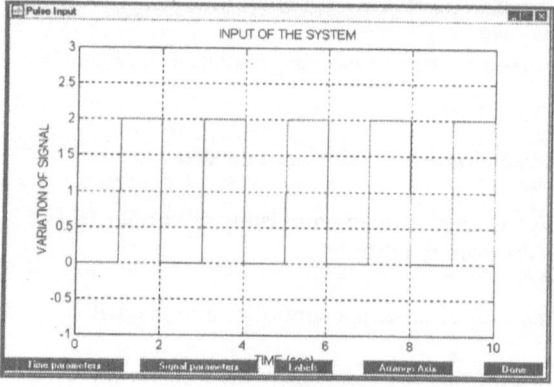

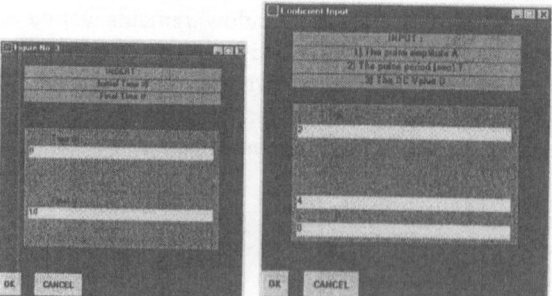

Figure 6

As can be seen, we have added more flexibility, allowing the user to use or create various types of inputs which are appropiate for the investigation of the control performance of the system (Figure 6).

- *Load from Library* is used to select a built-in signal from the library.

- *Last Saved* is used to select one of the previous saved inputs, while the current input of the system is under investigation. The user has the option to save an input when he (she) creates it (using the *Define New* menu option) as shown in Figure 6.

The *Control Parameters* menu is shown in Figure 7.

- *Controllability* displays a message to the user concerning the controllability of the open-loop system.

- *Open* and *Close* allows the user to investigate the behavior of the open-loop or the close-loop system calculating and displaying useful control variables.

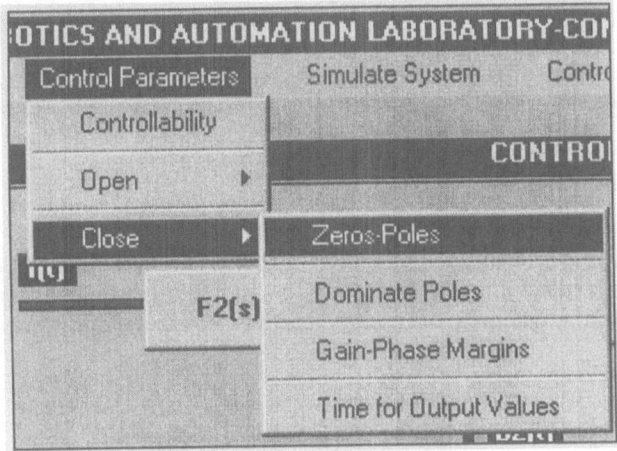

Figure 7

- *Zeros-Poles* option calculates and displays, in a pop-up window shown in Figure 8, the poles-zeros of the system. The user is also able to measure the real and imaginary parts of zero-poles by clicking the button *measure* (a crosshair will appear on the screen and the left mouse-button may be used to pick a zero or pole) in the window of Figure 8.

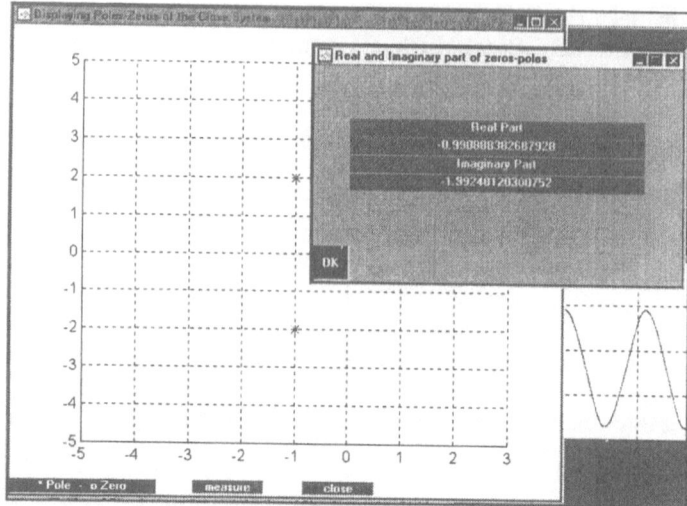

Figure 8

- *Gain-Phase Margins* calculates and displays gain-phase margins for open/close system.

- *Dominate Poles* investigates open/closed-loop system for the existence of dominating poles and displays the result in a pop-up window as shown in Figure 9 .

- *Time for Output Values* is used to pick points (using the mouse) of the system response (open/closed-loop, in the "INPUT SYSTEM" and the "OUTPUT of SYSTEM" respectively) and calculate the corresponding time and output values. The user is also able to calculate useful performance parameters, such as settling time, overshoot, rise time, peak time, steady state error etc.

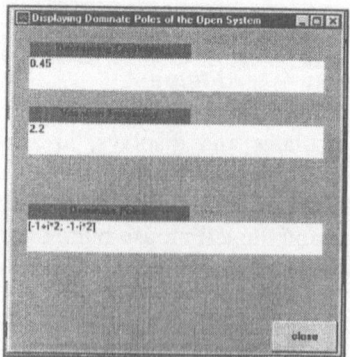

Figure 9

The *Simulate System* Menu is shown in Figures 10 and 11.

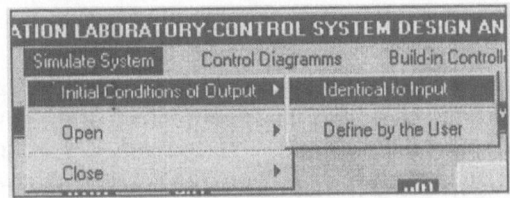

Figure10

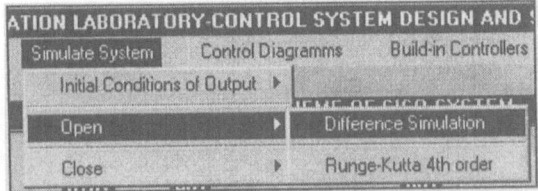

Figure 11

The Simulate menu enables the simulation of the current open/closed-loop system, calculating the system response for the currently defined input displayed in the *"INPUT SYSTEM"* window of the environment. The user can select either to use *Difference Simulation*, which via the use of an algorithm based on Taylor numerical integration, or *Runge-Kutta 4th order* numerical integration, enables the determination of non-zero initial conditions for the system, as shown in Figures 10 and 11.

The response of the system is displayed in the *"OUTPUT of SYSTEM"* window of the environment.

The *Control Diagrams* Menu shown in Figure 12, is more or less self-explanatory, with graphs displayed in their own windows. The user has the ability to view multiple graphs on different windows, at the same time, for the open-loop and close-loop system.

These windows have their own set of pull-down menus, which will be described later, in the plot-buttons presentation of the environment.

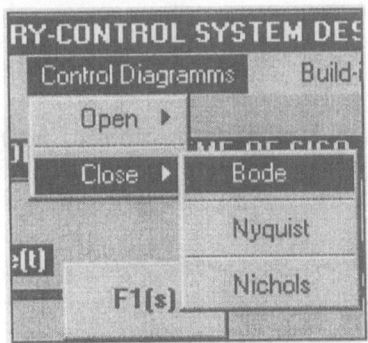

Figure 12

2.3   Input - Output Display Windows

These two windows with the label *"INPUT SYSTEM"* and *"OUTPUT of SYSTEM"* are permanent, as long as the central screen of the environment appears. The *"INPUT SYSTEM"* window displays the currently defined input for the system under investigation and the *"OUTPUT of SYSTEM"* window displays the last produced output from a simulation action, which stays alive until a fresh output occurs as a result of a simulation of open/close system, triggered by the user from Simulate System menu.

2.4   Group of Plot Buttons

The group of *Plot Buttons* is shown in Figure 13.

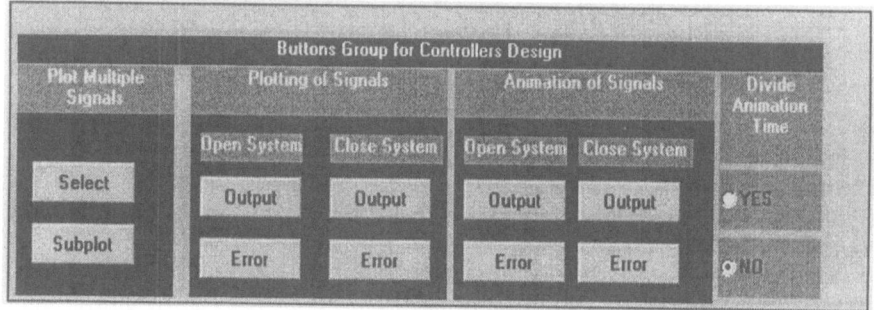

Figure 13

In the following paragraphs we present three subgroups of buttons:

- *The Plotting of Signals* subgroup is used for the plot of open/closed-loop system output and the plot of the error, calculated for the current defined input. When a plot is displayed, it appears in a window, as shown in Figures 14 and 15, which has its own set of pull-down menus. These menus allow various facilities. The user for example can enable the *Zoom Figure - On* option in order to "zoom in" feature, using the left mouse button and the right button to "zoom out".

The user can also arrange display properties of his (her) plot : fix labels or arrange the scaling of the axis (Figure 14), save the Figure (as a PS file in a predefined path).

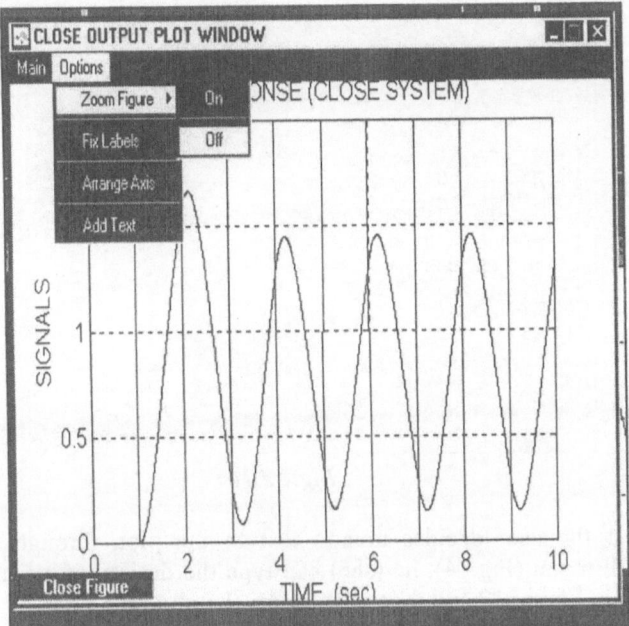

Figure 14

The user is able to resize his (her) plot in order to print it directly, using a dialogue box as shown in Figures 15 and 16, determining the desired printable height and width of the plot.

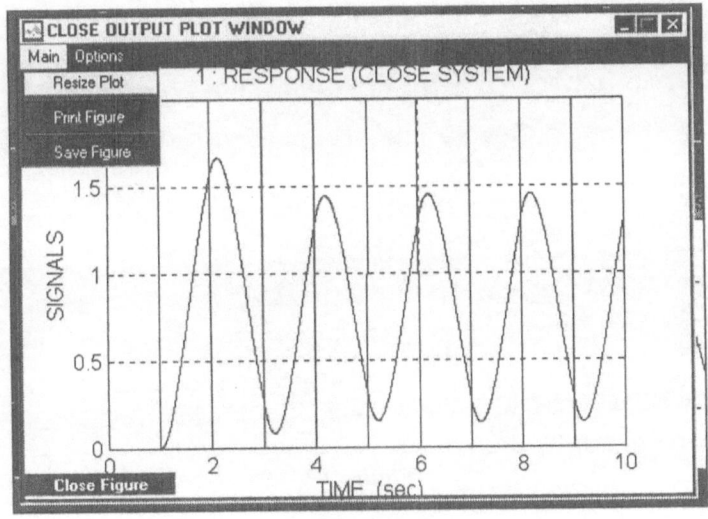

Figure 15

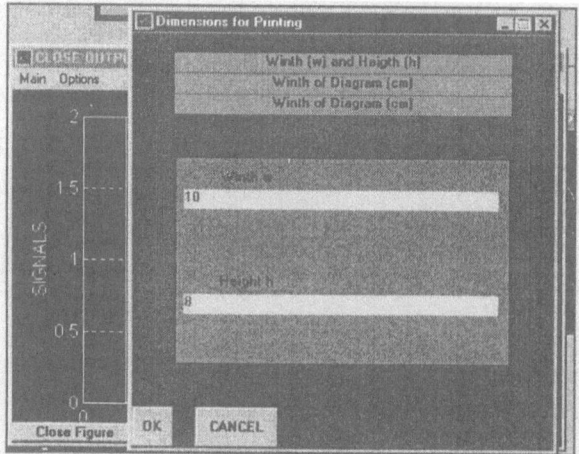

Figure 16

Finally, if the user decides to add text to the plot, through the *Add Text* selection of the menu (Fig.14), he (she) can type the desired string in the dialogue box, as shown in Figure 17 and then click inside the plot area.

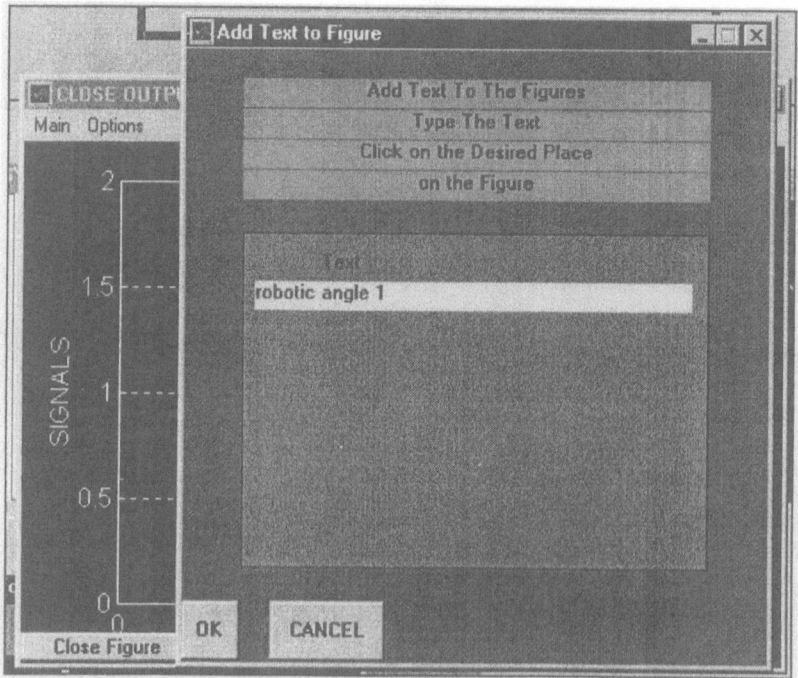

Figure 17

- *The Plot Multiple Signals* subgroup is shown in Figure 18.

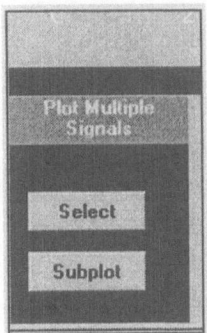

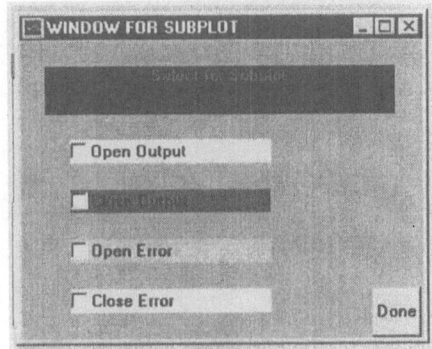

| Figure 18 | Figure 19 |

The user clicks on the *Select* button and using the dialogue box showed in Figure 19, he (she)'s able to select, the (open/closed-loop) system signals that he (she) is wants for the subplot.

Then, clicking the *Subplot* button, a pop-up window displays the desired subplot, as shown in Figure 20 (where the outputs to pulse input of both open and close system are selected).

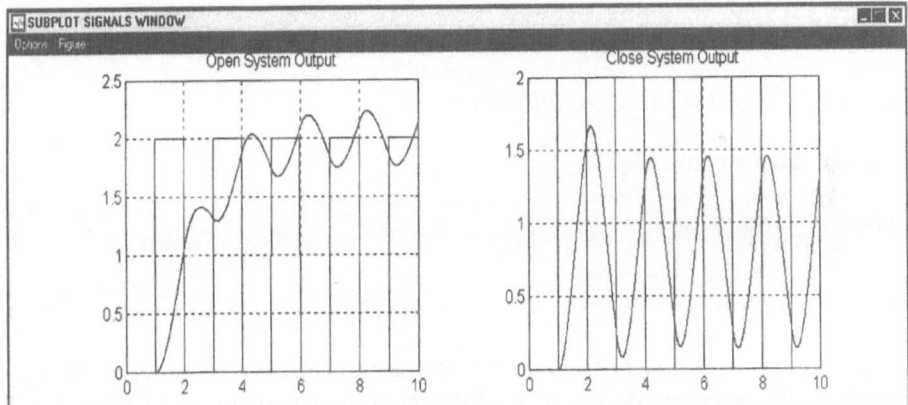

Figure 20

- *The Animation of Signals* subgroup is shown in Figure 21. These buttons combined with the radio button, for the type of requested animation (Divided animation time selection), perform the animation of selected (open/closed-loop) system response.

The user will decide if he wants to view segment by segment the animated response of the system and he will determine the time segments for this, through the dialogue box shown in Figure 22.

230

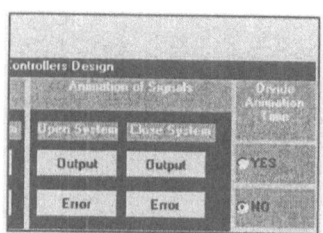
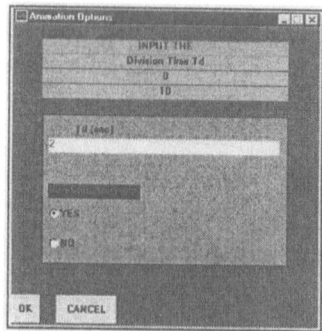

| Figure 21 | Figure 22 |

This option enables the user (student) to detect the nature of transient response and the steady-state response, through this animated version of signals, which are displayed in their own windows.

3   A Case Study

To illustrate the use of the COSAD environment and the variety of single-loop control problems, which can rapidly be studied using it, a representative example is presented.

Example :

Analytical Method of PID Controller Design [3]

Consider the plant given by:

$$G(s) = \frac{2}{s^2 + s - 2}$$

Open-loop System

The open-loop system has an unstable performance as we observe from the displayed step response of the system at the input-output windows of COSAD main screen, as shown in Figure 23.

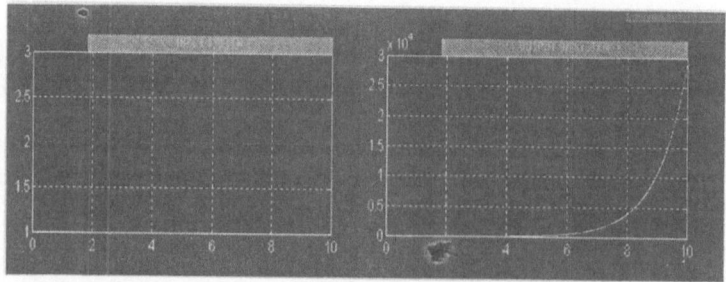

Figure 23

P - Controller

Introducing a P (proportional) controller (setting $F_1(s)$ and $F_2(s)$ to unity and $F_3(s)=k_p$) with a calculated proportional gain $k_p=4$ (calculating the transfer function of closed-loop system and properly determining the gain k_p, in order to achieve stability: $k_p > 2$), we succeed in stabilizing the closed-loop system , with a steady-state error about 0,8, as shown in Figure 24 where the user is able to view simultaneously (Figure 24a) the poles and the decreasing frequency of the close system (which explains the oscillatory behavior) and the output (Figure 24 b).

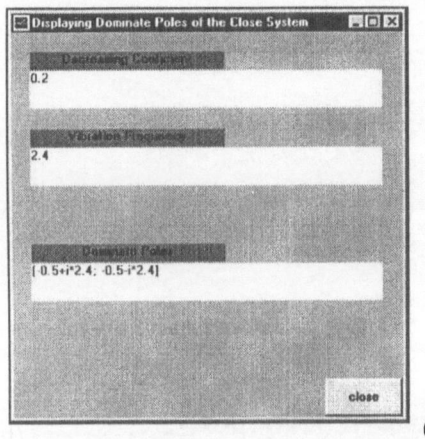

(a)

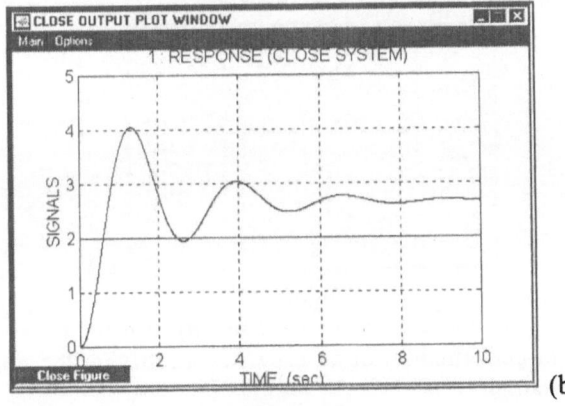

(b)

Figure 24

PI - Controller

Adding an integral term on the controller, with an Integral gain $k_i=1$ (in order to reduce the steady-state error of the closed-loop system) and simulating, one observes that the closed-loop system achieves zero value of steady-state error.

The user has the ability to view the results of plots in separate windows which can therefore be displayed simultaneously, as shown in Figure 25 (Closed-loop System Output Plot (a), Poles and Zeros Location (b), Bode Diagram (c),etc.).

232

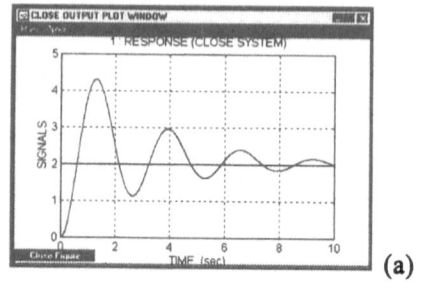 (a)

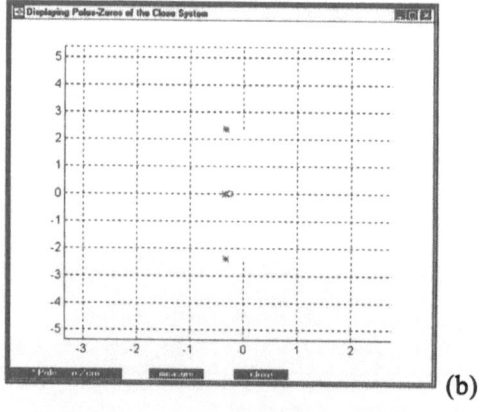 (b)

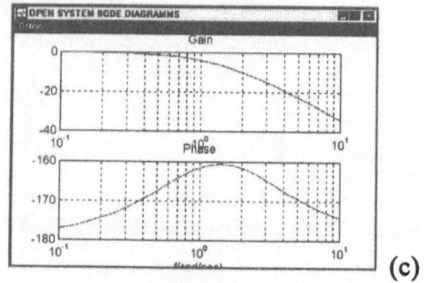 (c)

Figure 25

This feature allows the user (scientist, student) to make important remarks on the Control System performance, in an easy way (notifying the implication of the integral term for example).

PID - Controller

Finally adding a derivative term, with a derivative gain $k_d=8$, we achieve a stable behavior with zero value of steady-state error, but it is obvious that we have increased the system settling time.

The user displays simultaneously the system output (c), system poles (a), the dominating poles (b) and the Bode diagram (d), in order to see well the implication of the derivative term in the system behavior, as shown in Figure 26.

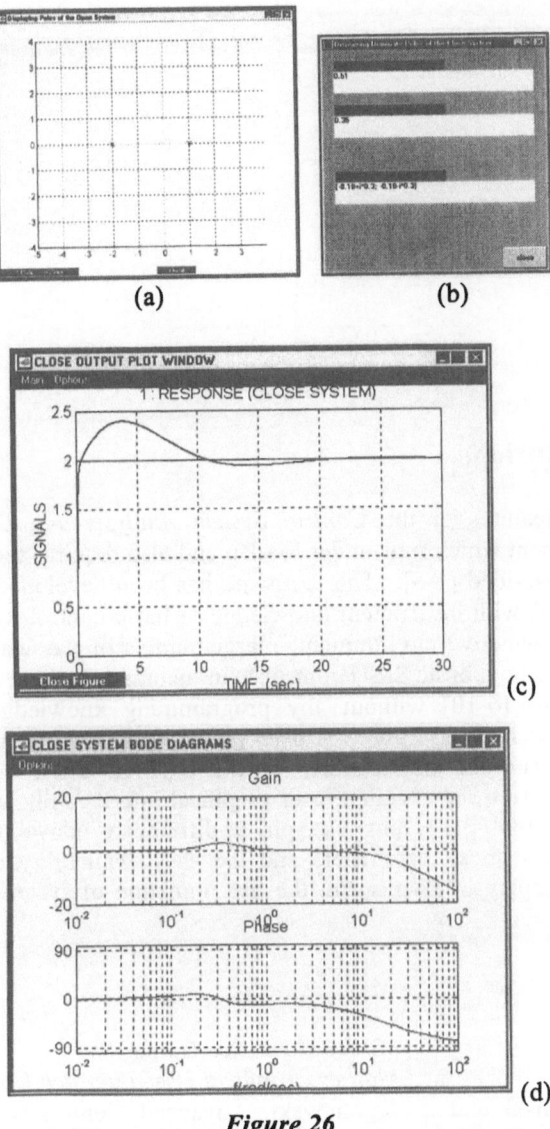

Figure 26

The user can experiment easily by changing the value of the derivative gain k_d and observing the system performance for a pulse system (with selected period T=10 sec) input.

For example, choosing $k_d=1$, we can achieve a satisfactory performance, as shown in Figure 27, where we have used the Subplot option of the COSAD environment, in order to include in the same plot the system output and the produced error from the reference input trajectory .

234

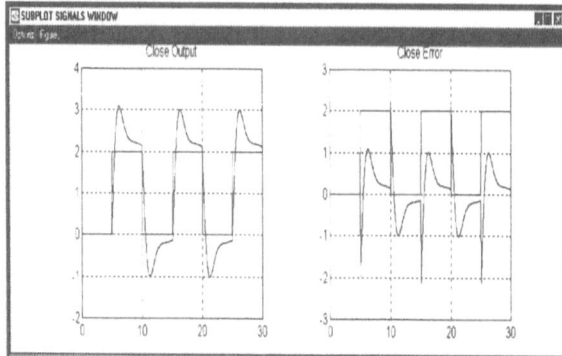

Figure 27

4   Conclusions

The main features of the *Control System Analysis and Design* (COSAD) graphic environment which runs under Matlab and also requires the Control System Toolbox were described [1-5]. The software has been developed so that the user (student, engineer) with insufficient knowledge of basic control theory, be able of carrying out, in a windows environment, a large number of classical control analysis and design studies on a basic SISO control loop including a plant, compensator and feedback elements [6-10] without any programming knowledge of Matlab. A representative case study example has been presented in order to illustrate both the easiness of use and the applicability of the COSAD environment. The main advantages, apart from the excellent user graphical interface for inputting data, are the availability of multiple response graphs in different windows, the evaluation of control and performance parameters and the easy availability of many options concerning the display of results and the determination of system parameters and inputs.

References

[1]   The Math Works Inc., *Matlab Reference Guide*,1993
[2]   The Math Works Inc., *Matlab Graphical User Interface Utili-ties*,1993
[3]   J. Móscinski and Z. Ogonowski, Advanced Control with Matlab and Simulink, *Ellis Horwood Publishing Company*,1995,
[4]   S. Nakamura, Numerical Analysis and Graphic Visualization with Matlab, *Prentice Hall PTR,1996*
[5]   Microsoft Corporation, The Windows Interface-An Application Design Guide, *Redmond*, WA, 1992
[6]   O.B. Sorensen and D. Atherton, Classical Control Using the Control Kit, *SAMS*, Vol. 21, pp. 293-303, 1995
[7]   K. Maekawa and G. K. H. Pang, Control System Design Automation for Mechanical Systems, *JINT*
[8]   A. A. Voda and I. D. Landau, A Method for the Auto-callibration of PID Controllers, *Automatica*, Vol. 31, No. 1,pp. 41-53, 1995

[9] S. G. Tzafestas (ed.),Applied Digital Control, *North Holland*, Amsterdam, 1985

[10] S. G. Tzafestas and J. K. Pal (eds),Real Time Microcomputer Control of Industrial Processes, *Kluwer*, Dordrecht/Boston, 1990

Appendix : COSAD MAIN SCREEN

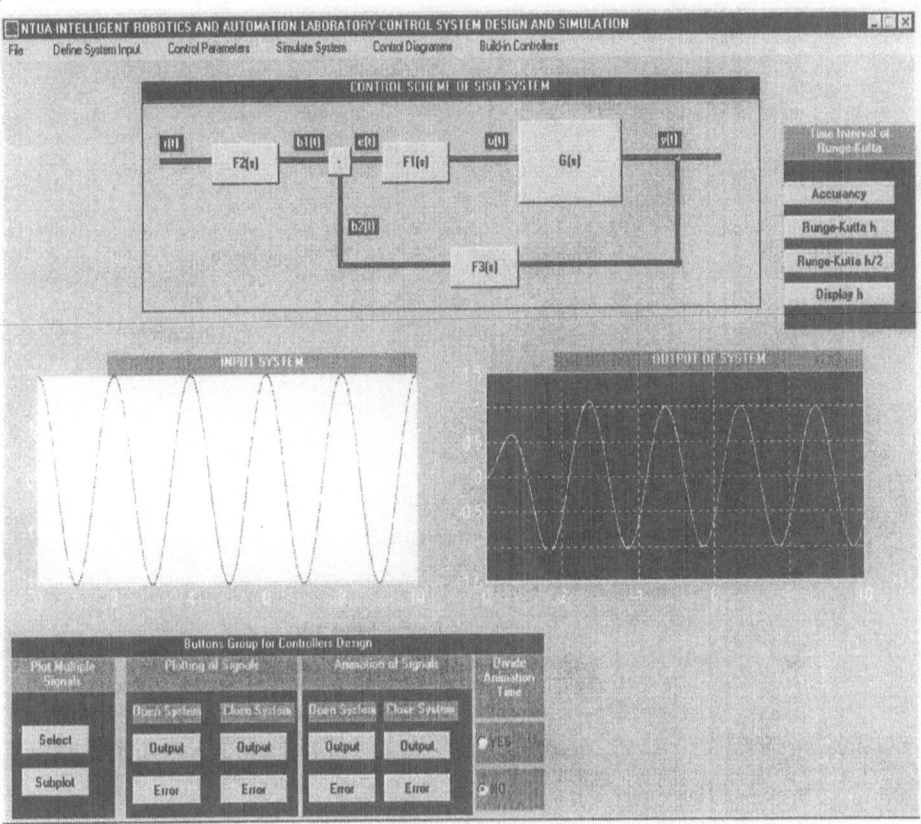

PART III

QUANTITATIVE FEEDBACK THEORY (QFT) CONTROL SYSTEM DESIGN

20

Smith Predictor for Uncertain Systems in the QFT Framework

M. Garcia-Sanz and J.G. Guillen

1 Introduction

Design of controllers for systems with a dominant time delay are notoriously difficult. In the last few decades many research efforts have focused that particular problem, especially in process control. The dynamic behaviour of many industrial processes contains time delays. The presence of time delays greatly increases the difficulty to achieve satisfactory performance of feedback controllers. For open-loop stable processes, the tracking problem can be improved substantially by introducing dead-time compensation. Perhaps the most popular scheme for those systems is the Smith Predictor Controller *SPC*, which was early introduced by O.J.M. Smith [1] in 1957. The *SPC* contains a model of the process with time delay in an inner loop (Figure 1), and can be easily implemented by using low cost digital micro-controllers.

On the other hand, speaking generally any mathematical model of a real process is based on a set of parameters q_i that are usually estimated between some lower and upper bounds q_i^- and q_i^+, due to the uncertainty that affects the description of the plant,

$$\mathbf{P} = \left\{ P(\mathbf{q}) \mid q_i \in \left[q_i^-, q_i^+ \right], \ i = 1,2,3,...l \right\} \tag{1}$$

In this context it is well known that the Smith Predictor technique may be very sensitive to model-plant mismatch, either in the time delay or in the rational part of the model. Under those circumstances considerable stability margins may not guarantee stability even with small modelling errors, [2], [3], [4] and [5].

An increasingly used engineering technique for robust control design is the Quantitative Feedback Theory (QFT), [6] and [7]. It is a frequency domain methodology that tries to achieve robust stability and robust performance over a specified region of model uncertainty.

Accordingly with these ideas, this paper presents two criteria for the design of a Smith Predictor Controller when the plant –rational part and time delay- is not precisely known. The first criterion is based on bandwidth frequency considerations and the second one introduces some guidelines to improve the design of the *SPC* by using the QFT technique. As a result, the present paper shows a two-step methodology of design, based on the above two criteria.

2 Structure and Basic Properties of the Smith Predictor

The basic control structure of the *SPC* is the model-based approach shown in Fig. 1, where $G(s) \exp(-s\theta)$ represents the real plant and $C(s)$ a conventional controller. A model of the open-loop process $\hat{G}(s) \exp(-s\hat{\theta})$ is used in an additional feedback loop in order to obtain an open-loop feedback signal that carries current and not delayed information.

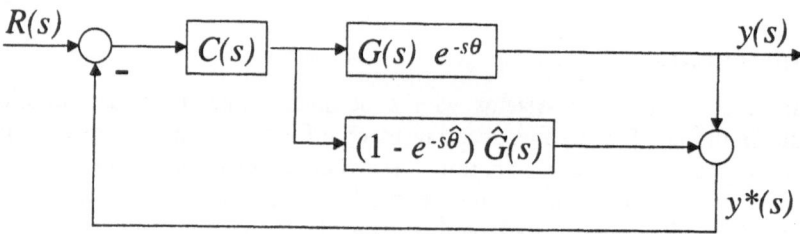

Figure 1. Smith Predictor Controller Structure

A simple analysis of that structure shows that, if $\hat{\theta} = \theta$ and $\hat{G}(s) = G(s)$, the time delay is eliminated from the characteristic equation. Thus the controller $C(s)$ can be synthesised without considering delay. This allows the designer to improve the closed-loop performance with respect to those achievable by a feedback regulator directly designed for the original plant. In this manner, the time response of the closed-loop system with a *SPC* will thus have the same shape as the response of the closed-loop system without the time delay compensated by $C(s)$; the only difference is that the output will be delayed by θ seconds. Indeed, in [8] it has been shown that the Smith Predictor structure provides a significant phase lead which justifies the above statement.

However, the real plant is never known exactly, and the actual behaviour and stability depend on all terms without cancellation. In this context, Palmor [3] and Marlin [9] introduce some rules to achieve practical and asymptotical stability and zero steady state offset with *SPC*'s without a perfect model.

3   Methodology of Design

The above *SPC* diagram can be rearranged in an equivalent structure, as it is shown in Figure 2.

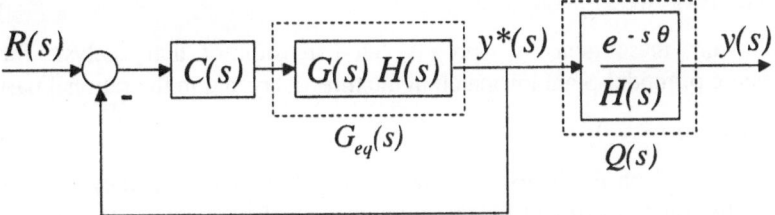

Figure 2. Equivalent Diagram of the Smith Predictor

where the expressions of the blocks are shown in the equations (2), (3) and (4).

$$H(s) = \left(1 - e^{-s\hat{\theta}}\right)\left(\frac{\hat{G}(s)}{G(s)}\right) + e^{-s\theta} \tag{2}$$

where, $G(s) \in \left[G(s)^-, G(s)^+\right]$ and $\theta \in \left[\theta^-, \theta^+\right]$

$$Q(s) = \frac{e^{-s\theta}}{H(s)} \tag{3}$$

$$G_{eq}(s) = G(s)\, H(s) = \left(1 - e^{-s\hat{\theta}}\right)\hat{G}(s) + G(s)\, e^{-s\hat{\theta}} \tag{4}$$

A previous analysis of the equivalent *SPC* structure showed that, if there is not uncertainty in the model, that is to say $H(s) = 1$, the time delay is eliminated from *y*(s)*. However, if there is some amount of model-plant mismatch, then the expression *H(s)* will be different from one. As a consequence, the previous control system behaviour will be affected by the uncertainty through *H(s)*, and particularly through the blocks $G_{eq}(s)$ and *Q(s)*. In this case, the selection of the model $\hat{G}(s)\exp(-s\hat{\theta})$ has a great impact on the set of values that *H(s)* adopts when the parameters of the model vary over the space of uncertainty. For that reason, it is necessary to analyse how to select the model of the *SPC*, and how can affect that choice in the control system behaviour.

The methodology of design proposed in this paper divides the study in two complementary steps. The first one analyses the influence of the selected model in the control system through the block *Q(s)*. The second step studies how affect the selection of the model in the design of the controller *C(s)* of the system $G_{eq}(s)$, by using the QFT technique.

First Step

The *H(s)* term that appears in the *Q(s)* block, out of the control loop (Figure 2), is the responsible of the deterioration of the system characteristics. Some years ago, a brief paper written by Santacesaria and Scattolini [10] introduced this problem and proposed a graphic solution for the simple case of model-plant mismatch in the time delay,

$$\hat{\theta} \neq \theta \quad \text{and} \quad \hat{G}(s) = G(s) \tag{5}$$

Now, following these ideas, we propose an extension of that analysis for the complete case of model-plant mismatch in the time delay and in the rational part,

$$\hat{\theta} \neq \theta \quad \text{and} \quad \hat{G}(s) \neq G(s) \tag{6}$$

The study is based on the analysis of the magnitude of the frequency response of $Q(j\omega)$ over the range of ω and for the complete set of parameter uncertainty. Any increase in the model-plant mismatch moves the resonance peaks to lower frequencies, reducing the bandwidth of the system. Accordingly with that idea, the *First Step* will search for the models of the plant $\hat{G}(s)\exp(-s\hat{\theta})$ that fulfil the desired bandwidth BW of the system, for the complete set of $|Q(j\omega)|$ over the space of uncertainty. The computer program that carries the *First Step* out is illustrated in Table I. It is a general procedure that can be used for any model of the plant, with any sort of rational part and any time delay, and for any kind of model-plant mismatch.

Table I. Outline for the *First Step* computer program

Item	*Operation*
1	Fix a desired closed-loop bandwidth BW
2	Select a model of the plant: $Mod_i = \left[\hat{G}(s)\exp(-s\hat{\theta})\right]_i$
3	Plot, for the selected model Mod_i, the magnitude of the frequency response of $Q(j\omega)$ for every plant of the space of parameter uncertainty.
4	If any $|Q(j\omega)|$ of the possible plants over the space of parameters uncertainty exhibits some amplification or attenuation about ±3 dB at frequencies lower than the desired bandwidth BW, then a deterioration of the system characteristics will be expected. Therefore, the selected model Mod_i must be rejected.
5	If every $|Q(j\omega)|$ of the possible plants over the space of parameters uncertainty does not exhibit some amplification or attenuation about ±3 dB at frequencies lower than the desired bandwidth BW, then no deterioration of the system characteristics will be expected. Therefore, the selected model Mod_i could be adopted.
6	Repeat from item 2 to item 5 for every possible models of the plant Mod_i selected over the space of parameter uncertainty.
7	Finally, a set of selectable models of the plant Mod_i that fulfils the desired bandwidth BW is obtained.

The procedure finds the set of models of the plant so that, if the *SPC* adopts one of them, the *Q(s)* block will not reduce the desired bandwidth BW. In this manner, any deterioration of the system behaviour due to the parameter uncertainty, within the frequency range (0 → BW), will be avoided.

Second Step

If there is a collection of possible models of the plant that fulfils the desired bandwidth BW (*First Step*), then an additional degree of freedom is still available in the selection of the *SPC* model. This additional degree of freedom will be used as a second criterion in order to improve the design of the *C(s)* controller by using the QFT methodology.

QFT templates are 'geometric' representations of the magnitude and phase uncertainty of the open loop transmission function *L(s)* in the Nichols chart. Since the compensator *C(s)* has no uncertainty, the whole uncertainty of *L(s)* is produced by the process. As it has been shown, the process that we want to control is $G_{eq}(s)$ -equation (4)-. In this context the open loop transmission function is given by,

$$L(s) = C(s)\, G_{eq}(s) \tag{7}$$

The aim of the *Second Step* deals with the analysis of the change suffered by the QFT plant templates when a *SPC*, with a particular model $\hat{G}(s)\exp(-s\hat{\theta})$, is included in the controller structure. In this case, QFT templates calculated for the same frequency and corresponding to two different models, can change their shapes drastically, producing different loop-shaping stages and hence very different control structures.

Table II. Outline for the *Second Step* computer program

Item	Operation
1	Select a representative model of the plant, $Mod_i = \left[\hat{G}(s)\exp(-s\hat{\theta})\right]_i$, from those that have successfully passed the *First Step* on.
2	Plot the QFT templates of the equivalent plant $G_{eq}(s)$ over the frequency range of interest.
3	Calculate the area of the templates.
4	Repeat from item 1 to item 3 for every possible model Mod_i selected over those who have successfully passed the *First Step* on.
5	Select the model that produces the smaller area of the templates set over the frequency range of interest.
6	As a result, the best model Mod_i of the plant, that is to say, the model that produces the smaller area of the templates set, is selected.

These ideas lead us to make the question that which is the optimal model of the plant that we have to choose within the set of models given above for the *First Step*, in order to improve the resulting templates, i.e. to make easier the loop-shaping. Initial investigations and the experience of many experimental simulations suggest that the smaller area the set of templates presents, the easier the controller design is.

The computer program that carries the *Second Step* out is illustrated in Table II. Again, it is a general procedure that can be used for any model of the plant, with any sort of rational part and any time delay, and for any kind of model-plant mismatch.

The model of the plant, selected for the *SPC* structure with the proposed methodology, will fulfil a desired bandwidth BW (*First Step*) and will present the less restrictive templates to the controller design stage (*Second Step*).

4   A Synthesis Example

For a simple example to clarify the above ideas, we consider the plant,

$$\text{Plant}: \ G(s)\,e^{-s\theta} = \frac{K}{\tau s+1}\,e^{-s\theta} \tag{8}$$

which captures the essential dynamics of many chemical, biological and industrial processes. Obviously, the mathematical model that describes the behaviour of that plant is the first order model with delay given by,

$$\text{Model of the Plant}: \ \hat{G}(s)\,e^{-s\hat{\theta}} = \frac{\hat{K}}{\hat{\tau} s+1}\,e^{-s\hat{\theta}} \tag{9}$$

For the present example, the parameters of the plant will adopt values that belong to the space of uncertainty given by,

$$\hat{K} \in [1,2];\ \hat{\tau} \in [1,2] \text{ and } \hat{\theta} \in [1,2] \tag{10}$$

First Step

As it was mentioned above, the *First Step* of the methodology is based on the analysis of the magnitude of the frequency response of $Q(j\omega)$ over the range of ω and for the complete set of parameter uncertainty. For the proposed plant and model (eq. 8, 9 and 10), the magnitude of the frequency response of $Q(j\omega)$ can be written as,

$$|Q(j\omega)| = \frac{1}{\sqrt{(1+M\,x)^2 + (M\,y)^2}} \tag{11}$$

where,

$$M = \left(\frac{\hat{K}}{K}\right)\left(\frac{1}{1+\omega^2\hat{\tau}^2}\right) \tag{12}$$

$$x = (1+\omega^2\ \tau\ \hat{\tau})\left[\cos\omega\theta - \cos\omega(\hat{\theta}-\theta)\right] - \omega\ (\tau-\hat{\tau})\left[\sin\omega(\hat{\theta}-\theta)+\sin\omega\theta\right] \quad (13)$$

$$y = (1+\omega^2\ \tau\ \hat{\tau})\left[\sin\omega(\hat{\theta}-\theta)+\sin\omega\theta\right] + \omega\ (\tau-\hat{\tau})\left[\cos\omega\theta - \cos\omega(\hat{\theta}-\theta)\right] \quad (14)$$

Now, fixing the desired closed-loop bandwidth to BW = 0.5 rad/sec (Item 1 of Table I) and running the computer program of the *First Step* (Items 2 to 7 of Table I), we can find the set of selectable models Mod_i of the plant that fulfils the desired specification, $\left|20\log_{10}\left|Q(j\omega)\right|\right| \leq 3$ db , over the range of frequency described by the bandwidth (0 → BW = 0.5 rad/sec).

The result of the *First Step* of the methodology for this example can be plotted as a 3D object, where the three axes represent the uncertainty of every parameter of the model (Figure 3). The set of models that fulfils the desired bandwidth and hence the *SPC* could adopt, are those which are allocated inside the 3D figure of the space of parameters, as it is shown in Figure 3.

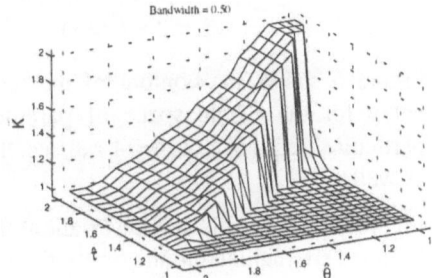

Figure 3. Parameters. First order model with delay. (BW = 0.50)

As a complement, the set of models outside the 3D figure represents the cases that do not fulfil the bandwidth specification..

Naturally, the larger the bandwidth specification is, the smaller the set of possible models is. Accordingly with that idea, Figures 4 and 5 show the 3D object obtained for a bandwidth specification of BW = 0.55 rad/sec and BW = 0.60 rad/sec respectively.

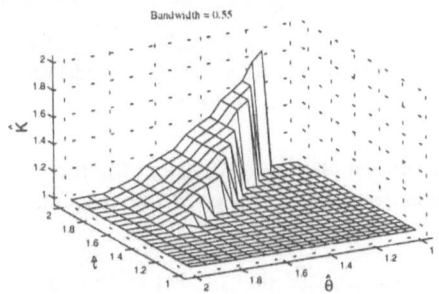

Figure 4. Parameters. First order model with delay. (BW = 0.55)

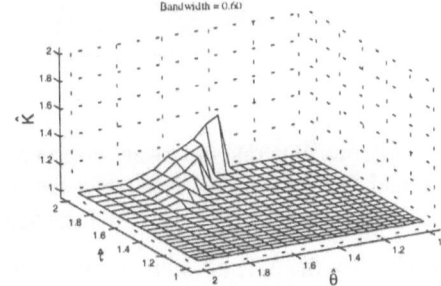

Figure 5. Parameters. First order model with delay. (BW = 0.60)

In the limit, if the bandwidth specification is larger enough, we can presume that there will be only one possible model.

246

Second Step

If there is a collection of possible models of the plant that fulfils the desired bandwidth of the first criterion, then an additional degree of freedom is still available in the selection the model of the *SPC*.

Assuming that premise, we select three models inside the 3D figure obtained for a bandwidth of BW = 0.50 rad/sec (Figure 3), given by,

$$\text{Case A} : (\hat{K} = 1; \ \hat{\tau} = 1.1157; \ \hat{\theta} = 2)$$
$$\text{Case B} : (\hat{K} = 1; \ \hat{\tau} = 2; \ \hat{\theta} = 1.9284) \tag{15}$$
$$\text{Case C} : (\hat{K} = 1; \ \hat{\tau} = 2; \ \hat{\theta} = 1)$$

Figure 6 shows the contour of the template of the original first order plant with delay for the whole space of parameter uncertainty (eq. 9 and 10) and for a particular frequency of $\omega = 1$ rad/sec. The three selected models (eq. 15) are shown as well.

Figure 7 presents the shape variation of the templates of $G_{eq}(s)$ when the models A, B and C are included in the Smith Predictor structure, also for a particular frequency of $\omega = 1$ rad/sec. Notice that the shape of the templates depends strongly on the selected model Mod_i of the *SPC*.

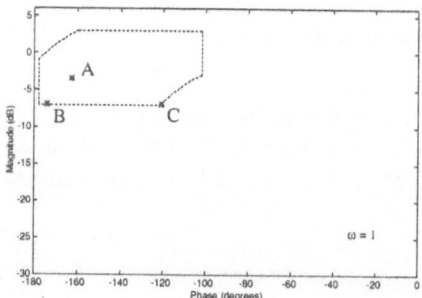

Figure 6. Template. First order model with delay. ($\omega = 1$)

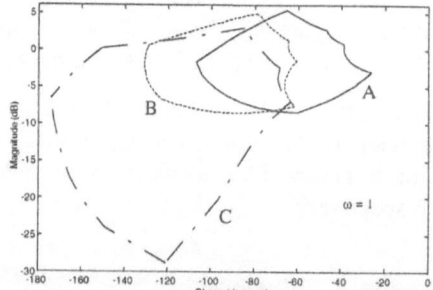

Figure 7. Templates. First order model with delay + Smith Predictor A, B or C

In order to illustrate the ideas proposed in the second criterion, we select the models A and C. Figures 8 and 9 show the templates for both cases over the frequency range of $\omega = [\,0.1 \quad 0.4 \quad 0.7 \quad 1\,]$ rad/sec.

Following the *Second Step* computer program, we can see that the set of templates of the case A presents a smaller area than the set of templates of the case C. For this reason and looking for the best controller design, we select the case A instead of the case C.

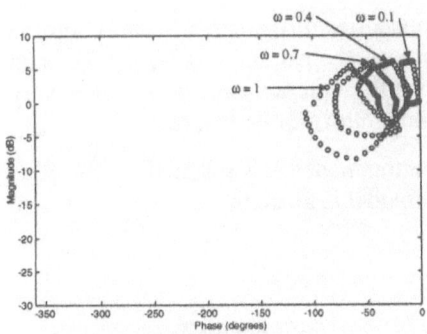

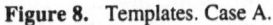

Figure 8. Templates. Case A.

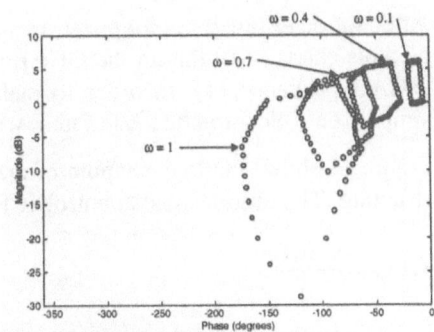

Figure 9. Templates. Case C.

To validate that choice, a complete design of the *C(s)* controller with the QFT technique, for both cases A and C, is presented. Equations (16), (17) and (18) exhibit the desired specifications for the control system. They are also shown in Figure 10.

$$\left| \frac{C(s)\,[G(s)\,H(s)]}{1+C(s)\,[G(s)\,H(s)]} \right| \le \text{Ws1}, \quad \text{where Ws1} = 1.2$$

(16)

for $\omega = [0.01\ 0.1\ 0.3\ 0.5\ 1]\,\text{rad/sec}$

$$\left| \frac{C(s)\,[G(s)\,H(s)]}{1+C(s)\,[G(s)\,H(s)]} \right| \le \text{Ws6}, \quad \text{where Ws6} = 0.5$$

(17)

for $\omega = 0.5\,\text{rad/sec}$

$$\text{Ws7a} \le \left| \frac{C(s)\,[G(s)\,H(s)]}{1+C(s)\,[G(s)\,H(s)]} \right| \le \text{Ws7b}$$

where : $\quad \text{Ws7a} = \left| \dfrac{0.08}{s^3 + 2.4\,s^2 + 0.84\,s + 0.08} \right|$

(18)

$$\text{Ws7b} = \left| \frac{0.0040\,(s^2 + 22\,s + 40)}{s^2 + 0.36\,s + 0.16} \right|$$

for $\omega = [0.01\ 0.1\ 0.3\ 0.5]\,\text{rad/sec}$

The specifications have been selected for a definite set of frequencies, also shown in the above equations. The boundary constraints of the control system, calculated taking into account the space of parameter uncertainty (eq. 10) and the mentioned performance specifications (eq. 16, 17, 18), are illustrated in Figures 11 and 12, for the cases A and C respectively.

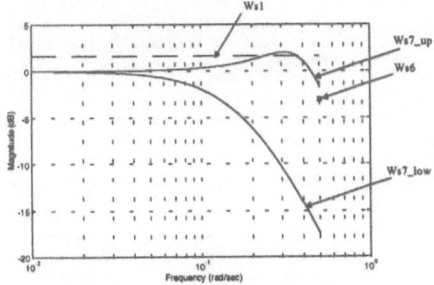

Figure 10. Controller Specifications.

248

After we determined the loop transmission boundaries of the cases A and C on the Nichols chart, according to the QFT procedure, we shaped $L(s) = C(s) G_{eq}(s)$ using a CAD package [11]. In order to make easier the comparison, we select a very simple controller structure k/s (a gain with an integrator) for both cases.

Figure 11 shows the loop shaping of $L(s)$ when the model A is selected for the *SPC* structure. The best-designed controller for that model is given by,

$$C(s) = \frac{0.1389}{s} \tag{19}$$

In addition, Figure 12 shows the loop shaping of $L(s)$ when the model C is selected for the *SPC* structure. The best-designed controller for that model is given by,

$$C(s) = \frac{0.1671}{s} \tag{20}$$

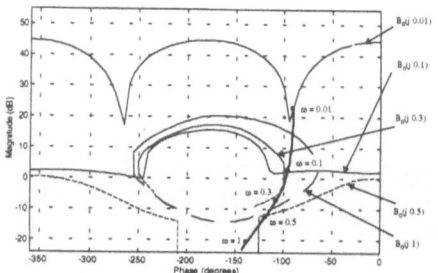

Figure 11. Loop-shaping. Case A.   **Figure 12.** Loop-shaping. Case C.

The controllers (22) and (23) are the best possible k/s expressions obtained with the CAD package [11] for the cases A and C respectively. Looking at the loop-shape obtained by both controllers, we can conclude that the first one fulfils the bounds constrains, while the second one does not. This means that the selected model of the plant, used in the *SPC,* not only modifies the templates and the loop shaping stage, but also can affect the controller structure.

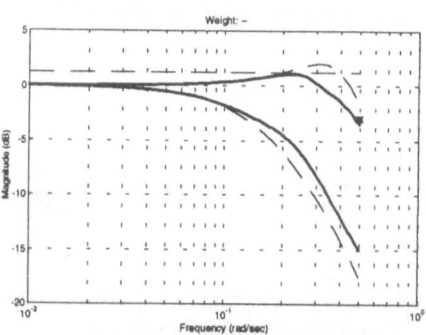
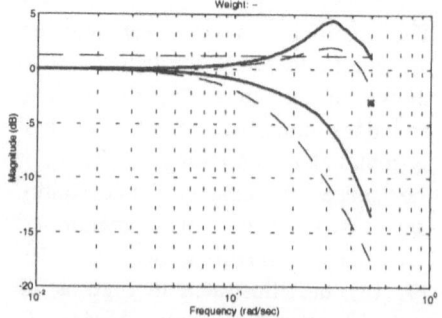

Figure 13. Analysis. Case A.   **Figure 14.** Analysis. Case C.

Finally, Figures 13 and 14 show the worst -over all uncertainty cases- closed loop response magnitude versus the specifications, again for the models A and C respectively. The dashed lines represent the performance specifications Ws1 and Ws7; the specification Ws6 is plotted as a simple point; and the solid lines show the response of the control system.

As one might expect, the first trial –model A- presents a good performance, always within the frequency specifications (Figure 13), while the response of the control system with the model C does not fulfil the required specifications (Figure 14).

5 Conclusions

It is well known that the Smith Predictor Controller *SPC* may be very sensitive to model-plant mismatch, resulting in a poor performance when model uncertainty is present. This paper introduces two criteria for the design of a *SPC* when the plant –rational part and time delay- is not precisely known.

The first criterion (*First Step*) is based on bandwidth frequency considerations. The procedure finds the set of models of the plant so that, if the *SPC* adopts one of them, the desired bandwidth BW will not be reduced by the effect of the parameter uncertainty.

The second criterion (*Second Step*) introduces some guidelines to improve the design of the *SPC* by using the QFT technique. That lead us to make the question that which is the optimal model of the plant that we have to choose in order to improve the resulting templates, i.e. to make easier the loop-shaping, and hence to present the less restrictive system to the controller design stage.

Acknowledgements--- The authors gratefully appreciate the support given by the Spanish 'Comisión Interministerial de Ciencia y Tecnología' (CICYT) under grant TAP'97-0471.

6 References

1. Smith, O.J.M., 1957. Closer control of loops with dead time. *Chem. Eng. Progr.*, **53**, 217-219.

2. Ioannides, A.C., G.J. Rogers and V. Latham, 1979. Stability limits of a Smith controller in simple systems containing a dead-time. *Int.J.Control*, **29**, 557-563.

3. Palmor, Z., 1980. Stability properties of Smith dead-time compensator controllers. *Int. J. Control*, **32**, 937-949.

4. Horowitz, I., 1983. Some properties of delayed controls (Smith regulator). *Int.J.Control*, **38**, 977-990.

5. Yamanaka, K. and E. Shimemura, 1987. Effects of mismatched Smith controller on stability in systems with time-delay. *Automatica*, **23**, 787-791.

6. Horowitz, I., 1991. Survey of quantitative feedback theory. *Int. J. Control*, **53** (2), 255-291.

7. D'Azzo, J. J. and C. H. Houpis, 1995. Quantitative Feedback Theory (QFT) Technique. In: *Linear Control System Analysis and Design*. 4th Ed. (McGraw Hill, New York), 580-635.

8. Astrom, K.J., 1977. Frequency domain properties of Otto Smith regulators. *Int.J.Control*, **26**, 307-314.

9. Marlin, T.E., 1995. *Process Control*. (McGraw-Hill, Inc.), 620-634.

10. Santacesaria, C. and R. Scattolini, 1993. Easy tuning of Smith Predictor in Presence of Delay Uncertainty, *Automatica*, **29** (6), 1595-1597.

11. Borghesani, C., Y. Chait and O. Yaniv, 1995. *Quantitative Feedback Theory Toolbox - For use with MATLAB*, 1st Ed. (The MathWorks Inc.).

21

Nonlinear QFT Based on Local Linearization

A. Baños, O. Yaniv and F.J. Montoya

1 Introduction

The basic idea of nonlinear QFT is to substitute an uncertain nonlinear plant by an equivalent linear family (ELF) of plants and attached set of disturbances. In this way, the non-linearity is transformed in two types of uncertainty to a linear equivalent counterpart: structured model uncertainty in the form of templates, giving by the set of equivalent linear plants, and unstructured uncertainty in the form of (input) disturbances on these linear plants. If in addition the original nonlinear plant has uncertainty, parametric or as disturbances, it adds an extra degree of uncertainty to the linear model giving larger templates and/or sets of disturbances.

Research of nonlinear QFT, initiated by Isaac Horowitz ([1,2]), has been centered in two possible ways to substitute such a nonlinear plant by an ELF and/or a set of disturbances: i) to associate a (usually hard to compute) LTI ELF to each acceptable output, and ii) to associate an (easier to compute) LTI ELF and a set of disturbances, to each acceptable output.

The first approach is hard in the sense that a systematic computation of templates can become problematical. It has been developed in [1,2] and extended in several papers, for example [3,4]. An analysis of this approach, including solutions to some difficulties, can be found in [5]. The second approach circumvents the problem of template computation, at the cost of loosing structure in the uncertainty, meaning, in general, more conservative results.

However, the second approach leaves a very open area, depending on the different techniques employed in computing ELF. The tractability/conservatively balance will depend on the ELF/set of disturbance interplay. Some different techniques, belonging to this general approach, have been developed in the literature ([6-8]).

2 Nonlinear QFT Based on Local Linearization

The main result of this work is a new technique belonging to the second type of approaches, as described in the last section, that is, associating an LTI ELF and a set of disturbances to each acceptable output. The difference with previous work is the way in which the set ELF is computed: using local linearization around an acceptable trajectory of the nonlinear system. Local linearization around a trajectory gives as direct result a linear time-varying set, being a "frozen" version of this set the final LTI ELF that is more convenient to use for design. Apparently, this approach is valid for an interesting class of problems. In this work we are mainly interested in showing an informal description of the technique, as well as a comparison with some previous QFT technique. A more rigorous theoretical foundation will be given elsewhere.

It should be noted that more traditional local linearization, usually about equilibrium points instead of system trajectories, can be easily integrated in this technique, since an equilibrium point $y(t) = y_e$ can always be expressed as a trajectory $\{y(t) = y_e, dy(t)/dt = 0, ...\}$. The technique is described in the following, using an example for illustrating its different steps. The example has been considered before in [5] and will be detailed in Section 3.

1. **Definition of the (time-varying) Equivalent Linear Family (ELF).** Let N be the original nonlinear plant. Define A_r as the set of closed loop acceptable outputs for the reference r. For each element $y(t)$ of A_r, define $P^{y(t)}$ as the time-varying local linearization of N around $y(t)$ (local linearization around a trajectory).

   Example (nonlinear RC circuit, details in [5]):

   $$N:u \rightarrow y, \quad y(t) + (1 + ay^2(t))y(t) = u(t), a \in [0,2]$$

   $$A_r = \{y(t)|0.9(1 - e^{-8t})u(t) \le y(t) \le 1.1(1 - e^{-12t})u(t)\}$$

   $$P^{y_a(t)}:u \rightarrow y, \quad \dot{y}(t) + (1 + 3ay_a^2(t))y(t) = u(t)$$

2. **Definition of the "frozen" LTI Equivalent Linear Family:** Our ELF is finally a "frozen" version of the above time-varying set. Define P^y as

   $$P^y = \{P^{y(t)}|t \in R\}$$

   Example (LTI ELF):

   $$P^{y_a}:u \rightarrow y, \quad \dot{y}(t) + (1 + 3ay_a^2)y(t) = u(t), \quad y_a \in [0,1.1]$$

3. **Definition of disturbances set.** To each $y(t)$ of A_r we associate a disturbance signal $d^{y(t)}$

   $$u(t) = N^I y(t) = u_{ELF}(t) - d^{y(t)}(t) \Rightarrow d^{y(t)} = (P^y)^I y(t) - N^I y(t)$$

   Example (disturbances set):

   $$d^{y(t)} = (P^{y(t)})^I y(t) - N^I y(t) = \dot{y}(t) + (1 + 3ay_a^2(t))y(t) - \dot{y}(t) - (1 + ay^2(t))y(t) =$$
   $$= 3ay_a^2 y(t) - ay^3(t)$$

The validation of the design for the original plants is based on the usual application of Schauder's fixed point theorem: (i) to each $y(t)$ of A_r associate a LTI P^y (the set given in 2.) and a disturbance $d^{y(t)}$ (the set given in 3.), (ii) design a controller for $\{P^y, d^{y(t)}\}$ that produces closed loop outputs in A_r, given as a result a mapping from A_r to A_r, and (iii) under some assumptions (the mapping must be continuous and A_r convex and compact) the controller is also valid for the original nonlinear plant.

3 Comparison with Other QFT Techniques

An interesting question is a comparison of this technique with other QFT techniques, overall with the "global" linearization method. Each technique provides different ELF's and sets of disturbances, as we will see in the next cases.

I. Nonlinear RC Circuit

An uncertain nonlinear RC circuit is used [5] for analysis of the global QFT approach. Here it is used for studying the local linearization approach and, in addition for comparing both approaches. Both the resulting ELF and the set of acceptable outputs are given in Table 1 (see details in Section 2).

	ELF	Set of disturbances
Global lin.	$P^{y_a} = \{\dfrac{y_a(s)}{u_a(s)}, u_a = N^I y_a, a \in [0,2]\},$ $ELF = \{P^{y_a}, y_a \in A_r\}$	$\varnothing$
Local lin.	$\{P^{y_a}(s) = \dfrac{1}{s+1+3ay_a^2},$ $y_a \in [0,1.1], a \in [0,2]\}$	$\{d^{y(t)} = 3ay_a^2 y(t) - ay^3(t),$ $y_a \in [0,1.1], a \in [0,2], y(t) \in A_r\}$

Table 1: ELFs and sets of disturbances for global and local approaches to QFT

Design #1 (Global linearization): A detailed design is given in [5]. Here we will only comment about the results.

Design #2 (Local linearization): Following the procedure discussed in Section 2, we obtain the ELF and the set of acceptable disturbances given in Table 1. Now, the problem is to track steps and also to reject disturbances, while keeping stability. The specification for tracking is given in the formulation of the design problem, but now we need to have an estimation of the worst-case disturbance due to the nonlinearity. This means to compute an upper bound for the disturbance magnitude

$$\{d^{y(t)} = 3ay_a^2 y(t) - ay^3(t), y_a \in [0,1.1], a \in [0,2], y(t) \in A_r\}$$

254

The result is given in Fig. 3.1, where a discretization of the acceptable output set is done using a ε-net ([5]). Finally, the disturbance rejection specification is adopted as a 95% rejection of disturbances upper bounded by the bound in Fig. 3.1.

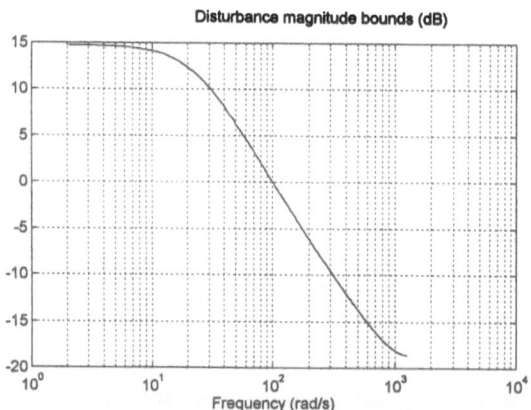

Fig. 3.1: Disturbance magnitude bounds

Then, using frequency domain specifications (tracking and relative stability), we compute templates and boundaries in the usual way. Worst-case boundaries for the nominal open loop function are given in Fig. 3.2.

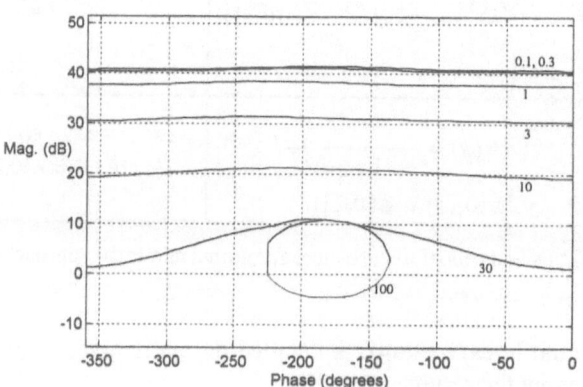

Fig. 3.2: Worst-case open-loop boundaries

Comparison of results: In Fig. 3.3, Bode diagrams for the compensators $G_1(s)$ and $G_2(s)$, corresponding to designs #1 and #2, are shown

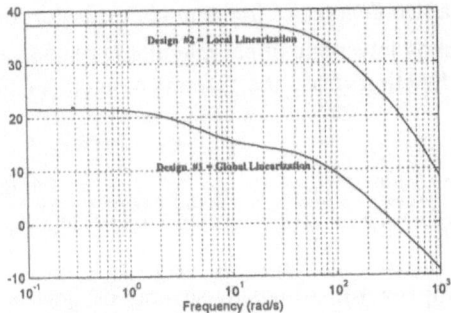

Fig. 3.3: Bode diagram of $G_1(s)$ and $G_2(s)$

A first result is the difference in terms of gain and bandwidth. It turns out that the "global linearization" technique is less conservative than the "local linearization" technique, as one would have expected. Time-domain simulations (Fig. 3.4) also confirm this, where it is clear that in the second design (Fig. 3.4.b), more feedback than needed is used for coping with the specifications.

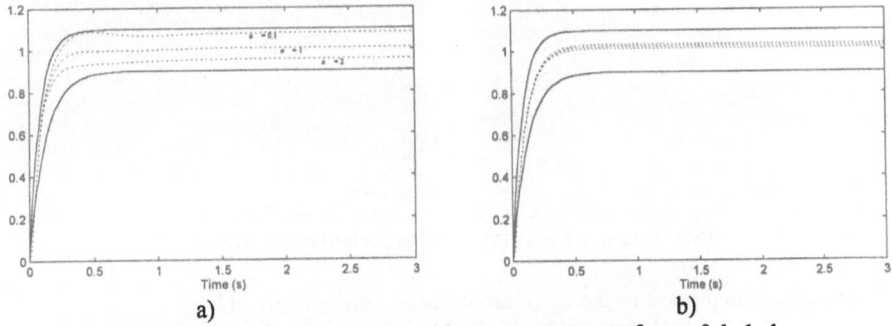

Fig. 3.4: Final nonlinear closed loop step response for a =0.1, 1, 1, using: (a) global linearization, (b) local linearization

B. Uncertain Van-der-Pol plant

In [9] some global QFT controllers are designed for an uncertain Van der Pol plant. For this plant the design is a bit more involved, since the resulting equivalent linear set contains unstable elements, and then the shaping of the feedback compensator is harder. In addition, the equivalent linear family exhibits a very large uncertainty for low-medium frequencies. Thus, a feedback controller with large gain/bandwith may be expected. We revisit here this design example, using both global and local approaches. The uncertain Van-der-Pol plant is given by:

$$\ddot{y}(t) + A\dot{y}(t)(By^2(t) - 1) + Ey(t) = Ku(t)$$

where the uncertain parameters A, B, E, and K belong to the intervals
$A \in [1,3]$, $B \in [1,4]$, $E \in [-2,-1]$, and $K \in [30,120]$

The goal is to design a two-degrees of freedom compensator for tracking steps with some tolerances, given by lower and upper magnitude bounds $a(\omega)$ and $b(\omega)$ of the closed loop transfer function, from the reference signal to the output. They are given by

$$a(\omega) = \left| \frac{4.8(s+100)}{s(s+2)(s+3)(s+4)(s+20)} \right|_{s=j\omega} \quad , \quad b(\omega) = \left| \frac{480(s+1)}{s(s+2)(s+3)(s+4)(s+20)} \right|_{s=j\omega}$$

Design #1 (global approach): Although several designs are made in [9], the computation of templates in that work is made without a systematic method. The validation of the design is then problematic, probably resulting in unnecessary overdesign. Here we made the design using a discretization of the set of acceptable outputs following the method given in [5]. Resulting templates are shown in Fig.3.5.

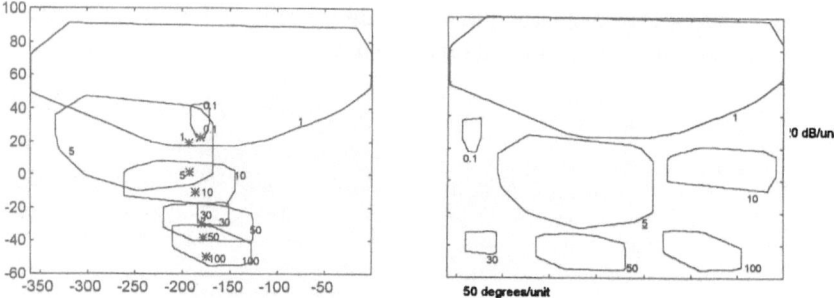

Fig. 3.5: Templates for the uncertain Van der Pol Plant
(Showing exact position -left- and relative size -right-)

where the nominal plant in the equivalent linear family is given by

$$P_0(s) = 15.3 \frac{(s+2.7)(s+80)}{(s+1)(s-1.3)(s+1.9)(s+100)}$$

Then, we compute tracking boundaries and also stability boundaries (6 dB) for the above templates. Worst-case boundaries are given in Fig. 3.6, where a shaping for $G(s)$ is also done. The design is given by

$$G(s) = 13.7 \frac{\left(1 + \dfrac{s}{7.9}\right)}{1 + \dfrac{2 \cdot 0.6}{700}s + \dfrac{1}{700^2}s^2} \quad , \quad F(s) = \frac{1}{\left(1 + \dfrac{s}{2.9}\right)\left(1 + \dfrac{s}{5.6}\right)}$$

and a simulation of the closed-loop step response is given in Fig. 3.7. Note that the closed-loop outputs do verify the specifications, but at a first sight using more feedback than needed, due to the high bandwidth of the feedback controller.

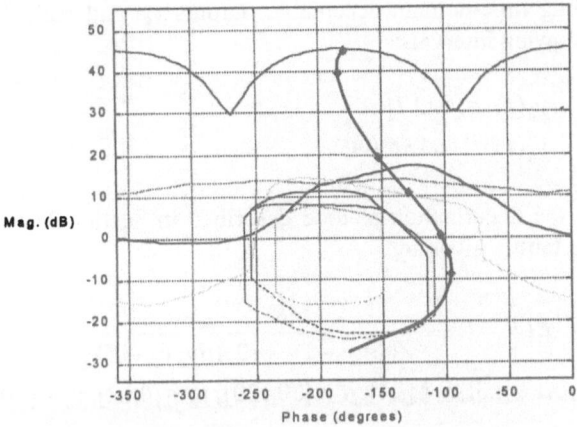

Fig. 3.6: Worst-case boundaries and loop shaping for
the uncertain Van der Pol plant

However, it should be pointed out that due to the characteristics of the Van-der-Pol plant, leading to unstable transfer functions in the ELF and large templates, the shaping of the feedback controller become very problematical and it become very hard, if not impossible, to obtain feedback compensators with less bandwidth. As a result, a modification of the specifications, allowing larger rise time and lesser overshoot, could be achieved with less control effort.

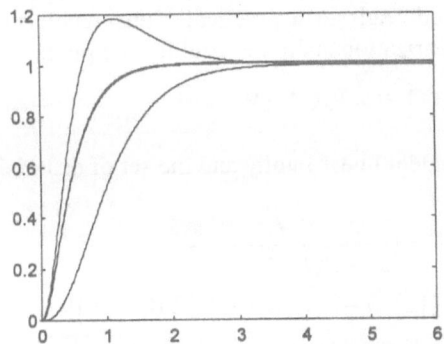

Fig. 3.7: Closed-loop time response for the uncertain Van der Pol plant

Design #2 (local approach): In the local approach, we first need to specify acceptable outputs in the time domain, since the linearization is about acceptable trajectories. The Van der Pol plant results in a second order linear plant, after linearization, thus it is necessary to specify not only bounds on the outputs, but also on its derivatives. For this end, we use the functions $a(\omega)$ and $b(\omega)$, as defined above. Acceptable outputs are defined in the frequency domain, as those signals bounded below and above by $a(\omega)$ and $b(\omega)$, and its having bounded derivatives.

The corresponding time-domain acceptable outputs y_a, and their derivatives take values in the following intervals:

$$y_a(t) \in [0,1.17]$$
$$\dot{y}_a(t) \in [-0.15,2.09]$$

Thus, following the linearization method described in Section 2, we arrived to the equivalent linear family given by

$$ELF = \{\frac{K}{s^2 + A(By_a^2 - 1)s + (2AB\dot{y}_a y_a + E)},$$
$$A \in [1,3], B \in [1,4], K \in [30,120], y_a \in [0,1.2], \dot{y}_a \in [-0.2,2.1]\}$$

and the set of disturbances given by

$$\{d^{y(t)} = \frac{AB}{K}(2\dot{y}_a y_a y(t) + (y_a^2 - y^2)\dot{y},$$
$$y_a \in [0,1.2], \dot{y}_a \in [-0.2,2.1], A \in [1,3], B \in [1,4], K \in [30,120], y(t) \in A_r\}$$

The observation of the set ELF reveals that the number of unstable poles varies along the set. This means that the number of crossings in the Nyquist or Nichols plot must change accordingly, and the shaping of a stabilizing feedback controller should be carefully executed ([6]). At this point, a direct application of the local nonlinear QFT method leads to a practically unsolvable problem. However, we may still consider linearization about the equilibrium point

$$(y_a(t) = 1, \dot{y}_a(t) = 0)$$

In this case, the equivalent linear family and the set of disturbances are given by

$$ELF = \{\frac{K}{s^2 + A(By_a^2 - 1)s + E},$$
$$A \in [1,3], B \in [1,4], K \in [30,120], y_a = 1\}$$

$$\{d^{y(t)} = \frac{AB}{K}(y_a^2 - y^2(t))\dot{y}(t),$$
$$y_a = 1, A \in [1,3], B \in [1,4], K \in [30,120], y(t) \in A_r\}$$

For this ELF, the number of unstable poles is invariant, then shaping of a feedback controller $G(s)$ is possible. Fig. 3.8 shows worst-case boundaries corresponding to tracking, relative stability (6 dB) and disturbance rejection (95%). Note that these boundaries are more demanding, overall at low frequencies, that in Design #1. As a result, we would obtain a similar feedback controller with a bit more gain/bandwidth.

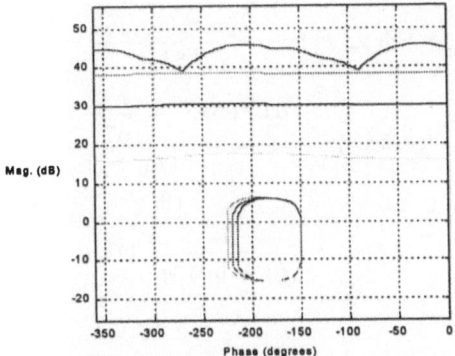

Fig. 3.8: Worst-case boundaries for the uncertain
Van der Pol plant (local linearization)

4   Example : pH Control

The dynamic model of pH process is taken from [10]. It is given by

$$\frac{dc_{H^+}(t)}{dt} = \frac{F_S c_{S_0} - u(t)c_{B_0} - (F_S + u(t))(c_{H^+}(t) - \dfrac{K_W}{c_{H^+}(t)})}{V(1 + \dfrac{K_W}{c^2_{H^+}(t)})}$$

$$y(t) = (pH) = -\log(c_{H^+}(t))$$

where there are some fixed parameters

$$V = 5.5 \; l$$
$$C_{S_0} = 0.05 \; mol \, / \, l$$
$$C_{B_0} = 0.1 \; mol \, / \, l$$
$$K_W = 10^{-14}$$

and the uncertain parameter

$$F_S \in [0.9, 1.1] \; l/min$$

Here the control input is u, which is limited to be in the interval $[0,1]$ l/min, and
he goal is to control the pH in the range $[3,11]$, with rise time in the order of 0.1
min to step changes. The linearization of the nonlinear dynamics with respect to the
equilibrium point $y_e \in [3,11]$, $d(y_e)/dt = 0$, is given after some simple computation
by (state-space model):

$$A(y_e, F_s) = -\frac{F_s + u_e(y_e, F_s)}{5.5}$$

$$B(y_e) = -\frac{0.1 + (10^{-y_e} - \dfrac{K_w}{10^{-y_e}})}{5.5(1 + \dfrac{K_w}{10^{-2y_e}})}$$

$$C(y_e) = -\frac{1}{10^{-y_e} \ln(10)}, \quad D = 0$$

where

$$u_e(y_e, F_S) = \frac{0.05 F_S - (10^{-y_e} - \dfrac{K_w}{10^{-y_e}}) F_S}{0.1 + (10^{-y_e} - \dfrac{K_w}{10^{-y_e}})}$$

The result is a set of first-order linear plants that are always stable and have a strong gain uncertainty. The effect of F_s is almost irrelevant in comparison with y_e. Specifications are translated to frequency domain as upper and lower bounds on closed loop frequency response magnitude (Fig. 4.1) :

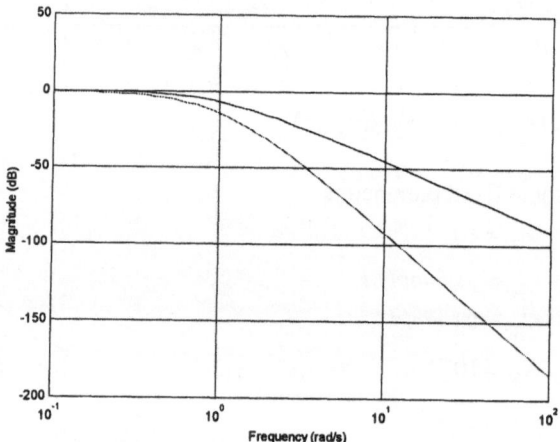

Fig. 4.1: Closed loop specifications as allowed variations on the Frequency response magnitude.

Tracking, stability and disturbance boundaries are mixed up to give worst case boundaries in Fig. 4.2, where a loop shaping is also shown. A time simulation of the resulting design is given in Fig. 4.3. The design performs well, except for the influence of the initial condition, that has not been taken account. The   uncertain parameter $F_{S0} \in [0.9, 1.1]$ is not significative compare to the nonlinearity.

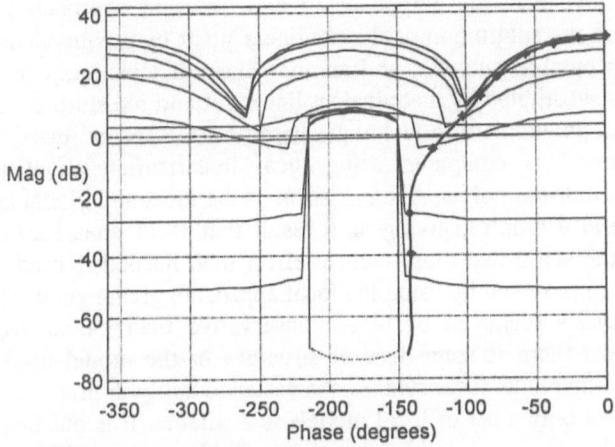

Fig. 4.2: Worst-case boundaries and loop shaping

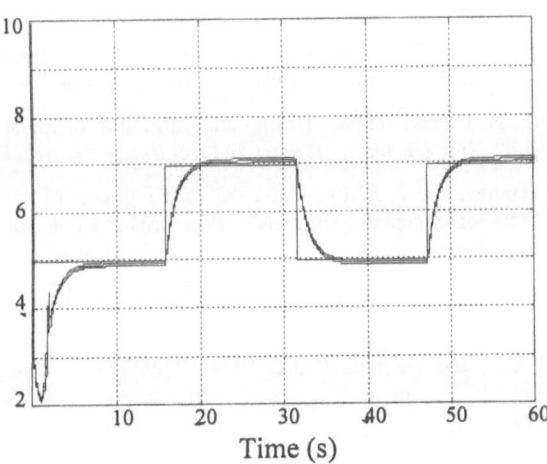

Fig. 4.3: Time-domain simulation of the closed loop response to changing pH reference (square wave)

Conclusions

Nonlinear QFT have been developed using two different approaches: a *global* approach based on the subtitution of the nonlinear plant by a equivalent linear set (depending of acceptable outputs), and another class of techniques in which the nonlinear plant is substituted by a equivalent linear set and a disturbances set. This work explores the potentials of a technique based on the second class, where the equivalent linear set is computed using local linearization about acceptable trajectories or equilibrium points. A comparison of the local and global approaches is made for several examples, giving as a result that local linearization is more conservative, in the sense that more control effort than needed is used. Finally, a pH control problem is solved by using the local approach, giving good results. As a result, local synthesis seems to be more conservative than global synthesis. A reason may be that there is some loss of structure in the model used for local synthesis. On the other side, local synthesis is easier to apply in practice, and is the unique option when only a set of local models is available. It is our belief that an integration of both techniques will be the most suitable one to achieve a solution with minimum control effort.

References

1. Horowitz, I., 1975, "A synthesis theory for linear time-varying feedback systems with plant uncertainty", *IEEE Trans. Automatic Control*, 20, 4:454-464.

2. Horowitz, I., 1976, "Synthesis of feedback systems with nonlinear and time-varying uncertain plants to satisfy quantitative performance specifications", *Proc. IEEE*, 64, 1:123-130.

3. Yaniv, O., and R. Boneh, 1997, "Robust LTV Feedback Synthesis for SISO Non-linear Plants", *Int. Journal on Robust and Nonlinear Control*, 7: 11-28.

4. Baños , A., and F. N. Bailey, 1988, "Design and validation of linear robust controller for nonlinear plants", *Int. Journal on Robust and Nonlinear Control*, 8:803-816.

5. Baños, A., F. N. Bailey, and F. J Montoya, 1997, "QFT design of nonlinear controllers by using finite sets of acceptable outputs", *Proc. Symposium on QFT and other Frequency Domain Methods and Applications*, University of Strathclyde, pp. 227-234.

6. Horowitz, I., 1991, "Survey of quantitative feedback theory (QFT)", *Int. J. Control*, 53, 2:255-291.

7. Oldak, S., Baril, C. , and Gutman, P. O., 1994, "Quantitative Design of a Class of Nonlinear Systems with Parameter Uncertainty", *Int. Journal of Robust and Nonlinear Control*, 4:101-117.

8. Yaniv, O., 1997, "Synthesis of LTV feedback around nonlinear MIMO systems", *Proc. Symposium on QFT and other Frequency Domain Methods and Applications*, University of Strathclyde, pp. 153-161.

9. Horowitz, I. and Shur, D., 1981, "Control of Uncertain Van der Pol Plant", *Int. J. Control*, 32, 2:199-219.

10. Klatt and Engell, 1995, "Nonlinear process control by a combination of exact linearization and gain scheduling", in Krener and Mayne (eds.), *Nonlinear Control Design 1995*, Pergamon Press, pp. 173-178.

22

Frequency Domain Control Structure Design Tools

E. Kontogiannis, N. Munro and S.T. Impram

1   Introduction

Very important parts of a multivariable design are the input/output (I/O) selection and I/O assignment or pairing, which both define the so-called control system structure design problem. I/O selection is the problem of choosing the variables that are the most appropriate to control particular outputs. A good physical understanding of the plant to be controlled is always the initial step towards the solution of this problem. However, in many cases the designer is still left with more than one choice from which he has to choose the best according to some criteria. This is the case in many real world applications, such as in aircraft and process control, where a number of variables are observable, but it is well known that only a subset of those equal to the number of inputs, can be independently controlled. Thus, the designer has the freedom to choose the outputs from the available measurements, and in doing so in a systematic way. A few criteria exist, which will be considered below, to help with this problem. Once such a choice has been made, it is then necessary to assign an input variable to the control of a particular output variable, i.e. to choose the appropriate I/O pairs. Most of the methods are scaling dependent, hence an initial I/O scaling is necessary before any of the aforementioned analytical tools are employed.

A systematic solution to these problems is obviously of paramount importance in the design of MIMO control systems. If an input is to be paired with an output over which it has a little or no influence, then clearly satisfactory control would be unattainable. Also, non-minimum phase characteristics, and/or high-levels of interaction, are common problems that could possibly be reduced, or even avoided by a more careful choice of the above. Obviously, decentralised control design methods like the Direct Nyquist Array (DNA), the Quantitative Feedback Theory (QFT) approach, and the Sequential Loop Closing (SLC) approach can benefit a great deal from a careful selection and pairing of the inputs and outputs.

2   Scaling Algorithms

Scaling is an essential part of the control systems analysis and design in any practical application. In MIMO systems scaling corresponds to the application of a diagonal transformation matrix, which changes the magnitudes of the variables within the system and normalises the effect of the inputs on each of the outputs, in terms of their magnitude (coupling). Scaling has a very significant effect in many areas of multivariable systems analysis and design, such as interaction analysis, conditioning of the problem, weighting function selection, model order reduction, etc.

A typical approach to scaling is to divide each variable by the maximum expected value or allowed change in the inputs, the disturbances, or the control error, hence, making each variable less than unity. There are also systematic methods of calculating an input/output scaling which improves the diagonal dominance of the system in some sense. Such methods are the Perron-Frobenius scaling and Edmunds' method which are described below.

2.1   Perron-Frobenius (PF) Method

This method was first introduced to control engineers by Mees [1-2] in the context of improving the diagonal dominance of multivariable systems. The aim of the algorithm is to maximise the worst dominance ratio for a particular order of inputs and outputs. The optimal input and output scaling matrices of a system G(s) are obtained from the elements of the Perron-Frobenius (PF) right and left eigenvectors, respectively. These are the eigenvectors corresponding to the largest eigenvalue of the corresponding comparison matrix, defined as $C(s) = |G(s)| |G_D^{-1}(s)|$, where $G_D(s)$ corresponds to the diagonal part of G(s). The restrictions on this method are that G(s) is assumed to be square and primitive. The latter is not a severe restriction; if all elements of G(s) are nonzero, then G(s) is primitive. If this is not the case, then usually G(s) is block diagonal, or block triangular, or at least can be made so by reordering the inputs and outputs. A dynamic scaling version of this method was later introduced [3] by Munro, and a version for uncertain parametric systems of interval or affine linear type has also been developed [4].

2.2   Edmunds' Method

This algorithm [5] sets all the row and column sums of the absolute value of a matrix to unity. It equalises and usually maximises the geometric mean of the row dominance and the geometric mean of the column dominance for any arrangement of inputs and outputs, which is the major benefit of the method. This algorithm will give the same scaling factors independently of the order of the inputs and outputs. For square systems it will ensure that the row and column dominance will be equal. It will always converge if the matrix is non-singular or if the matrix can be made non-singular by just changing the sign of its elements.

Example 1: Consider the 3×3 matrix G [5], where

$$G = \begin{bmatrix} 0.5 & -0.3 & 100 \\ 1 & 0 & 0.5 \\ 0.005 & 0.01 & 0.2 \end{bmatrix} \tag{1}$$

corresponds to a matrix of absolute values of the elements of the frequency response of a TFM at some frequency ω. The I/O scaling matrices are to be calculated using either method mentioned above.

Since element (2,2) is zero, the PF eigenvalue is infinite, and hence, the PF scaling fails. In order to overcome this problem, the system input/output pairing can be altered, such that no diagonal element is zero. Then, the PF scaling can be used and the optimum scaling can be found. However, this should not be considered as a major restriction, since a zero diagonal element is an indication of a bad selection of input/output pairs, and is something that one would take action against, especially if the design is to be carried out using a decentralised control method.

Going back to the example, the I/O pairs of G given in (1) can be rearranged as follows:

$$G_1 = \begin{bmatrix} 1 & 0 & 0.5 \\ 0.005 & 0.01 & 0.2 \\ 0.5 & -0.3 & 100 \end{bmatrix} \tag{2}$$

and the PF eigenvalue of the resulting matrix G_1 becomes 1.2558, which is less than 2 and hence diagonal, input and output scaling matrices (K and L, respectively) exist, which can be found to be:

$$K = \begin{bmatrix} 0.0195 & 0 & 0 \\ 0 & 0.8199 & 0 \\ 0 & 0 & 0.0100 \end{bmatrix}, L = \begin{bmatrix} 3.0563 & 0 & 0 \\ 0 & 84.397 & 0 \\ 0 & 0 & 0.7196 \end{bmatrix} \tag{3}$$

Applying the pre- and post-multiplying scaling matrices, the resulting system $G_s = LG_1K$ becomes:

$$G_s = \begin{bmatrix} 0.0597 & 0 & 0.0153 \\ 0.0082 & 0.6920 & 0.1688 \\ 0.0070 & 0.1770 & 0.7195 \end{bmatrix} \tag{4}$$

Note that G_s is diagonal dominant, even though G_1 was not.

Using Edmunds' method on the original matrix G, the pre- and post-multiplying scaling matrices can be found to be:

$$K = \begin{bmatrix} 0.7807 & 0 & 0 \\ 0 & 9.355 & 0 \\ 0 & 0 & 0.1031 \end{bmatrix}, L = \begin{bmatrix} 0.07402 & 0 & 0 \\ 0 & 1.202 & 0 \\ 0 & 0 & 8.469 \end{bmatrix} \tag{5}$$

and hence, the resulting scaled matrix $G_c = L\,G\,K$ becomes

$$G_c = \begin{bmatrix} 0.0289 & -0.2077 & 0.7634 \\ 0.9380 & 0 & 0.0620 \\ 0.0331 & 0.7923 & 0.1747 \end{bmatrix} \tag{6}$$

which can easily be made diagonal dominant by re-ordering.

Re-ordering now the matrix G_c, such that the resulting $G_{c,2}$ reflects the same input/output interconnections as the ones in G_s, where

$$G_{c,2} = \begin{bmatrix} 0.9380 & 0 & 0.0620 \\ 0.0331 & 0.7923 & 0.1747 \\ 0.0289 & -0.2077 & 0.7634 \end{bmatrix} \tag{7}$$

a direct comparison of the resulting diagonal dominance measures (row or column) is then available. The column dominance measures of the resulting matrices are 0.2558 for the PF scaling, and {0.0660, 0.2622, 0.3100} for the three columns of $G_{c,2}$ using Edmund's method. As expected, the PF method is giving a more dominant system $(0.2558 < \max\{0.066, 0.2622, 0.31\})$ than Edmunds' algorithm, since it is designed to maximise the worst dominance ratio. However, notice how small the magnitude of the element $G_s(1,1)$ became. This is a known problem of the PF scaling algorithm; sometimes in trying to make the dominance ratios equal, some diagonal elements can be made arbitrarily small.

3 Input/Output Selection Algorithms

The selection of the manipulated and controlled variables in a multivariable design is a very important task. For many systems, the number of measurements available is larger than the number of inputs to the system and the performance specifications can often be expressed in terms of a subset of these measurements. Therefore, knowing that the maximum number of outputs that can be independently controlled is equal to the number of inputs, the designer is left with the choice of selecting the outputs of the system from the range of measurements available.

A good physical understanding of the dynamic behaviour of the plant, as well as the knowledge of any practical considerations associated with it, are necessary for the right selection of the input/output variables. This stage should be the first towards the solution of the control structure design problem, and should be considered before any analytical tools are applied. The rules are to select those measurements which have a strong relationship with the outputs, and which can quickly detect a disturbance. The manipulations on the other hand should be selected such as to have a large effect on the selected outputs. If the plant is unstable, then the inputs must be selected such that the unstable modes are controllable, and the outputs must be selected such that the unstable modes are observable.

As suggested in [6], a formal analysis is possible using the model $y_A = G_A u_A + G_{dA} d$, where d denotes the disturbances within the system, and the subscript A stands for "all"; i.e., y_A stands for "all available measurements", which are considered as candidate outputs. Similarly, u_A corresponds to "all available manipulations", which are the candidate inputs. The model for a particular combination of the inputs and outputs is then $y = G u + G_d d$, where $G = S_O G_A S_I$ and $G_d = S_O G_{dA}$. The S_O and S_I are non-square permutation matrices (output and input, respectively) used for the selection of the desirable input/output pairs of all the possible combinations.

The number of r×c systems that can be selected from the available N×M full model

is given by $\begin{pmatrix} M \\ c \end{pmatrix}\begin{pmatrix} N \\ r \end{pmatrix} = \dfrac{M!}{c!(M-c)!} \dfrac{N!}{r!(N-r)!}$. If, for instance, M = N = 4 and r = c = 2, the number of possible 2×2 systems is 36, whereas if M = N = 10 and r = c = 5, the number of possible 5×5 systems is 63504. The combinatorial growth of the number of possible combinations is therefore evident. Performing even a very simple analysis procedure on each of these candidate systems can prove to be very time consuming. One way around this problem is to consider the full model (G_A) and calculate its (non-square) Relative Gain Array (RGA) [6], which together with some physical insight will reduce the number of combinations.

The non-square RGA is an extension of the classical RGA [7-9], which is briefly defined below.

3.1 Introduction to the Relative Gain Array

The Relative Gain Array (RGA) introduced by Bristol [7] over thirty years ago has found widespread use as a measure of interaction and as a tool for the control structure design problem.

Consider an n×n MIMO system with transfer function matrix G(s), and its inverse denoted by $\hat{G}(s) = G^{-1}(s) = [\hat{g}_{ij}(s)]_{i,j=1,\dots,n}$. The transfer function from input j to output i when all loops are open is $g_{ij}(s)$. The corresponding transfer function when all outputs except the ith one are tightly controlled will be denoted by $h_{ij}(s)$, and is given by $h_{ij}(s) = \dfrac{1}{\hat{g}_{ji}(s)}$ under the assumption of perfect control of all the other loops. Following this, the frequency dependent RGA matrix $\Lambda(s) = [\lambda_{ij}(s)]_{i,j=1,\ldots,n}$ can then be defined as:

$$\Lambda(s) = G(s) .* \hat{G}^{\top}(s) \tag{8}$$

where $.*$ denotes element-by-element multiplication (the Schur, or Hadamard product). Because a physical interpretation of the perfect control assumption is only meaningful at steady state[1], namely $\Lambda(0)$, many researchers have restricted the use of RGA to zero frequency.

For any square, non-singular matrix the RGA defined in Eq.8 has the following properties:

- It is input and output scaling independent; that is $\Lambda(K\,G\,L) = \Lambda(G)$, if K and L are diagonal matrices.

- All row and column sums are equal to one, i.e. $\sum_{j=1}^{n}\lambda_{ij} = \sum_{i=1}^{n}\lambda_{ij} = 1$.

- Any permutation of the rows or columns of G(s) results in the same permutations in the RGA matrix.

- If G(s) is triangular, or moreover diagonal, then $\Lambda(s) = I$.

The interested reader is referred to references [6,10-13] for a comprehensive study of the steady state and the dynamic RGA.

3.2 The non-square Relative Gain Array

The RGA is not only defined for square systems. Especially when the RGA is used on the full model of the system, $y_A = G_A\,u_A + G_{dA}\,d$; i.e. with all available inputs and outputs; it is generally non-square. The RGA can be generalised to non-square systems by the use of the pseudo-inverse in Eq.8. However, the non-square RGA does not have all the nice properties mentioned before. Furthermore, in the case where there are more inputs than outputs (full row rank), the RGA is independent of the output scaling only, and only the row sums are equal to one. On the other hand, in the case where there are more outputs (full column rank), the RGA is independent of the input scaling only, and only the column sums are equal to one. Nevertheless, it can be proved that the RGA of non-square matrices provides a useful tool of interpreting the information available in the singular vectors [6], which consequently can be used as an input/output selection algorithm as follows:

---

[1] If integral control is used, the assumption of perfect control at steady state is absolutely justifiable.

For the case of many candidate inputs (outputs), one may consider not using those inputs (outputs) corresponding to columns (rows) in the RGA whose sum of their elements is much smaller than unity.

Examining the RGA at different frequencies within the given bandwidth, the selection of the most important inputs and outputs can be made. Notice here that an input for instance, could be significant at steady state, but insignificant at higher frequencies, or vice versa. Hence, the RGA at different frequencies (not just the steady state) should be considered, before a selection is made.

Example [6]: Consider the following chemical plant at steady state, $G_A(0)$ say, which has 5 controlled outputs and 13 candidate inputs:

$$G_A^T(0) = \begin{bmatrix} 0.7878 & 1.1489 & 2.6640 & -3.0928 & -0.0703 \\ 0.6055 & 0.8814 & -0.1079 & -2.3769 & -0.0540 \\ 1.4722 & -5.0025 & -1.3279 & 8.8609 & 0.1824 \\ -1.5477 & -0.1083 & -0.0872 & 0.7539 & -0.0551 \\ 2.5653 & 6.9433 & 2.2032 & -1.5170 & 8.7714 \\ 1.4459 & 7.6959 & -0.9927 & -8.1797 & -0.2565 \\ 0.0000 & 0.0000 & 0.0000 & 0.0000 & 0.0000 \\ 0.1097 & -0.7272 & -0.1991 & 1.2574 & 0.0217 \\ 0.3485 & -2.9909 & -0.8223 & 5.2178 & 0.0853 \\ -1.5899 & -0.9647 & -0.3648 & 1.1514 & -8.5365 \\ 0.0000 & 0.0002 & -0.5397 & -0.0001 & 0.0000 \\ -0.0323 & -0.1351 & 0.0164 & 0.1451 & 0.0041 \\ -0.0443 & -0.1859 & 0.0212 & 0.1951 & 0.0054 \end{bmatrix} \quad (9)$$

For this system there are $\binom{13}{5} = 1287$ combinations of 5×5 systems, and 1716 combinations of 6×5. The column sums of the RGA corresponding to $G_A(0)$ are given below

$$\Lambda_\Sigma = [0.7675, 0.1459, 0.7282, 0.3978, 0.9481, 0.8548, 0.0000,$$
$$0.0108, 0.1775, 0.9417, 0.0268, 0.0003, 0.0005]$$

The largest five column sums, and hence the most favourable inputs, correspond to the 5th, 10th, 6th, 1st, and 3rd column of Λ_Σ. For this selection, the minimum singular value of the corresponding system, G_5 say, is $\underline{\sigma}(G_5) = 1.7257$, whereas $\underline{\sigma}(G_A) = 2.4473$. Hence, not much gain is lost in the input direction with the lowest gain by using only 5 from the 13 candidate inputs.

The aforementioned rule, however, should be used with some care. There are other factors that should be taken into account as well. For instance, the resulting system could have nasty RHP-zeros, or it could well be ill conditioned, whereas another I/O configuration, less appealing according to this criterion, could be less "problematic" later on in the design procedure. On the other hand, this rather crude initial analysis is the only way to avoid the combinatorial problem.

From the remaining combinations, a selection of the best input/output configuration can be made according to other analysis tools, which are further outlined below.

3.3 RHP-zeros

Even though the effect of RHP–zeros in multivariable systems is often not as serious as in the SISO case[2], they can limit the achievable performance of the control system design, especially when the right-half plane (RHP) zeros are lying within the closed-loop bandwidth [6]. Different input/output configurations produce different multivariable zeros. Hence, a general rule of thumb for such cases is to choose the configuration, which produces the least number of RHP zeros, located as far from the origin as possible.

3.4 Condition Number

The condition number $\kappa(\omega)$ of a plant $G(j\omega)$ is defined as:

$$\kappa(\omega) \overset{\Delta}{=} \frac{\bar{\sigma}(G(j\omega))}{\underline{\sigma}(G(j\omega))} \tag{10}$$

where $\bar{\sigma}(G(j\omega))$ and $\underline{\sigma}(G(j\omega))$ denote the maximum and minimum singular value of the system $G(s)$. A high condition number means that the plant in hand is ill conditioned, and hence, should be avoided if possible. The condition number is directly related with the functional controllability of the system, which is an indication of whether the outputs can be independently controlled. The larger the spread of the singular values at a range of frequencies, the more difficult it will be to achieve independent control of the outputs in this frequency range. Note that the condition number is scaling dependent.

4 I/O Assignment or Pairing Algorithms

The input/output pairing problem in multivariable systems design can be formidable. Even for relatively small systems, there are many distinct decentralised control system structures to choose from. For instance, for a 4×4 system there are 4! = 24 different pairings, for a 5×5 system 5! = 120, and so on. Thus, the need for

---

[2] The RHP-zeros in multivariable systems only apply in particular directions. Their deteriorating effect can often be moved to a given output, which could be less important to control well.

efficient algorithms, which will quickly eliminate structures that are inappropriate according to some criteria.

In the following three simple algorithms will be presented, which can be used to systematically calculate the best I/O pairings, while attempting to eliminate interaction as much as possible, and which will facilitate the use of a decentralised control design method.

4.1 The Relative Gain Array (RGA)

A brief introduction of the RGA was given earlier, and a version for non-square matrices was used to solve the I/O selection problem. In this section, the classical definition of the RGA for square matrices will be used, since the system is assumed that it has been squared down already in the previous step.

The RGA was initially given without any theoretical explanation, but was used very successfully especially in the process industry. As originally presented, the RGA involved only steady state considerations. More recently, several investigators have considered dynamic extensions of the RGA, and the implications of the RGA to the system stability, controllability and even performance has been proven and/or conjectured.

Based on the properties of the steady state RGA, the best I/O pairs can be found using the following guidelines: Relative gains in the range 0-1 indicate moderate interaction, with values of 0.5 being the worst. The variables, whose relative gains are closest to unity, should be paired. Negative values, or values much greater than one should be avoided. Specifically, relative gains greater than unity preserve the dynamic response but they reduce the effectiveness of the control action by reducing the loop gain. Pairs of variables with negative relative gains should also be avoided, since their dynamic response will be extremely poor.

At frequencies close to crossover, it can be proved [10] that for overall stability of a decentralised control scheme, the I/O pairings which make the system as close to triangular as possible should be chosen, which accordingly translates to its corresponding RGA being as close to the identity matrix as possible.

If the sign of a RGA element changes as s goes from 0 to ∞, then either the specific element has a RHP-zero, or the overall system, or just a sub-system, has a RHP transmission zero [10]. Any such zero may be detrimental for control, and therefore, pairings should be chosen such that the elements of the RGA are positive at all frequencies.

From the relationship between the RGA and the system condition number, it can also be proven [6] that if the plant's RGA contains large elements, the plant is ill conditioned and hence, the use of an inverting type of controller, like the one computed using the ALIGN algorithm [2], should be avoided.

4.2   The Performance Relative Gain Array (PRGA)

One of the main criticisms against the use of RGA has been its failure to predict one way coupling, because the RGA of a triangular TFM is equal to the identity matrix. To overcome this problem the PRGA [6,10], defined as

$$\Gamma(s) = diag[G(s)]G^{-1}(s) \tag{11}$$

was introduced. Note that the diagonal elements of the PRGA and RGA are the same, but the PRGA does not have the nice properties of the RGA. More specifically, the PRGA must be recomputed whenever the input/output arrangement in G is changed, it is input scaling independent but it is dependent on output scaling.

It can be proven [10] that for the low and intermediate frequencies, the diagonal elements of the PRGA (γ_{ii}) define a measure of the performance degradation in terms of the diagonal elements in the sensitivity function. For these frequencies it is desirable to have small elements in Γ. On the other hand, since $\gamma_{ii} = \lambda_{ii}$, for stability issues, it is required that the diagonal elements of Γ are as close to one as possible at frequencies close to crossover. The plots of the moduli the off-diagonal elements of the PRGA (γ_{ij}) give useful information about which pairs of inputs and outputs are expected to have large interaction.

4.3   Edmunds' Re-ordering Algorithm

This algorithm [5] attempts to reorder the inputs and outputs in order to maximise the diagonal (block diagonal) dominance. It first moves the most dominant elements to the principal diagonal of the matrix, and then iterates trying to improve the geometric mean of the dominance ratios by swapping pairs of inputs. The equivalent algorithm for block diagonal dominance has also been developed. A MATLAB function implementing Edmunds' scaling and reordering algorithm is available within the System Toolbox [14].

4.4   Recursive Bottleneck Assignment Problem

Given a complex matrix $G \in C_{n \times n}$, a positive matrix $D^c = \left[d_{ij}^c\right]_{i,j=1,...,n}$ called the column dominance matrix can be defined as follows:

$$d_{ij}^c \overset{\Delta}{=} \sum_{\substack{k=1 \\ k \neq j}}^{n} \frac{|g_{kj}|}{|g_{ij}|} \tag{12}$$

Similarly, the row dominance matrix can be defined. The ij-element of the column (row) dominance matrix corresponds to the new column (row) dominance measure of the jth column (row), when the rows (columns) i and j are interchanged. From this definition it is obvious that the column (row) diagonal dominance measures of the original matrix are given by the diagonal elements of $D^c(D^r)$. The calculation of the optimal permutation that minimises the worst column dominance measure can be expressed as a min-max problem as follows:

$$P_1 : \min_{\pi} \max_{i} d^c_{\pi(i),i} \tag{13}$$

where $\pi(i)$ is a (row[3]) permutation. An equivalent problem is the following: *"choose a set of n elements of D^c, called matching, such that there is exactly one element of the set in every row and column of D^c. Additionally minimise the maximum element in the matching"*. Problem P_1 is generally called the Bottleneck Assignment Problem (BAP). The solution of P_1 is not necessarily unique[4], and it usually only minimises the largest element of D^c. In order to minimise all column dominance measures (not just the worst), Bryant and Yeung [15] proposed the following recursive solution to the BAP (RBAP), based on graph theory.

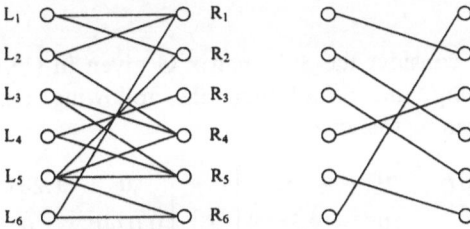

Figure 1: An example bi-partite graph and a corresponding matching

A bi-partite graph (Fig.1), G (V,B) say, has a node set V={L_1, ... , L_6, R_1, ... , R_6} and a branch set B. The nodes on the left represent the system outputs, whereas the nodes on the right represent the permuted outputs. A branch b_{ij} connecting two nodes (L_i, R_j) represents a permutation, i.e. the assignment of the original output i to the new output j, and the cost of such an assignment is d^c_{ij}.

Define now a matching **M** of the graph G(V,B) as a set of branches such that no two branches of this matching coincide with the same node and each node has a branch connected to it (Fig.1). Define also a cover **S** of G(V,B) as a set of branches such that each node has at least a branch connected to it. The following is an algorithmic solution to the recursive BAP [15]:

Step 0: Calculate the dominance matrix $D^c_n = D^c$, which is assumed to have distinct elements. Set k = n.

---

[3] Accordingly, $\pi(i)$ corresponds to a column permutation when the row dominance matrix $D^r = [d^r_{i,\pi(i)}]$ is considered.

[4] If the elements of the dominance matrix D^c are distinct, then the solution is unique.

Step 1: Sort the elements of D_k^c in ascending order to form a list, L_c say.

Step 2: Construct a minimum cover S_k for the matrix D_k^c. The last branch added to form the cover S_k is called the pivot.

Step 3: Supposing that the pivot is the branch b_{lm}, then delete the lth row and the mth column to form the new sub-matrix D_{k-1}^c. Set $k = k - 1$, and go back to Step 1. The min-max matching is the union of the pivot elements of all steps.

Two situations may arise in Step 3. If at each stage a min-max matching and a pivot can be found, then the cover S_k has a matching and the set of all the pivot elements is the solution. If at any stage a cover can not be found, the minimum cover does not contain a matching. Therefore, it has to be extended by adding one or more branches from the list L_c.

The three methods mentioned in this section will be further illustrated in the example below.

Example 2: Again, consider the 3×3 matrix G given in (1). Applying Edmunds' algorithm for scaling and re-ordering, the following input (K), and output compensators (L), can be found:

$$K = \begin{bmatrix} 0.7807 & 0 & 0 \\ 0 & 0 & 9.3549 \\ 0 & 0.1031 & 0 \end{bmatrix}, L = \begin{bmatrix} 0 & 1.2015 & 0 \\ 0.0740 & 0 & 0 \\ 0 & 0 & 8.4690 \end{bmatrix}$$

The RGA for the same matrix G can be found to be:

$$RGA(G) = \begin{bmatrix} -0.0024 & 0.0561 & 0.9463 \\ 1.0031 & 0 & -0.0031 \\ -0.0007 & 0.9439 & 0.0568 \end{bmatrix}$$

Using the guidelines given earlier, the elements of magnitude almost unity should be moved to the diagonal, i.e. the elements (2,1), (3,2) and (1,3) should become the diagonal elements. The permutation matrices that can bring these elements to the diagonal are:

Solution 1: L = [0 1 0; 1 0 0; 0 0 1], K = [1 0 0; 0 0 1; 0 1 0]

Solution 2: L = [0 1 0; 1 0 0; 0 0 1]

where, as before, K corresponds to a pre-, and L corresponds to a post-compensator matrix.

The column dominance matrix for the matrix G is given by:

$$D_3^c = \begin{bmatrix} 2.01 & 0.0333 & 0.007 \\ 0.505 & Inf & 200.4 \\ 300 & 30 & 502.5 \end{bmatrix}$$

and the corresponding ordered list L_c is : $L_c = \{d_{13}, d_{12}, d_{21}, d_{11}, d_{32}, d_{23}, d_{31}, d_{33}, d_{22}\}$.

Step 1: A minimum cover for the matrix $D_c^3 = [d_{ij}]_{i,j=1,\dots,3}$ is $S_3 = \{d_{13}, d_{12}, d_{21}, d_{11}, d_{32}\}$, and the corresponding pivot element is $d_{32} = 30$.

Step 2: Dropping the 3rd row and the 2nd column of the matrix, the matrix D_c^2 is formed, where

$$D_c^2 = \begin{bmatrix} 2.01 & 0.007 \\ 0.505 & 200.4 \end{bmatrix}$$

The matrix D_c^2 contains a minimum cover, which is $S_2 = \{d_{13}, d_{21}, d_{11}\}$. The pivot element in this step is $d_{11} = 2.01$.

Step 3: Dropping the 1st row and column, the remaining element $d_{23} = 200.4$ does not belong to the minimum cover. Hence, we go back to step 1 and increase the minimum cover by one element from the list L_c, forming $S_3 = S_3 \cup \{d_{23}\}$.

Now, performing the same steps again, it can be seen that S_3 now contains a matching and the set of pivot elements are: $\{d_{11}, d_{23}, d_{32}\}$. Hence, $\pi(1) = 1$, $\pi(2) = 3$, $\pi(3) = 2$ and consequently, the required row permutation matrix L is given by [1 0 0; 0 0 1; 0 1 0].

The previous two methods outperform the RBAP. This can be justified from the fact that both methods allow for simultaneous row and column operations in order to get the best dominance measure, whereas the RBAP considers only row or only column operations.

Conclusions

Some tools for the solution of the control system structure design problem have been presented in this paper. These tools are easy to apply, and provide a means of overcoming the combinatorial explosion of the possible I/O pairings, as well as improving the diagonal dominance of the system.

References

[1]   Mees A. I. 1981 Achieving Diagonal Dominance. *Systems and Control Letters*. Vol. 1 No. 2: 155-158.

[2]   Maciejowski J. M. 1989 *Multivariable Feedback Design*. Addison Wesley.

[3]   Munro N. 1987 Computer-Aided Design I: The Inverse Nyquist Array Design Method. In O'Reilly (Ed.), *Multivariable Control for Industrial Applications*, Stevenage: Peter Peregrinus, pp. 211-228.

[4]   Kontogiannis E. and Munro N. 1998 Robust stability conditions for MIMO systems with parametric uncertainty. To appear in Munro N. (ed) 1998 *Symbolic Methods in Control*, IEE Press, UK.

[5]   Edmunds J. M. 1998 Input and Output Scaling and Re-ordering for Diagonal Dominance and Block Diagonal Dominance. To appear in *IEE Proc. Control Theory and Applications*.

[6]   Skogestad S. and Postlethwaite I. 1996 *Multivariable Feedback Control : Analysis and Design*. Wiley, Chichester.

[7]   Bristol E. 1966 On a New Measure of Interaction for Multivariable Process Control. *IEEE Trans. Aut. Control* Vol. AC-11: 133-134.

[8]   Shinskey F. G. 1981 *Controlling Multivariable Processes*. Instr. Society of America.

[9]   Wang S. and Munro N. 1982 A Complete Proof of Bristol's Relative Gain Array. *Trans. Inst. M C* Vol. 4 No. 1: 53-56.

[10]   Hovd M. and Skogestad S. 1992 Simple Frequency-dependent Tools for Control Systems Analysis, Structure Selection and Design. *Automatica* Vol. 28 No. 5: 989-996.

[11]   Manousiouthakis V., Savage R. and Arkun Y. 1986 Synthesis of Decentralised Process Control Structures Using the Concept of Block Relative Gain. *AIChE Journal* Vol. 32 No. 6: 991-1003.

[12]   Nett C. N. and Manousiouthakis V. 1987 Euclidean Condition and Block Relative Gain: Connections, Conjectures, and Clarifications. *IEEE Trans. Aut. Control* Vol. AC-32 No. 5: 405- 407.

[13]   McAvoy T. J. 1981 Connection between Relative Gain and Control Loop Stability and Design. *AIChE Journal* Vol. 27 No. 4: 613-619.

[14]   Edmunds J. M. 1997 The System Toolbox Reference Manual. *Control Systems Centre Report 869*.

[15]   Bryant G. F. and Yeung L. F. 1990 *Multivariable Control System Design Techniques: Dominant and Direct Methods*. Wiley, Chichester.

23

Quantitative Pressure Controller Design for a Gas Recovery System

E. Boje

1 Introduction

This paper considers the design of a pressure controller to control the suction pressure of a gas compressor shown in Figure 1. The pressure is controlled by means of a valve (PV1) in the line between an upstream surge vessel and the (downstream) compressor suction. The design must maintain the compressor suction pressure within manufacturer's tolerances despite significant mass flow disturbances into an upstream surge vessel, illustrated in Figure 2. This practical application of the QFT design method considers the valve non-linearity, valve actuator and sensor rise times, the approximate effect of distributed gas dynamics, digital implementation issues and the use of a feed-forward signal from the surge vessel pressure measurement. Because of the fast dynamics, the design is required before equipment purchase and installation.

Quantitative feedback design theory (QFT) has been developed mainly in the aerospace industry where research funding and very stringent end-user requirements have driven the development. Not as much work has been reported in the literature on application in the process control industry.

The layout of the paper is as follows: Section 2 develops the model for the system under study. The feedback design and disturbance feedforward design is undertaken in Section 3. Section 3 also presents simulation results. Section 4 concludes the paper.

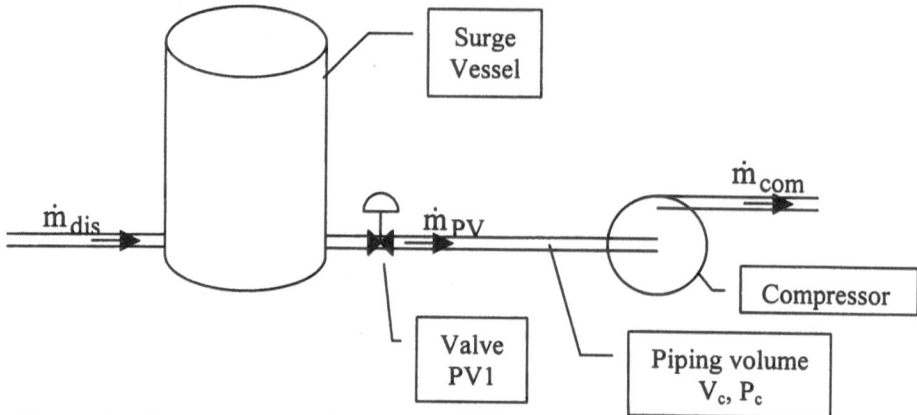

Figure 1 – Pressure control problem

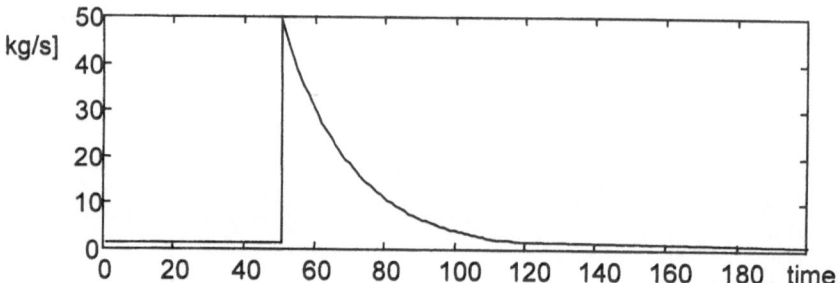

Figure 2 - Typical "worst case" $\dot{m}_{dist}$

2 Model

2.1 Pressure dynamics

The gas is assumed ideal and temperature effects negligible. For the surge vessel and compressor suction capacities (subscript s and c respectively),

$$\dot{P}_s = \frac{\mu RT}{V_s}\left(\dot{m}_{dist} - \dot{m}_{PV1}\right) = k_s\left(\dot{m}_{dist} - \dot{m}_{PV1}\right) \tag{1}$$

$$\dot{P}_c = \frac{\mu RT}{V_c}\left(\dot{m}_{PV1} - \dot{m}_{Comp}\right) = k_c\left(\dot{m}_{PV1} - \dot{m}_{Comp}\right) \tag{2}$$

where,

        P is the absolute pressure in Pa
        $\mu = 1/M$, the inverse of the molecular weight of the gas in kg/mol
        V is the volume of the respective capacity
        R is the (universal) gas constant
        T is the temperature in Kelvin

Numerical values of parameters used in this paper are summarised in Table 1. It is assumed that the pressure dynamics in the line are fast enough to be regarded as a single capacity and dead time given by (length)/(speed of sound):

$$T_d = \ell / c \tag{3}$$

The valve PV1, is an "equal percentage" valve and line pressure drops are negligible so that the installed behaviour is also equal percentage, with mass flow rate are given by,

$$\dot{m}_{PV1} \approx (KV) E^{(u/100 - 1)} \sqrt{\frac{(P_s - P_c)P_c}{T}} \tag{4}$$

$$= k_v E^{(u/100)} \sqrt{(P_s - P_c)P_c}$$

(if $(P_s\text{-}P_c) < P_s/2$, pressure drop less than half the upstream pressure. In eq(4),

KV   the valve CV in appropriate units

k_v the valve gain $\left(k_v = \dfrac{KV}{E\sqrt{T}} \right)$

E   equal percentage valve factor

u   the control signal in %

The downstream compressor is a constant (volumetric) displacement device so,

$$\dot{m}_{Comp} = \frac{\dot{m}_{demand} P_c}{P_c^{SetPoint}} \tag{5}$$

(i.e. approximately constant if the pressure is accurately controlled.)

SYMBOL	MEANING	VALUE
γ	C_p/C_v	1.33
μ	mol/kg	1/30E-3 for 30 molar gas
Z	Gas law factor	1
R	Gas Constant	8.314 (SI units)
$c = \sqrt{\gamma P / \rho}$ $= \sqrt{\gamma \mu R T}$	Speed of sound	340m/s at 40°C
P_s	Pressure in surge capacity	650-1000kPa
P_c	Pressure at compressor suction	600kPaA (desired set-point)
$T_d = \ell / c$	Dead time in suction header	0.15s
k_V	Valve gain (depends on valve size and type)	1.5×10^{-4} kg/s/kPa
E	equal % factor (depends on valve size and type)	50
V_c	Compressor header volume	5 m^3
V_s	Surge capacity volume	200m^3
ρ	Density at 0°, 101.33kPaA	0.95kg/m^3
τ_A	Valve actuator time constant	2s (hopefully!)
τ_S	Pressure sensor time constant	1s

Table 1 - Assumptions & Data

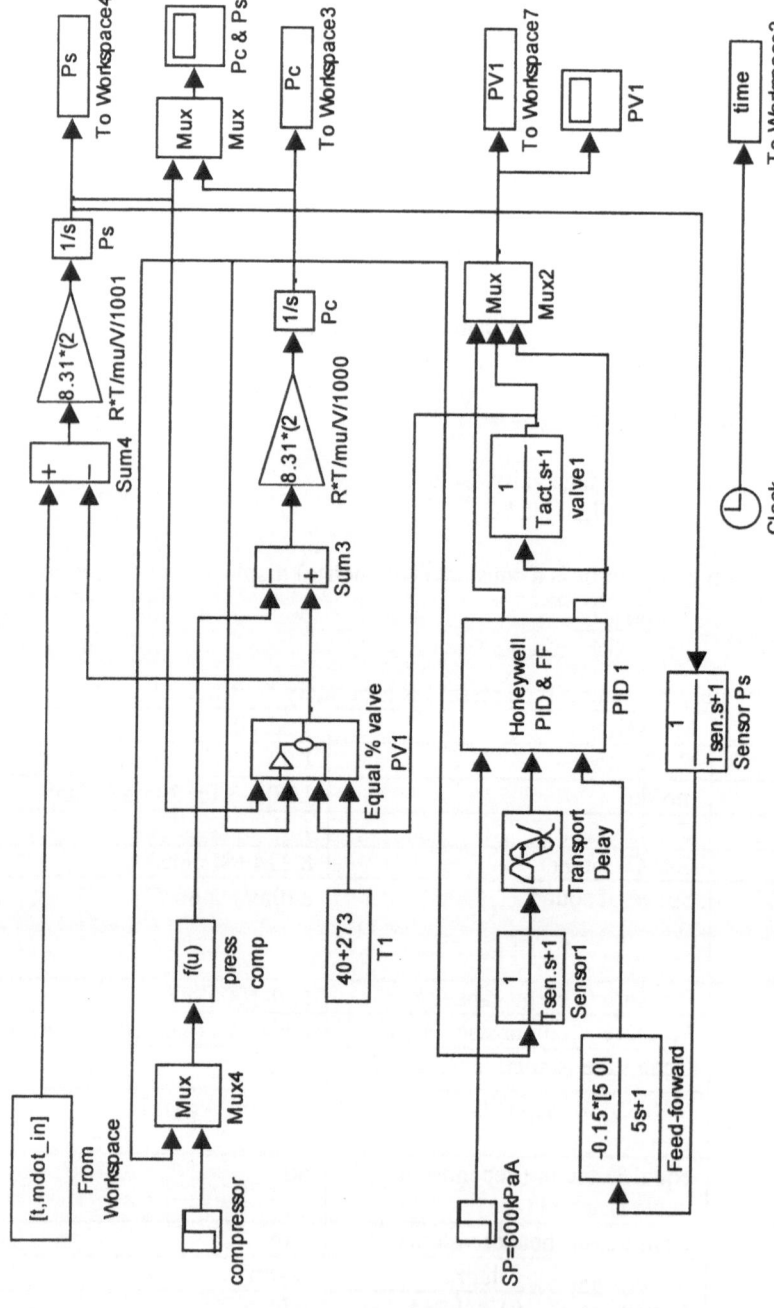

Figure 3 – Simulink model

There is some adiabatic temperature change across the valve. This and other temperature effects can be considered in a more detailed analysis. Changing gas composition and hence molecular weight could also be included if required.

The valve is assumed to have first order time dynamics. With slow recovery of the valve actuator, limit cycling of the valve may be expected by describing function analysis. The pressures are measured with fast sensors and are to be set up with no signal filtering.

The above non-linear model is implemented in Matlab-Simulink for testing the designed controller's response. The implementation diagram is shown in Figure 3.

2.2 Linearisation for feedback design

The only significant non-linearity in the above system is as a result of pressure behaviour across the valve and the valve's non-linear characteristic. Temperature effects and non-ideal nature of the gas could add additional non-linearity but these are secondary effects. As the pressure cannot change very rapidly because of the capacity of the surge vessel, and because the controller typically must be implemented using standard DCS or PLC tools, a robust linear feedback design approach is applied. The linearised models have structured uncertainty as a result of linearisation at varying operating points. In cases where the non-linearity is severe or the operating point changes result in fast variation of the linearisation, the non-linear QFT (Horowitz, 1982, Baños and Bailey, 1995) may be applicable. Clearly, in engineering application of linear design, any obvious static or dynamic gain scheduling would be applied before feedback design if there were sufficient benefit in terms of bandwidth reduction. As the pressure in the compressor suction is to be controlled, this is linearised around an operating point, P_s^*, P_c^*, u^* (where * refers to operating point, and Δ refers to change in variable) as,

$$\frac{d}{dt}\begin{pmatrix} \Delta P_c \\ \Delta P_s \end{pmatrix} = \begin{pmatrix} -k_c(a+b) & k_c c \\ k_s a & -k_s c \end{pmatrix}\begin{pmatrix} \Delta P_c \\ \Delta P_s \end{pmatrix} + \begin{pmatrix} 0 \\ k_s \end{pmatrix}\Delta \dot{m}_{dist} + \begin{pmatrix} k_c d \\ -k_s d \end{pmatrix}\Delta u \qquad (6)$$

with,

$$a = \frac{k_v E^{(u^*/100)}\left(2P_c^* - P_s^*\right)}{2\sqrt{\left(P_s^* - P_c^*\right)P_c^*}} \qquad\qquad c = \frac{k_v E^{(u^*/100)}P_c^*}{2\sqrt{\left(P_s^* - P_c^*\right)P_c^*}}$$

$$b = \frac{\dot{m}_{demand}}{P_c^{SetPoint}} \qquad\qquad d = k_v \frac{\ln(E)}{100}E^{(u^*/100)}\sqrt{\left(P_s^* - P_c^*\right)P_c^*}$$

This model is illustrated in Figure 4.

3 Design

In this section we will examine the quantitative design of the pressure controller. Section 3.1 considers the feedback design of the suction pressure loop. Section 3.2 examines the performance benefits available from using a feed-forward signal of the surge vessel pressure

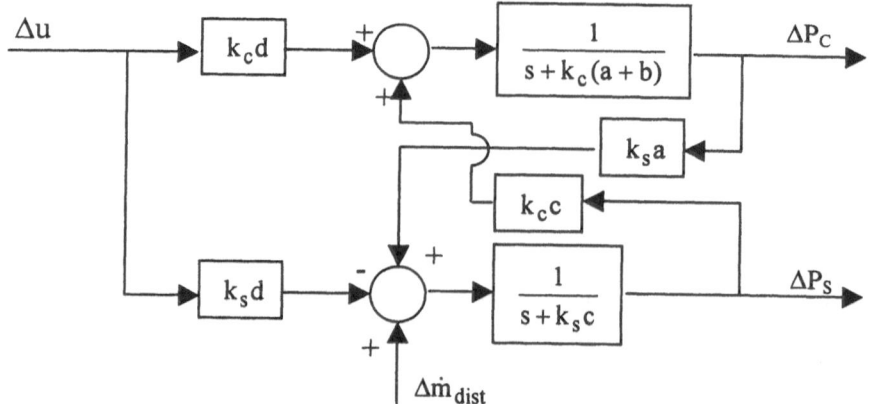

Figure 4 – Linearised Model

3.1 Quantitative Feedback Design of P_c

The client's design requirement is that the compressor suction pressure is never outside 50kPa of set-point for a given set of "worst case" disturbances at $\dot{m}_{dist}$ specified as a result of expected upstream plant operations. The feedback structure shown in Figure 5 will be used, with,

$$Q_d(s) = \frac{k_s k_c c e^{-sT_d}}{s^2 + (k_c(a+b) + k_s c)s + k_s k_c bc}$$
- Disturbance transfer function

$$Q(s) = \frac{k_c d s e^{-sT_d}}{s^2 + (k_c(a+b) + k_s c)s + k_s k_c bc}$$

$$\approx \frac{k_c d e^{-sT_d}}{s + k_c(a+b)}$$
- The "plant"

$$A(s) = \frac{1}{s\tau_A + 1}$$
- Valve actuator dynamics

$$H(s) = \frac{1}{s\tau_s + 1}$$
- Pressure sensor dynamics

Because of the low-passing nature of the surge vessel, with high gain feedback at low frequency (e.g. a PI controller), the transfer of this disturbance to the output,

$$T_{P_C/\dot{m}_{dist}} = \frac{Q_d}{1 + GAQH} = \frac{Q_d}{1 + L} \tag{7}$$

is band passing. For this design at least, nothing can be done about the physical plant hardware so the only way to maximise the disturbance rejection is to maximise the feedback loop bandwidth.

The design will be implemented in digital hardware and the approach of Eitelberg (1988) (see also Boje (1990) and Eitelberg and Boje (1991)) will be taken so that the specification of the sampling rate is an *outcome* of the design. Eitelberg shows that with respect to w-domain design $z = \dfrac{1 + wT/2}{1 - wT/2}$ and $w = \dfrac{2}{T}\dfrac{z-1}{z+1}$ where T is the sampling rate,

i) $Q_z(w) \approx Q_s(w)(1 - wT/2)$ \hfill (8)

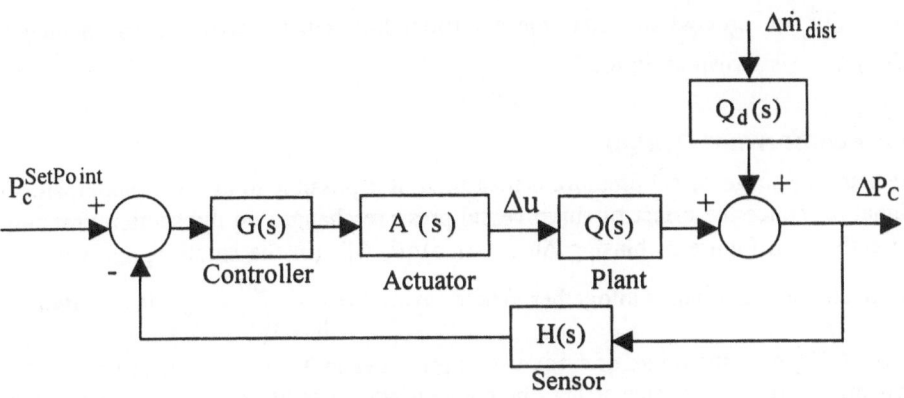

Figure 5 – Linearised model for feedback design

- the discrete plant model in the w-domain is approximately the same as the Laplace domain (continuous) plant model with the addition of a sampling effect (1-wT/2), and,
ii) specifications in the w-domain and in the s-domain have approximately equivalent performance.

Because of the very fast measurement signal and very fast valve actuator to be used, the design will be of a proportional and integral (PI) controller. This means that no derivative action will be allowed in the design. This constraint means that the controller phase is never more than 0°, a pretty realistic constraint in such a system where the basic engineering is properly done so that the plant capacity cannot be increased by feedback. (It may be necessary to relax this constraint somewhat during commissioning (tuning) if the process phase lag is more than contemplated here.) In addition to the constraint on controller structure, a single robust stability specification, $|1/(1+L)| < 4dB \ \forall \ \omega$, is used. This guarantees minimum gain and phase margins of 11dB and 55° respectively (based on linearisation of the theoretical model). The quantitative feedback design philosophy allows enhancement of the specifications as a result of design insight obtained from simulation and operating experience. We may then find it necessary to directly specify constraints on $\left| T_{P_C / \dot{m}_{dsit}} \right|$.

The design is performed using the Matlab QFT toolbox (Borghesani, *et al*, 1994) and is illustrated in Figure 6 along with the PI controller,

$$G = k_p \frac{(s + 1/T_i)}{s}, \text{ with reset time, } T_i = 5.9s \text{ and proportional gain, } k_p = 0.09$$

%/kPa and a sampling rate of 250ms. (9)

Notice that the non-minimum phase lag behaviour in the process and controller discretisation constrains the proportional gain (i.e. high frequency gain) to be quite low. For the process industry, the controller has rather fast reset. The achieved $\left|T_{P_C}/\dot{m}_{dsit}\right|$ is shown in Figure 7.

3.2 Feedforward Design

Because the surge vessel pressure is measured, it is possible to use this (anticipatory) signal to reduce the effect of surge vessel pressure changes on the suction pressure. Directly from Figure 5, biasing Δu by $-(c/d)\Delta P_s$ will (in the small signal analysis) eliminate the disturbance altogether. The ratio $c/d = \dfrac{50}{\ln(E)\left(P_s^* - P_c^*\right)}$ is uncertain in [0.02, 0.26] over the range of surge vessel pressures in Table 1. With valve actuator dynamics and pressure sensor lag, the compensation would only be effective at low frequency where the integrator of the PI controller can eliminate the disturbance. The worst case occurs when a significant inflow occurs with the surge vessel pressure low and the valve almost fully open. A compromise feedforward bias signal, $U_{bias}(s)$ = F(s) P_s(s) is used, with a low frequency washout,

$$F(s) = -k_F \frac{T_w s}{T_w s + 1}, \text{ with } k_F = 0.15 \text{ \%valve per kPa, and, } T_w = 5s. \quad (10)$$

3.3 Results

A typical inflow disturbance profile that results from a high pressure upstream process discharging into the surge vessel is shown in Figure 2. Non-linear simulation results are shown in Figure 8. The Matlab-Simulink model of Figure 3 includes correct implementation of a discretised PI "velocity" algorithm and non-linear valve model. Only with disturbance feedforward does the suction pressure satisfy (just!) the client requirement of ±50kPa of set-point. (It is expected the actual dP_s/dt in the surge vessel may be less rapid than the given data suggests because of distributed effects. In addition the compressor instrumentation will react to the measured pressure that will be a damped version of the calculated pressure.)

4 Conclusions

This paper has considered the quantitative design of a suction pressure controller for a gas recovery compressor. The controller maintains the compressor suction pressure within manufacturer's tolerances during significant mass flow disturbances into an upstream surge vessel. In this application, the design was required before equipment purchase. The QFT design method has allowed consideration of the valve non-linearity, approximate distributed gas dynamics, valve actuator and sensor dynamics, the effect of digital implementation and the use of a feed-forward measurement of surge vessel pressure. It also provides engineering insight, enabling design improvements to be made both during the design phase and during on-line commissioning as process knowledge improves and operating experience increases.

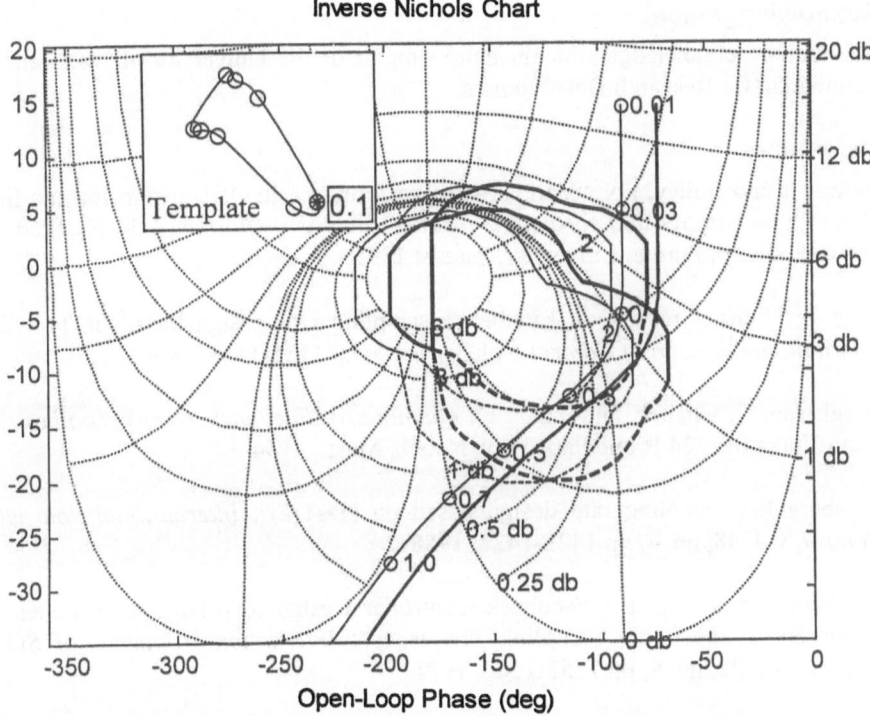

Figure 6 – Maximum bandwidth design with robust stability margin of 4dB

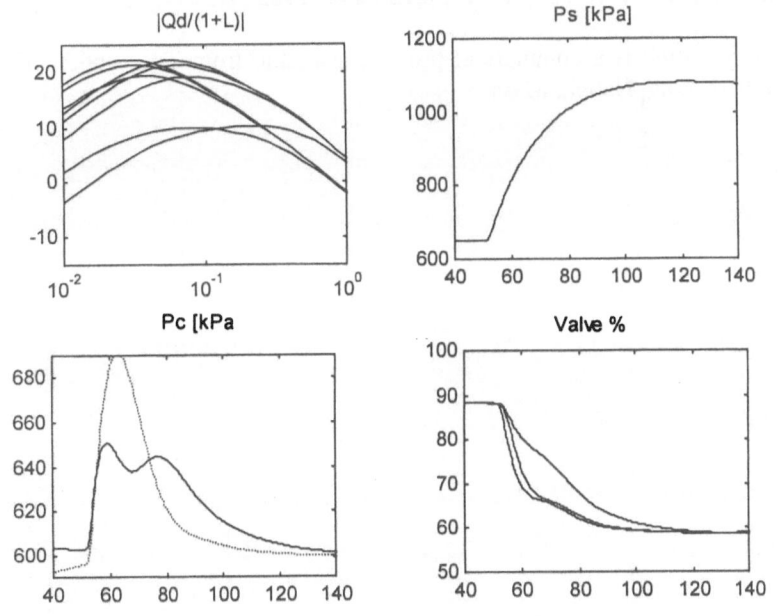

Figure 7 - Design results in frequency and time domains

Acknowledgement

The author acknowledges the financial support of the University of Natal and the Foundation for Research Development.

References

Baños, A and Bailey, FN, "Validation of performance in QFT design for non-linear plants", *Quantitative and parametric feedback theory symposium*, Eds. Nwokah ODI and Chandler P, Purdue University, August 1995.

Boje E, "Further results on Eitelberg's sampling rate design based on (1-sT/2)", *International Journal of Control*, vol. 51, no. 5, pp 1155-1158, 1990.

Borghesani C, Chait Y and Yaniv O, *Quantitative Feedback Theory Toolbox*, The MathWorks Inc., 24 Prime Park Way, Natick, Mass., 1994

Eitelberg E, "Sampling rate design based on (1-sT/2)", *International Journal of Control*, vol. 48, no. 4, pp 1423-1432, 1988.

Eitelberg E and Boje E, "Feedback Controller Design for Plants with modes and Disturbances above the Sampling Frequency", *International Journal of System Science*, vol. 22, no. 9, pp 1553-1562, 1991.

Horowitz, I, "Feedback systems with non-linear uncertain plants", *International Journal of Control*, vol. 36, no. 1, pp 155-171, 1982.

Matlab-Simulink is a commercial product available from The MathWorks Inc., 24 Prime Park Way, Natick, Mass.

PART IV

DRIVING SIMULATORS

24

Driving Simulators as Research and Training Tools for Improving Road Safety

E. Blana

1   Introduction

Driving simulators have been initially developed for military training purposes using the technology of flight simulators. To date, they are widely used not only for training but also for research. They have been used as research aids in a number of civil engineering, transport, psychology and ergonomics fields such as: innovative road design (e.g. testing the design of new tunnels, innovative highway design and road delineation, traffic calming); intelligent transport systems (e.g. new in-vehicle navigation systems, Head-Up-Displays, active pedals) and impaired driver behaviour (driving behaviour affected by drugs, alcohol, severe brain damage, fatigue); vehicle dynamics and layout (e.g. testing ABS, 4-wheel drive; interior design). They are also used in a number of countries for training professional drivers (1, 2). The primary thought behind these applications is the increase and enhancement of the provided road safety. This chapter will start by defining the problem of road safety in Greece and detail five case-studies of how driving simulators could be used both as research and training tools to improve Greek road safety standards.

2   Road accidents and their cost in Greece

Road accidents are the second cause of death in Greece (after deaths caused from malignant neoplasms of endothoracic organs) and the first cause of death for young people (between 25 and 34 years old). Between 1988 and 1995 there was a steady increase in the number of accidents (they increased 5% the last 8 years) and for 1995 they comprised the 91.9% of the total number of accidents in Greece (including falls; fire accidents; domestic and job-related accidents; rail and airplane accidents) (see Table 2-1).

One of the possible causes of the increased number of accidents in Greece the later years may be the increase of car ownership and the degree people use their cars. In 1980, there were 1.28 million cars and 120702 motorcycles (50 - >750 cc) in Greece, whereas by 1988 there were 2.21 million cars (72% increase) and 203582 (80% increase). To date, only in the area of Athens, there are 1.32 million cars (the fleet of cars increases almost 5% on average every year since 1991) whereas almost 40% of the total number of road accidents happens in the metropolitan area of the city of Athens.

Table 2-1 Total number of accidents in Greece

Year	Total	Road	Other
1988	27288	87.3%	12.7%
1989	26767	88.7%	11.3%
1990	25384	89.3%	10.7%
1991	25664	90.5%	9.5%
1992	26488	91.1%	8.9%
1993	26797	90.1%	9.9%
1994	26034	91.8%	8.2%
1995	25572	91.9%	8.1%

Table 2-2 Accidents and Casualties by severity: GR: 1988-1995

Year	Total Accid.	Total Casualt.	Casualties		
			Fatal (%)	Serious (%)	Slight (%)
1988	23832	35132	1784 (5.08)	3902 (11.11)	29446 (83.82)
1989	23737	35408	2016 (5.69)	4014 (11.34)	29468 (83.22)
1990	22672	33389	1986 (5.95)	6930 (20.76)	27515 (82.41)
1991	23230	33854	2013 (5.95)	3883 (11.47)	27958 (82.58)
1992	24125	34781	1995 (5.74)	4000 (11.50)	28786 (82.76)
1993	24365	34652	2008 (5.79)	3278 (9.46)	29367 (84.75)
1994	23892	34144	2076 (6.08)	3387 (9.92)	28681 (84.00)
1995	23492	34043	2149 (6.31)	3491 (10.25)	28403 (83.43)

Accidents as well as their casualties are classified into three categories: fatal, serious and slight. In Greece, every year and for the period 19888-1995 happened approximately 24000 road accidents with 35000 casualties. About 9 people die in every 100 accidents each year, or approximately 6 people every day!! (see Table 2-2). The problem of road safety has been addressed occasionally from the state using various measures but it has been proved very difficult either to implement, monitor, evaluate and/or enforce any of these measures. Possible reasons could be the impatience of people to «see the results» (therefore not enough time for implementation, monitoring and evaluation) and the unwillingness of people to understand and face the fact that «it can also happened to them». The cost of road accidents is tremendous not only in monetary terms but also in political and social terms (5).

Road accidents cost a vast amount of money to the Greek state either direct or indirect (in terms of decreasing population, loss of labour force, compensation to accident casualties but also to their relatives in case of a fatal accident, hospital expenses, police/fire-brigade expenses and other administrative costs) as well as to insurance companies. The average cost of an accident varies according to the type of collision (see Table 2-3). The most expensive type of accident to pay off is when a vehicle crashes to a solid object and probably this is the reason why the collision damage waiver insurance has a very high premium in Greece.

Table 2-3 Average cost by type of collision for vehicles and casualties

Type of collision	Cost (dra)
Head-on	1,598,291
Side	1,247,714
Rear	1,383,038
With solid object	3,617,084
With pedestrian	618,056

Since every year approx. 24000 accidents happen on the Greek roads, the state and the insurance companies have to pay on average 41 billion drachmas (120 million ECU) per year as compensation to the accident victims. On the other hand hospital cost also vary according to the type of collision (see Table 2-4). The most expensive hospital cost is for head-on collisions and the average cost is 320,000 drachmas. This means that for the 34000 casualties per year, the state and the insurance companies pay off at minimum approximately 11 billion drachmas (32 million ECU) to the hospitals (in 1994 prices). In the above calculations, the cost of calling the police and the fire-brigade and pay for their expenses, the cost of time-loss of other road users waiting for the road to clear, the cost of absence from work, the cost of emotional and psychological disturbance are not included.

It becomes obvious that road accidents has become the number one problem in Greece and the number one priority for the state and the insurance companies. Driving simulators could significantly contribute to the decrease of road accidents even if they could only be used as demonstrative and training tools. The following section will detail five case-studies which will explain how simulators could benefit the Greek nation by conducting experiments, proving training and demonstrating road and traffic related issues which are of particular importance and applicability to the Greek driving standards.

Table 2-4 Average hospital cost per type of collision

Type of collision	Average hospital cost per accident (dra)
Head-on	526,933
Side	320,387
Front-side	349,621
Rear	234,797
With parked vehicle	239,525
With solid object	348,201
With pedestrian	289,774
Run-off	252,852
Overturning	303,767

3 The use of driving simulators in Greece as research and training tools for improving road safety

Driving simulators is a totally new field of technology, development and applications in Greece, either for research or training purposes. However, it is considered economically feasible and beneficiary (for the Greek state, the Greek society and the Greek private businesses) to develop and use driving simulators for the improvement of road safety.

3.1 Case study 1: Driving simulators and road environment

About 40 % of fatal road accidents happen every year on motorways, 20% on rural roads and 40% on any other Greek road. The percentage of fatal accidents on motorways has been remained almost constant along the period 1988-1995, however there was a small increase during the years 1989 and 1990. On the other hand, serious and fatal accidents have a different distribution according to the different type of roads compared to the fatal accidents. For the serious accidents, approximately 30% happen on motorways, 20% on rural roads and 50% on any other road, whereas the respective percentages for the slight accidents are 18%, 15% and 67% (see Table 3-1 and Table 3-2) (4). Greek motorways have been improved significantly the last 10 years. Major alternations to their alignment, environment and furniture are happening the last 5 years and most of the motorways look like continuous work zones today. On the other hand, the rural and urban road network has not been improved significantly, although traffic calming measures has been introduced to a number of municipalities of Attica and major cities of Greece the last 4 years.

Table 3-1 Road accidents by severity (fatal) by road type: GR: 1988-1995

Year	Total	Fatal (%)		
		M/w	Rural	Other
1988	23832	627 (39.78)	338 (21.45)	611 (38.77)
1989	23737	738 (42.56)	362 (20.88	634 (36.56)
1990	22672	724 (41.07)	333 (18.89)	706 (40.05)
1991	23230	673 (38.44)	349 (19.93)	729 (41.63)
1992	24125	698 (39.64)	373 (21.18)	690 (39.18)
1993	24365	694 (38.47)	397 (22.01)	713 (39.52)
1994	23892	700 (38.19)	437 (23.84)	696 (37.97)
1995	23492	750 (39.54)	433 (22.83)	714 (37.64)

Table 3-2 Road accidents by severity (serious and slight) by road type: GR: 1988-1995

Year	Total	Serious (%)			Slight (%)		
		M/w	Rural	Other	M/w	Rural	Other
1988	23832	866 (30.43)	677 (23.79)	1303 (45.78)	3420 (17.62)	2994 (15.43)	12996 (66.96)
1989	23737	886 (30.87)	670 (23.34)	1314 (45.78)	3487 (18.23)	2811 (14.69)	12835 (67.08)
1990	22672	825 (29.10)	691 (24.37)	1319 (46.53)	3288 (18.19)	2707 (14.98)	12079 (66.83)
1991	23230	765 (27.03)	632 (22.33)	1433 (50.64)	3338 (17.90)	2827 (15.16)	12484 (66.94)
1992	24125	790 (26.49)	633 (21.23)	1559 (52.28)	3497 (18.04)	3004 (15.50)	12881 (66.46)
1993	24365	697 (29.26)	588 (24.69)	1097 (46.05)	3634 (18.01)	3241 (16.06)	13304 (65.93)
1994	23892	751 (30.83)	661 (27.13)	1024 (42.04)	3652 (18.61)	3240 (16.51)	2731 (64.88)
1995	23492	712 (28.74)	682 (27.53)	1083 (43.72)	3512 (18.37)	3061 (16.01)	12545 (65.62)

Driving simulators could be used to investigate various parameters that affect road safety on Greek roads

* driver behaviour along the work zones on motorways
* driving Greek «habits» on motorways (e.g. driving on the emergency lane)
* driving behaviour along the toll zone on motorways
* driving Greek «habits» on rural roads (single carriageway roads with no medians) (e.g. driving on the left of the lane and force other drivers to illegal overtaking)
* construction «habits» on Greek urban roads for reducing speed (e.g. constructing road humps of any type, anywhere using no design guidelines).

The above areas of applications are only some of which a driving simulator could be used. Public authorities (e.g. the Ministry of Environment and Public Works) and private businesses (construction companies) could significantly benefit from the use of a driving simulator to investigate driver behaviour on the Greek roads.

3.2 Case study 2: Driving simulators and type of accidents

It has been demonstrated earlier that the cost of road accidents varies significantly according to the type of collision (see section 2). It would be extremely beneficial to investigate driver behaviour under extreme road environment and traffic conditions and monitor driver avoidance actions. Driving simulators provide an inherently safe environment. There is no endangerment to the driver or other road users under these critical driving conditions. Drivers can survive a crash and learn from their mistakes. In particular accidents involving two vehicles or vehicle hitting a pedestrian or animal constitute on average the 75% of all accidents. On the other hand, a non-negligible percentage (13%) are run- off-the-road accidents (see Table 3-3) (4).

Possible research but also training areas where a driving simulator could be used are::
* monitoring, analysing and evaluating avoidance manoeuvres
* monitoring, analysing and evaluating driver behaviour under critical traffic and road environment conditions
* evaluating the effectiveness of safety barriers and safety bollards located by the edge of the road
* evaluating the effectiveness of hazard warning signs on tight bends (e.g. chevron signs)
* risk-perception studies, especially when driving on curves
* gap acceptance studies
* drivers' training under traffic and road environment scenarios of high risk

These type of studies would mostly benefit insurance companies, the Ministry of Environment and Public Works (Division of Traffic Signs) and private companies that make traffic signs and safety fences.

3.3 Case study 3: Driving simulators and traffic violations (Greek driving «culture»

There are approximately 1.5 million traffic violations every year in Greece. Almost 50% of them relate to illegal parking. Traffic police register 24 different traffic violations and according to the seriousness of the violation, takes the proper action. The following table (Table 3-4) shows the five most commonly occurred traffic violations on Greek roads. These are speeding, red-light violation, illegal overtaking, moving on the oncoming traffic lane and moving wrongly on a one-way street (4).

Speeding constitutes on average the 35% of traffic violations (it has been increased almost 30% since 1988), red-light violation the 20%, illegal overtaking the 12%, moving on the opposite lane the 7%, moving wrongly on a one-way street the 7% and any other traffic violation the 19% (see Table 3-4). It is well known that speeding is one of the major factors that can cause a road accident as well as the violation of «priority» signs (e.g. traffic lights, «give-priority» sign, «stop» sign). Usually the derived accident is fatal or serious.

Table 3-3 Road accidents by type of accident: GR: 1988-1995

Year	Total	Type of accident (%)				
		Collis 1*	Collis 2**	Run-off	Over-turn	Other
1988	23832	18984 (79.66)	1071 (4.49)	2872 (12.05)	804 (3.37)	101 (0.42)
1989	23737	18853 (79.42)	1126 (4.74)	2881 (12.14)	754 (3.18)	123 (0.52)
1990	22672	17706 (78.10)	1147 (5.06)	2977 (13.13)	766 (3.39)	76 (0.34)
1991	23230	17765 (76.47)	1197 (5.15)	3042 (13.10)	961 (4.14)	265 (1.14)
1992	24125	18694 (77.49)	1158 (4.80)	3001 (12.44)	1087 (4.51)	185 (0.77)
1993	24365	18843 (77.34)	1253 (5.14)	3120 (12.81)	943 (3.87)	206 (0.85)
1994	23892	18159 (76.00)	1258 (5.27)	3309 (13.85)	1036 (4.34)	130 (0.54)
1995	23492	17733 (75.49)	1143 (4.87)	3523 (15.00)	1015 (4.32)	78 (0.33)

*collision of a vehicle with pedestrian, other vehicle(s) and or animal
** collision of a vehicle with a fixed object (e.g. wall, tree, fence)

A Greek study using questionnaires (6), showed that although 85% of the Greek drivers believe in the existence of speed limits on rural roads (including motorways), there is a significant percentage of drivers (30%) believing that speed limit compliance does not lead to a reduction of accidents (this percentage almost match with the percentage of drivers who are caught by the police for speeding). Although 80% of the drivers asked, stated that they obey the speed limit most of the times, they also stated that they believe that only 10% of the «other» drivers obey the speed limit. Greek drivers believe that the three most important reasons for not complying with the speed limit is: a) the speed limit is unrealistic; b) they want to reach their destination as soon as possible and c) there are no police cars in sight whereas they believe that «other» drivers do not comply with the speed limit because a) they want to reach their destination as soon as possible; b) they are no police cars in sight and c) they do not believe that speeding is dangerous.

These differences in their opinion about how many drivers obey the speed limit most of the times when talking about themselves and the «other» drivers or which are the reasons for not complying with the speed limit implies in a way that they are afraid of taking the responsibility of braking the law and the consequences this action has to themselves and the other road users and accuse the «other» drivers. Choosing their speed on curves depends on how far they can see, the road surface condition, the apparent severity of the curve, the presence of hazard warning signs (e.g. chevrons).

In addition, besides speeding behaviour, Greek drivers on rural areas tend to believe that roads belong to them, therefore they force other (non-local) drivers either to unnecessary, illegal and dangerous overtaking and/or to move to the opposite lane for some distance (until they let them to complete the overtaking manoeuvre and reenter the main stream). On the other hand, Greek drivers on urban roads tend to believe that red-light violation can save them some minutes (e.g. when violating more than 2-3 traffic lights in a row, since most of them are synchronised), and they ignore stop signs (they believe that other drivers would stop in any case since other drivers would not like to get involved in an accident). However, the reality is always different, thus the observed number of these types of violations and the respective type of accidents is very high on Greek roads.

Combining the results from the different studies and statistical data from the Greek police, driving simulator experiments could be conducted relative to the Greek standards. The Greek driving «culture» could be monitored and analysed and solutions could be found that would address the problem of this different behaviour. Simulators then, at a second stage, could be used both for training but also for demonstrative purposes, e.g. drivers could realise the effects of this improper behaviour (without losing their lives, but still remember the feeling of being at high risk and crash).

3.4   Case study 4: Driving simulators and drinking-and-driving

Drink-and-drive accidents have been increased the last years in Greece. A 1993 study showed that the percentage of accidents involving drunk drivers was 7.5% (3) whereas by 1995 this percentage has been increased to 12.5% (a 66 percent increase) (7). Although drink-and-drive accidents are only a small percentage of road accidents, they are responsible for the 43% of all fatal accidents (for the year 1993).

However as Table 3-5 shows, only a small percentage of drivers (not being involved in an accident) are examined (on average 4.5%) every year since 1985 with the exception of year 1993 (a «don't drink and drive» campaign started from the mass media) and the last two years. In addition for the majority (approx. 60%) of these drivers the results of the breath-test is not known. On the other hand, for those drivers whose test results were known, almost 40% of them had BAC more than the legal. The legal blood alcohol level (BAC) for Greece is $0.5^0/_{00}$ (or 0.5 gr/lt) (3).

Table 3-4 The five most import traffic violations: GR: 1988-1995

Year	No of traffic violations*	Speeding (%)	Red light violation (%)	Illegal overta king (%)	Moving on the opposite lane (%)	Moving wrongly on a one-way street (%)
1988	183290	54526 (29.75)	36662 (20.00)	22691 (12.38)	14840 (8.10)	20355 (11.11)
1989	167377	48272 (28.84)	42089 (25.15)	14802 (8.84)	14507 (8.67)	18941 (11.32)
1990	206526	84137 (40.74)	42988 (20.81)	21011 (10.17)	15260 (7.39)	18889 (9.15)
1991	227236	102779 (45.23)	44000 (19.36)	29155 (12.83)	16766 (7.38)	16938 (7.45)
1992	240668	97060 (40.33)	52093 (21.65)	31461 (13.07)	21532 (8.95)	17839 (7.41)
1993	188408	72686 (38.58)	41133 (21.83)	23366 (12.40)	19262 (10.22)	14902 (7.91)
1994	171526	70024 (40.82)	36425 (21.24)	21285 (12.41)	16551 (9.65)	12474 (7.27)
1995	191092	70805 (37.05)	39675 (20.76)	23105 (12.09)	14649 (7.67)	12675 (6.63)

* Traffic violations that affect directly driver's road safety (including speeding, illegal overtaking, moving on the opposite lane, «give-priority» violation , driving too close to the front vehicle, drunk driving, red light violation, violation of policemen orders, moving illegally on a one-way street, not moving on the right side of the road, violation of pedestrian priority and driver distraction

A Greek study showed that on accidents involving two vehicles, drivers with BAC between 0.5 gr/lt and 1.0 gr/lt are 5 times more risky to get involved to an accident, drivers with BAC between 1.0 gr/lt and 1.5 gr/lt are 6 times more risky, whereas drivers with BAC greater 1.5 gr/lt are 14 times more risky compared to a sober driver. It seems that increasing the BAC, the risk index follows an exponential distribution. For single vehicle accidents, the respective risk index according to the various BAC vary significantly: for the first category (BAC between 1.0 gr/lt and 1.5

gr/lt) the risk index is 4.5 more than of a sober driver, for the second category (BAC between 1.0 gr/lt and 1.5 gr/lt) the risk index is 8 times more whereas for the last category (BAC >1.5 gr/lt) is 21 time more compared to a sober driver (7).

Driving simulators could be used both for research and demonstrative purposes in the case of drink-and-drive accidents. Various experiments in the simulator could demonstrate to young drivers up to the age of 34 (whether they are novice or experienced) the effect that alcohol has to their driving behaviour and its consequences. A similar study (8) has been conducted in the University of Montreal in Canada. A pedagogical program was developed using a driving simulator for dissuading alcohol-impaired driving in youth. It was proved that the goal of disseminating knowledge and modifying emotional responses can be achieved by using simulation techniques. On the other hand, an experiment using drunk subjects could verify that indeed the legal limit of 0.5 gr/lt BAC is the right one (or not) for the Greek people, since most of studies had been based on police data only.

Table 3-5 BAC results for drivers involved in an accident: GR: 1985-1993

Years	Involved drivers /examined drivers (%)	BAC $<0.5^0/_{00}$	BAC $>0.5^0/_{00}$ (% of known results)	Unknown results (% of examined drivers)
1985	34435/1036 (3.0%)	306	191 (38.4%)	539 (52.0%)
1986	31295/1063 (3.4%)	258	197 (43.3%)	609 (57.2%)
1987	30430/1145 (3.8%)	437	255 (36.8%)	450 (39.4%)
1988	34498/1989 (5.8%)	336	285 (36.8%)	368 (37.2%)
1989	33285/1441 (4.3%)	402	259 (39.1%)	780 (54.1%)
1990	32187/1453 (4.5%)	352	240 (40.5%)	861 (59.2%)
1991	34227/1958 (5.7%)	314	229 (42.1%)	1415 (72.2%)
1992	36494/1932 (5.3%)	313	219 (41.1%)	1400 (72.4%)
1993	36917/2584 (7.0%)*	268	237 (46.9%)	2079 (80.4%)

*by 1993 the mass media began to report on the subjects, therefore law enforcement increased!

3.5   Case study 5: Driving simulators and training

The current situation in Greece concerning driving training and education is the following:
- No schools exist for training driving instructors
- No theoretical lessons are available (relative to the operation of the vehicle)
- No first aid lessons are provided

- The learner gets no driving lessons on motorways, driving at night or generally under difficult traffic and environmental conditions (e.g. rain, fog)
- The learner is not taught driving techniques but only what is required by the examiners
- The examiners of the candidate drivers have no special training
- Only about 1/40 of the learners and 1/90 of the instructors wear their seatbelts during the lesson (research conducted in Athens in 1992).

Driving simulators could be widely used in Greece for training purposes, not only to train candidate and novice drivers but also to train the driving instructors and the examiners. They could also be used to train professional drivers (e.g. coach and buses drivers, trucks drivers).

4   Conclusions

This chapter introduced the severe problem of decreased road safety in Greece and showed ways of using driving simulators both for research, training and demonstrative purposes to improve road safety in Greece (and not only). The high acquisition cost of a driving simulator could be pay off if by its use we could avoid the cost of hospitalise 100 people out of the 32000 who are injured every year or the cost of insurance claims for 100 damaged vehicles by any type of collision.

References

1. Allen, R.W., Klein, R.H. and Ziedman, K (1979). Automobile research simulators: a review and new approaches. Transportation Research Board 706, pp 9-15.
2. Blana, E. (1996). A survey of driving research simulators around the world. ITS Working Paper 481. Institute for Transport Studies. University of Leeds. England.
3. Papadopoulos, J. (1996). Accidents: Prevention is possible. Supreme Confederation of large families of Ellas (A.S.P.E.).
4. Greek Police Statistical Data, 1995.
5. Mintsis, G., Taxilaris, C., and Petropoulos, J. (1994). Contribution in the estimation of the costs of road accidents with injured persons. Proceedings of the 1st Panhellelinic Congress on Road Safety. Thessaloniki, 28-29 March.
6. Blana, E. (1994). Comparison of Greek and English drivers' attitudes and behaviour regarding speed. MSc (Eng) Thesis. Institute for Transport Studies. University of Leeds. England.

7. Georgiopoulos, E. (1997). Investigating the effect of alcohol to drivers' risk factor. Diploma Thesis. National Technical University of Athens. Greece.

8. Bergeron, J., Perraton, F., Paquette, M. , Joly, P., Laviolette, E. and Godin, L. (1996). Application of driving simulation techniques for the dissuasion of alcohol-impaired driving. Proceeding of the Driving Simulation Conference 1997. September 8-9, Lyon, France.

25

Critical Judgements on Feasible Emergency Manoeuvres: A Comparative Study Between Test Track and Driving Simulator

J. Fréchaux and G. Malaterre

As many other research institutes or companies which use driving simulators, INRETS have started to study validity problems over the last few years.

The question is : What kind of driving activity should be studied and using which simulator ?

In our approach, we try to answer this question :

(a) studying different tasks involved in driving activity, from the more basic ones to the more complex ones (involving decision makings and interactions with intelligent traffic).

(b) isolating the different technical variables of the simulator (image quality, width of the visual field, transport delay...) and testing their importance in relation to the task studied, i.e. the way they influence the driver activity.

As a result, we could understand, for each task studied, the role of different characteristics of the simulator in order to make it more efficient.

Our criteria of validity are comparisons between performance results in both situation, actual driving and simulator, and moreover, comparison between the psychological mechanisms the drivers rely on.

In our validation program, which has been under progress for 2 years, we firstly examined two tasks involving perceptual activities : estimations of distance and speed (1). Concerning the speed experiment, after comparing simulator data to test track ones (2), we noted that subjects had no major problem to estimate speed, despite the large dispersion on the simulator.

On the other hand, subjects have difficulties to estimate distances longer than 80 metres (they tended to under produce distances), probably because of the poor image resolution on our station, which is the case on most of the low cost simulators. Texture affected the subject's responses for both experiments. The width of the visual field, which was only studied in the speed experiment, also affected the subjects' performance. On the other and, the others variables (road markings, sound) have no effect on subjects' performance.

This chapter concerns a more complex task which is the estimation of feasible emergency manoeuvres by the driver (braking and swerving). This task implies both distance and speed estimations. As we have seen above, subjects seem to have difficulties to assess long distances which is necessary to evaluate the moment to trigger the manoeuvre at high speeds. We can make the hypothesis that subjects probably tend to get closer to the obstacle and thus to trigger the manoeuvre later than test track subjects would do. This task implies actions on vehicle controls too. The resulting questions is : does the simulator driver regulate the system as a driver in an actual situation ?

2   Description of the experiment

The work presented here is a replicate of an experiment carried out on a test track by Malaterre in 1987 (3). The aim of this research was to determine if subjects placed in a simulated emergency situation (the subjects never actually performed the manoeuvre), were able to perceive that above a certain speed swerving remained possible nearer to the obstacle than braking. This is a complex judgement. It relies on a good perception of Time to Collision but also on a good awareness of the dynamic properties of the vehicle driven.
We used a similar procedure on the simulator.
24 subjects participated in this experiment. 14 males and 10 females ranging from 21 to 40 years of age, each with more than 100,000 kilometres of driving experience.
The experimental variables were the following :
. gender (between subjects),
. manoeuvre : braking or swerving (within subjects),
. speed : varying from 40 to 120 km/h, 8 levels (within subjects),
. width of the visual scene (specific to simulator) : 64° or 25° (between subjects).
The visual scene was projected on a flat screen 3,20 m x 2,40 m providing a 64° view angle. It was computed by a Silicon Graphics Reality Engine 33 MHz, and displayed by a video projector Sony VPH 1272QM. The system was interfaced with a fixed base vehicle mock-up.
The subjects who drove the simulator were invited to indicate by pressing a switch the last moment beyond which the manoeuvre would not be possible. The obstacle was simulated by plastic cones and the subjects were instructed to drive between them after they had pressed the switch, and to keep their speed unchanged.

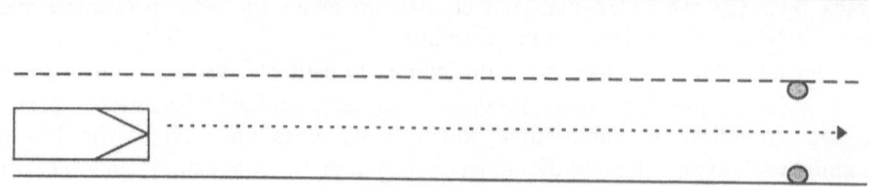

Figure 1 : experimental design for the experiment on feasible emergency manoeuvres.

3   Results

3.1   Theoretical longitudinal acceleration

The data were converted into theoretical longitudinal acceleration for both manoeuvres, in order to make the comparison possible. They were collected for speeds varying from 40 to 120 km/h.

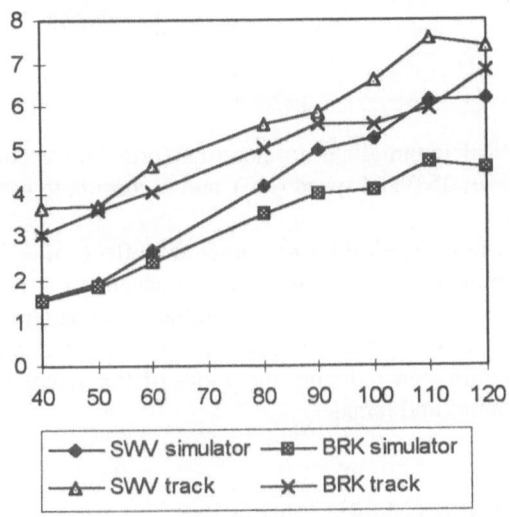

Figure 2 : theoretical longitudinal acceleration ($\gamma = S / 2 \times TC$; with S for speed and TC for Time to Collision) according to situation (simulator and test track), manoeuvre (swerving and braking), and speed (km/h).

The results exhibit similar tendencies on simulator and test track. In both cases swerving was estimated possible closer to the obstacle than braking for comparable speeds, which fits with the physical model. On the other hand, whatever the manoeuvre, accelerations corresponding to drivers' estimates are significantly lower on the simulator. This result reveals a tendency to adopt larger safety margins on simulators, maybe because drivers are not so confident in their estimates in this situation than in actual driving. Women perceive the advantages of swerving as well as men, but initiate both manoeuvres farther from the obstacle. This last point is consistent with previous results which have shown that women tend to keep larger margins (4). Contrary to classical results in simulation, standard deviations are not significantly higher on the simulator. Probably because they were already high in test track situation (range = 5).

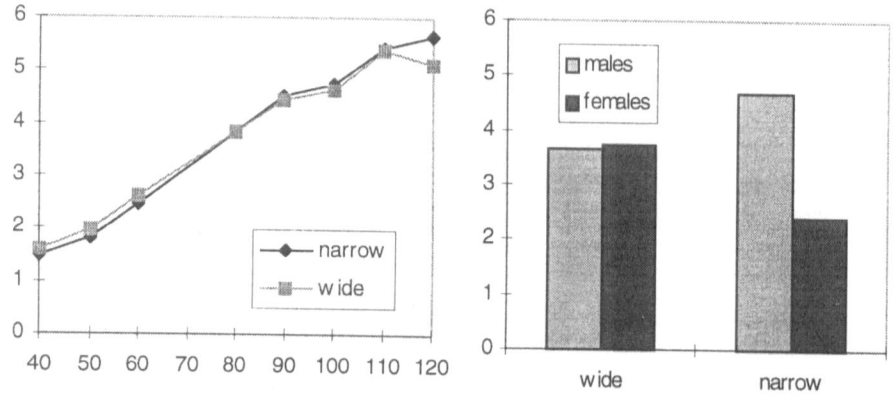

Figure 3 : theoretical longitudinal acceleration (ordinate) according to the visual field width (64° versus 25°) and speed (left), and according to gender (right).

Surprisingly, visual field width does not show any effect on estimates (anova). On the other hand, women's theoretical longitudinal acceleration is twice smaller with narrow visual field than men's, i.e. they initiate manoeuvres farther from the obstacle. This result between males and females is hard to explain. We know that vection sensation is affected reducing the width of the visual field but this effect should affect both males and females.

3.2   Feasible / non feasible comparison

As in the test track study we computed theoretical transversal acceleration which should result of the actual manoeuvre. We have also taken into account a reaction time (400 ms) for braking.

We then computed numbers of estimates corresponding to non feasible manoeuvres, according to admissible acceleration thresholds. For braking we have fixed it at 8 m/s², which correspond to the maximum performance of a vehicle in good condition, on dry road and wheels blocked. For swerving, we have chosen two different thresholds, 5 and 6 m/s². It is to be noted that in normal driving conditions, drivers use rarely more than 3 or 4 m/s².

TRACK			SIMULATOR		
braking $\gamma > 8m/s^2$	swerving $\gamma > 5m/s^2$	swerving $\gamma > 6m/s^2$	braking $\gamma > 8m/s^2$	swerving $\gamma > 5m/s^2$	swerving $\gamma > 6m/s^2$
N= 81/384	N= 130/384	N= 99/384	N= 0/768	N= 102/768	N= 75/768
21,1 %	33,9 %	27,8 %	0 %	13,3 %	9,8 %

Table 1 : numbers of estimates corresponding to non feasible manoeuvres, according to admissible accelerations thresholds (32 runs per subject and manoeuvre).

Compared to the test track situation, the number of estimates corresponding to non feasible manoeuvres is lower on the simulator. In accordance with the theoretical longitudinal acceleration we have seen above, this result reveals that on the simulator, drivers tend to adopt larger margins. Subjects seem not to be confident in their judgements, may be because of the short sight distance of the visual scene and / or because of the lack of sensation of motion. It should be stressed that without these references the drivers adopt a more cautious behaviour.

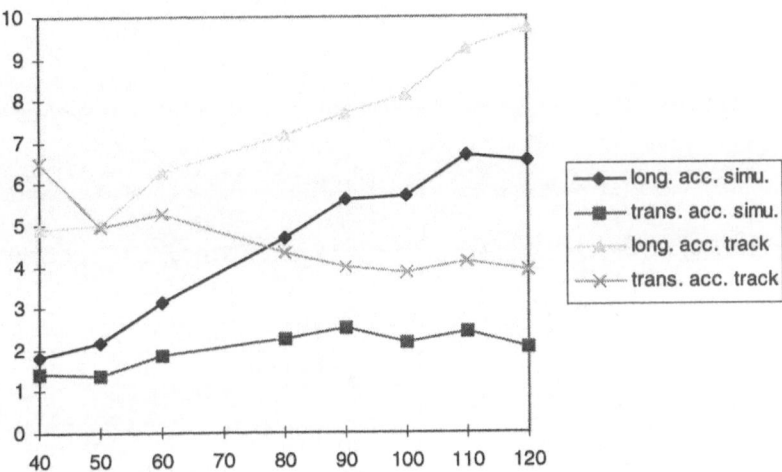

Figure 4 : longitudinal and transversal accelerations (ordinate) in which estimates would have resulted, according to speed and situation. A reaction time (400 ms) was integrated to make it closer to reality

Longitudinal and transversal accelerations were computed according to speed and averaged over subjects. Here again, despite scale differences between the simulator and the test track, the results show the same tendencies. As for braking, effective acceleration tends to increase with speed whereas transversal acceleration remains constant in both situations.

3.3 Research of the functions : Time = F (speed) and physical comparison

As in the test track and for each subject, we adjusted linear functions : time = F(speed) and made a comparison with the physical model.
The parameters of the linear function are :
- a : slope lane : the acceleration value. Threshold chosen by subjects,
- b : reaction time
Concerning the physical model parameters are :
For braking :

$\gamma = 8m/s^2$, so TC = S/16. The line is a linear function : y = ax + b, with a = 1/16 and b = 0,4.

For swerving :

$\gamma = 6m/s^2$. The line is a linear function with a = 0 and b = 1,45.

Thus, for each subject's data we plotted two lines :
- one resulting from the best adjustment (solid line),
- one corresponding to the physical model (dotted line).

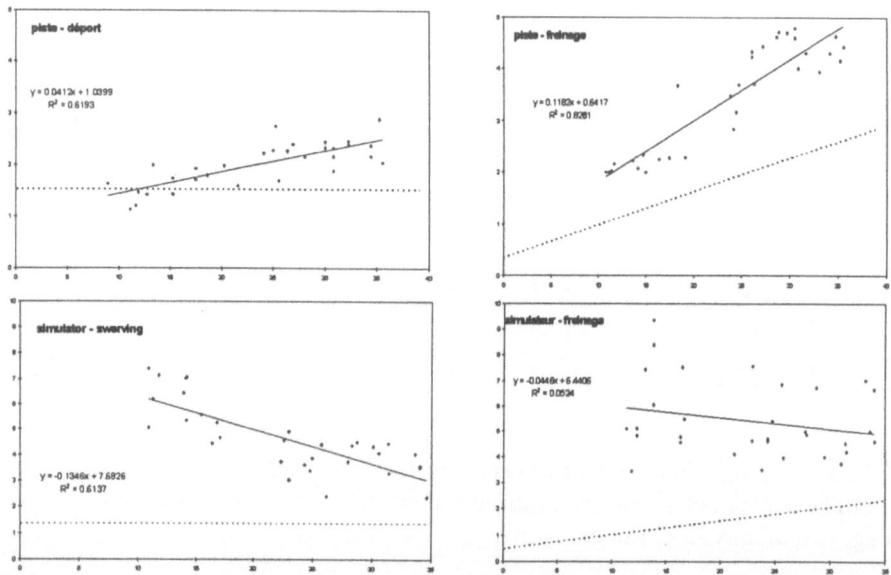

Figure 5 : linear function adjustment (solid line) and physical model (dotted line) comparison. Two representative examples for each situation and each manoeuvre.

On test track, Malaterre made a typology of the subject's functions obtained according to manoeuvres. 4 classes have been found :

Type 1 : steep slope for braking and gentle (or null) one for swerving (in accordance with the physical model),

Type 2 : steep slope for both braking and swerving (examples shown above),

Type 3 : null slope for both braking and swerving,

Type 4 : gentle slope for braking and steep one for swerving.

On simulator, as opposed to the test track, we note that simulator TCs decrease with speed increase. This means that the more speed increases the more the subjects initiate their manoeuvres late, which is not a normal driver behaviour. This could be explained by the short distance of visibility provided by the station. We saw at the beginning of this chapter that subjects have had difficulties to estimate distances longer than 80 metres. They have probably initiated their manoeuvres only when they were in the comfortable visibility zone, whatever the speed. And at high speed, subjects have less time to evaluate the situation and decide to trigger the manoeuvre. We can make the hypothesis that distance does not increase proportionately with speed because it was already large for low speeds (40 and 50 km/h) and that the short distance of visibility does not allow to subjects to chose distance proportionally to speed. As a result, time to collision decreases with speed increase.

4   Conclusion

Results show the same tendencies on the driving simulator and on the test track. As many other experiments have already shown, there is a scale difference between the two situations. Generally drivers adopt a more cautious behaviour on simulator. On the other hand, when we analyse more closely the results, perceptive mechanisms involved in this task seem to be different, probably because subjects have difficulties to perceive objects at long distances.

Two other experiments have already been carried out. In the first one, we studied actual braking, in order to stop at a nominated point or not. Results will be compared to Spurr's (5) and Newcomb's (6) ones, which have been obtained on a test track. It will also be interesting to compare the results with the experiment which has been presented in this chapter. We have seen that both braking and swerving estimates were affected by the short distance of visibility of the visual scene. Braking at a nominated point will be particularly difficult for the subjects because of the visual scene and also because of the absence of motion feedback needed to perform the task. This experiment was carried out on two fixed-base simulators (SIM^2 - INRETS and SHERPA - PSA), which will enable us to compare them and particularly to evaluate the influence of the visual motion feedback (pitch and roll included in the model, but only visually restituted) provided by the simulator of PSA on subjects performance. A second experiment has been just carried out and concerns headway regulation. Here again comparison with results already obtained will be interesting.

References

1. Malaterre G, Fréchaux J 1997. How to assess and improve validity on a driving simulator ? ISATA. Symposium on Automotive Technology & Automation. Simulation, Diagnosis and Virtual Reality Applications in the Automotive Industry including Supercomputer Application. Italy, Florence:137-143

2. Denton G.G 1966 A subjective scale of speed when driving a motor vehicle. *Ergonomics*, 9, 3:203-210

3. Malaterre G, Peytavin J F, Jaumier F, Kleinmann A 1987 L'estimation des manoeuvres réalisables en situation d'urgence au volant d'une automobile. Rapport n°46

4. Evans L, Wasielewski P, 1982 Risky driving related to driver and vehicle characteristics. *Accident Analysis & Prevention*, 15, 2:121-136.

5. Spurr R.T, 1971 Driving behaviour during braking. Symposium on Psychological aspects of driver behaviour, 1.2, the Netherlands, Swov

6. Newcomb TP, 1981 Driver Behavior during braking. SAE Technical Paper Series, n°810832

Acknowledgements
The authors thank gratefully Stéphane Espié, Jacky Robouant and Gilles Rousseau who developed the images and software necessary for this experiment.

26

A Historical Perspective of the Use of Driving Simulators in Road Safety Research

D. Pollock, S. Bayarri and E. Vicente

1 Introduction

The use of driving simulators in their present form is a relatively recent phenomenon in behavioural research. This technology evolved from simple mechanical and video devices by adapting newer technologies. Driving simulators provide a task which mimics real driving while at the same time allowing for a level of experimental control which would be impossible in a real traffic environment, enabling researchers to assess driving performance under dangerous conditions and test the effects of new in-vehicle and roadway technologies before they are actually implemented.

This chapter describes the evolution of the use of driving simulators in traffic and road safety research during the past twenty five years and will concentrate on two specific aspects of this development. First, the different types of simulators that have been employed as research tools during this period will be analysed. Second, the research topics that have been studied using this technology will be examined.

1.1 Methodology

The methodology used to quantitatively analyse this evolution was based on an extensive literature search of the articles included in the American Psychological Association's database, PsycLit. The articles were selected according to the following three-step process. First, a literature search was carried out in PsycLit covering the period between 1974 and 1997. The search profile used was "SIMULAT* AND DRIV*" applied to all the fields included in this database. As a result of this first step, 420 references were uncovered. This search strategy offered the advantage of being exhaustive, but the inconvenience of including many articles that do not deal with the topic of driving simulation. Thus, the second step entailed examining each reference individually in order to eliminate those articles which did not deal with driving simulation. In addition, the principal journals that are included in PsycLit were consulted directly to be certain that the last year (1997) was complete. The result of this second step was 155 references. The qualitative analysis of the types of driving simulators is based on these articles, as well earlier articles cited within these studies. The results of this analysis provide a classification of the different kinds of simulators employed, based on the type of visual display, degree of fidelity with regard to the physical environment, and degree of interactivity between the driver's input or control operation and the simulator's output.

Because the analysis of the apparatus that are referred to as a driving simulators revealed great diversity, a further step was taken to specify exactly what type of instrument would be considered a driving simulator in the analysis of the research topics. The final criteria for selection were that the articles be empirical studies and that they include at least a steering task. The studies that employed only a partial simulation of driving sub-tasks, but did not include a steering task, were excluded. The resulting 115 articles were analysed in order to provide a classification of the topics which have been studied using this research tool, including most frequent applications, classic areas of research and recently emerging areas of investigation.

2 The Evolution of Driving Simulator Technology

Researchers' interest in studying the effects of different variables on driving behaviour led to the development of a variety of devices that imitate a real driving environment. Harms [1] observes that despite their common purpose, driving simulators differ considerably with respect to basic design and technical specifications. The evolution that driving simulators have undergone from the first mechanical display devices of the 1960's and 70's to the modern full-scale high fidelity moving base versions of the 90's has been a process of continual growth and improvement made possible by the incorporation of the latest advances in information technology and hydromechanics. In the following sections, the progress achieved in driving simulation technology during the past thirty years will be described.

2.1 Earliest Devices

The earliest devices used to simulate driving were simple apparatus and fall into the category that Wachtel [2] describes as "home-grown" devices which are typically designed and built in university laboratories and used to support faculty and student research. One of the primary concerns of the developers of the first simulators was the creation of a visual display that represents the driving environment as realistically as possible. Because the first research in this field was carried out before computer technology became widespread, the problem of creating a realistic visual display was solved in a variety of curious and often ingenious manners. The pioneers in this field relied on one of two means of visual reproduction of the driving scene--motion picture display or mechanical display (often using an 'external viewpoint'). Another important aspect that had to be addressed by simulator designers was response-system realism [3], which includes both the degree to which drivers' control actions correspond with real driving behaviour and the interactivity between these actions and the scene presented in the visual display. Thus, in order for the simulation to be realistic, the vehicle controls must mimic the real ones and when the driver operates these controls the scenario simulation model must change the visual display accordingly. These concepts will be addressed as they apply to the different types of simulators.

The first driving simulators were mechanical devices that relied on one of four means of depicting the driving scenario: 1) a scale model car circuit, 2) a moving belt display, 3) a shadow projection display, and 4) a closed circuit TV-terrain model display. The model car set-up is the least realistic of the four in terms of visual display. An example of this apparatus is found in Currie [4] consisting of a track and model cars set up on a platform. In this simulation the subject controls one of the model cars by operating a steering wheel, and accelerator and brake pedals. The task involves driving around the track and reacting to three event cars which are controlled by the experimenter.

The moving belt simulator is another of the earlier attempts to present the driver with a visual environment similar to that encountered in the real world. This type of mechanical simulation appears in a study carried out by Regina et al. [5]. The apparatus in this study includes a real automobile chassis which stands at one end of five adjacent movable belts which form a 70-ft. loop. The belt system simulates two driving lanes, the centreline, and roadside scenery. An optical lens system mounted on the front of the car made the scale model of the roadway appear life-size to the driver. In the driving task the driver interacted with a lead car by accelerating, decelerating, and braking as the situation required. Additional sensory input included motor noise, wind effects and speedometer readings. Because the model car controlled by the subject was attached to the belt that depicted a lane of the roadway, steering was not a part of the driving task. Compared to the scale model car circuit, the moving road simulator is an improvement in terms of visual display realism, however, the interaction between the driver's control operations and the visual scene is inferior. Both of these devices reproduced an external viewpoint.

The shortcomings associated with the earliest mechanical displays led to new and more complex solutions to the problem of visual display and response-system realism. In the shadow projection simulator, also referred to as a point light source simulator, the visual display was presented on a rear-projection screen located in front of the mock-up. The visual display was formed as a refracted image produced when illumination from a 25-watt point light source was passed through a transparent Plexiglas disk. Roadway scenes painted on the Plexiglas disk, as well as objects placed on the disk surface, made up the projection source from which the visual image was produced. Simulated movement was produced when the disk was moved beneath the stationary point light source. Movement of the disk was controlled through the driver's manipulation of the controls located in the mock-up [6, 7].

Lastly, the closed circuit TV-terrain model display simulator creates the visual scene using a small scale model of a roadway, one or more movable cameras, and television projector or monitor. In this simulation the driver's control operations are translated into camera movements and the resulting roadway scene is viewed on a TV screen. A good example of the use of a closed circuit TV-terrain simulator appears in Blaauw's study [8]. Among the mechanical displays, this simulation technique represents the most satisfactory solution to the twofold problem of creating a realistic visual display and a response-system that corresponds to real world driving. This system of reproducing vehicle movement was first used in aviation simulation.

In addition to the mechanical display technique, the other early simulators used a motion picture display. In these simulators, which were primarily used in the 1970's [9, 10, 11, 12, 13, 14], a film taken on a real roadway is projected in front of the driver. The subject is instructed to operate the controls (steering wheel, brake, and accelerator) as if he or she were driving along the projected roadway. This type of task is called open-loop simulation because the driver's control actions have no effect on the driving scene, that is there is no interactivity between the driver's behaviour and the visual image in the display. However, the driver's control actions are recorded in some way (for example, on a chart recorder) and later analysed. Some motion picture display simulators, for example the AETNA Driving Trainer used in Edwards et al. [12], incorporate interactivity between the driver's control actions and the filmed scene with regard to the *speed* with which the roadway advances. In another study, Ward et al. [15] devised a curious form of driver-scenario interactivity in which a car-following task was carried out by means of a light-weight laser which was affixed to the steering wheel so it could be used to track the lead vehicle.

Because of their better visual realism, motion picture simulators are especially appropriate for studying such variables as visual perception and eye movements [9,14, 13, 11]. Of course, their lack of response-system realism makes them much less suitable for studying driving performance variables. Although this type of simulation was prevalent during the 1970's and in subsequent years fell into disuse as more advanced technology took its place, it has not disappeared altogether. This visual display technique is still employed today [16, 15, 17], although to a much lesser extent. In contrast, the mechanical means of producing visual driving images that were first used have been abandoned altogether; replaced by the computer generated image displays that will be discussed in the following section.

2.2 Desktop Simulation

The developments in information technology were adopted by simulator designers during the mid- to late 1970's. These new advances offered researchers an inexpensive means to greatly improve driving simulation. For the first time, simulation developers in small university laboratories could have the best of both worlds--visual display realism, as well as response-system realism. While the early computer generated images came nowhere close to the visual quality of a film representation, they were a great improvement over the mechanical versions. It was this gain together with the ability to translate the driver's control actions into the motion represented in the driving scenario that made this development so advantageous.

The first desktop simulators appeared in the literature during the late 1970's [18, 19, 20]. These simulators included the following characteristics. The visual display consisted of elemental 2D or 3D computer generated graphics. The road scene was presented to the driver via a typical CRT monitor or TV screen placed at eye level on a table or desk--thus, the name desktop simulator. Usually the driver control environment included only the main car controls--steering wheel, accelerator and brake pedals--and the driver sat in an ordinary chair. However, in some cases a more realistic physical environment was achieved using a car mock-up or real car cockpit [18, 19, 21, 22]. In other cases simpler versions of the desktop simulator were employed in which the only control mechanism was a computer mouse used to track the simulated road scene [23 and 24].

Desktop simulators have been used extensively in recent years as research tool [22, 23, 24, 25, 26, 27, 28, 29, 30] and are also widely used for driver assessment, education and training. The popularity of these systems stems from their versatility and low cost. These simulators can be programmed in-house to conform to the specific needs of each research project, creating a simulation that is custom-made. Although the desktop simulator represented a step forward in driving simulator technology, it still left some simulation needs unresolved--for example, that of real physical motion, higher quality 3D graphics and a more complete driver control environment.

2.3 Intermediate Level Simulators

In parallel to the development of the desktop simulator other researchers were creating more sophisticated systems that also relied on computer generated graphics. The simulators that fall into this category incorporate one or more of the following characteristics which put them a step above the desktops: 1) high quality 3D graphics and scene complexity; 2) a visual display that is projected on a large screen or is viewed through a special lens that provides a wider field of view; 3) the driver control environment is a real car, a genuine car cockpit or a realistic mock-up with a complete set of controls and instrumentation; 4) a limited degree of real physical motion.

The first driving simulator with these characteristics appeared in the mid-1970's. It was a moving base simulator developed at the Virginia Polytechnic Institute and State University (VPI & SU) by Walter W. Wierwille, a pioneer in driving simulator technology, and his collaborators [31, 32, 33, 34, 35, 36, 37, 38, 39, 40]. For its time, this driving simulator was quite an advanced device. It provided a high degree of fidelity in vehicle handling that was achieved via a four-degree of freedom hydraulically actuated physical motion system co-ordinated with a dynamic visual scene. In addition, this simulator incorporated vibration cues and an audio system that simulated sounds for rolling resistance, engine/drive-train noise, tire screech on severe braking, and tire squeal on severe cornering. Although the VPI simulator reproduces some of the physical motion cues associated with real driving, this device does not create the large range of motion of the modern moving base simulators that will be described later. Nevertheless, this device represented a giant leap in simulator technology when it was built and it is a forerunner of the present-day moving base simulators.

Among the fixed base simulators in this category is the Systems Technology Incorporated (STI) driving simulator which first appeared in the literature in the 1980's [41, 42, 43 22]. Advanced aspects of this simulator include a completely instrumented cab and an especially complex visual display. The road display consisted of three components: a CRT image optically combined with two slide-projected images. One slide image was a traffic sign and the other was a horizon scene used as a background. Both the sign and background images were horizontally deflected by a servo-controlled mirror which was moved proportionately to the vehicle heading. In addition, a Fresnel lens was mounted in front of the CRT monitor in order to provide magnification and to collimate the road image to create the illusion of distance. Other examples of driving simulators employed in the 1980's that incorporate one or more of the advances listed above are the systems that appear in the following studies: Reid et al. [44] use a large screen projection of the visual scene; Godthelp [45] included a visual scenario that was projected on large screens providing a 120° horizontal field of view; and Drory [46] used a simulated truck cabin.

The improvements of the 1980's became standard equipment in the new generation of intermediate level fixed base simulators of the 1990's. Carsten and Gallimore [47] call this class of devices "medium cost" simulators and they provide a list of the features typically found in them: "the provision of a full-sized and complete vehicle, with all the normal controls operational; the use of real-time animation to create a scene that is projected in front of the driver; construction of the simulator around a specialised visual simulation workstation (costing perhaps £100,000); and the lack of a moving base to subject the driver to gravitational and inertia forces".

These authors are the developers of a driving simulator at the University of Leeds which conforms to these specifications. In addition to these features, the Leeds simulator provides the driver with several other types of feedback, like steering wheel forces and instrument panel output [47]. Other driving simulators in current use that fit Carsten and Gallimore's description are: the simulator at the TNO Institute for Perception, which was used in Korteling's [48] research on the effects of ageing on driving performance; the driving simulator at the Traffic Research Centre (TRC) in Groningen, which was employed in Van-Winsum and Heino's [49] study of time-headway; and the simulator used by Horne and Reyner [50] to study the effects of different treatments on driver sleepiness.

2.4 High Fidelity Moving Base Simulators

The high fidelity moving base simulators represent the top of the line in driving simulation. These devices are patterned after simulators that were originally developed for the aviation industry [2]. Their pricetag ranges in the tens of millions of dollars and only a handful of these systems exist in the world [2]. A general description of these simulators is offered by Wachtel [2]:

> These high end devices typically include projected computer-generated imagery (CGI) displays which provide fields of view from the driver's eye position of 180 to 360°. In addition, they incorporate sophisticated motion systems which permit angular rotation as well as longitudinal and lateral translation to simulate vehicle movements (p.3).

Of the high fidelity moving base simulators, the one that appears most often in the literature dedicated to traffic and road safety research is the VTI Simulator of the Swedish Road and Traffic Institute. Examples of research carried out using the VTI simulator can be found in (51, 52, and 53). Other simulators included in this category are the Daimler-Benz simulator in Berlin, the United States National Driving Simulator (IDS) at the University of Iowa. These apparatus enable researchers to create the most realistic simulated driving experiences possible. Having completed the classification of different devices that have been used in traffic and road safety, the following sections will be dedicated to the analysis and discussion of the research topics that have been investigated during the past 24 years (1974-1997) using driving simulators as a research tool (limited to those that include at least a steering task).

3 Research Topics

The technological aspects are not the only component of driving simulation research that has evolved during the past 24 years. The topics that have been investigated using this methodology have undergone their own transformation. In order to better observe the evolution of the specific objectives that have interested researchers in driving simulation during this period, the research topics that were investigated in the articles being reviewed were analysed. This analysis includes a broad classification of general topics (see table 1) and a more detailed classification of subtopics (see table 2). These classifications were not formed quantitatively (i.e. based on key words or descriptors), but qualitatively (i.e. based on the article's contents).

3.1. Main Topics

TABLE 1
Frequency of Main Topic by Year of Publication

Main topic	1974-1976	1977-1979	1980-1982	1983-1985	1986-1988	1989-1991	1992-1994	1995-1997	Total
Driver	13	11	7	3	8	14	11	20	87
Vehicle	0	0	1	0	1	1	4	4	11
Environment	2	0	0	1	1	3	1	0	8
Simulation	2	1	2	0	1	0	3	0	9
Total	17	12	10	4	11	18	19	24	115

The general categories include the three dimensions that constitute the traffic system: driver, vehicle, and environment. In addition, the category of "simulation" was included to refer to the those articles that concentrate solely on methodological aspects, such as: validation studies, research concerning techniques used to measure dependent variables, and investigation centred on simulator ergonomics. These categories are exclusive in that each article can pertain to only one category. The results of this classification are shown in Table 1. The driver was the main topic of 76% of the articles, followed by the vehicle (10%), simulation (8%), and the environment (7%).

3.2 Subtopics

In contrast to the previous classification which is composed of four exclusive categories, the subtopics that make up this classification are not exclusive. In this classification an article may be included in more than one category if the research involves more than one subtopic. For example, a study dealing with the effects of mobile telephone use upon attention would be classified in the category of "psychological processes" for studying attention as well as the "vehicle" category because mobile telephones are considered a part of the new technology within the vehicle.

TABLE 2
Frequency of subtopics

Main topic	Subtopic	Frequency
Driver	Alcohol	22
	Drugs (not including alcohol)	19
	Psychological processes	31
	Individual differences	15
	Transitory states	9
	Illnesses/handicaps	12
	Psychophysiology	12
	Driving performance	6
	Driver education	3
Environment	Environment	8
Vehicle	Vehicle	11
Simulation	Simulation	9
	TOTAL	157

The classification of subtopics resulted in the nine driver related categories listed in Table 2. The remaining main topics: environment, vehicle, and simulation, were not subdivided because relatively few articles dealt with these topics. Regarding this point, it is important to mention that the reason these topic are under represented in this sample of articles may be due to the fact that these areas are more technical and less related to human factors than the majority of the studies referenced in PsycLit. In order to locate additional simulation studies concerning these topics, perhaps it is necessary to consult specialised journals and conference proceedings. The results of this analysis show that the driver related subtopics most frequently studied are psychological processes (27%), followed by alcohol (25%), and drugs other than alcohol (22%). Moreover, within the category of psychological processes, attentional mechanisms are by far the most widely studied.

3.2.1. Classic Topics

In addition to being the most frequently investigated, these three principal subtopics form the core of driving simulation research. Because the large number of articles dedicated to these questions is maintained throughout this 24 year period, they may be considered classic topics in driving simulation research (see graph 1).

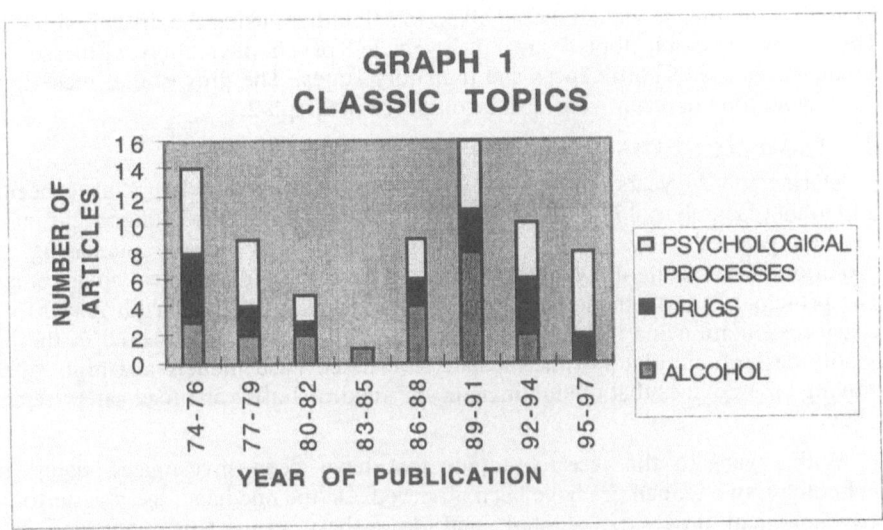

GRAPH 1
CLASSIC TOPICS

It is interesting to note that Sivak's [54] review of the literature dedicated to traffic psychology presents results that are very similar to these. In his survey, alcohol is the principal topic studied within the field of traffic and road safety, followed by drugs other than alcohol, and psychological processes. Thus, these classic topics not only represent the traditional applications of driving simulation methodology in the field of traffic psychology, but also constitute the principal topics within the broader field of traffic and road safety in general.

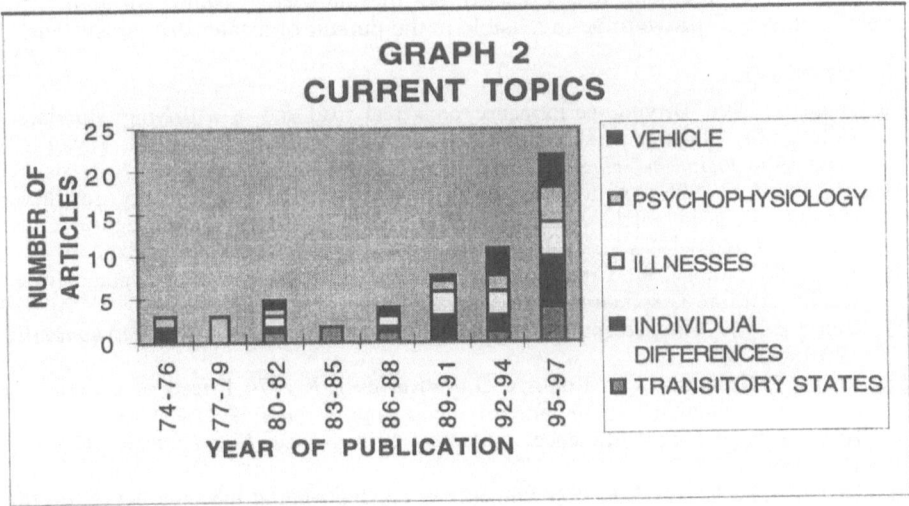

GRAPH 2
CURRENT TOPICS

3.2.2. Current Topics

Although the classic topics are quantitatively dominant, there exists another group of topics that in spite of being quantitatively less significant have received increasing attention in recent years. Information regarding these current topics is especially interesting because they reveal the new problems that need to be solved, as well as the means that are being taken to solve them using the driving simulator. These new research topics are: the vehicle, psychophysiology, illnesses and handicaps, individual differences and transitory states. The growth that these topics have undergone in recent years can be observed in Graph 2.

4   Conclusions

During the 24 years that this survey covers, driving simulators have been an important research tool in the field of traffic and road safety. In addition, during the past ten years the use of this methodology has been steadily increasing. The evolution of this technology has been marked by enormous technical advancements. The principal improvements are related to the degree of realism with regard to the visual presentation and the response system. This progress is observed in the high quality desktop simulators, intermediate level fixed base models and high fidelity moving base devices that predominate in the field of traffic and road safety research today.

With regard to the research topics that have been investigated using this technology, two tendencies have been observed. On the one hand, the classic topics--psychological processes, alcohol, and drugs--have consistently attracted much attention throughout the period studied. However, recently an ever increasing amount of research has been dedicated to new topics--vehicle, psychophysiology, illnesses and handicaps, individual differences and transitory states--which reveal the direction that this area of research is taking currently.

In summary, driving simulators have proved to be an invaluable tool for studying those driving related variables that would be difficult or impossible to investigate by any other means. This methodology combines the advantages of a high level of ecological validity, experimental control, and precision in driver performance measurement. It is expected that the knowledge gained through these investigations will prove to be invaluable in the pursuit of a safer driving system.

References

[1]   Harms L 1996 Driving performance on a real road and in a driving simulator: Results of a validation study. In Gale A G, Brown Y D, Haslegrave C M, Taylor S P (eds) 1996 *Vision in Vehicles -V.* Elsavier, Amsterdam, pp. 19-26

[2]   Wachtel J A 1996 Applications of appropriate simulator technology for driver training, licensing and assessment. In Gale A G, Brown Y D, Haslegrave C M, Taylor S P (eds) 1996 *Vision in Vehicles-V.* Elsavier, Amsterdam, pp.3-11

[3]   Schiff W, Arnone W, Cross S 1994 Driving assessment with computer-video scenarios: More is sometimes better. *Behav Res Meth* 26(2):192-194

[4]   Currie L 1969 The perception of danger in a simulated driving task. *Ergonomics* 12(6):841-849

[5]   Regina E G, Smith G M, Keiper C G, McKelvey R K 1974 Effects of caffeine on alertness in simulated automobile driving. *J Appl Psychol* 59(4):483-489

[6]   Hagen R E 1975 Sex differences in driving performance. *Hum Factors* 17(2):165-171

[7]   Dott A B, McKelvey R K 1977 Influence of ethyl alcohol in moderate levels on the ability to steer a fixed-base shadowgraph driving simulator. *Hum Factors*, 19(3):295-300

[8]   Blaauw G J 1982 Driving experience and task demands in simulator and instrumented car: A validation study. *Hum Factors* 24(4):473-486

[9]   Allen J A, Schroeder S R, Ball P G 1974 Effects of head restriction on drivers' eye movements and errors in simulated dangerous situations. *J Appl Psychol* 59(5):643-64

[10] Baron M L, Williges R C 1975 Transfer effectiveness of a driving simulator. *Hum Factors* 17(1):71-80

[11] Beideman L R, Stern J A 1977 Aspects of the eyeblink during simulated driving as a function of alcohol. *Hum Factors* 19(1):73-77

[12] Edwards D S, Hahn C P, Fleishman E A 1977 Evaluation of laboratory methods for the study of driver behavior: Relations between simulator and street performance. *J Appl Psychol* 62(5):559-566

[13] Ceder A 1977 Drivers' eye movements as related to attention in simulated traffic flow conditions. *Hum Factors* 19(6):571-581

[14] Allen J A, Schroeder S R, Ball P G 1978 Effects of experience and short-term practice on drivers' eye movements and errors in simulated dangerous situations. *Percept Mot Skills* 47(3, Pt 1):767-776

[15] Ward N J, Parkes A, Crone P R 1995 Effect of background scene complexity and field dependence on the legibility of head-up displays for automotive applications. *Hum Factors* 37(4):735-745

[16] McKnight A J, McKnight A S 1993 The effect of cellular phone use upon driver attention. *Accident Anal Prev* 25(3):259-265

[17] Nilsson T, Nelson T M, Carlson D Development of fatigue symptoms during simulated driving. *Accident Anal Prev* 29(4):479-488

[18] Witt H, Hoyos C G 1976 Advance information on the road: A simulator study of the effect of road markings *Hum Factors* 18(6):521-532

[19] Donges E 1978 A two-level model of driver steering behavior. *Hum Factors* 20(6):691-707

[20] Baxter J, Harrison J Y 1979 A nonlinear model describing driver behavior on straight roads. *Hum Factors* 21(1):87-97

[21] Mouran R R, Herman M, Moussa-Hamouda E 1980 Direct looks and control location in automobiles. *Hum Factors* 22(4):417-425

[22] Lenné M G, Triggs T J, Redman J R 1997 Time of day variations in driving performance. *Accident Anal Prev* 29 (4):431-437

[23] Sivak M, Flannagan M J, Sato T, Traube, E. C., Aoki M 1994 Reaction times to neon, LED, and fast incandescent brake lamps. *Ergonomics* 37(6):989-994

[24] Liu Y 1996 Quantitative assessment of effects of visual scanning on concurrent task performance. *Ergonomics,* 39(3):382-399

[25] Brouwer W H, Waterink W, Van-Wolffelaar P C, Rothengatter T 1991 Divided attention in experienced young and older drivers: Lane tracking and visual analysis in a dynamic driving simulator. Special Issue: Safety and mobility of elderly drivers: Part I. *Hum Factors* 33(5):573-582.

[26] Jancke L, Musial F, Vogt J, Kalveram K T 1994 Monitoring radio programs and time of day affect simulated car-driving performance. *Percept Mot Skills,* 79(1, Pt 2), Spec Issue 484-486

[27] Gianutsos R 1994 Driving advisement with the Elemental Driving Simulator (EDS): When less suffices. *Behav Res Meth* 26(2):183-186

[28] Briem V, Hedman L R 1995 Behavioural effects of mobile telephone use during simulated driving. *Ergonomics* 38(12):2536-2562

[29] Dorn L, Matthews G 1995 Prediction of mood and risk appraisals from trait measures: Two studies of simulated driving. *Eur J Pers* 9(1):25-42

[30] Desmond P A, Matthews G 1997 Implications of task-induced fatigue effects for in-vehicle countermeasures to driver fatigue. *Accident Anal Prev* 29(4):515-523

[31] Wierwille W W, Fung P P 1975 Comparison of computer-generated and simulated motion picture displays in a driving simulation. *Hum Factors,* 17(6):577-590

[32] Wierwille W W, Guttmann J C, Hicks T G, Muto W H 1977 Secondary task measurement of workload as a function of simulated vehicle dynamics and driving conditions. *Hum Factors* 19(6):557-565

[33] Hicks T G, Wierwille W W 1979 Comparison of five mental workload assessment procedures in a moving-base driving simulator. *Hum Factors* 21(2):129-143

[34] Casali J G, Wierwille W W 1980 The effects of various design alternatives on moving-base driving simulator discomfort. *Hum Factors* 22(6):741-756

[35] Muto W H, Wierwille W W 1982 The effect of repeated emergency response trials on performance during extended-duration simulated driving. *Hum Factors* 24(6):693-698

[36] Wierwille W W, Casali J G, Repa B S 1983 Driver steering reaction time to abrupt-onset crosswinds, as measured in a moving-base driving simulator. *Hum Factors* 25(1):103-116

[37] Skipper J H, Wierwille W W 1986 Drowsy driver detection using discriminant analysis. *Hum Factors* 28(5):527-540

[38] Frank L H, Casali J G, Wierwille W W 1988 Effects of visual display and motion system delays on operator performance and uneasiness in a driving simulator. *Hum Factors* 30(2):201-217

[39] Rogers S B, Wierwille W W 1988 The occurrence of accelerator and brake pedal actuation errors during simulated driving. *Hum Factors* 30(1):71-81.

[40] Imbeau D, Wierwille W W, Wolf L D, Chun G A 1989 Effects of instrument panel luminance and chromaticity on reading performance and preference in simulated driving. *Hum Factors* 31(2):147-160

[41] Ranney T A, Gawron V J 1986 The effects of pavement edgelines on performance in a driving simulator under sober and alcohol-dosed conditions. *Hum Factors* 28(5):511-525

[42] Gawron V J, Ranney T A 1988 The effects of alcohol dosing on driving performance on a closed course and in a driving simulator. *Ergonomics* 31(9):1219-1244

[43] Gawron V J, Ranney T A 1990 The effects of spot treatments on performance in a driving simulator under sober and alcohol-dosed conditions. *Accident Anal Prev* 22(3):263-279

[44] Reid L D, Solowka E N, Billing A M 1981 A systematic study of driver steering behaviour. *Ergonomics* 24(6):447-462

[45] Godthelp J 1985 Precognitive control: Open- and closed-loop steering in a lane-change manoeuvre. *Ergonomics* 28(10):1419-1438

[46] Drory A 1985 Effects of rest and secondary task on simulated truck-driving task performance. *Hum Factors* 27(2):201-207

[47] Carsten O M J, Gallimore S 1996 The Leeds Driving Simulator: A new tool for research in driver behaviour. In Gale A G, Brown I D, Haslegrave C M, Taylor SP (eds) 1996*Vision in Vehicles-V* . Elsavier, Amsterdam, pp.11-19

[48] Korteling J E 1994 Effects of aging, skill modification, and demand alternation on multiple-task performance. *Hum Factors* 36(1):27-43

[49] Van-Winsum W, Heino A 1996 Choice of time-headway in car-following and the role of time-to-collision information in braking. *Ergonomics* 39(4):579-592

[50] Horne J A, Reyner L A 1996 Counteracting driver sleepiness: Effects of napping, caffeine, and placebo. *Psychophysiology* 33(3):306-309

[51] Lovsund P, Hedin A, Tornros J 1991 Effects on driving performance of visual field effects: A driving simulator study. *Accident Anal Prev* 23(4):331-342

[52] Alm H, Nilsson L 1994 Changes in driver behaviour as a function of handsfree mobile phones: A simulator study. *Accident Anal Prev* 26(4):441-451

[53] Alm H, Nilsson L 1995 The effects of a mobile telephone task on driver behaviour in a car following situation. *Accident Anal Prev* 27(5):707-715

[54] Sivak M 1997 Recent psychological literature on driving behaviour: What, where and by whom? *Appl Psychol* 46(3):303-310

27

Multimodal Driving Simulation in Realistic Urban Environments

S. Donikian, G. Moreau and G. Thomas

1 Introduction

Reproducing the real multimodal traffic of a city, as completely as possible, implies the simulation of autonomous entities like living beings [1]. Such entities are able to perceive their environment, to communicate with other creatures and to execute some actions (drive a car or walk in the street for example). To perform a simulation composed of a large set of dynamic entities "living" and interacting in a complex environment, we need to implement different models: environment models, mechanical models, motion control models, behavioural models, sensor models, geometric models and scenario.

Databases for virtual environments are often restricted to the geometric level, when they must also contain physical, topological and semantic information. Accordingly, we have developed VUEMS, a Virtual Urban Environment Modelling System. The main aim of VUEMS is to build a realistic virtual copy of a real city (Rennes in France) in which we would perform driving simulations. VUEMS produces two complementary outputs: the 3D geometric representation of the scene and its symbolic representation used by sensors and behavioural entities. From our point of view and in accordance with some psychological studies, different paradigms are required to describe a behavioural model (the brain part of a complete entity). This model should be both cognitive and reactive, treating flows of data to and from its environment. To describe realistic behaviours, we have proposed to use a formal model based on Hierarchical Parallel Transition Systems (HPTS). To integrate all these models we have proposed a simulation platform : GASP. This platform takes into account real time synchronization and data communication between cooperative processes distributed on an heterogeneous network of workstations and parallel machines.

In this paper, we present the software environment (VUEMS, behavioural model, GASP), a couple of models (tram, car driver) and some applications.

2 Software Environment

To perform modelling and simulation of virtual urban environments, we need models of the environment and of dynamic entities (geometry, mechanics and behaviour). Therefore we have developped our own tools (see in figure 1); it provides us modularity and flexibility of development and simulation. We first present the two modelling tools which generate the environment and the behavioural models. Then, we introduce the simulation platform.

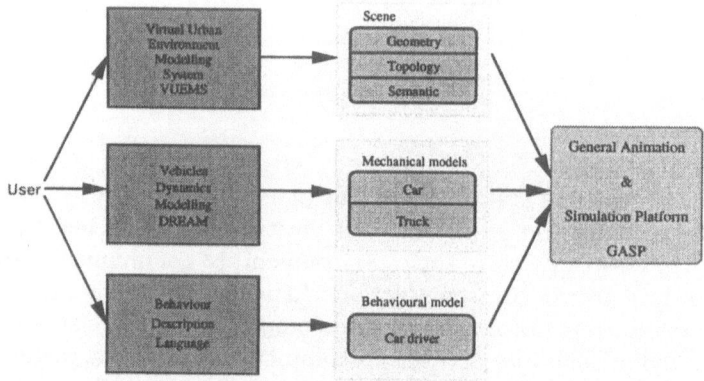

FIGURE 1. Overall view of the SIAMES Simulation Environment.

2.1 VUEMS: Virtual Urban Environment Modelling System

Informations required for driving simulation are of different kinds: the road network, with its geometry (road-shape), its rules (road-signs) and its environment (buildings, parks, ...) [2, 3]. This is sufficient for driving simulation in which vehicles are all driven by a user in the loop. Since autonomous vehicles are added in the simulation, other kinds of information become necessary [4, 5]. We cannot perform in real time the simulation of human vision, therefore it is impossible to build in-line topological and semantic models of the environment. Those informations must be represented in the database to be used for the emulation of human vision. The database contains different kinds of informations: on the road, on the road network, and on the city (name of streets, quarters, particular buildings, squares). As far as we know there is no normalization of the design of elements like a round-about or a crossroads. Each element of the thoroughfares in a urban environment is unique, but it is possible to classify them in a little number of categories. This allows us to describe the structure of complex crossroads (see in figure 2).

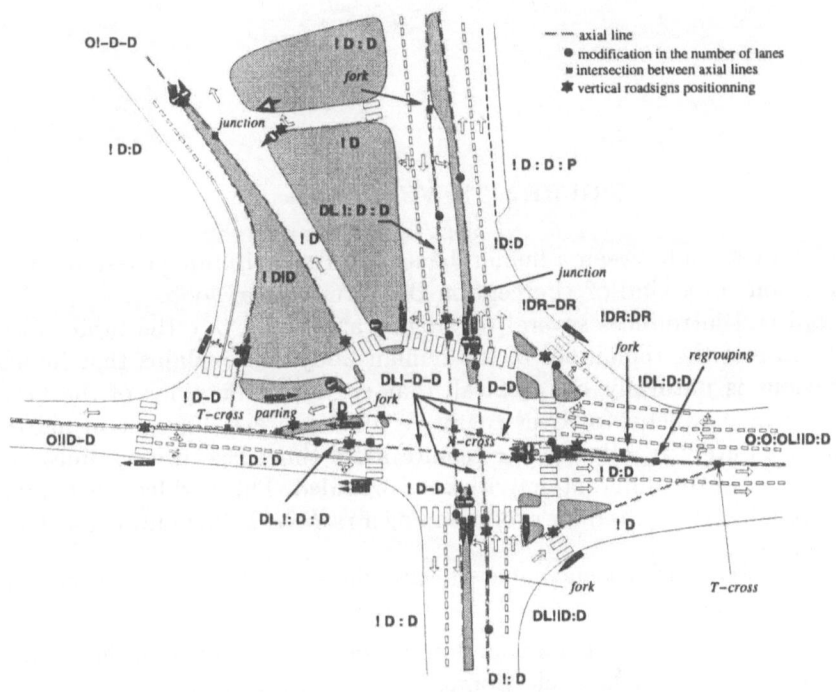

FIGURE 2. A usual crossroads near the university and its representation.

VUEMS (Virtual Urban Environment Modelling System) [6] enables to build a virtual copy of thoroughfares of real cities. It uses, as inputs, different kinds of information: cartographic databases, scanned maps of roadways, traffic lights organisation and synchronization. After the interactive description of the road-network, VUEMS (see in figure 3) produces two complementary outputs: the 3D geometric representation of the scene (including automatic texturing) and its symbolic representation (geometric, topological and semantic levels) used by sensors and deliberative agents.

2.2 Behaviour Modelling

Our goal being to add some dynamic entities "living" in a virtual environment, we need to simulate the behaviour of virtual creatures [7]. Behavioural animation consists of a high level motion control of dynamic objects [8, 9], which offers the ability to simulate autonomous entities like organisms and living beings [10]. Psychological studies [11] have shown that

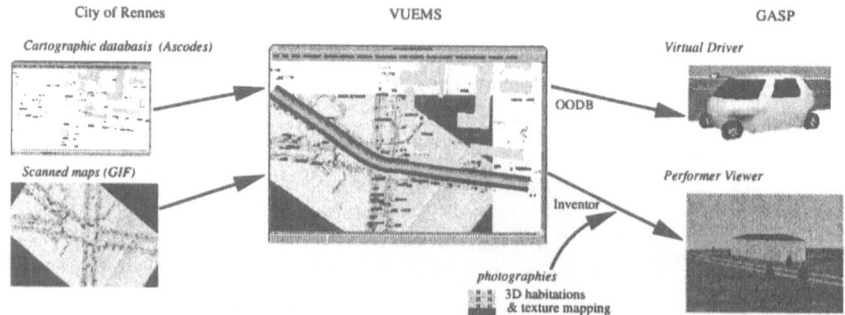

City of Rennes　　　　　　　VUEMS　　　　　　　GASP

Cartographic databasis (Ascodes)　　　　　　　　　Virtual Driver

OODB

Scanned maps (GIF)　　　　　　　　　　Performer Viewer

Inventor

photographies
3D habitations
& texture mapping

FIGURE 3. The VUEMS Structure.

the interactions between a human being and its environment (see in figure 4) are done in a kind of "Perception-Decision-Action" loop.

Lord [12] introduces several paradigms about the way the brain works and controls the remainder of the human body. He explains that human behaviour is naturally hierarchical, that cognitive functions of the brain are run in parallel. Moreover cognitive functions are different in nature: some are purely reactive, other require more time. Executions times and frequencies of the different activities are provided. This had lead us to state that paradigms required for programming a *realistic* behavioural model are:

- reactivity, which encompasses sporadic or asynchronous events and exceptions,

- modularity in the behaviour description, which allows parallelism and concurrency of sub-behaviours,

- data-flow [13, 14], for communications between different modules,

- hierarchical structuring of the behaviour, which means the possibility of preempting sub-behaviours on transitions in the meta-behaviour, as a kind of exception or interruption. It means also that sub-behaviours can notify the meta-behaviour of their activity.

- frequency handling for execution of sub-behaviours. This provides the ability to model reaction times in perception activities.

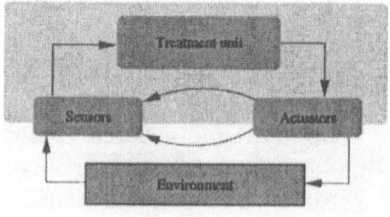

FIGURE 4. The human organism and its environment.

Therefore we have presented the HPTS formalism [8] which consists of a reactive system which is composed of a hierarchy of concurrent state machines (possible behaviours). Each state machine of the system can be viewed as a black-box with an In/Out data-flow and a set of control parameters. Though state-machines may be coded directly with an imperative programming language like C++, we have decided to build a language [15] for behaviour description. Otherwise the problem is that it quickly becomes quite difficult to update a complex state machine and therefore to reuse it in future developments. Moreover the code of the transition systems becomes unreadable or inefficient. This is why we propose a language that allows the description of both the hierarchical parallel state machines and their associated data-flows. This language is compiled and efficient C++ code for GASP is generated. The change of a transition condition is thus quite easy. Despite the benefits of the language approach, the description of behaviour remains quite difficult to people who are not computer scientists. Therefore we have started the building of a graphical tool in order to allow behavioural specialists to test their models in an interactive simulation. This project is currently underway and is more than a simple graphical tool for drawing state machines, it also includes a graphical description of the dataflows and the associated integration functions.

2.3 GASP: a General Animation and Simulation Platform

To perform a simulation composed of a large set of dynamic entities evolving and interacting in a complex environment, we need to implement different models: environment models, mechanical models, motion control models, behavioural models, sensor models, geometric models and scenarios. In a system mixing different entities defined by different kinds of models (descriptive, generator and behavioral), it is necessary to take into account the explicit management of time, either during the specification phase (memorization, prediction, action duration) or during the execution phase (synchronization of objects with different internal times). Nevertheless, in a simulation, all simulated entities do not require the same level of realism and by way of consequence the same computation time. Then, it is interesting to mix different motion control models in a same system to benefit from advantages of each motion control model; GASP [16] intends to answer to this requirement, using an object oriented programming methodology.

As we have to simulate universes with a great number of entities, a lot of CPU resources is required. So, in order to reduce the computation time we need to distribute these entities over a network on different computers or on different processors in the same machine. Our simulation platform manages data communication between cooperative processes distributed on an heterogeneous network of workstations and parallel machines, furthermore it takes into account real time synchronization between modules with very different calculation frequencies.

326

The main objective of GASP is to give the ability to simulate different entities composed themselves of different modules in different hardware configurations, without any change for the animation modules. When someone specifies a module, he does not have to make any hypothesis on the network location of other modules he must interact with. Nevertheless, he must be able to name them, and for that, modules are structured in a simulation tree (see in figure 5). Each module of a simulation is a specialisation of a class named *PsSimulObject*. The *PsSimulObject* class can be viewed as the container of a computation function $Y = F(X, CP)$, where X is a set of inputs, Y a set of outputs and CP a set of Control Parameters. X and Y determine the data-flow from and to other objects. Each object has its own frequency and is activated periodically to calculate its new state. At each simulation step, the new input values are used to compute the outputs. This requires to connect each input of the object to an output of another object. This data dependency can be static or dynamic, as we cannot know at the beginning of a simulation, which objects might interact later. To take into account dynamic data dependencies, the number of inputs of an object can change during the simulation, unlike the one of outputs.

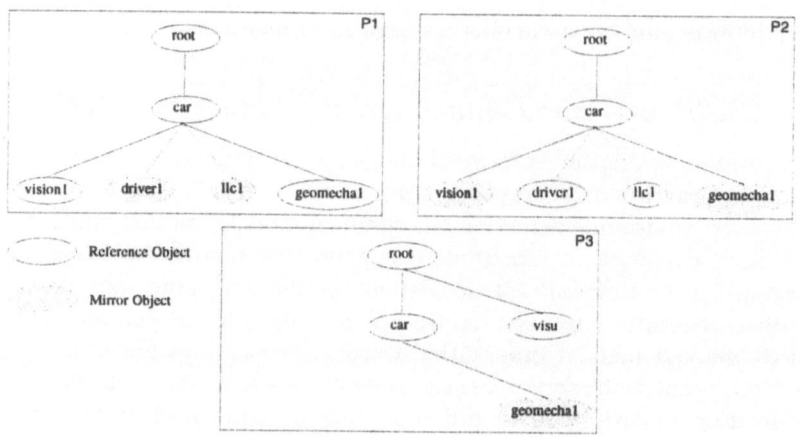

FIGURE 5. Simulation sub-trees on different processes.

A configuration file is used for each simulation to define which dynamic objects are used and on which hardware. As several processes can be used, this file describes first which processors are used, and then each process is named and located on a processor. As the modules of an entity can be located in separate processes, the location of each module is specified in the configuration file. During the simulation, inputs of an object must be supplied by values of outputs. Rather than to define specifically how each reference object must send the new values of its outputs to interested reference object, it has been preferred an automatic mechanism which is based on a client/server mechanism. Each time the inputs of objects of a process require the value of the outputs of another one reference object, an object which contains only the outputs and control parameters of the reference object is created for this process: we call it a mirror object. The continuous communication between two agents can be managed by a two steps mechanism: firstly, the reference object communicates to its mirror the new value of its outputs and control parameters; secondly, the object interested by outputs or control parameters of another object can contact the embodiment of this object in its own process. As each reference object runs at its own internal frequency, the data-flow communication channel must include all the mechanisms to adapt to the local frequency of the producer and of consumers (over-sampling, sub-sampling, interpolation and extrapolation). With the intention of minimising communications between processes, the frequency of the communication between a reference object and each of its mirrors is computed especially for each mirror. The data-flow communication between the distributed processes is presented in figure 6.

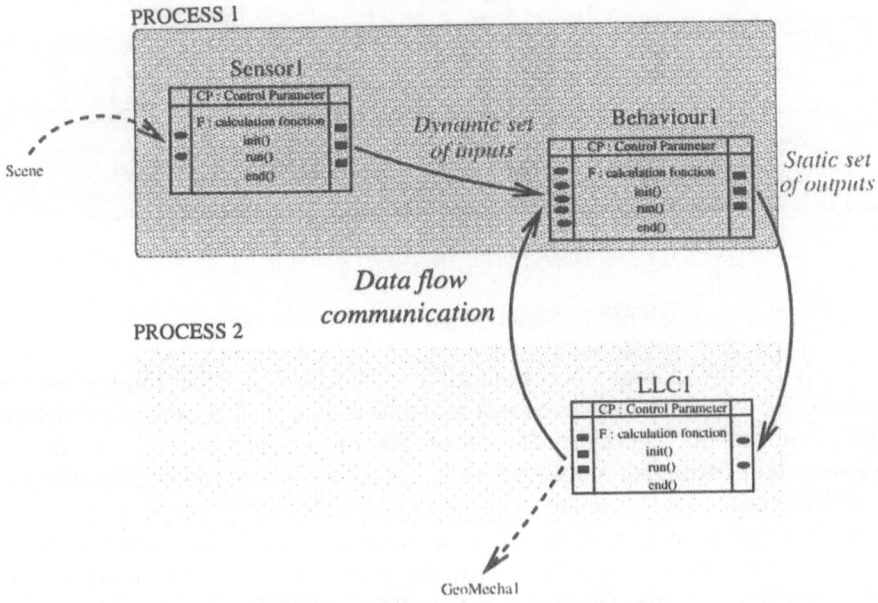

FIGURE 6. Data-flow communication between distributed processes.

3 Introducing Artificial Life in Virtual Environments

Now that we have defined our framework for simulating traffic in a urban environment, we must include simulation modules. Several modules have been defined, but we will introduce here only the virtual car driver and the tram, which is one of our most recent development. Thanks to the modular architecture of GASP, we have been able to integrate all modules in the same application. Two applications will be presented: one deals with multimodal traffic in the city of Rennes, whereas the other is a real-case study of the implementation of tram tracks in the city of Nantes.

3.1 Simulation modules

In this section we describe the automated car driver model in a urban environment that has been implemented whithin GASP. We successively discuss the three components of the "Perception-Decision-Action" loop. The global architecture of the system is presented in figure 7. It shows the different modules used in the simulation and the data flows between them.

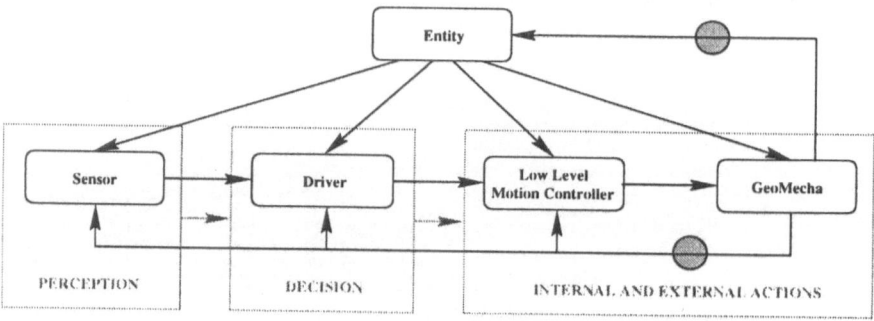

FIGURE 7. Architecture of the virtual vehicle.

In the realm of real-time animation or simulation, it is impossible to completely simulate human vision and the building of a mental model of the environment. Therefore, the automatic driver gets a local view of its environment through a sensor which is in fact a filter of the whole environment database. Two different types of objects are taken into account in the sensor: static objects (buildings, road signals, traffic lights) and dynamic objects (cars, trucks, bicycles). Objects that would be hidden by closer objects are eliminated thanks to a Z-buffer algorithm.

The car driver decisional model simultaneously performs three different activities: traffic handling (i.e. following the other cars, possibly overtaking them), road network following (i.e. following the road, changing lane, taking turns in crossroads) and traffic lights and road signals handling (i.e. adapting speed to the situation). These activities are coded in our behaviour description language [15] with several state machines. An overview of the decisional model is shown in figure 8. This figure also presents a detailed view of the traffic lights handling state-machine.

The goal of the decisional model is to produce a target point and an output action with parameters for the low-level controller. These actions

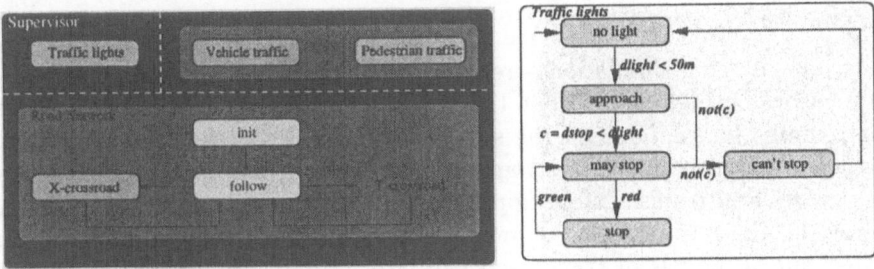

FIGURE 8. Architecture of the decisional model of an automated car driver in a urban environement and detailed view of the traffic light handling state-machine

include a normal free driving mode at a desired speed, a following mode and different breaking modes. The goal of the low-level controller is to produce a guidance torque, an engine torque and a brake pedal pressure as inputs for the mechanical model. The mechanical aspect of the car is modelled with DREAM [17], our rigid and deformable bodies modelling system. By means of Lagrange's equations, DREAM computes exact motion equations in a symbolic form for analysis and then generates numerical C++ simulation code for GASP.

Adding trams and tramways requires several new modules in GASP, and updates in Vuems: trams are moving along tramways and are controlled by special traffic lights and stop at tram stations. Tram traffic lights are themselves controlled by triggers disposed on the tramway in order to give a priority to trams. These traffic lights are connected to normal traffic lights. The architecture of the virtual tram is presented in figure 9. There are four different types of triggers placed along the tramway. The LD trigger is placed 200m before the crossroads, the CD and RAZ triggers are just before the entrance of the crossroads and the VUT trigger signals that the tram has passed the crossroad. Triggers generate events that are taken into account by the tram traffic lights controller. Its behaviour is handled by a hierarchical state machine described in our behaviour modelling language presented in section 2.2. The state machine is quite complex, it requires about 70 states.

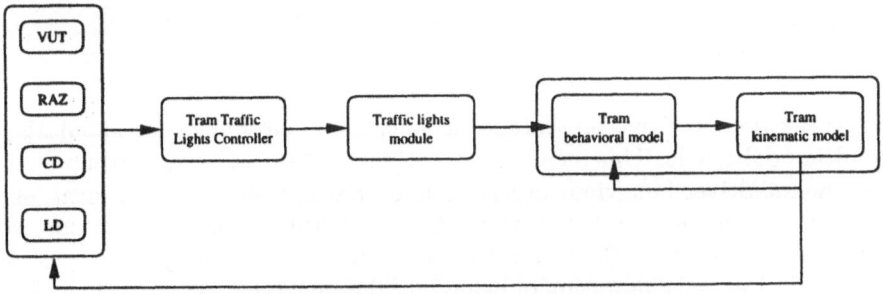

FIGURE 9. Architecture of the virtual tram.

3.2 Applications

By now, driving simulations are commonly limited to cars and trucks interactions on highways. Urban traffic has a higher degree of complexity, as it requires interactions on the same thoroughfare between not only cars, trucks, bicyclists and pedestrians, but also public transportation systems as busses and trams. As our approach is modular, we have started to integrate all these transportation modes into GASP, our simulation platform. Mechanical models of trucks and cars are available, as well as kinematic models of bicyclists and pedestrians. Behavioural models of car and tram drivers, as well as a biker have been described, and we are still working on the behaviour of the pedestrian which is more complicated as he is not constrained to stay on the thoroughfare but, unlike the others, he can wander about everywhere in the city.

FIGURE 10. Views of a multimodal simulation

Our goal is to perform multimodal simulations of the traffic in realistic urban environments; As mentioned in section 3.1, we have recently integrated a model of tram to perform some simulations on a real case study (Croix Bonneaux Crossroads in the city of Nantes). This crossroads is in fact a round-about which is crossed by two tramway tracks which merge inside the crossroads (cf figure 11). We have been asked to study this crossroads to evaluate possibilities of deadlocks due to a high traffic demand.

4 Conclusion

In this paper we have presented our software environment for simulating multimodal traffic. This includes: realistic modelling of virtual urban environments, driver behaviour description, scenario, simulation platform, use of dynamic models and motion control algorithms. The applications we have presented prove the interest of the system; we are currently working on validation and calibration of behavioural models thanks to psychological and statistical studies. Many studies have been performed by psychologists

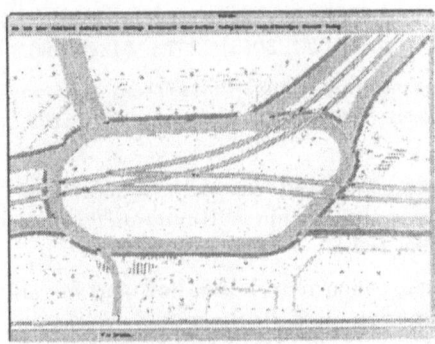

FIGURE 11. Simulation of a tram in the city of Nantes

to analyse the human behaviour during the driving task, but very few of them have focused on pedestrians and bicyclists. In order to analyse the behaviour of bicyclists and pedestrians, we have decided to make some experiments with walkers and bikers immersion in the virtual environment. We are currently buying an immersion equipment including an image wall and a Silicon Graphics Onyx-2 and we already own a car cockpit to undertake these studies.

Applications are not restricted to simulation in urban environments. Work is underway in the realm of high traffic flows on highways (thousands of vehicles, 6km highway section). Within the DIATS project[1], we have been given the opportunity to compare our results to macro-models and statistical models of vehicles flows.

References

[1] Stéphane Espié. Modular driving simulation and traffic simulation. In *Driving simulation conference*, Lyon, France, September 1997.

[2] D.F. Evans. Ground vehicle database modeling. In *Real Time Systems'94*, Paris, France, 1994.

[3] B.E. Artz. An analytical road segment terrain database for driving simulation. In *DSC'95*, pages 274–284, Sophia Antipolis, France, September 1995.

[4] Y.E. Papelis and S. Bahauddin. Logical modeling of roadway environment to support real-time simulation of autonomous traffic. In *SIVE95: the First Workshop on Simulation and Interaction in Virtual Environments*, pages 62–71, University of Iowa, Iowa City, U.S.A., July 1995.

---

[1]DIATS is a project sponsored by the European Community. It aims at defining and studying some ATT (Advanced Transport Telematics) scenarios on interurban highways, in order to bring more efficient management of the existing road network. http://www.soton.ac.uk/~trgwww/diats/diats.htm

[5] S. Bayarri, M. Fernandez, M. Perez, and R. Rodriguez. Virtual reality for driving simulation. *Communications of the ACM*, 39(5):72–76, May 1996.

[6] S. Donikian. Vuems: a virtual urban environment modeling system. In *Computer Graphics International'97*, Hasselt-Diepenbeek, Belgium, June 1997. IEEE Computer Society Press.

[7] B. Blumberg P. Maes, T. Darrell and A. Pentland. The alive system: Full-body interaction with autonomous agents. In *Computer Animation'95*, pages 11–18, Geneva, Switzerland, April 1995. IEEE.

[8] S. Donikian and E. Rutten. Reactivity, concurrency, data-flow and hierarchical preemption for behavioural animation. In E.H. Blake R.C. Veltkamp, editor, *Programming Paradigms in Graphics'95*, Eurographics Collection. Springer-Verlag, 1995.

[9] Xiaoyuan Tu and Demetri Terzopoulos. Artificial fishes: Physics, locomotion, perception, behavior. In *Computer Graphics (SIGGRAPH'94 Proceedings)*, pages 43–50, Orlando, Florida, July 1994.

[10] N. I. Badler, C. B. Phillips, and B. L. Webber. *Simulating Humans : Computer Graphics Animation and Control*. Oxford University Press, 1993.

[11] Hanspeter A. Mallot. Behavior-oriented approaches to cognition : theoretical perspectives. *Theory in biosciences*, 116:196–220, 1997.

[12] R. G. Lord and P. E. Levy. Moving from cognition to action : A control theory perspective. *Applied Psychology : an international review*, 43 (3):335–398, 1994.

[13] B.M. Blumberg and T.A. Galyean. Multi-level direction of autonomous creatures for real-time virtual environments. In *Siggraph*, pages 47–54, Los Angeles, California, U.S.A., August 1995. ACM.

[14] O. Ahmad, J. Cremer, S. Hansen, J. Kearney, and P. Willemsen. Hierarchical, concurrent state machines for behavior modeling and scenario control. In *Conference on AI, Planning, and Simulation in High Autonomy Systems*, Gainesville, Florida, USA, 1994.

[15] Rémi Cozot and Guillaume Moreau. Fast simulation models for an effective aircraft training. In *Simtect'97 Advancing Simulation Technology*, Canberra, Australie, 1997.

[16] Stéphane Donikian, Alain Chauffaut, Thierry Duval, and Richard Kulpa. Gasp: from modular programming to distributed execution. In *Computer Animation'98*, Philadelphia, USA, June 1998. IEEE Computer Society Press.

[17] Rémi Cozot. From multibody systems modelling to distributed real-time simulation. In ACM, editor, *American Simulation Symposium*, New Orleans, USA, 1996.

28

An Architecture for Optimal Management of the Traffic Simulation Complexity in a Driving Simulator

M. Fernández, G. Martin, I. Coma and S. Bayarri

1 Introduction

Low-cost technologies associated to the 3D graphics and simulation fields have made possible a rising interest in using driving simulators for experimental purposes in the areas of road safety and human factors research. This trend is confirmed by the increase in the number of publications that reference this kind of apparatus as the source of their data.

The driving simulator offers a high degree of experimental control in relation to the real world but it also improves the degree of ecological validity that can be obtained in classical laboratory tests. Of course driving simulator data can not be directly treated as real data and the influence of the measurement device (the simulator itself. in this case) and its environment must be considered as a distorting factor.

However, the driving simulation systems have achieved an acceptable degree of realism in providing a restitution for movement (mobile platforms), visual (3D image generators using realistic texturing) and audio stimuli. as well as a good reproduction of the mechanical and dynamical vehicle behaviour (see in figure 1 the generic structure of a typical driving simulator).

The previously pointed increment in the use of driving simulation devices has also enlarged the range of experiments that can be considered. The experiments include frequently the measures that involve the control and representation of a complex scenario surrounding the driver. This aspect has received recently a higher research interest than other parts of driving simulation technology and it is currently one of the main research topics in this field. There have been a good number of contributions in recent years [1][2] but we are still in the way to achieve an 'ideal traffic simulator'.

The traffic in a driving simulator has to accomplish a couple of basic requirements [3]:

- A good degree of control is needed upon some of the vehicles in order to produce or measure certain reactions in the driver.

- Second, it is necessary to have a good degree of naturalism and randomness for the configuration of ambient traffic in order to provide a feeling of immersion to the driver.

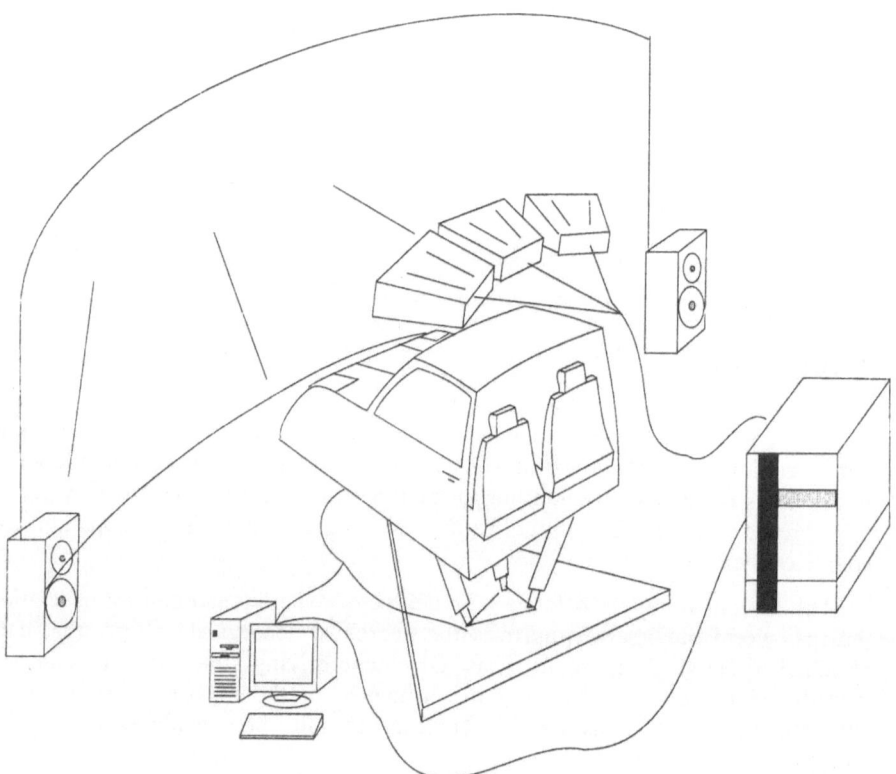

Figure 1: Schema of Driving Simulator Components

In this paper we will present an analysis of traffic simulation techniques oriented towards driving simulation. Our aim is to clarify the possible approaches to traffic simulation in order to achieve the two main requirements earlier indicated.

In the next paragraph we will evaluate different approaches to traffic simulation coming from different application fields and we will extract some properties that allow us to reuse some of these developments in our problem domain.

The degree of naturalism that we will be able to achieve will have something to do with the number of vehicles involved in the simulation, so we will need to control the computational cost in order to maintain the performance in low-medium cost platforms.

To deal with the problems related to computational cost of the simulation we will have to evaluate the specific conditions in which we will use the simulated traffic, especially those related to 3D real-time graphics.

2 Traffic and Driver Models

Traffic and driver modelling is a matter that has been addressed by several research areas, mainly due to its high impact in our society and to the complexity of the problem. Hence, traffic engineers, psychologists, robotics engineers and computer science engineers have tackled the problem from their own perspectives producing a wide range of approaches to traffic and driver modelling.

Driving simulators contribute with new requirements to this generic modelling needs, but of course some of the work formerly carried out in other areas can be evaluated and reused to comply with some of the new necessities.

It is clear that one basic requirement in driving simulators is to describe the behaviour of individual vehicles involved in the traffic flow. Other important issue to consider in this case is the visual aspect of the evolution of the vehicles. This 'naturalism' is necessary in order to avoid non-desired distractions to the driver. And, as we pointed out previously, it will be important to have a certain degree of control upon the behaviour of the vehicles in order to guarantee the repeatability of experimental conditions.

To achieve some of those conditions a simulation model has to be complete and detailed [4]. *Complete* means that the model should include all the aspects involved in the process of driving, and the term *detailed* refers to the accuracy of the model when explaining the decisions that traffic elements make, the description of the information they need to make these decisions and the specification of their consequences.

In order to be able to answer these questions we have started analysing driver models from the point of view of the psychological approach that can provide us with a decomposition of the driving task. Following this idea, we can consider the well-known model that divides the driving task in three decision levels[5]:

- *Strategical Level*, in charge of long and medium term goals (route planning, timing selection, etc.)

- *Tactical Level*, related with the short term decision making (selection of proper lane, speed selection, etc.)

- *Operational Level*, in charge of the control of motor actions related to low level behaviour (lane tracking, acceleration control, etc.)

Other authors have followed an approach based on a detailed description of a wide range of driving situations and subtasks. The situation is characterised by a set of elements involved in it and a set of events and consequences associated to the action [6]. This abstract approach can help us to understand the driving task but does not offer an easy way to get solutions in a computational approach.

Another kind of models that contribute to the natural modelling of driver behaviour are the those ones based on motivational aspects of the driver during the driving process which affect the subject behaviour: "level of risk", "intentions", etc. These models can provide new clues about some of the elements which have to be considered when we select the parameter to generate a random distribution of driving entities.

Other category of traffic models is the one developed by traffic engineers. Those models collect the expertise in the management and simulation of high rates of traffic densities and flows. Within the different subcategories of these traffic simulation techniques, the microscopic models are the ones suitable for our purposes because those models describe the behaviour of individual vehicles. These models have addressed mainly the tactical and strategical levels [7, 8, 9]. In general, the operational level has not been addressed in this field because in most of the cases the aim of this kind of simulation is oriented toward studies in the configuration of the overall traffic flow and not towards describing in detail how the driver really performs the actions.

Finally, we will make reference to driver models developed in the field of robotics and automated cars. This area has provided some interesting contributions at tactical and operational levels in the scope of the Automated Highway System Program (AHS). The program is oriented towards intelligent vehicles able to drive autonomously. Within the program some important work has been developed in situation awareness [10]. Also important contributions have been made in the operational level by describing operational controllers to track lanes and to control speed [11].

As a conclusion from the previously listed works developed in different fields, we can say that it is possible to reuse some parts of each one of these approaches. From the psychological area we can extract the driving task decomposition that allows to tackled the traffic simulation from several levels of complexity that can solved independently. Also from the psychological approach it is possible to extract the parameter characterisation of the driver in terms of motivational behaviour. From the traffic engineering approach we will reuse the way of simulating the traffic as an aggregate of vehicles, which will allow the control of high traffic densities. Finally, the robotics approach can provide the operational controllers to describe in a realistic fashion the vehicles' movement.

The other important feature to take into account in driving simulators is the need of real-time performance. The traffic has to be simulated without an important overload contribution to the computational cost of the whole process. For instance, if we require an acceptable frame rate of about 20-25 frames/second the time that we have available for the whole simulation (dynamic model of the car, evolution of the traffic, 3D graphics processing, etc.) is about 40 milliseconds. In a multiprocessing system the traffic simulation part may be not a problem if we can dedicate one processor for this purpose, but in medium-low cost systems it is necessary to reduce the computational load in order to preserve most of the processing time for drawing purposes. As a consequence of this, usually the traffic simulation has to be simplified by reducing the amount of vehicles involved in the scenario or by reducing the accuracy in the behaviour simulation (cinematic simulation at operational level instead of dynamic simulation that can be more realistic), or by using both simplifications. Both of this options may lead to a failure in achieving the desired 'naturalism' in the appearance of the traffic scenario.

To minimise the effects of this reduction in the computation power available we can take advantage of the perceptual limits of the visual presentation. The basic idea is that not all the traffic environment is visible simultaneously to the driver, nor with the same detail. That suggest that we can use some techniques similar to the one used in 3D real-time graphics to reduce the geometric complexity of the scene. In the interactive graphics domain it is common to use culling techniques based on visibility properties and also level of detail management based on several criteria, like the distance from the objects to the viewer [12]. The transposition of this technique to the traffic management conveys the reduction of the quality and accuracy of the traffic simulation based on visibility and distance criteria. The authors followed this approach in previous works when using a combination of microscopic and macroscopic traffic simulation to manage the dynamics complexity in a large urban environment with high traffic density [13]. In the next chapter we will present an architecture that generalises this idea, including at the same time the proper mechanisms to guarantee an adequate scenario control.

3 Management of Traffic Simulation in a Driving simulator

Based on the ideas presented in the previous paragraph, the architecture that we propose is organised as a set of hierarchical objects that will cover the different levels of reasoning and acting in which we have decomposed the driving task. An additional level will be introduced in order to manage the degree of complexity of the behavioural representation and computation for each vehicle within the simulation scenario. As we have earlier said, the selection of the complexity (accuracy) needed to simulate an aspect of the behaviour will be based on the visibility state and distance of the vehicle in each simulation step.

The objects in charge of describing the vehicle behaviour are called **behavioural servers** and will act upon the **vehicle object** after this one has requested its services (see figure 2). The **vehicle object** encapsulates a set of parameters that define the specific behavioural characteristics of each individual vehicle. Examples of these parameters are desired speed, maximum acceleration rate, degree of impatience, level of risk, etc. The vehicle object also contains a set of attributes which store information describing the current state of the vehicle: spatial position, orientation, speed, lane membership, etc. The behavioural server will be allowed to modify only some of these attributes as a function of the server level in the control hierarchy (strategical, tactical, operational) and will also be influenced in its actions by the current state of the vehicle as well as by the internal parameters defining the 'personality' of the vehicle.

Other important issue that will contribute to conform the general behaviour of the vehicle object is the way in which it retrieves information about its environment.

338

In order to perform this task, the vehicle has access to a bunch of **environment sensors**, with different degrees of complexity, to which it can be subscribed. Generally the sensors are selected controlling at each moment depending on the type of servers that are controlling the vehicle. The sensors can also be configured to introduce some level of error in their measurements. That property will contribute to achieve a more naturalistic behaviour generating vehicles that can make wrong decisions based on imperfect sensors. The degree of inaccuracy is also a function of some of the parameters describing the internal state of the driver: degree of fatigue, blood alcohol concentration, etc.

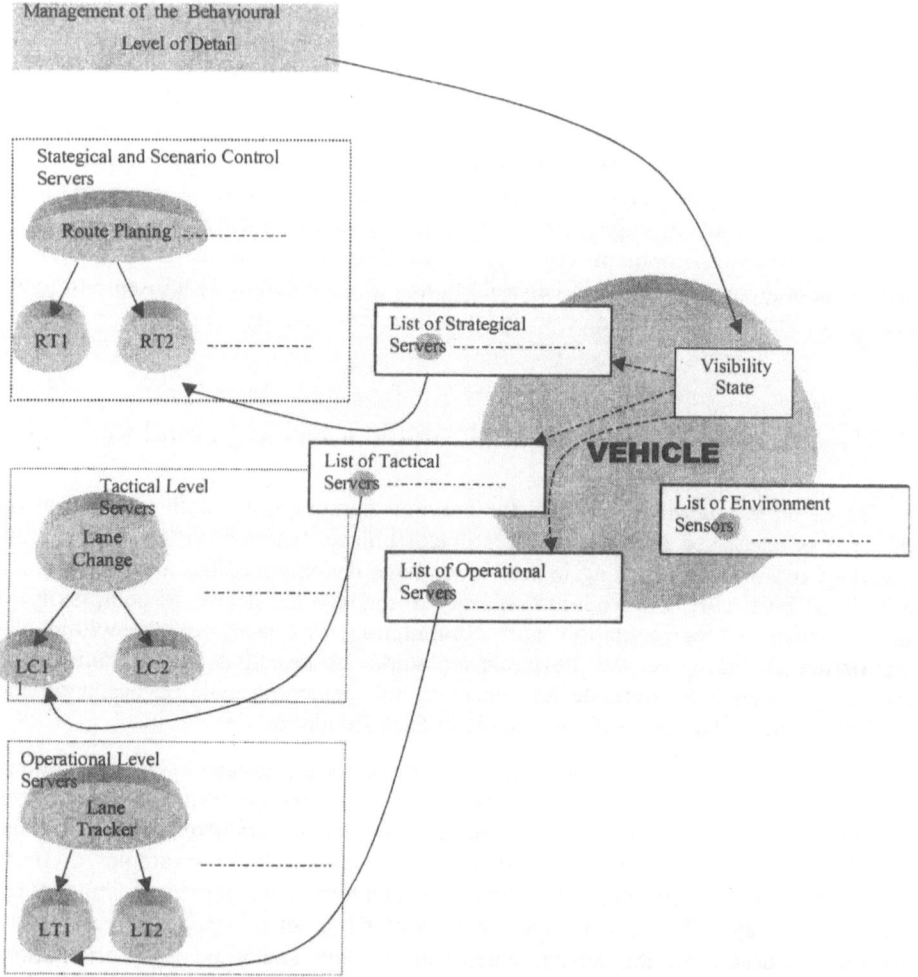

Figure 2: Architecture of the Traffic Simulation Management

The **behavioural servers** will use the attributes and state stored in the vehicle object to make decisions and establish the evolution of the state of the vehicles as well as to update the attributes that they are allowed to modify in each simulation step.

There are three different levels of behavioural server, each one to represent one of the levels in which the driving task is decomposed. As we have previously indicated, a higher level of decision has been introduced to be in charge of managing the complexity level used in every simulation step to describe the vehicle behaviour.

The **Behaviour Level-of-Detail Manager,** as we can see in figure 2, is on the top of the decision hierarchy. To better understand the way it works we will take a brief look at the flow of a real-time 3D graphics application. We can see the main loop in these applications as consisting of three main processes or functions, that are executed in this order:

1. *Application Process*: it is in charge of updating the data structures of the application, including every simulation computations (in our case that includes traffic simulation).

2. *Culling Process*: it performs a traversal of the hierarchical tree representing the geometrical description of the scene. During this traversal a test computation of the areas of the scenario that are inside of the viewing frustum, defined by the position and orientation of the viewer, is made. The objects outside the viewing area are rejected for the draw process, while the ones that are inside are re-evaluated and the proper level-of-detail for its representation is selected based on the distance to the viewer. As a result of this process we obtain a display list containing all the objects to be rendered in the draw process.

3. *Draw Process*: It is in charge of performing the render of the scene to be visualised. This process uses as input data the display list obtained from the culling process. This list is usually sent to the 3D dedicated hardware for rendering.

The manager of the behaviour level of detail (BLOD) will work in relation to the overall culling process in order to avoid a duplication of the visibility and distance tests. If we analyse the sequence of processes presented above, we will realise that in principle it is not adapted to our objective, since the application process is called previously to the culling process. To overcome that, we have assumed the compromise of loosing some accuracy in the management of behavioural computation detail by using in the application function the results of the culling process of the previous frame, so that there will be a delay of one frame time between detail test and behaviour computation. If we consider a normal frame rate of 25 frames per second that means an error of about 40 milliseconds, which is acceptable.

We have modified the classical culling process in order to consider the lanes, links and intersections (elements defined at the topological level of the correlated simulation database [14]) as nodes that can be visible or invisible. In addition, a label containing the information about the distance from the viewer is assigned to a number of parts dividing these elements in sections. In the following application process the manager will read these values and will assign the proper state of complexity to each vehicle based on the lane, link or intersection where it is located. Then the proper object is selected from the list of behavioural servers to request its computation services. Several levels of detail are defined for each layer of behavioural servers, and also for each task performed by a specific server.

The **strategical level servers,** as we introduced before, are related with the long and medium term driver's decision making process. They will be in charge of performing actions such as route planing. This level of service is also appropriate to performs tasks concerning the scenario control, because a way of controlling the actions of the vehicles is to establish their long-medium term goals. In this level we have included some control objects, as a *behavioural path-follower* which allows to define the spatial and behavioural evolution of a vehicle within the scenario. The hierarchy of behavioural servers will make that decisions taken by the strategical server will force actions at inferior levels, i.e. if the path follower decides to take an exit it will force the tactical lane change server to look for a lane that is adequate to reach the desired exit. At the same time, if the tactical lane change server has found the change to the target lane to be safe, it will force the vehicle to select a proper operational lane change server to perform the action.

The different servers are fully independent and only have to accomplish with the interfaces defined for each one of the basic tasks represented. This level of encapsulation allows a high degree of flexibility and freedom to select the model to represent each one of the tasks. For instance, we can develop a lane change server based on a classical ruled approach and at the same time design for the operational level a fuzzy controller to perform the action. It is also possible to have different approaches for servers at the same level and control different vehicles using one or another approach.

4 Results, Discussion and Future Work

The architecture that we present here has been already implemented and it is being used to carried out assessment test of the ergonomics of Automatic Cruise Control Systems at Nissan Cambridge Basic Research Centre in Massachusetts, USA. The scenario prepared for this test includes a combination of urban and interurban environments. The whole scenario has about 16 square kilometres, with about 10 miles of 3-lane highway, about 5 miles of an state road, and a small town with about 30 streets. The whole scenario involves more than three hundred vehicles and the system is running in a SGI Octane with a single R10000 processor. The experiment is running at an average frame rate of 20 frames/second (see figure 4).

The architecture that we have presented allows us to maintain the cost of the traffic simulation (using a good degree of realism for the operational and tactical servers -at the highest BLOD-) below the 10 % of the total cost of the simulation. In figure 3 it is shown a comparison between the time required for traffic simulation when using the behavioural level of detail management and without it. Another point to be stressed is the good performance obtained in the scenario control tasks due to the hierarchical structure of servers. The behavioural path-follower server has been able to control in a natural and precise way the vehicles used in the experiment. The vehicles were forced to cut in front of the driver at precise time headway in order to assess the driver's reaction while driving using the ACC system.

We can conclude that this architecture makes compatible a good degree of naturalism, using a large number of vehicles, with an acceptable computational cost.

At the same time the hierarchical architecture guarantees a good degree of control upon specific vehicles in the scenario. Future work will be oriented towards the extension of basic tasks considered by the system, and to the generalisation of the server interfaces.

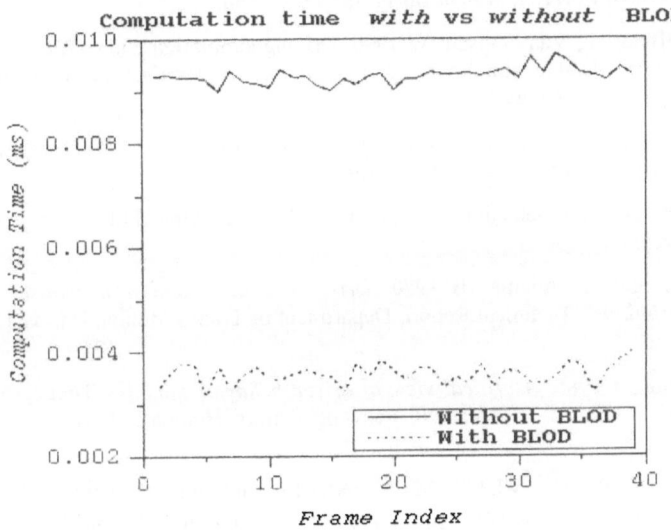

Figure 3 Computation Times with and without Complexity Management

Figure 4: Views of Experiment Scenario at NISSAN CBR

5 REFERENCES

[1]   Cremer J, Kearney J, Papelis Y, Romano R 1994. The software architecture of scenario control in the Iowa driving simulator. *Proceedings of the 4th Computer Generated Forces and Behavioral Representation..*

[2]   Wolffelar P, Van Winsum W 1996. Driving simulation and traffic scenario control in the TRC driving simulator. *Symposium on the Design and Validation of Driving Simulators. Valencia.*

[3]   Bickenbach H 1995 Training Design and Scenario Control. *Proceedings of the Driving Simulation Conference. DCS'95.France.*

[4]   Reece D 1992. Selective Perception for Robot Driving. PhD thesis, Carnegie Mellon University.

[5]   McKnight J, Adams B 1970. *Driver education and task analisis volume1: Task descriptions.* Technical Report, Department of Transportation, National Higway Safety Bureau.

[6]   Michon, J 1985. *A critical view of driver behavior models: What do we know, what should we do?.* L. Evans and R. Schwing, editors. Human Behaviour and Traffic Safety. Plenum.

[7]   Gipps G P 1986. MULTSIM: A model for simulating vehicular traffic on multilane arterial roads. *Mathematics And Computers in Simulation 28.* Pp. 291-295

[8]   Barcelo J et al. 1996. Microscopic trafficc simulation for ITS analysis. *Contribution to the 25th Anniversary of Centre de Recherche sur les Transports, Universite de Montreal.*

[9]   Yang Q 1997. A Simulation Laboratory for Evaluation of Dynamic Traffic Management Systems. PhD Thesis, Massachusetts Institute of Technology.

[10] Sukthankar R 1997. Situational Awareness for Tactical Driving. PhD Thesis, Carnegie Mellon University.

[11] Pomerlau D. 1995. RALPH: Rapidly adapting lateral position handler. *Proceedings of IEEE Intelligent Vehicles.*

[12] Funkouser T A 1993. Database and Display Algorithms for Interactive Visualization of Architectural Models". PhD. Thesis, University of Berkeley.

[13] Bayarri S, Fernández M, Pérez M 1996. Virtual Reality for driving simulation. *Communications of ACM.* Vol. 39, N° 5. Pp 72-77.

[14] Evans D F 1992.Correlated Database Generation For Driving Simulation. *Proceedings of IMAGE VI Conference.*

29

Software Challenges for High Fidelity Driving Simulators

Y.E. Papelis

1 Introduction

The cost of hardware typically employed in high fidelity driving simulators has been steadily declining. This decline has made it possible to design and construct driving simulators at a fraction of the cost of what was possible within the last few years. This progressive reduction in hardware costs has made simulators affordable for a variety of disciplines for which simulation has been too expensive to be beneficial. Such disciplines include transportation related human factors, training, highway safety and medicine. The uses of simulators within these new fields have brought new requirements on the capabilities of the software that accompanies these simulators. Unfortunately, unlike hardware whose cost has been steadily declining due to rapid technical advancements, software costs have not kept pace with this trend. Furthermore, the commercial availability of software customized to driving simulation applications is minimal, primarily because of the small market for such a vertical application. These two factors have lead to a plethora of driving simulators at various cost categories that utilize the latest, most advanced hardware available, yet fail to deliver supporting software to enable effective utilization of the device.

This chapter provides an overview of functional requirements and associated challenges posed upon advanced driving simulators. The focus is on disciplines for which advanced simulators are finally within reach due to lowering hardware costs.

2 Overview of Driving simulation usage

Uses of driving simulators can be broadly divided in these areas: human factors experimentation, training, medical research and virtual proving ground prototyping.

Human factors is a very general term, even when limited to transportation, that includes the study of the driver when interacting with the vehicle, the outside environment or the overall transportation infrastructure. In the past, simulators have been used to investigate numerous human factors issues including advanced in-vehicle devices such as heads-up or heads-down displays, anti-lock braking systems, or new highway concepts. [1, 2]

Training applications often utilize simulators to reduce the risk of training in an actual environment, or when training is necessary for situations that are hard to recreate in real-life. Simulators can also be useful in compressing training time, for example by presenting the trainee with virtual environments whose meteorological conditions vary in ways that in real life would not be achievable in short spans.

Medical applications of driving simulators include the study of the effects of drugs, or medical conditions on the ability to drive (or lack of such ability).

Virtual proving ground prototyping [3] is a relatively new use of advanced driving simulators and refers to the utilization of a driving simulator in lieu of an actual model for the purpose of conducting engineering design of a vehicle or a vehicle component. Virtual proving ground requires that simulators utilize dynamic models that provide engineering level fidelity with regards to their modeling of the vehicle system. Such fidelity provides performance data in enough detail to allow comparisons between design options exclusively in software. Often times, actual hardware components are integrated with the software dynamic model providing hybrid "hardware-in-the-loop" systems that can test the design of either the hardware component or vehicle dynamic model as simulated in software. Using such tools design engineers can modify key design parameters and almost instantaneously obtain feedback of the performance of the vehicle under actual usage conditions, while driven by a human operator. This is in contrast to off-line simulations that are not interactive and thus cannot predict the reactions generated by the closed loop system that includes the human in the loop.

Each of the aforementioned uses of driving simulation presents its own challenges to the various software components. The first three, which we will collectively refer to as human sciences experimentation, provide the most challenge for the software that interacts with the user and/or operator, as opposed to the virtual proving ground applications which present the most challenge for the software that is part of the underlying simulator device (such as the high fidelity vehicle model) but is not in direct interaction with the user.

Virtual proving ground applications are still the domain of advanced, one-of-a-kind devices that often utilize motion bases or high fidelity cueing systems. Due to their high overall cost (relative to other simulators), the portion incurred due to customized and specialized software is small and has a lower impact to potential users. In contrast, human sciences researchers can often conduct scientifically valid and useful research on more affordable (from the hardware standpoint) devices. Unfortunately, software costs often dominate the cost such devices despite their advanced hardware capabilities. The result is often complete lack of software, or software whose capabilities do not make full use of the available hardware. The derived effect is severe limitations to the usefulness of these devices. This is unfortunate, because it is within the human sciences experimentation field where simulators have the potential of becoming wide spread, effective tools. With these factors in mind, this chapter concentrates on the challenges for simulator software in the human sciences applications.

3   Human Sciences Experimentation Requirements

There are several challenges in the effective use of advanced simulators for scientific research. These are best identified by looking at the steps involved in using such devices for training and/or human factors applications.

In training applications, an instructor or the simulator itself presents the trainee with a series of lessons. These often involve special events to test the trainee's readiness, or gradual increase in the skill level required to successfully complete a lesson.

Such requirements sound straightforward but a simple decomposition into more detailed technical requirements yields a variety of challenges. Let us first consider the issue of lessons or exercises. It is almost a necessity for an instructor or trainer to have the ability to generate different exercises. Generating different exercises may involve developing different scenery such as new roads or intersections, or may involve utilizing existing scenery but creating a more challenging environment, maybe by injecting heavy traffic in a winding rural road. Doing either task often requires a person that is not necessarily trained in software development to communicate to the device their instructions for creating the new scenery of dynamic environment. The ability to author lessons is a key capability that can be very complex if not handled by very powerful and innovative software.

In addition to authoring exercises, it is reasonable to expect the system to provide facilities to catalogue, test and store exercises for later use. This implies the existence of some database management system designed using a structure and interface that is familiar to trainers. Trainees often have multiple sessions on a simulator requiring the capability for the system to remember students, keep track of their performance and even be able to adjust the curriculum to match the student's learning progress.

During the actual training, someone has to evaluate the performance of the trainee. An instructor can do that, although this limits the instructor to only supervising a single student so a form of automatic evaluation would be another key requirement. Automatic evaluation software should also be programmable to allow the trainer to customize the evaluation. Additional requirements include the ability of the system to provide instantaneous feedback to the student in case of serious mistakes, or for purely for instructional purposes. Additional related requirements include the ability of the system to provide support to the instructor for monitoring multiple students, or allowing multiple students to interact in the same virtual environment.

Generally, in human factors or medical applications, a hypothesis is put to a test by running several subjects through the same or similar conditions, then collecting data on their behavior and/or performance and then using statistical techniques to validate the hypothesis. A key requirement is the ability of an experimenter to develop the situations or events that the subject is exposed during the simulator trial. This action is parallel to the development of lessons in training applications. Since multiple subjects are often involved in such studies, the system should also be able to keep track of subjects, their exposure conditions and collect data for each exposure. Keeping track of the subjects and the conditions assigned to them (i.e., subject 1 is presented condition A, subject 2 is presented condition C etc.) is conceptually a simple task, yet in practice involves extensive logistical problems and when done manually poses the risk of human error. For example, if a single operator has to start the simulator, select the exposure while at the same time providing instructions to the subject, the risk is high that mistakes could be made in selecting the proper startup files. The risk becomes even higher when the workload increases by something as simple as a subject failing to show up. One would consider it a requirement for an advanced simulator dedicated to such work to provide software that facilitates experimentation and minimizes the risk of human error under most conditions.

Once data is collected it has to be analyzed through statistical means for hypothesis validation. Due to the real-time constraints, most systems usually collect raw data during the simulation. This data may include the position of the vehicle, its velocity, the control inputs and other similar parameters collected at some frequency. However, the data required by experimenters is defined at a completely different domain and includes behavioral descriptions as opposed to engineering level variables. In few cases, the two may coincide and it is simple to extract the data required by the experimenter. Other times however, this conversion is not straightforward. For example, the reaction time to an incursion maneuver is a variable often used in research studies. Deriving this variable requires knowing when the other vehicle caused the incursion and its position at that time. Other similar variables include lane deviation (how far from the center of the lane is the driver positioning the car), following distance (how far behind a lead vehicle is a driver driving), number of lane changes etc. The process of extracting these higher level variables from the raw data produced by the simulator is often referred to as data reduction, and is another key requirement for driving simulator software.

In looking at these requirements, one can notice certain commonalties in the software components for the various disciplines. We believe that all such requirements can be categorized into four groups: User Management, the ability of the system to understand the concept of a whole experiment or training session as opposed to simply starting and ending the simulator hardware; Scene Authoring, the ability of the system to allow users to develop the static scenery component of the simulator's virtual environment; Scenario Authoring, the ability of the system to allow users to define the dynamic elements of the virtual environment; and Data Reduction, the ability of the system to provide useful data to the experimenter after a session. Note that these four areas are closely related and often overlap, but the decomposition is useful in looking at the unique requirements of each group.

3.1 User Management

In almost all uses of simulators, (with the exception of demonstrations or entertainment sessions) there is a need to maintain some connection and continuity among drivers dictated by an experiment, a training session or some similar higher level activity for which the simulator acts as a tool.

Depending on the complexity of the simulator and the particular usage pattern, there can be several items that have to be properly coordinated and matched against each driver before a drive begins. These include anything from configuration parameters and names of data files to mundane issues of proper VCR tape labels. Experienced software developers have probably already recognized this as a configuration management problem, something that is not terribly challenging from a technical standpoint, yet is extremely important in any project.

The details that require tracking vary between training and research applications but in general terms, the user management software should be able to keep track of all the associations between drivers and their parameters within the context of an experiment. Multiple experiments should also be tracked as individual units. As experiments take place, individual experiment entries should be updated to reflect additions or changes to the baseline configuration. Such changes may include the actual names of data files produced by the simulators or other salient information such as notes about non standard events or miscellaneous log entries. All of this information can be naturally organized in a hierarchy of the form:

```
Experiment:
      Subject:
            Initial Simulator Configuration
            Other non simulator data (i.e., questionnaires, video tape IDs etc.)
      Subject:
            Initial Simulator Configuration
            Other non simulator data (i.e., questionnaires, video tape IDs etc.)
      etc ...
```

A related requirement is the ability of the simulator to help the operator follow the stored configuration of an experiment during the sequence of drives. The goal of this is to minimize the potential for human error. For example, the simulator should provide graphical tools that force the operator to specify which subject is about to drive and perform extensive error checking to ensure that no duplicates or missing data exist and that each data file is properly tagged and associated with the correct subject. Graphical user interfaces and software to achieve such functionality are not complicated, but are rarely part of driving simulators as they are employed today.

3.2   Scene Authoring

Scene authoring refers to the development of the static portions of the virtual environment presented by the simulator to the driver. In driving simulator applications, this scenery usually consists of roadways, vehicles, signage, traffic lights vegetation and adjacent structures such as buildings, sidewalks etc. Note that we make a distinction between the visual appearance of the scenery and its behavior or evolution that takes place during a simulator session. Scene authoring refers to the construction of the visual models but not the development of dynamic events or interactions within the environment. Figure 1 illustrates a driving simulator scene taking place in a city setting.

Scene construction is a challenging task, even for experienced users. A large factor contributing to this complexity is the stringent requirements for real-time display of the scenery which forces developers to utilize complicated data structures for structuring the 3D world. The sheer number of details that a developer has to provide is often intimidating even for scenes of modest complexity. Another contributing factor to the complexity is that in designing scenes appropriate for driving simulator applications, one must at least be familiar with road construction guidelines, sign placement rules and numerous other rules and specifications that should be applied to roadway design, at least for scenes that are supposed to mimic the real world. Scene development under these conditions is not well suited for users of driving simulators that have no expertise or interest in real-time scene modeling or civil engineering

It is primarily because of these limitations that the majority of driving simulators do not provide tools to allow users to develop their own scenes. Instead, the simulators are delivered with large scenes already developed and the users are limited to using what is available. Additional scenes can always be developed when required, but not by the users, and often with substantial cost and time involvement. This is unfortunate since providing the ability to the user to develop scenes of their liking would make simulators much more effective and attractive as tools.

Figure 1 – Illustration of Driving Simulator Scenery.

Therefore, one of the key requirements for scene generation software is simplicity of use, and efficiency. To circumvent the problem of having to specify numerous details, scene generation tools should operate at a high level, maybe allowing the user to combine smaller but existing scenes into larger scenes suitable to the user's needs. This approach, often referred to as tile based scene generation has the potential advantage of ensuring that larger scenes represent valid roadways, assuming that the smaller building blocks are already valid roadways. The technical approach to achieving such functionality is still subject of ongoing research and development [4, 5], but we consider it to be a key requirement, especially for simulators with high fidelity visual systems who can benefit the most from rich, realistic scenes.

3.3 Scenario Authoring

Scenario authoring refers to the development of the dynamic aspects of the virtual environment of the simulator. The term includes the specification of the motion of other cars, trucks, pedestrians, traffic lights, changes in environmental conditions, or any other change in the environment that takes place during the simulator's execution.

This term often includes various initial conditions of the environment such as the position of the own vehicle or time of day upon start up. Scenario control is a term often used to describe the software responsible for simulating the environment and producing the movements of all entities in the virtual environment.

The most basic form of scenario control software consists of specifying paths that entities follow at a velocity that is independent of the driver's actions. With enough patience and time, one can create fairly complicated dynamic scenes consisting of large number of vehicles and other dynamic elements. The biggest disadvantage of this approach is that the scenario is not reactive to the driver. For example, if the driver stops and the following vehicle is not programmed to stop at exactly the same time, a collision will occur. Given that no two drivers will ever drive a particular route in exactly the same way, use of non-reactive scenario control is not a viable option, especially for high fidelity simulators.

A variety of techniques exist for addressing the need for consistency in the events presented to a series of drivers independent of their timing. One of the simpler methods includes programming the exact path of each entity but also specify that entity's distance to the driver at each point on the path. This ensures that no matter how slow or fast the driver travels, everything will happen the same way when viewed relative to the driver. There are still disadvantages to this method, for example if the driver comes to a complete stop, so does everything else in the virtual world, something that is not consistent with reality. Also, if a scenario consists of more than a few vehicles, defining all the paths involves quite a lot of manual effort, much of which involves trial and error.

Another technique, that is often used in high fidelity simulators, is the development of autonomous driver models which are then responsible for controlling each entity at runtime. These driver models utilize rules and decision making similar to human drivers as far as reacting to other vehicles, navigating the virtual environment or following the rules of the road, and can often produce behaviors that cannot be differentiated against a human driver. A rich scene populated with multiple autonomous entities looks very realistic and be extremely immersive to the driver. The negative aspect of this scenario control approach is that the autonomy of the individual entities makes it harder to force specific events to happen, especially when these events are not normal. For example, forcing one of the scenario vehicles to ignore a STOP sign may be needed for testing the simulator's driver reaction time, but the autonomous behaviors should not do it as it violates the rules of the road.

In recent years there have been extensive research [6] in techniques that merge the autonomy of scenario, which is required for realism, with the controllability and determinism of events, which is required for repeatable scientific studies. The underlying implementation techniques vary but the general idea is to use dynamically evaluated rules that force autonomous entities to behave in non-standard ways at the proper time. Different authoring paradigms exist, some using scripts written is custom designed languages, while others using more graphical means for defining the rules and conditions. Tools utilizing scripted languages are generally harder to use but can be very powerful since they allow arbitrary programs to be developed in their custom languages. Graphical tools are generally easier to use

Clearly, an effective scenario authoring and control software is necessary in high fidelity simulators. The most important requirement in such a system is the proper matching of features to the requirements of the simulator user. For example, a scenario system using a scripted language may not be the most appropriate for users who are not experienced in developing software. A graphical tool providing high level constructs for developing scenarios and include the ability to re-use existing scenarios would be more appropriate in that case. Users experienced in programming however, may prefer a powerful scripting language or a graphical tool that provides powerful features and extensive capabilities. This tradeoff between usability and power of a tool is rather common in popular commercial software packages.

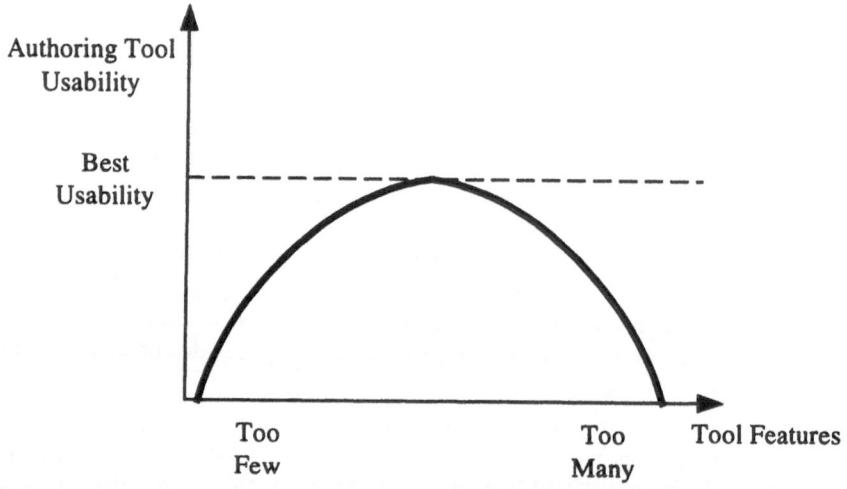

Figure 2 – Authoring Tool Usability vs. Features.

Figure 2 illustrates a curve showing the usability of tools versus the amount of features (or power) of a given tool. Clearly, too few features can severely limit the usability of a tool. At that point, adding more power to the tool increases usability to a point where too much power overwhelms the user and reduces the tools' usability. Note that each application domain has its own curve with different slopes that depend heavily on how much standardization exists for scenarios within the application domain. The more standardized the scenarios, the fewer customization features are needed in authoring tools. As an extreme example of this tradeoff consider flight simulators whose scenarios are often limited to taking off and landing sequences, thus require very limited authoring capabilities for new scenarios. In driving simulation applications, where there is great need to produce a variety of scenarios, it is important to provide tools that are a match to the expectations and capabilities of the users. By providing the right amount of power and features, these tools can maximize their usability and in turn the usability of the simulators. Better yet, borrowing from practices often employed by commercial software packages, tools can present different complexity to the user based on the user's expertise and background.

3.4 Data Reduction

Data reduction refers to the process by which raw data produced by the simulator in real-time is transformed into data that is useful for further analysis by users of the simulator. The complexity of this transformation depends on the complexity of the reduced variables and the original state of the raw data.

Typically, simulators store data records describing the state of the own vehicle multiple times per second. The sampling frequency varies but often ranges between 10 and 60 Hz. Higher sample rates may be necessary for specialized applications but that does not change the basic requirements of the data reduction software.

If the simulator does not operate using a fixed time step the data must contain enough information about recording time drift to allow reconstruction of the original timing. This is usually the first step taken by data reduction software followed by removal of any anomalies in the data. Anomalies are typically introduced by censor noise, software glitches, or plain failures of the simulator hardware to keep track with the delivery of data. As simulator technology has evolved over the last few years, such problems are minimized and are often eliminated. Following this data conditioning step, calculations are made to derive the necessary variables. It is in this step where one sees the highest variability in requirements.

Variables that are directly related to the raw data and only refer to the state of the vehicle, independent of anything else in the virtual environment, are easier to produce. Assuming that the state of the own vehicle along with all control inputs are saved in the raw data stream, examples of this class of variables include velocity, engine RPM, brake pedal presses, accelerator pedal presses, steering wheel reversals etc. There are two key related requirements. The software should document how such calculations are made including threshold or coefficient values. The second requirement is the ability of the user to change or fine tune these threshold values. For example, consider calculating the number of brake presses during a simulator run. To determine a brake press, one has to pick a threshold pressure or displacement beyond which the simulator driver has to depress the pedal before the software will count one press. Knowing the threshold value is necessary for publishing such results or for providing the opportunity to fellow researchers to validate a conclusion by repeating an experiment. If this information is not known, no one else can repeat a research study and any results produced have limited usefulness for the community. The ability to change the parameters is useful in facilitating use of the reduction for multiple purposes, for example in looking at hard braking as opposed to light braking.

When reduced variables refer to the own vehicle state in relation to some other quantity, data reduction becomes more complicated. For example, if instead of looking at the number of brake presses for the whole duration of the drive, one wants to compute the number of brake presses while driving a particular segment of road, it is necessary to have a notion of the virtual environment in the form of roadways, and be able to identify the particular segment of road at data reduction time. Other examples of relative variables include lane deviation (position of the vehicle measured relative to the roadway), following distance (position of the vehicle measured relative to another object), or count of STOP sign violations (vehicle velocity used in the near STOP signs). Within this group, variables that are in relation to quantities that are static across a drive or the whole experiment are easier to define and calculate than variables that depend on varying quantities. For example, it is simpler to calculate the velocity of the own vehicle at a given intersection or after a fixed amount of time since the start of the drive as opposed to when passing another vehicle. Variables that are centered around interactions with other vehicles pose the biggest challenge as the software has to search for that occurrence which is error prone due to the variability of the subject's behaviors. As the richness of the simulator's virtual environment increases so is the potential complexity of deriving complicated variables or measuring the performance of the driver while in the simulator. There are great challenges for the development of techniques that allow rapid development of data reduction specifications that include complicated, inter-related variables that are comparable across simulators. This is the subject of ongoing research and no elegant solutions exist. At a minimum, driving simulators should provide data reduction tools that have the basic ability of conditioning data, producing basic variables with well documented and tunable thresholds and have the ability to export data for further analysis by customized software.

4 Conclusion

This chapter provided an overview of technical challenges in high fidelity driving simulators. These challenges originate from the ever widening use of driving in various disciplines that can now find simulators affordable due to decreasing costs.

5 References

[1]   Bloomfield, J. R., Buck, J.R., et.al. 1994 Human Factors Aspects of the Transfer of Control from the Automated Highway System to the Driver. In: Technical Report for FHWA, under Contract DTFH61-92-C-00100.

[2]   McGehhee, D.V., Dingus, T. A., Papelis, Y.E., Bartelme, M.J. 1995 The Use of Specialized Scenes and Scenarios on the Iowa Driving Simulator for the Evaluation of Rear-End Crash Avoidance Performance *TRB meeting*, Washington, DC, 1995.

[3]   Haug, E. J., Cremer, J. Papelis, Y. E., Solis, D. A., Ranganathan, R. 1998 Virtual Proving Ground Simulation for Vehicle Design. ASME Design Automation Conference, submitted.

[4]   Kearney, J., Allen, S., Bahauddin, S., et. al., 1996 Tile-based Scene Modeling for Driving Simulation. In: *Proceedings of the IMAGE IV Conference*, Scottsdale, Arizona.

[5]   Papelis, Y. E., Bahauddin, S., 1998 Rapid Development of Domain Specific Correlated Databases Using Parametrized Tiles. In *Proceedings of the 1998 IMAGE IV Conference*, Scottsdale, Arizona, to appear.

[6]   Scenario 1996; Workshop on Scenario and Traffic Generation in Driving Simulation, December 6-7, 1996, Orlando, Florida.

PART V

INDUSTRIAL ROBOT ANALYSIS, DESIGN AND CONTROL

30

Trends in Robot Control : Autonomous Behavior and Remote Control – the VR Contribution

Ph. Coiffet

1 Introduction

In the huge amount of litterature about robot control few papers precise the basic assumptions on which their contribution relies as connected to a robot definition and vocation. That point is not important when the commented control level is concerned with an improving of robustness or precision, and so on... for a mechanism, trajectories of which are known in advance. The problem becomes more conspicuous when it deals with the motion or behavior to be respected in order to perform a given task in an ill-known environment. So an introduction will be dedicated to the basic assumptions from which behavior autonomy demands to be studied.

In a next section, the various ways covering all robotic researches will be presented, allowing to guess the different ideas people developped believing to approach, either a behavior autonomy or a good remote mastering of robots. Analyzing these approaches, a new possibility for autonomy can be examined that takes under consideration some forgotten features of environment. The way to go toward some selfawareness will be exposed underlining the difficulties.

In Teleoperation, the two main traditional drawbacks preventing from an extensive use of such a system deal with a poor friendly using possibility from one hand, and a not less poor information feedback quality from the slave site to the master one, from another hand. VR techniques can renew the teleoperation concept through the creation of a virtual universe, well adapted to man behavior from one side, and also well adapted to control machines from another side. VR world becomes a new adaptive and intelligent interface between the human operator and his real workplace. Difficulties and ways to overcome them will be discussed in both cases of a remote short distance and long distance control. The conclusion that can be proposed is the idea that a perfect teleoperation system must pass through some capacity of the controlled machine to be autonomous in some extent. The point deals with the level of autonomy that has to be aimed at. And the basic problem exposed at the begining is once more again raised.

2 Control and autonomous behavior

2.1 Basic statements in order to understand the fundamental demands to build robots

2.1.1 Declaring something about an ideal robot control assumes that a clear definition and a clear goal can be given to the machine named robot. In any useful industrial or practical application, the definition and goal problems do not exist because tasks to be performed as well as environment characteristics drive the designer to precise drawing up of specifications. From these last ones, a suitable machine using available tehnologies can be made and meet specifications. It is only after this step that the decision to name or to not name the machine as "robot" is made, in function of trade policy or discussions about advertizing or used technologies or and so on... [1].

But, when some softwared properties such as intelligence, skill, behavior autonomy, awareness etc., generally reserved to human beings appreciation, have to be implemented on a machine, all things become fuzzy about robot definition and goal, resulting in a difficulty to clearly understand what way is really tracked by specialists. So, it seems interesting to notice some elements that might gather the approval of most people and would allow to highlight the dispute and to detect new ways about a better robot control.

STATEMENT 1: Robot is a material machine which owns a behavior evocating this one of a living being (belonging to animals class) to a human observer.

This statement is not a robot definition because terms such as: "living being behavior" or "evocating" remain very fuzzy and complex, but it is a kind of axiomatic assumption that will be heavy in its consequences. It is now supposed that this first statement is accepted.

STATEMENT 2: A living being is mainly noticeable by its complexity.

No one seems able to definitively give a correct definition of life [2][3]. Anatomy and physiology of the least living being is so complex that all characteristics are not understood, and no one was able to create a living being without an extraordinary complex genetic base. So, robot cannot be designed (excepted through a biogenetic way) as a copy of any living being.

STATEMENT 3: Living beings are also noticeable by their diversity.

From microbe to human being, all animals are alive. To make a robot, must one imitate one given living being rather than another one? There is no answer to that question.

CONCLUSION 1: It must be faced to a contradiction: from one side robot has to exhibit an animal behavior, and from another side, life, source of this behavior, cannot be imitated. Consequently, it is necessary to seek inside manifestations of life, a part of which being the animal behavior, for what features could be tentatively synthetized.

Among the main consequences of life allowed through a material behavior, there are phenomena devoted to survival (that constitutes the main deep motivation of the individual for acting).

So, the question becomes now: how survival is possible from the origine of time or from the origin of the apparition of a living kind? and what consequences on the robot designing?

2.1.2 An evolution image based upon the system theory. Considering the classical Darwin's evolution theory [4][5][6] (carefully preventing to adopt any philosophical position in this paper), it can be detected a basic point for the future robot designing with the notion of mutual interaction and transformation of the individual through its exchanges with its environment.

The "classical" Darwin's theory says that only a kind of "unilateral" transformation takes place, i.e. only the individual is modified in order to get the best adaptation to survival in a given environment. In fact, other explanations can be proposed: A system is globally composed with:

a) a structural body,
b) an internal functioning system,
c) an input system through which the environment can act on the system,
d) an output system through which the system can act on the environment.

Input and output systems have two features: the signal "shapes" they receive or transmit and the signal "natures" i.e. the nature of the material support. Four processes can take place, all of them generating both an environment and a system modifications (figure 2.1):

1. *a regulation process.* The system modifies its input signals to get a desired output signal. There is modification neither of the system structure nor of the internal functioning. But the environment is submitted to a transformation coming from the system. This scheme represents the final current state of the evolution for the system or the living being.

2. *an adaptive process.* To get a given output the environment modifies the input nature and can also, through this way, modify the system structure

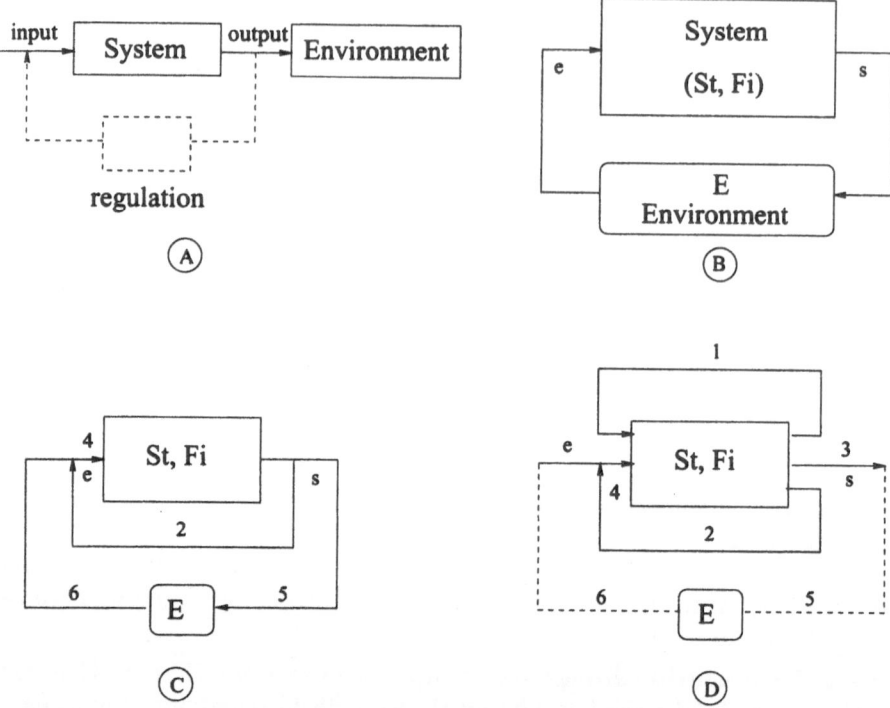

Fig. 2.1 A: regulation. System makes an input signals transformation into desired output signals. In case of output perturbations due to environment effect, outputs are regulated via inputs modifications based upon difference of desired outputs with actual outputs. **B:** adaptation. System will be named adaptive if, to maintain desired output s, environment modifies input e, but also could modify structure S_t and/or internal functioning F_i. **C:** Possible modifications generated by external signals and making to grow the system complexity. s(signal) modifies E(nature) via 5. E(nature) modifies e(signal) via 6. e(signal) modifies (St, Fi)(nature) via 4. (S_t, F_i)(nature) modifies e(nature) via 2 and s(nature) via 5. (signal) means signals features for a given signal nature (support nature). **D:** possible modifications coming from internal complexity. (S_t,F_i)(nature) modifies (S_t,F_i) nature via 1. (S_t,F_i)(nature) modifies s(nature) via 3 and e(nature) via 2.

and/or the internal functioning. At the final state of the evolution, that adaptive process can be shown on a relatively long period. The two previous processes can be find on man-built machines. The two next processes occurred only during the steps of evolution generating strong transformations in a far past time.

3. *a complexity growing process.* The system output signals shapes modified the environment nature. The environment nature could modify the system input signals shapes. The system input signals shapes could modify the nature of the system structure and of the internal functioning. The system structure, in return, could transform the input signals nature as well as the input signal nature. All that modifications can be interpreted (with

few additional explanations such as attractivity) as a source of system growing complexity.

4. *a selfmodification process.* When the internal complexity reaches a high level, it can be understood that it can generate a selftransformation of its nature, inducing another transformation of the input and output systems nature, resulting in return in an environment transformation.

Considering these processes as possibly real during the living beings evolution, two main conclusions can be drawn, interesting the robot concept.

CONCLUSIONS 2:

a) Survival of a living being as exhibited by its reality and by its behavior finds its possibility into a very long period of mutual living being-environment transformations, resulting in a kind of behavior stability guided by regulation and adaptivity processes. Consequently, if a robot has to evocate a living being behavior, it cannot be designed without the design of an adapted environment. Putting the peculiar case of human being aside, to conceive a robot "alone" is a non-sense procedure, a no meaning fact.

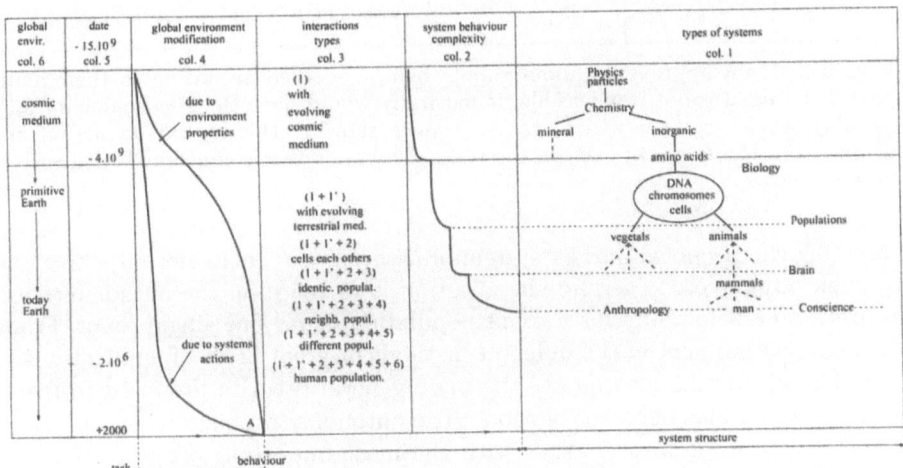

Fig. 2.2 Image of the natural systems evolution.

b) Secondly, as shown in figure 2.2, a summary image of the evolution allows to point out three different ways or goals in robot studies. Either the objective is only to perform some tasks, or it is to get some behaviors, or it can be to build a structural system exhibiting some properties that can be proved via behaviors or via tasks performing.

2.1.3 Robotic researches organization. Taking into account the previous analysis, a better understanding of robot control contributions can be completed as shown in figure 2.3. In fact, reading a paper, a first step can be done

360

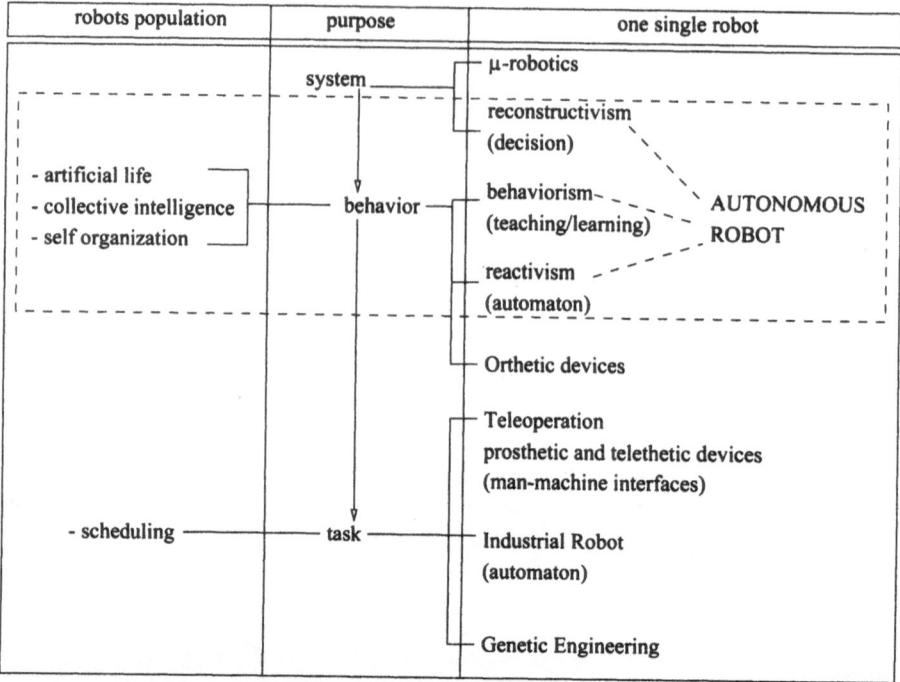

Fig. 2.3 Main approaches about robots being classified according to their main goal. If an ideal robot was feasible, it naturally would have the good behavior and suitably would execute wished tasks. Arrows showing this logical deduction are displayed.In the frame are shown the scientific ways directly concerned with search of autonomy.

inserting the proposal into a "system objective" or a "behavior objective" or a "task objective". A second classification item relies on the author interest in proposing something for a robot population or for one single robot. From there, an assignment of the different methods or great fields of applications is possible. Figure 2.3 also quotes the main general theories proposed to reach a real autonomous behavior of robot. The autonomy problem seems the most interesting one because it deals with the most important property that researchers would master. It opens a kind of breaking in the scientific progress and is directly related to the huge human potentiality coming both from his body and his brain. It is the reason why the control level considered along this paper is this one related to an autonomy capacity.

2.2 Robotic properties as inherited from evolution from living beings

It was explained in the previous section that the living being behavior, primarily devoted to its survival into a specific environment, was an inheritance from the evolutionary process bringing out a mutual being-environment adaptation. Manifestation of survival passes through a living being behavior, itself expressed by a capacity to perform some actions and tasks on or into the surrounding adequate environment.

2.2.1 Task and action definitions. An easy method to define a task consists in considering a starting environment state and a desired final one, and to consider these two states are stable relative to the desired environment transformation. The task is the process that performs the global environment transformation.

But the process uses a succession of steps that leave stable the environment at each step without achieving the global transformation. These steps are actions. They are themselves composed with a succession of motions generated by forces. So, the two main questions are: How to execute an action? and, How to chain actions in order to perform a task?

2.2.2 Automatic and reactive execution of actions or "Genetic dexterity". Considering the kinematical properties of a mechanism owning joints, it is obvious it can potentially perform motions and successions of motions. Now, adding efficient actuators and some internal regulation control level, it can effectively perform these successions of motions. If external sensors and reflex loops are added, it can perform a variety of actions in automatic mode depending on the environment characteristics.

For a living being, that capacity to perform a number of actions (potentially useful to its survival in a given environment) is a genetic one, something it has at birth. That property will be named: "dexterity". In a similar manner, making a robot with some kinematics, actuators, internal sensors associated to regulation loops and simple external sensors associated to reflex loops, allows, taking into consideration adapted environment properties, to easily program an automatic performing of various actions. Its dexterity relative to an adapted environment will be so defined and will be considered as a genetic inheritance (given by the human designer).

2.2.3 Chaining of actions generating a task performance. Any robotician is able to correctly achieve the previous stage, and, in matter of control, some of them spend time to improve (via researches in mechanics and in automatic control) the robot behavior inside every action of a desired pack [7]. Of course, it is a good thing, but a more difficult problem is in the methods able to ensure the actions chaining in order to reach a task execution, if possible exhibiting real behavior autonomy properties. "Skilfulness" can be chosen to name the property corresponding to that actions chaining demonstration.

So, the question is now: from where or from whom can come the automatic chaining of actions? Adopting a multiagent point of view, it can be thought of an intervening of one or several agents among the following list, in order to drive the robot from an action to the suitable next one.

1– the human operator. The case classically occurs in teleoperation although, very often, the guiding goes down up to motions and actions execution. This case is not considered now. Another situation is met through a programming elaborated by man. But, in fact, that program becomes a robot property when it runs. The last possibility is a direct request from man during a task execution indicating the next action. Conclusively, although man is the robot designer and its basic programmer, it is of poor interest to consider him as an active agent during the actions chaining inside a perspective of control proving an autonomous behavior.

2– the environment which has a number of properties linked, for instance, to its geometry, topology, gravity presence, etc. and, which is able to influence the robot behavior. As a simple instance, in a flipper game, whatever the path of the ball, it always terminates its trajectory into the same hole. Easily dividing that whole trajectory considered as a task completion into a number of actions, the actions chaining is only due, whatever the actions and the ball-flipper-bumpers interactions, to the gravity effect, that is to say to an environment property. (In this instance, man intervention to ensure, from time to time, some actions chaining is clear).

3– the interactions. They will have a major role for the robot guidance. They can be of two kinds:

– natural interactions when a robot part is in physical contact with either another part of itself or a part of the environment. Motions issued from natural interactions are only depending on mechanics laws. So, it is necessary to have an a priori model of the contact events in order to establish the conformity of a desired motion with the real motion.
– artificial interactions. They rely on sensors presence detecting at various distances (including contact) some features of the environment. Implemented by the designer, either on the robot or on the environment, they can provide an information linked to the next or future motion or action.

So, interactions can play two roles:
The first one is compulsory and occurs inside an action. It can involve natural or/and artificial interactions. Interactions effects are belonging to the "dexterity implementation" process and allow the automatic performance of any action. They are part of the exhibition of the genetic robot properties (under conditions robot moves in an adequate environment).
The second role copes with their possible implication into the actions chaining process. In this case they are a candidate agent to solve that problem.

4– the robot itself that owns, firstly, the basic components quoted previously, and, consequently, owning an adequate dexterity to achieve actions. Programs can be added to it in order it plays a role in the actions chaining. And the basic problem consists in designing such programs. The responsability distribution into the various agents in matter of decision to select an appropriate series of actions driving to a task execution is strongly depending, not only on the task (that can be known in advance as a goal), but mainly on the knowledge level of the environment. In fact, five practical cases will occur as indicated in figure 2.4 that allows to be aware of the problem when working on a solution.

The main conclusion of figure 2.4 examination is the following:

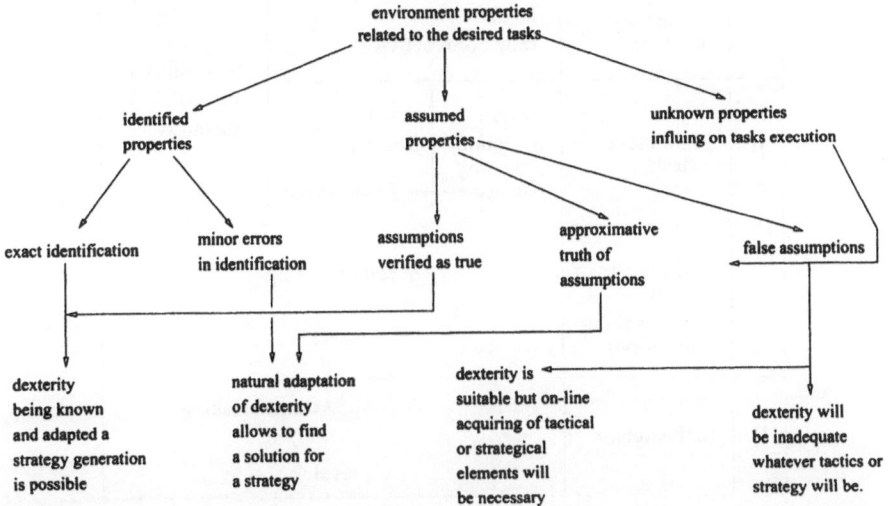

Fig. 2.4 The various levels of knowledge on the environment oblige to generate various strategies for getting actions chaining.

STATEMENT 4: If environment features in relation with the robot dexterity implementation are neither known in advance nor acquirable on line, any task performance will be impossible, excepted by chance. That evidence is sometimes forgotten...

2.3 Control levels

As depicted on figure 2.5 , in order to get an ideal autonomous robot, control can be dispatched into four main levels: control corresponding to exhibit robot dexterity (level A), control devoted to the actions chaining exhibiting skilfulness (level B), control dedicated to get a selfcapacity of project (level C), at last, control to get selfawareness (level D).

2.3.1 Control level A for dexterity exhibition. It consists in automation of separated actions. It was already explained this level can be considered as mastered. It is composed with an internal regulation level and external reflex or reactive loops that are adaptive to some characteristics of the environment. So, the designer can prepare the suitable automata for each type of action included into an a priori list. That control can be named: zero level control. Although a lot of methods to achieve that zero level are proposed in litterature for a long time, the only difference that can be found among them deal with time or precision or robustness performances. So, it seems not useful to insist on them here [8].

2.3.2 Control level B for skilfulness or actions chaining exhibition. With that B level one's reaches the heart of the main robot control researches because, up to date, the results remain partially questionable.

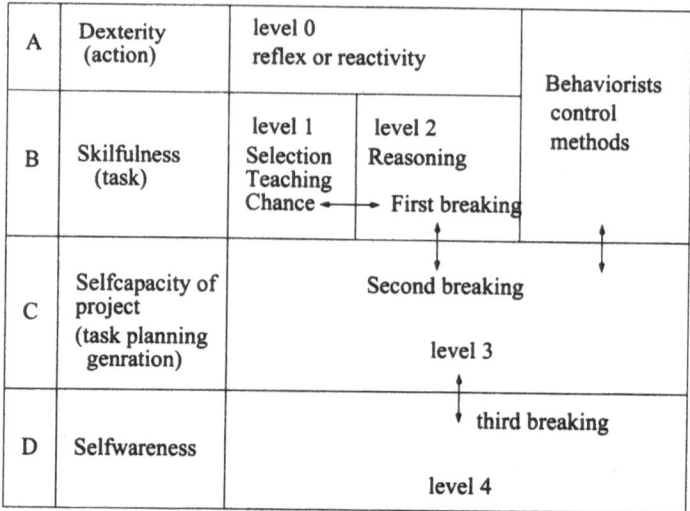

Fig. 2.5 Control levels and breakings in methods.

Putting provisorily aside the behaviorism approach, two main ways (sometimes connected) can be distinguished. What make these ways different is not only the basic principles of the methods themselves but their efficiency when they are implemented. One way can work but it seems dificult to go further on in autonomy with it. The other way, potentially better from an autonomy point of view, was for a long time deceiving in its results.

General way of the first control level: learning/teaching.
Although the number of teaching/learning methods is very large, [9] [10] [11] [12] the result is very often the same one. Machine can reproduce, directly or indirectly, what was taught.The object of the teaching/learning can be in the field of software or in a material field. So, the actions chaining can find a solution with the teaching/learning procedures. However, it is a difficult discussion to know the real autonomy level that can be got with these procedures because depending a little on the teaching methods. Any way, the robot behavior will be this one of an automaton, even when some variety in the actions chaining could be observed (behaviorism way is not considered here).

Improving the first control level.
To make the actions chaining it can be advantageous to take under consideration two other phenomena, often neglected in robot control design, and reducing the role of teaching/learning.

1– robot "specialization state".
In the succession of actions leading to a task performance, it is obvious that nature of every action cannot be any. So, the execution of one kind of action forbids to chose some actions natures as the next one. So, after every action, it can be considered that the robot is in some state of specialization favoring some types of actions as the next one and forbidding other types of actions.

On this base, before using a teaching procedure to link two successive actions, using of a "selection filter" can be interesting. From another side, if the choice for a next action is seen as a chosen answer to a set of requests coming from the environment or from the robot, it is possible, in function of the peculiar situation, that the set of requests will be reduced to one request. With this opportunity, the selection filter is enough to choose the next action, and the teaching procedure becomes not compulsory. So, about the whole series of chainings, some ones could only call for the selection process while other ones could use a teaching process after the selection filter.

2– interactions properties and attractors.

Considering a new time the next action choice as a selected answer to a list of requests or possibilities, several requests can be suitable relative to the final goal that is aimed at in the task performance. In other words, the only requests to be avoided are these ones leading to a sure failure of the task execution. Under these circumstances, it can be possible to adopt chance as a matter of choice of a request. Two phenomena can occur increasing the success probability. The first one copes with the fact that the choice by "releasing" takes place after the selection step. The second one is based upon the possible "reinforcing effect" of the environment through the robot-environment interactions, and through the presence of "attractors" pulling the robot on some specific and positive ways for the task execution.

This reinforcing effect has been observed in Nature with insects societies such as bees or ants [13]. Whatever the individual behavior, multiplicity of interactions and creation of new interactions through presence and creation of new attraction zones make the insects societies compulsory converging toward some specific tasks executions.

May be that phenomenon could be transposed to the case of a single robot under some conditions to be fulfilled by the environment. Up to date it remains a question. In conclusion about the first control level, depending on applications, it is possible to adequately combine the three previous stages: Selection process (suitable for some actions chainings), Selection process and Teaching process (for other actions chainings), Selection process and Releasing (for other actions chainings).

General way for the second control level: reasoning.

Instead of adopting the track of teaching/learning, people worked since the first times of Artificial Intelligence appearing on another way to ensure the actions chaining, a way based upon reasoning and planning generators.

A planning generator can be designed off line [14] [15]. In this case, after every action, it will provide the next one. However the action execution and the signal of end of action must be in total conformity with the forecasts. The Artificial Intelligence (AI) procedure is only located into the phase of planning generation, and is not really useful in robotics for which a hand-made generator is often easy to build. A planning generator could be made on line [16] and would become extremely interesting. But difficult problems arise:

1. robot must know the goal, the possible actions, the environment characteristics in terms of actions, the past actions it performed and after what action it is now [17]. If these knowledges are effective, the AI problem to generate an actions chaining is the same as previously for the off line planning generation case.

2. the deep problem deals with translation of named objects involved into actions, into physical features the robot is able to acquire and to match with the objects names. That fits with the general problem of environment recognizing, relatively unsuccessfully attempted through visual images interpretation, for a long time [18]. As an instance, if the present action is defined with: "grip the jug", it is necessary to recognize the jug. While the general problem is far from a suitable solution, several simplifications can be adopted solving peculiar cases.

    a) In the previous instance, if the jug is the only object of the environment: "grip the jug" becomes: "grip object" without need of information about why it is named: this jug. "Grip object" can become an automatic but adaptive procedure thanks to simple sensors. So, the first simplification consists in attributing to each object a simple recognizing mark, either artificial, or natural (specific feature), leading to transform the environment recognizing into an easy identification process.

    b) Instead of taking under consideration the final goal, a task analysis will show that the task performance can be divided into compulsory steps (every one including several actions) that become subgoals. Robot will have to only take in charge the reaching of each subgoal involving few actions and simplifying the planning generation problem.

    So, in every application,accepting some restrictions relative to the general case, an actions chaining based upon reasoning can be possible, even if the general case cannot be solved.

Combination of the first and second control level.

Way through learning/teaching and way through reasoning could be combined, reducing contribution, therefore difficulty, in reasoning approach design.

In fact, at the first level, after the Selection process, sometimes a teaching/learning stage could be used, and sometimes a reasoning stage, eventually followed with a releasing stage, could be proposed, in function of the local problem to be solved about the request choice, inside a task performance, to get the actions chaining. Although the control choices strongly depend on the application, this method provides an approach means to keep a good robot autonomy inside the execution of a given task.

Behaviorist approach.

Behaviorist do not cut dexterity from skilfulness in their attemps to give an autonomous behavior to robots. They try to make in one shot control levels 0,1 and 2 through imitation techniques [19].

In living organisms, it is well known that the sensorymotor systems are born with strong possibilities of selflearning and selfcoordination, considering the only body (in its growing) as well as the body interacting with the environment. A sensory education (autonomous or guided) leads to not only dexterity but to reflex behaviors (although complex) when facing some situations. So, a large part of actions chainings can be selflearned and become reflex. Consequently, sensory coordination in task performance is the main topics studied by behaviorists.

2.3.3 A and C control levels to get a capacity to conceive projects and then to get selfwareness. It can be understood that there is a kind of breaking between level 1 and level 2, the first one being based on teaching and the second one on reasoning. With level C a new problem is raised. One's passes from a real task performance to an appreciation about a feasability of a succession of tasks that can lead to a real execution of a complex project. Basically it is possible to reduce the problem to an appreciation of feasability of one task.

Mostly this kind of problem is studied by cogniticians whose goal is to reproduce some human brain properties. To be efficient solutions call for two connected levels:

1– an adequate description of knowledges necessary to solve the problem i.e. knowledge of environment, knowledge of task, knowledge of robot dexterity [20].

2– a clear correspondance or mapping of the description used by the reasoning level with the description of reality as making possible the physical execution.

This second point is not correctly solved even if a solution will probably emerge soon. Assuming this solution at disposal, an interesting consequence will arise : robot will be able to estimate its capacities to perform a task and to find a tasks and actions organization to execute a project. It does not mean it will be able to invent or to discover a project but to appreciate whether a project is feasible or not. It would begin to have a knowledge of itself which is a decisive progress. From that point, to get the faculty to propose a new project as well as to reach level D with a consciousness of its existence remain focal points about which approximately nothing is known in scientific terms, the main contribution coming from philosophy and biology that still remain far from the scientific demands [21][22].

2.4 Conclusion

In the presentation of a possible methodology to generate robot control levels in order to get some behavior autonomy an analysis of some important features of living beings was first proposed. Results of the analysis allowed to better situate the set of contributions in matter of robot control and to know the present limits and possible ways of progress. Because mastering of robot autonomy is not a fact, another solution in robotics consists in leaving man in the control loop in order to compensate weaknesses of robot control. This method is named teleoperation. In spite of man presence all things are not solved as explained in the next section.

3 Human remote control

Teleoperation is an old technical and scientific research domain and was initiated by needs of human protection to dangerous radiations when manipulating radioactive objects or products.

A first progress took place in 1947 when Goertz at Argonne Laboratory (USA) discovered the master-slave techniques still largely used in nuclear activities at present time. After a period during which the slave-master links were only mechanical, the same Goertz powered electrically each articulation of both master and slave parts, then allowing to have a single electric cable connecting the two subsets. From that time (1954) any distance separating slave from master was made possible and applications could be extended to other things than manipulators, for instance vehicles. Later, the cable could be suppressed and replaced with radio or other electromagnetic waves transmission systems [23].

Another decisive step for teleoperation progress occured in years 1980 when a computer was introduced in order to aid man in piloting and to give some automation possibilities in elementary tasks performings [24].

Unfortunately, whatever endeavours of researchers, the two next drawbacks were not sufficiently decreased and remain a handicap to a satisfying using even nowadays. Indeed, using a teleoperation system is a difficult work, needing important training, long time for every task execution and skill. The two big drawbacks are: first, a lack of friendly using of the master station in relation with its consequences into the slave universe, and secondly, relatively imperfect sensory feedbacks from the slave universe to the human operator making difficult his analysis of the real slave situation (in spite of presence of a lot of cameras in the slave environment and of force feedbacks during an object manipulation).

To improve the usual systems a new technique appeared since about 1990 that could transform the teleoperation general problem into a telework problem. It deals with Virtual and Augmented Reality. The new expected improvements will be due to the possibility of using a VR system as a buffer or as an interface between the human operator and the slave, in such a way the work to be done becomes well adapted to man dexterity while the slave control system can also be well adapted to a work performed with a machine.

To explain the point in details it seems good to start with some short explanations about VR systems.

3.1 Remind on VR systems principles

As shown on figure 3.1 a VR system cannot be thought without a human operator because all main components are in connection with him and designed

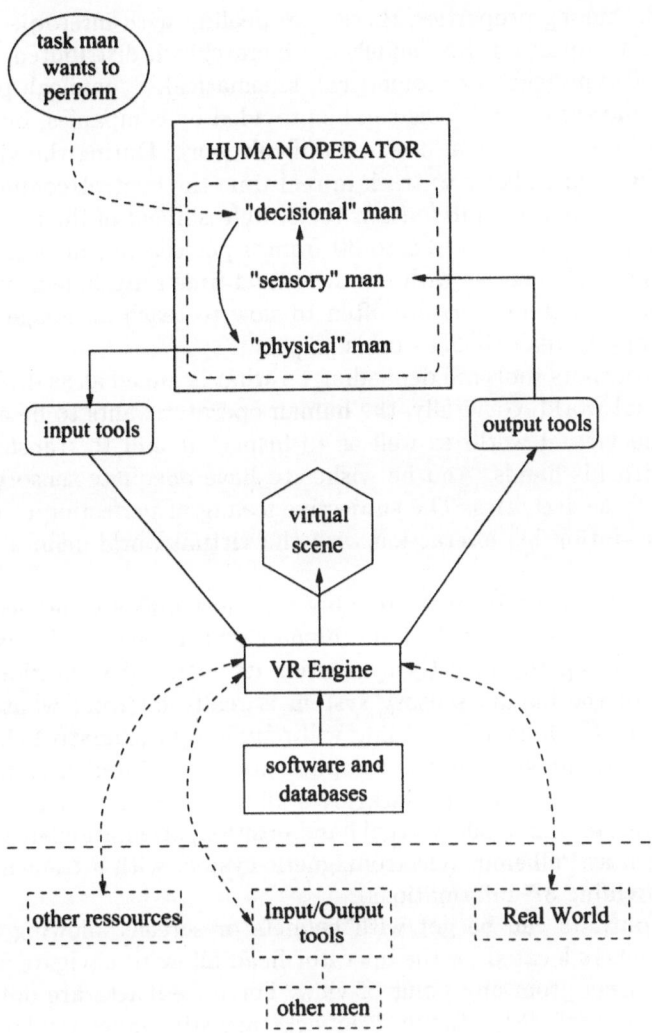

Fig. 3.1 Basic functioning of a general VR system.

in function of his features [25]. Throwing a glance to the functioning, anything starts with a man decision to do something in the virtual world. The decision is translated into a physical modification of a part of the human body. This modification constitutes the command order. Through input tools that order is going to modify something in the virtual world, and that last modification is going to generate signals adequately transmitted to the man sensory system through output tools. Man sensory system becomes the control organ of the virtual world evolutions.

To build a VR system it is therefore necessary to make VR universes, input tools and output tools. Making a VR world starts with graphical modellings respecting the specifications of the properties of objects present in the virtual world. Among properties, these ones dealing with interobjects relations are the most important. So, models are hierarchical, distributed, and coping with several types such as geometrical, kinematical, dynamical, physical etc. [26]. Some databases and tools can be provided by companies, but to build a virtual world is generally a long and difficult work. During the virtual world using, a VR engine allows to track in real time the control/command indications coming from the input/output tools. Refreshment of the rendering must reach a frequency of about 25 to 30 frames per second, in order to keep a continuous visual effect as with a movie. Real time refreshment of scenes is a problem because computers are often to slow to reach an image complexity corresponding to realistic 3D worlds [27].

Inputs/outputs tools are depending on what is aimed at as desired actions in the virtual world. Generally, the human operator wants to be able to navigate in the virtual world as well as to inspect it and to transform it with tools or with his hands. And he wishes to have adequate sensory feedbacks during all these activities. The subjective feeling of perfection or of operator satisfaction during his interaction with the virtual world defines the quality of his immersion.

The immersion problem is probably the most difficult one because it assumes that a correct model of the human sensory system is achieved and available allowing to correctly synthetize its artificial excitations. In fact, knowledge of the human sensory system is really far from what is needed, and consequently, immersion feeling will remain approximative [28].

Classical inputs systems use computer mouse or joystick or trackball. A peculiar attention is given to datagloves allowing to animate a virtual hand from the operator hand [30]. Virtual hand position/orientation are got through trackers such as Polhemus (electromagnetic system with a transmitter and a receiver providing 3D informations).

Visual outputs can be got with helmets or screens allowing 3D images display. Trackers located on the operator head allow to navigate and look at the virtual scene from any point of view. Force feedbacks are obtainable on manual input tools [31]. Tactile feedbacks are still under studies while 3D sound is mastered [29].

As it can be seen on figure 3.1, a VR system can be considered either as a close system devoted to one single user, or as an open system connected to external items. In the first case the basic utility is for improving simulations quality or for training, individual game... [32]. The second case can be divided into great fields of applications. When a VR system is connected to an information network it can be considered as representative of the future multimedia networks implementation. Information Highways are a possible illustration [33]. When a VR system is connected to several users who can collaborate inside the same virtual world, teaching to several students or concurrent engineering gives an idea of the class of applications. At last, a new feature can occur when a VR system is connected to a real world. VR system becomes a control/command station to remotely master machines. Teleoperation or Telepresence or Telework is a subsequent application, and more precisely the subject of this paper [34][35].

3.2 Computer aided teleoperation and its large variety of ways

The important amount of litterature on teleoperation can give a feeling of disorder in methods to improve that technology because, sometimes, the same terms are used to name very different things. The expression "shared control" is a good instance commented in [36].

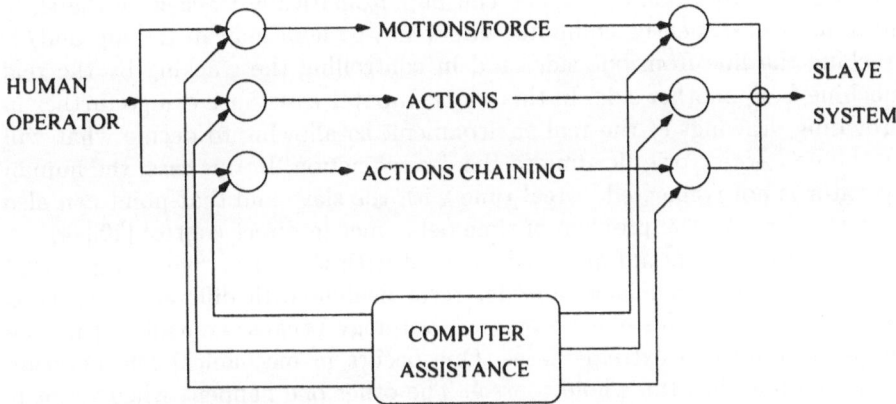

Fig. 3.2 Computer assistance possibilities showing the large variety of ways to go from a permanent and low level man intervention up to a great autonomy of the slave system in task performing.

In fact, a way to highlight the question consists in coming back to some notions exposed in part 1 of the paper. Previously, three main levels were defined in a task analysis: the application of forces or motions, succession of which composing actions, and actions chaining allowing the task performance. As shown on figure 3.2, the role of computer assistance can be for the human operator or for the slave system or for both, to contribute to one or several levels relative to the generation of motion/forces, or of actions, or of actions chaining. This simple description exhibits 63 basic classes of possible assistance. And, of course, each class can be relied on a lot of different means depending on available technology. VR could give one of the best results because it is the most advanced technology.

Indeed, this number of classes is too high because, mostly, help to motion at the operator side, for instance, is also help to determine what motion is suitable, leading to an indirect help to determine actions and actions chaining... But the estimation indicates the problem complexity level.

At basic control level, the first aid that can be given through a computer is tuning of servoing in the slave-master link, and, in case of real images transmitted via video cameras, servoing of cameras motions in direction of zones on which the operator is interested in attaching his gaze [37]. This former computer assistance helps the operator both in performing smooth motions and in performing the adequate motions. Another basic and classical help to motion is using of passive or active compliance when touching surfaces or assembling parts through sensory force feedbacks. As soon as something is superimposed on image in order to easy control, the information under question must act on both sides of the teleoperation system: human operator and slave machine. "Virtual fixtures" are an instance of this type of aid [38][40]. Some guidelines can be displayed in the scene, lines representing the adequate motion to be performed. Thus, the operator can directly track the line with a virtual image of his worktool, or the line can be automatically tracked by the slave machine. Consequently, computer intervenes to help man in tracing and/or tracking the line from one side, and in controlling the tracking by the real machine from another side. In this way computer assistance can go further in providing drawings of the real environment, so allowing to decide what will be the best trajectories leading to the desired action. In this case, the human operator is not connected in real time with the slave and that point can also help to overcome the problem of time delay met in direct control [39][44].

After the general aid approach covered with the terms "virtual fixtures" or/and virtual mechanisms, another term dealing with different realities is "shared control". This expression is a tautology because control will not be shared in only two extreme cases. One occurs in mechanical teleoperation for which man has the whole control. The other one happens when robot is fully automated as in industrial robots. As the real teleoperation situation is exactly between the two extreme cases, it is obvious that control is shared, a part being dedicated to man (eventually aided by computer) and the other part to robot (compulsory aided by computer).

So, the question is not related to shared control itself but to the choices leading to a sharing of control for man and slave.

For instance in [41] shared control only means the possibility for the operator to manipulate a not rigid robot in order to get stable contacts with the environment while compliance properties make autonomous a suitable behavior of the slave robot. In [42] shared control is understood at an action level. The human operator has the responsability to trigger a slave action through a single gesture. The authors explain the system functioning in terms of "action units". For the authors of [43], shared control has another meaning. The task being known in advance, it is decomposed into independant possible subtasks. For every subtask there are several available modulus depending on different criteria. To perform the task, there are several chainings of subtasks, and every subtask can be of different kind. Shared control means that an automatic control chaining is made taking into consideration a weighting of various solutions. So, shared control is a weighted contribution of several modulus present in the computer (the weighting depending on a state model that intervenes as a supervisor).

Conclusively, examining recent contributions to improve teleoperation, it can be seen some general trends about a computer aid.

There are two basic problems:

1. the decisional problem including two levels: What motion must be performed now? or What action must be performed now?
2. the execution problem: How the selected motion will be performed? or How the selected action will be performed?

Both problems raise the question of real time additionally to this one of physical suitability. So, time can be divided into two periods:

a) a period to decide, and it is wished that this period can be as short as possible. It is the meaning of real time when specific informations are displayed during the teleoperation task in order to choose the best next motion or action.
b) a period for motion or action execution. Here two possibilities occur:
   – either the execution can be automatically done by the slave after a triggering coming from the operator. The execution operation is not exactly in real time but delayed by characteristics of communication system. This approach can be related to teleprogramming. It is then difficult for the operator to intervene during the motion or action execution, excepted in stopping the execution at a point that takes into account the transmission delay.
   – or the operator wants to master in real time any behavior of the slave, "swallowing" the delay, and permanently guiding the slave evolution, mainly because he wants to take advantage of force feedbacks (necessarily usable in real time). That point is a very difficult challenge [44].

So, summarizing, general progresses were done in matter of aid to the operator at a decisional level (through improving detection of pertinent information and proposal of strategies), and in matter of aid to the slave control (through automation of some motions and actions).

3.3 Originality of VR contribution

In computer assistance, even using graphical or visual means, the goal consists in providing to the operator realistic views of the slave scene. From that, the operator has to generate a tactical or strategical method to perform the desired task. That task is only known or displayed in the operator brain.

VR techniques can bring out a contribution in this situation, for instance in improving scene realism, in giving possibilities of superimposing various informations, and finally in immersing the operator into a virtual scene which will be a pertinent copy (from a task performance point of view) of the real world. In such a way the VR contribution can be considered as located on a "classical" way of teleoperation progress. But another way does exist, based upon more original properties offered by VR: the world and task transposition way.

Indeed, whatever the classical computer aids, man has to face:

– an image of the real world possibly including a very complex system and making difficult not only its understanding itself but also the brain activity that has to deduce from the world understanding what adapted motions or actions to be performed to progress toward the task execution.

– a control/command system (arm or joystick for instance) that makes not very natural generation of his suitable gestures.

– an impossibility to control in parallel several slave systems, mainly if they are not similar, even if they must perform the same task.

The ideal situation would be, for the operator, to execute natural gestures (for instance with his hands) in a not complex natural environment to achieve not complex tasks. If considered as a series of actions, it must be noticed that tasks through a clever decomposition are never complex. Only context and constraints associated to tasks are complex. VR techniques allow to go forward on this direction because:

– natural or realistic environment means: environment reacting to man as a real one, and it is the VR goal.

– in this environment, man can directly act with his hands or with a tool manipulated with his hands as in the usual real world.

– a task being a succession of not complex actions corresponding to not complex environment transformations, an adapted environment for desired tasks can be built and could be not very complex. So, the control/command chain composed with man-master station-slave world could be replaced with the chain: man-VR world-slave world that offers:

- a better potential solution concerning the ergonomical problem of man-system relations and interactions;
- an increasing of the number of possible sensory feedbacks (figure3.3);
- a replacement of the man-slave world relations problem with this one of the VR world-slave-world transformations (figure3.3).

3.4 A method taking advantage of VR originality: the hidden robot concept

Looking at a task to be performed by a slave robot in a slave environment, the purpose is neither the robot nor the present environment. The purpose is a future environment state expressed through its transformation. In other words, only that transformation is interesting. But such a transformation includes two aspects:

1. one deals with the transformation functionality or nature that can be described with an action name or a series of actions such as: go to, insert, grip, cut and so on.
2. the second one defines the environment objects or the environment locations where those actions take place.

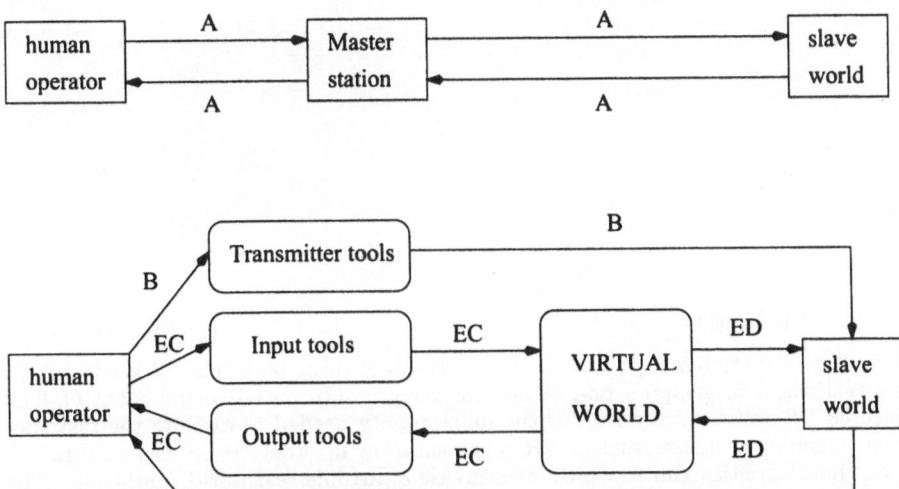

Fig. 3.3 Comparing classical and VR teleoperation control. In classical approach, whatever the data they track loop A. With VR, that basic loop can be replaced with four loops: B,C,D,E. Loop C deals with man-virtual world interactions. Loop D deals with virtual world- real world exchanges. Loop E deals with orders reaching the slave world through the virtual world and information coming from the slave world and reaching man through the virtual world. Loop B can be a part or identical to loop A. Overlapping of loops E,C,D indicates a separation of the informations flows into a part only regarding the man-VR world relations, a part only regarding the VR world-Slave world relations, and a part regarding the three components.

Consequently, Virtual reality techniques offer the possibility to build a world keeping an only functional copy of the real environment, adapted to the desired task transformations. And the task part devoted to locations must be prepared as another transformation involving the virtual world and the real world (figure 3.4).

So, it can be said that the Virtual World will be an Intermediary Functional Representation (IFR) between real world and man [45]. It is easy to see that, in any IFR, the first object to be disappeared will be the "picture" of the system moving the real operational tool. For instance a gripper could be represented with another gripper moved by man in the VR world, and more precisely with a virtual hand itself because robot will be replaced with man. The robot functionality will be transferred in the virtual world to the man functionality.

As any executive real machine, and more peculiarly real robot, will be replaced with man, robot picture can never appeared in virtual world during a task execution. So, it is legitimating that principle: the hidden robot concept.

In this approach, the teleoperation problem is dispatched into two main parts:

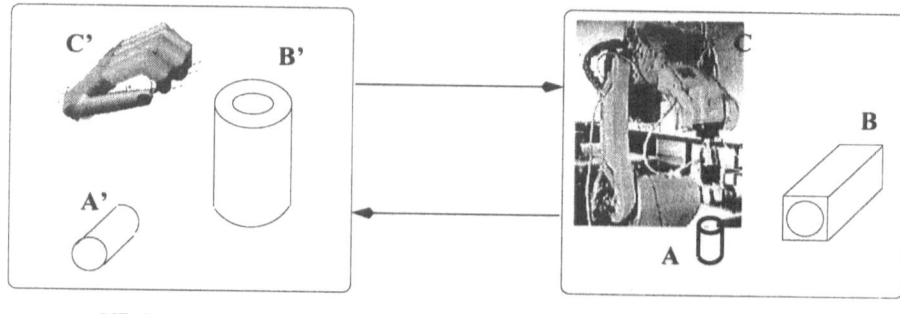

VR World   Real World

Fig. 3.4 Functional task representation (or IFR principle). The task consists in the real world in gripping part A and inserting it into part B using robot C. The selected VR representation keeps the functionality needed to execute the task. So, it is possible to choose such a VR representation in order to be well adapted to man characteristics and to cancel or decrease a possible real world complexity. The executive problem becomes the geometrical and topological correspondances of the VR world with the real world (direct and reverse transformations as needed for control).

1. choice of the IFR and construction of the man-virtual world relations and interactions.
2. building of the virtual world-real world relations and transformations.

The first point involves the general know-how in matter of VR if the second point is forgotten during an instant. However, that first point will have to take under consideration all constraints imposed by the virtual world-real world links. Among them, for instance, there are: communication delay, real robot velocity and any element related to tactical or strategical tasks performance aspects [46].

The second point is concerned with the general problem of calibration [47] and of errors recovery [48] in order to get a correct coincidence of what happens in the virtual world and in the real world.

Some experiments such as these ones described in [49] are instructive because they consist in remotely controlling, in parallel, four different robots located at four different places (in Japan and in France), performing the same puzzle assembly tasks.(figure3.5). With different robots controlled in parallel a common IFR is imperative and enhances the interest of the method. However, the experiments show three problems (not taking into account the usual technical difficulties):

1– *the safety problems.* How to be sure of what happens in the real world and how to discover a safe strategy? Those difficulties were solved through adding three levels of feedbacks: the basic one is only devoted to the virtual scene evolution under man action. A second level reintroduces a graphic modelling of virtual robots but the robots virtual joints motions were directly controlled with the robots real joints evolutions. At last, a video conference system provides images from the real scenes. Using the three information lev-

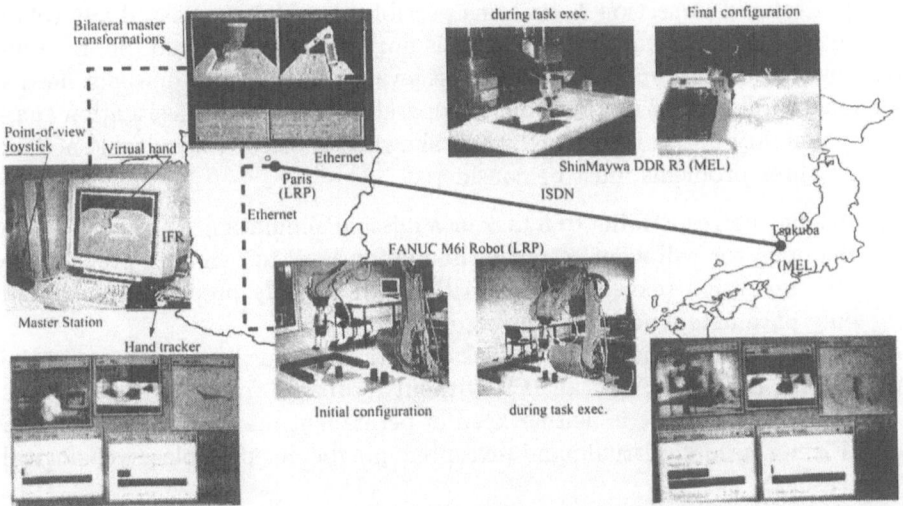

Fig. 3.5 An experiment of remote control of several robots through a VR station and using the hidden robot concept. On left the operator and the virtual scene (main control/command level); Above :synthetic robots controlled by remote real robots (second control level for safety); bottom right: real teleconference images (third control level for safety)..

els allowed for the operator to be sure of the real events and to adapt his orders to the real situation.

2– *the time delay problems.* Although tentatively solved through several methods, they had a great influence at a strategic level and the time delay-strategy relations remain ill elucidated.

3– *the robots autonomy problems.* A careful calibration was done in order to get all robots gripping the suitable part at the prepared in advance suitable spot, only allowing to have a light difference in time, compensated via the operator supervision strategy. If every part position in each robot environment was random (as wishable), a less or more complicated recognizing step should have been necessary making longer the mean time of picking, and increasing the time discrepancy of robots, making longer the strategy elaboration time, and finally decreasing the interest for the operator to elaborate a gripping gesture in the virtual world, while an easy push-button procedure or a program instruction could have had the exact same effect. That point is important and raises the open question of the relation of man intervention methods with the robot autonomy level.

4 Conclusion

In teleoperation, being supposed presence of a computer assistance such as this one possibly provided by VR techniques, at what level should man intervene? at motion/force level?, at action level? at actions chaining level?

As depicted in section 1, it seems possible to make automated any robot action while the actions chaining level is not correctly mastered. So, man intervention seems compulsory at this last level. And this level does not need a manual complex hand motion from the operator. It just needs a situation analysis possibility and a triggering signal allowing to select the suitable action. In fact three problems must be considered:

1. this one corresponding to a task or a mission simulation. During this step the operator will want to reach the motion level to carefully prepare the task execution (excepted when actions are already preprogrammed and only parameters must be delivered).
2. this one corresponding to the task execution. Either it can be trust in the actions execution, and an IFR without feedbacks from the real world is enough. And there is neither need of permanent manual interventions in IFR nor need of virtual hand (excepted, maybe, for possible psychological reasons);
3. or the operator will wish to intervene for corrections during an action execution. In this case, it is useful to forecast a possible intervention at motion level, but it makes imperative to have a good understanding of the real scene and not only of its representation in the virtual world. That means, the "hidden robot" cannot be permanently hidden...

In conclusion, while in section 1 it was demonstrated that a long way remained to be done before reaching an interesting autonomy level for robot, in section 2 it was shown that keeping man in the general robot control loop is not a simpler solution to master robots, even using the new VR techniques.

However, it is good to keep optimism because the distance up to the final goal becomes clearer every time a new idea is introduced.

References

1. Coiffet, P. (1996): *Robots: définitions et classifications*, Les techniques de l'ingénieur Publ., Paris, R7700, April 1996.
2. Farmer, J.D., Belin, A. (1990): *Artificial life: the Coming Evolution*, Cambridge University Press.
3. Heudin, J.C. (1994): *La Vie Artificielle*, Editions Hermès, Paris.
4. Gould, S.J. (1982): *Is a new and General Theory of Evolution Emerging?* in J.Smith Ed.: Evolution Now, A Century After Darwin., San Francisco, W.H.Freeman, 129-145.
5. Gribbin, J. (1992): *A la poursuite du Big-Bang*, Flammarion Publ., Paris.
6. Bartusiak, M. (1993): *Trough a Universe Darkly: A Cosmic tale of Ancient Ethers, Dark Matter, and the Fate of the Universe*, Harper Collins Publ.
7. M'sirdi, N.K., Manamani, N., Nadjar-Gauthier, N. (1998): *Methodology for Control of Legged Robots with Fast Gaits*, Proc. of ECPD Conference on Robotics and Active Systems, Moscow, August 1998.
8. *Handbook of Robotics* G.Nof Ed., Wiley Publ., New York, 1997.

9. Lin, L.J. (1992): *Self-Improving reactive Agents Based on Reinforcement Learning, Planning and Teaching*, Machine Learning, Kluwer Academic Publ., Boston, **8**, 293-321.

10. Maes, P., Brooks,R.A. (1990): *Learning to Coordinate Behaviors*, Proc. of AAAI'90, Boston, 1990; Morgan-Kaufmann Publ., Boston, 696-702.

11. Sehad, S., Touzet,C. (1994): *Apprentissage par enforcement por l'acquisition de comportements en robotique*, Actes des journes NSI'94, 67-70, Chamonix, Mai.

12. Sutton, R. (1988): *Learning to Predict by the Method of Temporal Differences*, Machine Learning, Kluwer Academic Publ., Boston, **3**, 9-44.

13. Drogoul, A., Ferber, J., Corbara, B., DFresneau, D. (1992): *A Behavioral Simulation Model for the Study of Emergent Social Structures*, Toward a Practice of Autonomous Systems. MIT Press. P.Bourgine and F.Varela Eds., 161-170.

14. Nilson, N.J., (1988): *Principes d'intelligence artificielle*, Cepadues Publ., Toulouse.

15. Farreny H., Ghallab, M. (1987): *Elments d'intelligence artificielle*, Hermès Publ., Paris.

16. Kaelbling, L.P., Rosenschein, S.J. (1990): *Action and Planning in Embedded Agents*, In Maes P. Ed.: Designing Autonomous Agents, Theory and Practice from Biology to Engineering and Back. Elsevier Sc. Publ., Amsterdam, 35-48.

17. Maes, P. (1990): *Situated Agents Can Have a Goal ?*, In Maes P. Ed.: Designing Autonomous Agents, Theory and Practice from Biology to Engineering and Back. Elsevier Sciences Publ., Amsterdam, 49-70.

18. Faugeras, O., Hebert, M. (1986): *The representation, recognition and locating of 3D objects*, Int. J. of Robotics Research, **5**(3).

19. Berthouze, L., Kuniyoshi, Y. (1998): *Emergence and categorization of Coordinated Visual Behavior through Embodied Interaction*, Autonomous Robots and Machine Learning, **17**, 1-15.

20. Tzafestas, S.G. (Ed.) (1997): *Computer Assisted Management and Control of Manufacturing Systems*, Springer Verlag, London.

21. Eccles, J.C. (1992): *Evolution du cerveau et cration de la conscience*, Flammarion Publ., Paris.

22. Changeux, J.P., Ricoeur, P. (1998): *La nature et la rgle*, Odile J. Publ., Paris.

23. Vertut, J., Coiffet, P. (1984): *Teleoperation- Technology Evolution*, Series: Robotics. Vol.3A. Prentice Hall Publ.

24. Coiffet P., Vertut, J. (1985): *Computer Aided teleoperation*, Series: Robotics, Vol.3B, Prentice Hall Publ.

25. Burdea, G., Coiffet, P. (1994): *Virtual Reality Technology*, J.Wiley and Sons Publ., New York.

26. Foley J., Van Dam, A., Feiner, S., Hughes, J. (1990): *Computer Graphics. Principles and Practice*, Addison-Wesley Publ., 175 pp.

27. Funkhouser, T., Sequin, c., Teller, S. (1992): *Management of Large Amounts of Data in Interactive Building Walkthroughs*, Proc. of the 1992 Symp. on Interactive 3D Graphics, ACM, 11-20.

28. Fuchs, H., Bishop, G. (1992): *Research Directions in Virtual Environments*, Computer Graphics, **26**(3).

29. Pimentel, K., Teixeira, K. (1993): *Virtual Reality: through the New Looking Glass*, Windcrest, Mc Graw Hill, New York, 301 pp.

30. Bouzit, M. (1996): *Conception et mise en oeuvre d'un gant de donnes retour d'effort pour la tlmanipulation d'objets virtuels et rels*, Thesis. University Pierre et Marie Curie, Paris, 29 Nov.

31. Burdea, G., (1996): *Force and Touch Feedback for Virtual Reality*, John Wiley and Sons Publ., New York.

32. Itoh, M., Inagaki, T. (1996): *Design of Human Interface for Situation Awareness*, Proc. of 5th IEEE Int. Workshop on Robot and Man Communication, 478-483, Tsukuba, Japan.
33. Ohya, J., Ebihara, K., Kurumisawa, J., Nakatsu, R. (1996): *Virtual Kabuki Theater: Towards the Realization of Human Metamorphosis Systems*, Proc. of 5th IEEE Int. Workshop on Robot and Human Commnication, 416-421, Japan.
34. Sheridan, T.B. (1992): *Telerobotics, Automation and Human Supervisory Control*, The MIT Press, Cambridge, USA.
35. Hirzinger, G., Brunner, B., Dietrich, J., Heindl, J. (1994): *ROTEX: the First Remotly Controlled Robot in Space*, Proc. of IEEE Int. Conf. on Robotics and Automation, 2604-2611.
36. Kheddar, A. (1997): *Téléopération basée sur le concept du robot caché*, Doctorat Thesis, Université Pierre et Marie Curie (Paris VI), Paris, 19 Dcember 1997.
37. Kuspriyanto, P. (1981): *Contribution l'amlioration des performances d'un systme de tlopration par asservissement adaptatif d'une camra sur l'outil manipul par le bras eslave*, Doctorat Thesis, Universit de Montpellier (France).
38. Rosenberg, L.B. (1993): *The use of Virtual Fixtures to Enhance Operator Performance in Time Delayed Teleoperation*, Technical report AL-TR-1993-XXX-USAF Armstrong Laboratory, XPAFB OH.
39. Kosuge, K., Murayama, H., Takeo, T. (1996): *Bilateral Feedback Control of Telemanipulators via Computer Network.]*, Proc. of IEEE/SRJ IROS'96, **3**, 1380-85, Osaka, Japan.
40. Joly, L.D., Andriot, C. (1995): *Imposing Motion Constraints to a Force Reflecting Telerobot through Real Time Simulation of Virtual Mechanisms*, Proc. of IEEE ICRA'95, **1**, 357-362, Nagoya, Japan.
41. Bejczy, A.K., Kim, W.S. (1990): *Predictive Displays and Shared Compliance Control for Time-Delayed Telemanipulation*, Proc. of IEEE IROS'90, 407-412, Tsuchiura, Japan.
42. Michelman, P., Allen, P. (1994): *Shared Autonomy in a Robot Hand Teleoperation System*, Proc. of IEEE IROS'94, **1**, 253-259, Munich.
43. Douglas, A., Xu, Y. (1994): *Real Time Shared Control System for Space Telerobotics*, JIMT Reprint.
44. Spong, M.W. (1993): *Communication delay and Control in Telerobotics*, Journal of the Robotic Society of Japan, **II**(6), 803-810.
45. Kheddar, A., Tzafestas, C., Coiffet, P. (1997): *The Hidden Robot Concept-High Level Abstraction Teleoperation*, Proc. of IEEE/RSJ IROS'97, **3**, 1818-1824, France.
46. Kheddar, A., Coiffet, P., Kotoku, T., Tanie, K. (1997): *Multirobot Teleoperation-Analysis and Prognosis*, Proc. of IEEE ROMAN'97, 166-171, Sendai, Japan.
47. Kim, W.S. (1996): *Virtual Reality Calibration and Preview/predictive Displays for Telerobotics*, Presence. Teleoperators and Virtual Env., **5**(2), 173-190.
48. Kheddar, A., Tanie, K., Coiffet, P. (1998): *Detection of Discrepancies and Sensory Based Recovery from a Class of Discrepancies in Virtual Reality Based Telemanipulation Systems*, Proc. of IEEE ICRA'98, **4**, 2877-2883, Belgium.
49. Kheddar, A., Tzafestas, C., Coiffet, P., Kotoku, T., Tanie, K. (1998): *Multi-robot Teleoperation Using Direct Human Hand Actions*, RSJ Int. Journal of Advanced Robotics, **11**(8), 779-825.

31

Robust Control of a Non-Holonomic Underactuated SCARA Robot

J. Mareczek, M. Buss and G. Schmidt

1 Introduction

Underactuated robots (UAR) have been a recent topic of interest [1–6]. Control strategies for UAR apply for example as emergency control in case of actuator failure. In future cost reductions could be a motivation to build robots with fewer actuators than joints for example by replacing actuators with holding brakes. Especially the related reduction of mass could open a new field of applications in space robotics. Beside various applications, control of UAR contributes to understanding of nonlinear and nonholonomic systems.

It is well–known that UAR are with second-order nonholonomic constraints [1]. Most control algorithms for UAR are based on *strong inertial coupling* between links [2], others assume a holding brake [7] or periodic inputs to control both, actuated and non–actuated joints simultaneously by exploiting chaotic system behavior [3].

When implementing a typical feedback linearizing control law to UAR, the overall system behavior is quite sensitive to parameter estimation errors. Classical methods for control of nonlinear systems [8] use Lyapunov methods for achieving global asymptotic stability if the uncertainties satisfy the matching condition, i.e. enter the system through the same channel as the control input. For the UAR under consideration here, the matching condition is not satisfied. The approach presented in [9] takes uncertainties into account when calculating a globally stabilizing control Lyapunov function recursively. However, this method requires strict feedback form, which is often not achievable for underactuated systems.

In this chapter we present a global and robust switching control law for the underactuated SCARA-type robot R2D1 (2 rotational-degrees-of-freedom, 1 drive). The robot R2D1 has a DD-motor in the first and a holding brake in

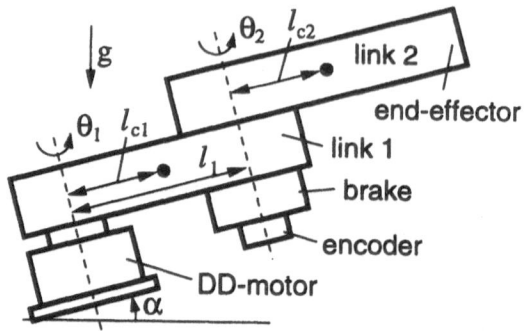

Figure 1.1: Schematic of R2D1

the second joint, see Figure 1.1. The rotational plane can be inclined against the gravitational field at an angle α between $-\pi/6 \leq \alpha \leq 0$.

The end–effector positioning of R2D1 is performed in two major steps. First, the cartesian coordinates of the desired position are transformed into joint space coordinates and the unactuated joint is controlled to its desired joint angle. In the second step, the holding brake is applied and the actuated joint is controlled. Here the challenging first step of unactuated joint position control will be discussed.

2 Dynamic model

The dynamic equations of R2D1 are

$$M(\theta)\ddot{\theta} + n(\theta,\dot{\theta}) + g(\theta) + f(\dot{\theta}) = \tau , \qquad (1)$$

where $\theta = [\theta_1, \theta_2]$ are the joint rotational angles, $\tau = [\tau_1, \tau_b]$ is the motor/brake torque, $n = [n_1, n_2]$ computes the coriolis and centrifugal torque, $g = [g_1, g_2]$ is the gravitational torque, $M = [m_{i,j}]$, $i, j \in \{1,2\}$ is the inertia matrix and $f = [f_1, f_2]$ is the frictional torque. With the brake at joint 2 fixed a brake torque τ_b can be generated; otherwise $\tau_b \equiv 0$. Using the abbreviations $s_\alpha = \sin(\alpha)$, $c_2 = \cos(\theta_2)$, $s_2 = \sin(\theta_2)$, $s_{12} = \sin(\theta_1 + \theta_2)$, the dynamic parameters of (1) are given as

$$
\begin{aligned}
m_{11} &= A + 2Bc_2 & n_1 &= -B(\dot{\theta}_2^2 + 2\dot{\theta}_1\dot{\theta}_2)s_2 \\
m_{12} &= m_{22} + Bc_2 & n_2 &= Bs_2\dot{\theta}_1^2 \\
m_{21} &= m_{12} & g_1 &= -Cs_\alpha s_1 + g_2 \\
m_{22} &= m_2 l_{c2}^2 + I_2 & g_2 &= -Ds_\alpha s_{12} ,
\end{aligned}
$$

with

$$
\begin{aligned}
A &= m_1 l_{c1}^2 + m_2(l_1^2 + l_{c2}^2) + I_1 + I_2 & B &= m_2 l_1 l_{c2} \\
C &= g_0(m_1 l_{c1} + m_2 l_1) & D &= g_0 m_2 l_{c2} ,
\end{aligned}
$$

where the physical parameters are estimated in standard units to $l_1 = 0.300$, $l_{c1} = 0.206$, $l_{c2} = 0.092$, $I_1 = 0.430$, $I_2 = 0.127$, $m_1 = 10.2$, $m_2 = 5.75$.

This yields $A = 1.55$, $B = 0.16$, $C = 37.5$, $D = 5.20$, $m_{22} = 0.18$. A viscous and static friction model for both joints established from experiments is $f_i = a_i \operatorname{sign}(\dot{\theta}_i) + b_i \dot{\theta}_i$, $i \in \{1,2\}$, where $\operatorname{sign}(0) = 0$ and $a_1 = 2.3$, $a_2 = 0.32$, $b_1 = 0.15$, $b_2 = 0.01$.

3 Position control of the unactuated joint

In this section control of the unactuated joint without use of the holding brake (i.e. $\tau_b \equiv 0$) is proposed. The non–collocated linearization (NCL) partial decoupling stategy of [2, 4, 7] is adapted, together with a PD-controller for stabilization of the linearized system.

3.1 Non–collocated linearization

NCL is used to compensate the nonlinearities in the dynamics of the 2nd joint with the result that in the non–perturbed case, i.e. the case of exactly known physical parameters, a 2nd order integrator results. Solving the 2nd row of (1) for $\ddot{\theta}_1$ and substituting the result into the first row one obtains

$$m_{12}^* \ddot{\theta}_2 + n_1^* + g_1^* + f_1^* = \tau_1 \qquad (2)$$
$$m_{12} \ddot{\theta}_1 + m_{22} \ddot{\theta}_2 + n_2 + g_2 + f_2 = 0 \qquad (3)$$

with the abbreviations

$$
\begin{aligned}
m_{12}^* &:= m_{12} - \tfrac{m_{11}}{m_{12}} m_{22} & g_1^* &:= g_1 - \tfrac{m_{11}}{m_{12}} g_2 \\
n_1^* &:= n_1 - \tfrac{m_{11}}{m_{12}} n_2 & f_1^* &:= f_1 - \tfrac{m_{11}}{m_{12}} f_2 .
\end{aligned}
\qquad (4)
$$

Applying the *computed torque* method for the 1st joint, the motor torque results as

$$\tau_1 = v_2 \tilde{m}_{12}^* + \tilde{n}_1^* + \tilde{g}_1^* + \tilde{f}_1^* , \qquad (5)$$

with v_2 denoting the new control input and $\tilde{m}_{12}^*, \tilde{n}_1^*, \tilde{g}_1^*, \tilde{f}_1^*$ as estimated dynamic terms. Inserting (5) in (2), solving for $\ddot{\theta}_2$ and inserting the result in (3) yields

$$
\begin{aligned}
\ddot{\theta}_1 &= -\tfrac{1}{m_{12}}\left((v_2 \tilde{m}_{12}^* + \Delta)\tfrac{m_{22}}{m_{12}^*} + n_2 + g_2 + f_2\right) \\
\ddot{\theta}_2 &= \tfrac{1}{m_{12}^*}(v_2 \tilde{m}_{12}^* + \Delta) ,
\end{aligned}
\qquad (6)
$$

where $\Delta n_1^* = \tilde{n}_1^* - n_1^*$, $\Delta g_1^* = \tilde{g}_1^* - g_1^*$, $\Delta f_1^* = \tilde{f}_1^* - f_1^*$ and $\Delta(\boldsymbol{\theta}, \dot{\boldsymbol{\theta}}) = \Delta n_1^* + \Delta g_1^* + \Delta f_1^*$ denote estimation errors. In case of *ideal parameters estimates*, (6) simplifies to the completely decoupled and linearized system

$$
\begin{aligned}
\ddot{\theta}_1 &= -\tfrac{m_{22}}{m_{12}} v_2 - \tfrac{n_2 + g_2 + f_2}{m_{12}} \\
\ddot{\theta}_2 &= v_2 .
\end{aligned}
\qquad (7)
$$

Hence the 2nd joint is a double integrator system with the 1st joint dynamics as a non–holonomic constraint. The latter is 2nd–order non–holonomic as the highest order of non–integrable state derivative is 2 [1].

The feedback linearization law (5) exists only in case of *strong inertial coupling*, i.e. $\tilde{m}_{12} \neq 0$. R2D1 is designed such that $0.02 \leq \tilde{m}_{12}(\theta_2) \leq 0.34$, hence the two links are strongly inertially coupled in the entire configuration space.

3.2 PD–control of the linearized 2nd joint

The resulting double integrator of the 2nd joint is stabilized by a PD–controller

$$v_2 = k_p e_2 + k_d \dot{e}_2 \, , \tag{8}$$

with $e = [e_2, \dot{e}_2] = [\theta_2^d - \theta_2, -\dot{\theta}_2]$ as the control error and θ_2^d as a piecewise constant desired angle. From the 2nd row of (7) one obtains in the *non–perturbed* case the homogeneous error differential equation

$$\ddot{e}_2 + k_d \dot{e}_2 + k_p e_2 = 0 \, , \tag{9}$$

with the control parameters $k_p > 0, k_d > 0$. In the non–perturbed case the steady–state error goes to zero, whereas in the *perturbed case* the error differential equation is given by

$$\ddot{e}_2 + \bar{k}_d(\theta_2)\dot{e}_2 + \bar{k}_p(\theta_2)e_2 = z(\boldsymbol{\theta}, \dot{\boldsymbol{\theta}}) \, , \tag{10}$$

with the abbreviation

$$\gamma(\theta_2) = \frac{\tilde{m}_{12}^*}{m_{12}^*} = \frac{|\tilde{\boldsymbol{M}}|m_{12}}{|\boldsymbol{M}|\tilde{m}_{12}} \tag{11}$$

and $\bar{k}_d(\theta_2) = k_d\gamma$, $\bar{k}_p(\theta_2) = k_p\gamma$, $z(\boldsymbol{\theta}, \dot{\boldsymbol{\theta}}) = -\boldsymbol{\Delta}/m_{12^*}$. The right hand side of (11) may cause a non–zero steady–state error, which can be reduced by increasing the value of k_p.

Understanding the dynamic behavior of the non–perturbed system is important for analyzing the stability of the total system. Therefore the next section discusses the zero–dynamics and stability of the non–perturbed system.

3.3 Zero–dynamics and stability

In case of ideal parameter estimates, $e_2(\infty) = v_2(\infty) = f_2(\infty) = 0$. Consequently the steady state dynamics of the total system are governed by the *zero–dynamics*

$$\ddot{\theta}_1 = -\frac{n_2 + g_2}{m_{12}} \, . \tag{12}$$

Figure 3.1 shows trajectories of the zero–dynamics for two characteristic inclination angles α. Depending on the initial values and the control parameters the system approaches a stable or an unstable trajectory. While the velocity component of trajectories in Figure 3.1 (a) with $\alpha = 0$ and positive (negative) θ_2^d converge in the upper (lower) plane asymptotically to zero, the trajectories

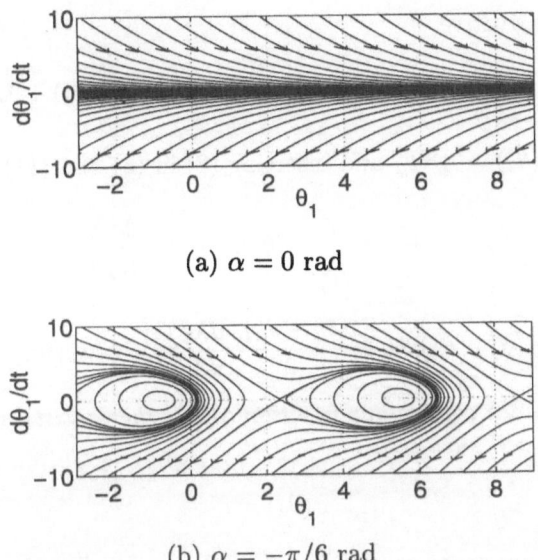

(a) $\alpha = 0$ rad

(b) $\alpha = -\pi/6$ rad

Figure 3.1: Zero–dynamics, $\theta_2^d = \pi/4$ rad.

in the lower (upper) half plane are unstable. In case of non–zero inclination angles α, cyclic motions may occur, see Figure 3.1 (b).

The controlled 2nd joint is asymptotically stable if $k_d, k_p > 0$. However, this does not ensure stability of the whole system as the 2nd–order non–holonomic constraint may be unstable. Another possible reason for instability results from the motor torque limitation which may cause the loss of output controllability.

A third reason for instability may occur due to *parameter perturbation* since the non–collocated linearized system remains nonlinear and coupled with the first joint dynamics. The non–homogeneous error differential equation (10), which depends also on θ_1 and $\dot{\theta}_1$, shows that the control error e_2 is coupled with the motion of the 1st joint. Therefore the resulting control parameters $\bar{k}_p$ and $\bar{k}_d$ as well as the disturbance term z is a function of all states and position control of the 2nd joint may become unstable. This motivates a stability analysis of the NCL–PD–controlled 2nd–joint, in case of physical parameter perturbations using a Lyapunov function.

3.4 Lyapunov function for the NCL–PD–controlled 2nd joint

With the position dependent parameter $\lambda_3(\theta_2) > 0$ and the constant parameter $\lambda_4 > 0$, a candidate Lyapunov function for the 2nd joint is given by

$$V_2 = \frac{1}{2} \left[\lambda_3(\theta_2)(\theta_2^d - \theta_2)^2 + \lambda_4 \dot{\theta}_2^2 \right] , \tag{13}$$

with its time derivative

$$\dot{V}_2 = -\lambda_3(\theta_2)(\theta_2^d - \theta_2)\dot{\theta}_2 + \frac{1}{2}\left(\theta_2^d - \theta_2\right)^2 \dot{\lambda}_3(\theta_2) + \lambda_4\dot{\theta}_2\ddot{\theta}_2 . \qquad (14)$$

Considering $\dot{\lambda}_3(\theta_2) = \frac{d\lambda_3}{d\theta_2}\dot{\theta}_2$ and inserting (8) in (6) and (11) one obtains

$$\dot{V}_2 = \sigma + \lambda_4\dot{\theta}_2\left(-\gamma(\theta_2)k_d\dot{\theta}_2 + \frac{\Delta}{m_{12}^*}\right), \text{ with} \qquad (15)$$

$$\sigma = -\lambda_3(\theta_2)\left(\theta_2^d - \theta_2\right)\dot{\theta}_2 + \frac{\left(\theta_2^d - \theta_2\right)^2}{2}\frac{d\lambda_3}{d\theta_2}\dot{\theta}_2 + \lambda_4\dot{\theta}_2\gamma(\theta_2)k_p\left(\theta_2^d - \theta_2\right) .$$

Lemma 1 *Given a continuous function $\gamma(\theta_2)$, there exists a positive definite function $\lambda_3(\theta_2) > 0$ such that*

$$\sigma(\theta_2) = 0 . \qquad (16)$$

The corresponding proof can be found in the Appendix. Applying Lemma 1 to (15) yields

$$\dot{V}_2(\boldsymbol{\theta}, \dot{\boldsymbol{\theta}}) = \frac{\lambda_4}{m_{12}^*(\theta_2)}\dot{\theta}_2\left(-\tilde{m}_{12}^*(\theta_2)\dot{\theta}_2 k_d + \Delta(\boldsymbol{\theta}, \dot{\boldsymbol{\theta}})\right) . \qquad (17)$$

In the *non-perturbed* case, (17) simplifies to $\dot{V}_2 = -\lambda_4\dot{\theta}_2^2 k_d \leq 0$, hence V_2 is a Lyapunov function for the 2nd joint in the whole state space.

Because (17) is globally indefinite in the *perturbed* case, the state space region in which V_2 is a Lyapunov function reduces to

$$\mathcal{G} = \left\{\boldsymbol{\theta}, \dot{\boldsymbol{\theta}} \mid \dot{V}_2(\boldsymbol{\theta}, \dot{\boldsymbol{\theta}}) < 0\right\} . \qquad (18)$$

It is clear by inspection of (17), that the higher the value of k_d, the more negative is $\dot{V}_2$, hence the larger is $\mathcal{G}$. Additionally there may exist small values of $\dot{\theta}_2$, ($|\dot{\theta}_2| < 0.2$ rad/s) such that $\dot{V}_2 \geq 0$. Experiments have shown that the region in state space where this may happen is stable due to friction.

4 Robust Stabilization of R2D1

In order to find a robust stability region in state space, first for the non-perturbed case the region $\mathcal{G}$ is determined and controlled to be *quasi-invariant* using a variable structure controller. Taking velocity limitations of the motor and an additional stable region of the 2nd joint into account, a stability region $\mathcal{S}$ is obtained from $\mathcal{G}$. Finally the most conservative region $\mathcal{S}^*$ with regard of physical parameter perturbations is determined with the assumption of *contractivity*.

4.1 Quasi–invariant, nominal stability region for the 2nd joint

The standard Lyapunov stability analysis for determining an invariant region for the 2nd joint dynamics requires to find a closed contourline of V_2, completely inside the negative definite region $\mathcal{G}$ of $\dot{V}_2$. Trajectories starting within this invariant region cannot leave it and converge asymptotically to the origin as shown in Figure 4.1 (a).

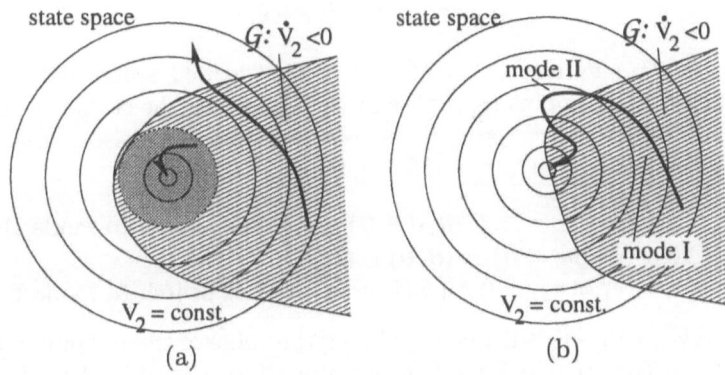

Figure 4.1: Invariant regions: (a) Invariant, asymptotic stability region, (b) Quasi–invariant, asymptotic stability region

For the NCL–PD controlled R2D1 it is not possible for the Lyapunov function defined above to find a closed contourline inside $\mathcal{G}$. One possible configuration of V_2 and $\dot{V}_2$ is sketched in Figure 4.1 (b). A trajectory starting in $\mathcal{G}$ converges towards the origin because V_2 is negative definite. However, it may leave $\mathcal{G}$ and may even get unstable outside of $\mathcal{G}$.

In order to prevent system instability, a variable structure control law is proposed, which makes R2D1 a simple *hybrid, discrete–continuous system* with 2 control modes. In the interior of $\mathcal{G}$ the system is NCL–PD controlled (mode I). When the sytem leaves the stable area mode II is activated, consisting of a damping controller for both joints with the brake switched on. The only purpose of mode II is to force the system trajectory back into the stable region, where mode I is reactivated, see Figure 4.1 (b). This has the effect of making $\mathcal{G}$ *quasi-invariant*.

Definition 1 *Given the region $\mathcal{G}$ in which $V_2 > 0$ and $\dot{V}_2 < 0$ suppose a system with initial state located in the interior of $\mathcal{G}$. If trajectories, leaving $\mathcal{G}$ can be forced to return by a suitable variable structure control law, then $\mathcal{G}$ is quasi–invariant. If in addition the values of V_2 at the reentry points form a decreasing sequence, then the quasi–invariant region is asymptotically stable as shown in Figure 4.1 (b).*

4.2 Enhanced control strategy

Next an enhanced control strategy to achieve a quasi–invariant region $\mathcal{G}$ for R2D1 is proposed. In the interior of $\mathcal{G}$, mode I is active, hence the brake is off and the 2nd joint is NCL–PD controlled. In mode II the brake is switched on while the 1st joint is simultaneously position controlled by a sliding mode controller, ensuring that the unstable motion of R2D1 is strongly damped. The active damping control law is

$$\tau_1 = \tau_{max} \, \text{sign} \, (e_1 + \dot{e}_1) \, , \tag{19}$$

where $e_1 = \theta_1^d - \theta_1$ and $\dot{e}_1 = \dot{\theta}_1^d - \dot{\theta}_1$. The desired states θ_1^d and $\dot{\theta}_1^d$ are chosen such that the point $(\theta_1^d; \dot{\theta}_1^d)$ is in the interior of the stable region of the zero–dynamics.

The enhanced control strategy (cf. Figure 4.2) is given by:

IF	(mode = I) $\wedge$ $(\boldsymbol{\theta}, \dot{\boldsymbol{\theta}} \notin \mathcal{G})$	THEN switch to mode II
ELSE IF	(mode = II) $\wedge$ $(\boldsymbol{\theta}, \dot{\boldsymbol{\theta}}) \in \mathcal{G}$ $\wedge$	
	$(e_1 = \dot{e}_1 = 0) \wedge (\Delta V_2 < 0)$	THEN switch to mode I

Here, $\Delta V_2 = V_2 - \bar{V}_2$ with $\bar{V}_2$ (V_2) being the value of the Lyapunov function at the escape (reentry) point. The meaning of $\Delta V_2 < 0$ is that the system looses energy when reentering $\mathcal{G}$. A formal argument for this is given in the next section.

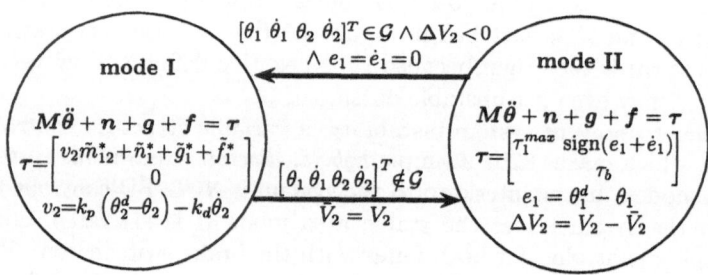

Figure 4.2: Control law of R2D1 described by a hybrid automaton

Applying the above enhanced control strategy, R2D1 becomes a hybrid control system and can be described by the hybrid automaton of Figure 4.2. In mode I the system is NCL–PD controlled, while it changes to mode II when the trajectory leaves $\mathcal{G}$. At this point the value of the Lyapunov function and the position of the 1st joint is stored. In mode II the 1st joint is position controlled to θ_1^d by the sliding mode control (19), while the second joint is damped by applying the brake. Once the system trajectory reenters $\mathcal{G}$ and the 1st joint reaches its desired position θ_1^d and $\Delta V_2 < 0$ holds, mode I with NCL–PD control becomes active again.

In order to confirm asymptotic stability of the controlled R2D1, it is shown first, that the enhanced control strategy renders $\mathcal{G}$ quasi–invariant. This can be checked by observing the value of V_2 (the sign of ΔV_2) at the switching points.

In the following the time intervals when mode I is active are described by $[t_I^{b,i}, t_I^{e,i}]$, $i \in \{0, 1, 2, \cdots\}$, with the upper index b for the start time and e for the end time of the i–th time interval. The set of all start times is given by $\mathcal{T}_I^a = \{t_I^{b,0}, t_I^{b,1}, \cdots t_I^{b,i}, \cdots\}$. Similarly the time intervals in mode II are described by $[t_{II}^{b,i}, t_{II}^{e,i}]$, and the set of all end times is denoted by $\mathcal{T}_{II}^e$. It is always assumed that $t_I^{e,i} < t^* < t_I^{b,i+1}$ holds, with t^* as the time when the energy of the 2nd joint in mode II has decreased below the escape value $\bar{V}_2$, i.e. $\Delta V(t^*) = 0$.

Theorem 1 *Assuming a sufficiently strong brake moment τ_b, the control of mode II forces the system trajectories to return into the quasi–invariant stability region $\mathcal{G}$ with $\Delta V_2 < 0$ in finite time.*

Proof: The sign of $\dot{V}_2$ changes from negative to positive if the system leaves $\mathcal{G}$. As V_2 is continuous, there exists a time t^* in mode II, when $\dot{V}_2$ becomes negative again, i.e. $\Delta V_2(t^*) \leq 0$. Inserting this condition in (13) results in

$$2\Delta V_2 = \lambda_4 \left(\dot{\theta}_2(t^*)^2 - \dot{\theta}_2(t_I^e)^2 \right) + \lambda_3(\theta_2(t^*))e_2(t^*)^2 - \lambda_3(\theta_2(t_I^e))e_2(t_I^e)^2 . \quad (20)$$

With t_b as the braking time, i.e. the time duration when the brake is first applied until zero velocity of the 2nd joint, and $\dot{\theta}_2(t_b) = 0$ it is assumed without loss of generality that $t^* = t_b$. It follows from (17) that $\dot{\theta}_2(t_I^e) \neq 0$, consequently (20) reduces together with $e_2(t_b) = \theta_2^d - \int_{t_I^e}^{t_b} \dot{\theta}_2 dt$ to

$$2\Delta V_2 = -\lambda_4 \dot{\theta}_2(t_I^e)^2 + \lambda_3(t_b)e_2(t_b)^2 - \lambda_3(\theta_2(t_I^e))e_2(t_I^e)^2 .$$

This means that a braking time $t_b > 0$ exists such that $\Delta V_2(t_b) < 0$ holds. ∎

Next we will show, that the control of mode II leads the system trajectories back into $\mathcal{G}$ while the energy of the 2nd joint decreases. The existence of a Lyapunov function within the modes is not sufficient for asymptotically stable behavior of the whole system. Rather than observing the system energy in the modes, one has to make sure that the sequence of system energy values at the switching points is monotonously decreasing. In what follows $\langle x_i \rangle$ denotes the sequence of values x_i.

Theorem 2 *The 2nd joint is asymptotically stable, if the sequence $\mathcal{F}_I = \langle V_2(\mathcal{T}_I^a) \rangle$ decreases strictly monotonously and if there exists a Lyapunov function for the 2nd joint in mode I.*

Proof: A formal proof of a related problem is given in [10] using *pseudo–Lyapunov functions*. Therefore the proof is only briefly sketched here.

In contrast to mode I, where the controlled 2nd joint is asymptotically stable as $\dot{V}_2 < 0$, in mode II the energy of the 2nd joint may also increase. Growing of energy is prevented at switching points, if the energy value at reentry to mode I proves to be lower than the energy when the system leaves mode I. In the case of R2D1 this is assured by Theorem 1.

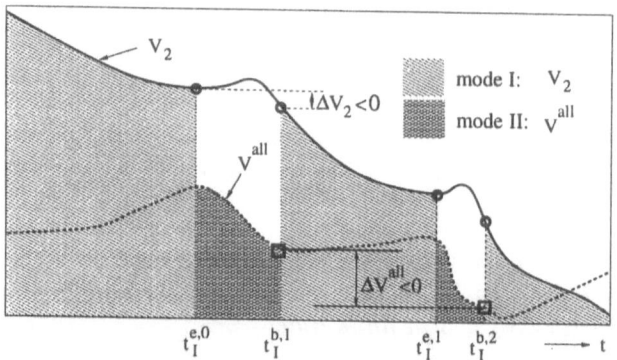

Figure 4.3: Lyapunov functions for mode I: $V_2(t)$ and for mode II: $V^{all}(t)$.

Further clarification may be obtained by Figure 4.3, with V^{all} as a Lyapunov function for the total system in mode II. While $t < t_I^{e,0}$, mode I is active. At time $t \geq t_I^{e,0}$, mode II is active as $\dot{V}_2 \geq 0$, hence the system leaves $\mathcal{G}$. The energy of the 2nd joint is decreased by ΔV_2 at the reentry time $t_I^{b,1}$. ∎

4.3 Stability of the total system R2D1

A stability region for the whole system is obtained by the intersection of the stability region $\mathcal{G}$ of the 2nd joint with a stability region of the 1st joint. The latter may be determined by a Lyapunov stability analysis. However due to the initial value and control parameter dependent, possibly unstable zero–dynamics, the resulting stability region is too conservative and not useful for control purposes.

Nevertheless, a stability result is obtained by a velocity and hence energy limitation of the 1st joint. This procedure also allows to take into account control input limitations of the motor. The velocity limitations are chosen such that the corresponding region contains only stable regions of the state space. This is achieved by considering the zero–dynamics. For $\alpha \neq 0$ there exists a velocity limitation $|\dot{\theta}_1| \leq \dot{\theta}_1^{max}$ such that the 1st joint follows stable (cyclic) trajectories.

In the case $\alpha = 0$ with θ_2^d being positive (negative), the whole upper (lower) phase plane is stable, hence the stable velocity region for the 1st joint is determined by $\dot{\theta}_1 \geq 0$ ($\dot{\theta}_1 \leq 0$). In the following we assume $\alpha \neq 0$ without loss of generality.

Augmenting the supplemental narrow stability region for the 2nd joint

$|\dot{\theta}_2| \leq \dot{\theta}_2^{limit}$, mentioned earlier in Section 3.4 and restricting the result to $|\dot{\theta}_1| \leq \dot{\theta}_1^{max}$ leads to the stability region $\mathcal{S}$ for the whole system as

$$\mathcal{S} = \left\{ \boldsymbol{\theta}, \dot{\boldsymbol{\theta}} \mid \left(\dot{V}_2(\boldsymbol{\theta}, \dot{\boldsymbol{\theta}}) < 0 \vee |\dot{\theta}_2| < \dot{\theta}_2^{limit} \right) \wedge |\dot{\theta}_1| < \dot{\theta}_1^{max} \right\} .$$

Remark 1 *The augmentation of $\mathcal{G}$ by $|\dot{\theta}_2| \leq \dot{\theta}_2^{limit}$ compensates the effect of a possibly too small brake torque as $\dot{\theta}_2(t_I^e)^2$ grows and therefore the braking time t_b may be increased (see the proof of Theorem 1). In this supplementary region, the 2nd joint may follow small residual and bounded oscillations around the desired angle such that strictly speaking, the condition for asymptotic stability is not fulfilled.*

Remark 2 *One assumption of Theorem 1 is that the 2nd joint velocity $\dot{\theta}_2(t_{II}^{b,i})$ at the entry point of mode II must not be zero. Therefore in the special case of $\dot{\theta}_2 = 0 \wedge |\dot{\theta}_1| > \dot{\theta}_1^{max}$, the controller of mode II cannot force the system to fulfill the switching condition $\Delta V_2 < 0$. However, as the brake remains active until $\dot{\theta}_2 = 0$ holds, the system is driven back into the stability region $\mathcal{S}$, not necessarily being inside $\mathcal{G}$. Therefore this case has to be treated seperately in the switching rule.*

As already mentioned in Remarks 1 and 2, $\mathcal{S}$ is rendered quasi–invariant by the enhanced control strategy such that in addition to asymptotic stability of the 2nd joint, the energy of the 1st joint is bounded. This leads to the following stability result.

Corollary 1 *The total system is stable on $\mathcal{S}$ if the 2nd joint is asymptotically stable with respect to Theorem 2 and if a Lyapunov function V^{all} exists for the total system in mode II such that $\langle V^{all}(\mathcal{T}_{II}^e) \rangle$ decreases monotonously.*

Proof: Referring to the definition of $\mathcal{S}$, the 1st joint velocity and hence also its energy is bounded in mode I. When leaving mode II, the states of the 1st joint are located at the point $(\theta_1^d, \dot{\theta}_1^d)$ chosen to be close to a stable cycle of the zero–dynamics. Hence, with e_2 being sufficiently small, the whole trajectory is attracted by this stable cycle. As the 2nd joint converges asymptotically towards the desired angle, e_2 converges to zero and the total system is stable. This means that only the energy values of the system at the points when it leaves mode II must form a decreasing sequence. These points are marked in Figure 4.3 by small squares. A Lyapunov function for the total system is given by

$$V^{all} = \frac{1}{2} \left(e_1^2 + \lambda_1^{all} \dot{e}_1^2 + \lambda_3(\theta_2)\theta_2^2 + \lambda_2^{all} \dot{\theta}_2^2 \right) ,$$

with $\lambda_1^{all}, \lambda_2^{all} > 0$. The corresponding time derivative is

$$\dot{V}^{all} = -e_1\dot{\theta}_1 + \lambda_3(\theta_2)\theta_2\dot{\theta}_2 + \lambda_1^{all}\dot{\theta}_1\ddot{\theta}_1 + \lambda_2^{all}\dot{\theta}_2\ddot{\theta}_2 .$$

When the motion of the joints are damped, $\dot{\theta}_1\ddot{\theta}_1 < 0$, $\dot{\theta}_2\ddot{\theta}_2 < 0$ holds. Therefore $\dot{V}^{all} < 0$ with sufficiently large values $\lambda_1^{all} > 0$ and $\lambda_2^{all} > 0$.

Combining the last two terms in V^{all} results in $V^{all} = \frac{1}{2}\left(e_1^2 + \lambda_1^{all}\dot{e}_1^2 + V_2\right)$. With the values of V_2 at the switching points forming a strictly decreasing sequence and $e_1(\mathcal{T}_{II}^e) = \dot{e}_1(\mathcal{T}_{II}^e) = 0$, $\langle V^{all}(\mathcal{T}_{II}^e)\rangle$ is a monotonously decreasing sequence, which concludes the proof. ∎

4.4 Robust stability region

The derived *nominal, quasi-invariant* stability region $\mathcal{S}$, is now reduced to a robust (with regard to physical parameter perturbations) stability region $\mathcal{S}^*$.

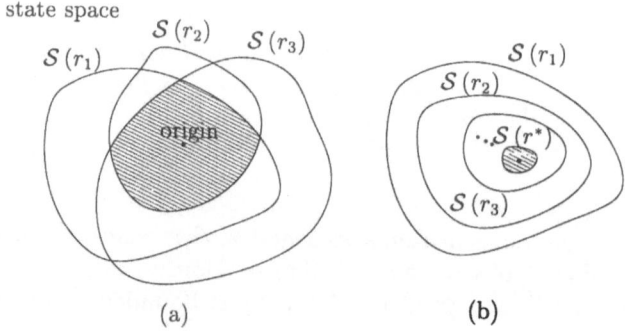

Figure 4.4: Conservative stability region: (a) Intersection method, (b) Contractivity

With the Lyapunov function $\dot{V}_2$ depending on physical parameter estimation errors, the stability region $\mathcal{S}$ depends on the estimation errors also. Therefore a relative parameter estimation error is defined by $x_i := (1 + r_{x_i})\,\tilde{x}_i$, with x_i as a physical parameter. All relative parameter errors can be represented compactly in the error vector r as

$$r = \left[r_{l_1},\ r_{l_{c1}},\ r_{l_{c2}},\ r_{I_1},\ r_{I_2},\ r_{m_1},\ r_{m_2},\ r_{a_1},\ r_{b_1},\ r_{a_2},\ r_{b_2}\right].$$

For R2D1 the absolute values of the maximal relative errors are assumed to be bounded by $|r_{l_1}^{max}| \leq 0.03$, $|r_{l_{c1}}^{max}| \leq 0.03$, $|r_{l_{c2}}^{max}| \leq 0.03$, $|r_{I_1}^{max}| \leq 0.03$, $|r_{I_2}^{max}| \leq 0.03$, $|r_{m_1}^{max}| \leq 0.03$, $|r_{m_2}^{max}| \leq 0.03$, $|r_{a_1}^{max}| \leq 0.1$, $|r_{b_1}^{max}| \leq 0.1$, $|r_{a_2}^{max}| \leq 0.1$ und $|r_{b_2}^{max}| \leq 0.1$.

Definition 2 *System (6) with respect to the error vector r is denoted by $\Sigma(r)$. Furthermore the relative errors are assumed to be bounded. The resulting error space is denoted by $\mathcal{R}$. The set of systems $\Sigma(r)$ is called robustly stable, iff all $\Sigma(r)$ are stable for all $r \in \mathcal{R}$.*

For each system $\Sigma(r)$ there exists a corresponding stability region $\mathcal{S}(r)$. The most conservative and therefore robust stability region is obtained by the intersection of all individual regions $\mathcal{S}(r)$ as illustrated in Figure 4.4 (a).

Definition 3 *If the different regions* $S(r)$ *do not intersect as illustrated in Figure 4.4 (b) and shrink in size with* $\|r\|$ *increasing, then* S *is said to be contractive with respect to* $\|r\|$.

In the case of ideal parameter estimates, hence if $r \to 0$, S is the whole state space as $\dot{V}_2$ is globally negative semidefinite. The larger the relative errors, the larger the absolute values of the error component Δ in (17) and the more differs the fraction $\tilde{m}_{12}^*/m_{12}^*$ of (15) from its nominal value 1. This has the effect of shrinking the negative definite region of $\dot{V}_2$, hence reducing the extension of S. However, the origin is a fixed point of this contraction. Numerical studies have shown that increasing $\|r\|$ causes a contraction of S with the origin as a fixed point. However, the exact relationship between $\|r\|$ and the form of S is not yet clear and seems to be a topic worth further research.

Assumption 1 *It is assumed for* R2D1 *that* $S(r)$ *is contractive.*

In what follows, the robust stability region is denoted by S^* and the corresponding error vector is denoted by r^*. Clearly S^* is the smallest region in state space from all possible regions $S(r)$. Therefore, one has to find the error vector r^*, minimizing the state space extension of $S(r)$. To solve this problem numerically, $\dot{V}_2$ is calculated with a fine grid of the state space $\mathcal{I} \subset \mathbb{R}^4$. The number of points in state space where $V_2 > 0$ is used as a cost function J for an optimization routine. Clearly, the error vector maximizing J renders S most conservative. The optimization problem is given by

$$r^* = \max_r J(r, k_d, \theta, \dot{\theta}) \approx \max_r \sum_{\mathcal{I}} \text{sign}\left(\dot{V}_2\left(r, k_d, \theta, \dot{\theta}\right)\right) . \qquad (21)$$

As the cost function $J(r, k_d, \theta, \dot{\theta})$ is not continuous, a genetic algorithm (GA) as given in [11] is applied, using a crossover propability of $p_c = 1.0$, a mutation probability of $p_m = 1/176$ and a number of coding bits $b = 16$ for each component of the error vector r. With $i = 1, 2$, the searching area is determined by

$$\mathcal{I} := \left\{\theta, \dot{\theta} \mid \theta_i \in \{-\pi, -3/4\pi, \cdots, \pi\}, \; \dot{\theta}_i \in \{-5.0, -4.9, \cdots, 5.0\}\right\} .$$

The control parameters were set to $k_d = 21$, $k_p = 221$, the inclination angle is $\alpha = -\pi/6$. Using these values, the GA obtains after 150 generations, each consisting of 30 individuals, the most conservative error vector

$$r^* = 10^{-2}[3.0, \; 3.0, \; 2.7, \; -1.4, \; -3.0, \; 0.3, \; 2.9, \; 6.4, \; -1.5, \; -7.1, \; 4.1] .$$

This vector is used for the computation of the stability region S^*. The exact stability region is wider than S^*, being conservative as assuring stability for all possible error vectors of the error space.

5   Experiment

The following experiment was performed with $\alpha = -30°$, $k_p = 221$, $k_d = 21$, $\dot{\theta}_2^{limit} = 0.12$ rad/s, $\dot{\theta}_1^{max} = 10$ rad/s in order to validate the proposed approach. The desired value θ_2^d is a square wave function with amplitude $\pi/2$ rad and a period of $10s$, i.e. every $5s$ the command value and the transient behavior are changing. The command value θ_1^d for the 1st joint in mode II is computed to be the equilibrium point of the zero–dynamics given by $\theta_1^d = \pi - \theta_2^d$.

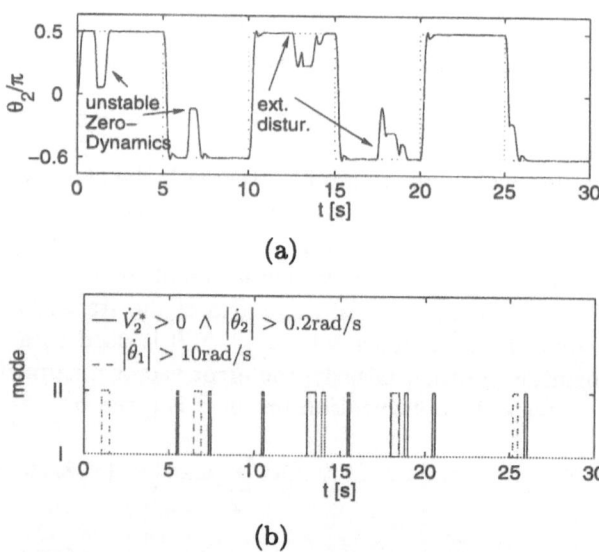

(a)

(b)

Figure 5.1: Sample experimental results

During the first 10s, Figure 5.1 (a) shows the case of unstable zero–dynamics. In the 2nd period of the square wave function ($10s \leq t \leq 20s$), the system is disturbed twice by hand. However, the enhanced control startegy is capable of stabilizing the system. During the last period ($20s \leq t \leq 30s$) the system approaches stable trajectories of the zero–dynamics without the need of mode II control.

The corresponding system mode is shown in Figure 5.1 (b) by the solid line and the dashed line. Each line determines a characteristic instability reason: the dashed line indicates changing mode in case of $|\dot{\theta}_2| > 10$rad/s, i.e. the system leaves the stability region S because of exceeding the joint velocity limit due to unstable zero–dynamics. The case of $\dot{V}_2^* > 0 \wedge |\dot{\theta}_2| > 0.2$rad/s (solid line) indicates instability of the 2nd joint control due to physical parameter perturbations.

Performing the same experiment with equal system and PD–control parameters but without the enhanced control strategy causes instability from the very beginning. A video clip of R2D1 is available in [12].

6 Conclusions

A globally stabilizing and robust position control method for the underactuated SCARA type robot R2D1 was presented. Inclination of the rotating plane of the robot enables investigations with and without graviational influence.

The presented derivation of the non–collocated linearization takes physical parameter perturbations into account. It was shown, that the PD–controlled system can be unstable due to these perturbations. This motivated the introduction of a stability region and an enhanced control strategy, rendering the stability region quasi–invariant. Stability of the enhanced control strategy is shown.

Under the assumption of contractivity, a conservative and robust stability region was determined by an optimization problem solved by a genetic algorithm.

The enhanced control strategy enables the asymptotic stabilization of the unactuated joint with respect to output limitations, physical parameter perturbations and the unstable internal dynamics of the uncontrolled 1st joint. The proposed control strategy may be applied as an emergency control in space robotics, referring to the examples mentioned in the introduction. However, invariant regions in state space for the uncontrolled joints, hence control of nonlinear non–minimum phase systems remains a widely open topic for future research.

References

[1] G. Oriolo and Y. Nakamura, "Control of Mechanical Systems with Second-Order Non-holonomic Constraints: Underactuated Manipulators," in *Proceedings of the IEEE Conference on Decision and Control*, (Brighton, England), pp. 2398–2403, 1991.

[2] M. W. Spong, "Partial Feedback Linearization of Underactuated Mechanical Systems," in *Proceedings of the IEEE/RSJ/GI International Conference on Intelligent Robots and Systems IROS*, (München), pp. 314–321, 1994.

[3] T. Suzuki, M. Koinuma, and Y. Nakamura, "Chaos and Nonlinear Control of a Nonholo-nomic Free-Joint Manipulator," in *Proceedings of the IEEE International Conference on Robotics and Automation*, (Minneapolis, Minnesota), pp. 2668–2675, 1996.

[4] J. Mareczek, M. Buss, and G. Schmidt, "Comparison of Control Algorithms for a Nonholonomic Underactuated 2-DOF Robot," in *Proceedings of the IEEE/ASME International Conference on Advanced Intelligent Mechatronics AIM'97*, (Tokyo, Japan, Paper No. 96), 1997.

[5] J. Mareczek, M. Buss, and G. Schmidt, "Robust Global Stabilization of the Underactuated 2-DOF Manipulator R2D1," in *Proceedings of the IEEE International Conference on Robotics and Automation*, (Leuven, Belgium), pp. 2640–2645, 1998.

[6] M. Bergerman and Y. Xu, "Robust joint and Cartesian control of underactuated manipulators," *Transactions of the ASME: Journal of Dynamic Systems, Measurement and Control*, vol. 118, pp. 557–565, Sep. 1996.

[7] H. Arai and S. Tachi, "Position Control of a Manipulator with Passive Joints Using Dynamic Coupling," *IEEE Transactions on Robotics and Automation*, vol. 7, pp. 528–534, August 1991.

[8] F. Najson and E. Kreindler, "On the Lyapunov Approach to Robust Stabilization of Uncertain Nonlinear Systems," in *International Journal of Robust and Nonlinear Control*, (Haifa, Israel), pp. 41–63, 1996.

[9] R. Freeman and P. Kokotović, *Robust Nonliner Control Design*. Berlin: Birkhäuser-Verlag, 1 ed., 1996.

[10] M. S. Branicky, "Stability of switched and hybrid systems," in *Proc. 33rd IEEE Conf. Decision Control*, (Lake Buena Vista), pp. 3498–3503, 1994.

[11] D. Goldberg, *Algorithms in Search, Optimization, and Machine Learning*. Addison-Wiley Publishing Company, Inc., 1989.

[12] *http://www.lsr.e-technik.tu-muenchen.de/movies/dd32.mpg*. Institute of Automatic Control Engineering LSR, Technische Universität München.

Appendix

Proof of Lemma 1: Inserting $\dot{\lambda}_3 = d\lambda_3/d\theta_2 \, \dot{\theta}_2$ in (16), three cases have to be distinguished: $\dot{\theta}_2 = 0$, $\theta_2^d - \theta_2 = 0$ and $(\dot{\theta}_2 \neq 0) \wedge (\theta_2^d - \theta_2 \neq 0)$. For the first two cases the assumption (16) follows directly. After some elementary manipulations, the 3rd case is given by

$$\frac{d\lambda_3}{d\theta_2} - \frac{\lambda_3}{\theta_2^d - \theta_2} = -\lambda_4 k_p \frac{\gamma(\theta_2)}{\theta_2^d - \theta_2} \; . \tag{22}$$

This is an ordinary linear 1st order differential equation of the form $d\lambda_3/d\theta_2 + a(\theta_2)\lambda_3 = f(\theta_2)$ with $a(\theta_2) = -\left(\theta_2^d - \theta_2\right)^{-1}$ and $f(\theta_2) = -\lambda_4 k_p \gamma(\theta_2)/\theta_2^d - \theta_2$. The solution may be determined by means of *variation of constants* as $a(\theta_2)$ is continuous on the interval, chosen to be without loss of generality $\mathcal{I} = [0; 2\pi[\setminus \theta_2^d$. The continuity of $f(\theta_2)$ follows from the continuity of $\gamma(\theta_2)$ on this interval. Applying variation of constants, the solution is given by

$$\lambda_3(\theta_2) = e^{-A(\theta_2)} \left(\int e^{A(\xi)} f(\xi) d\xi + c \right) , \tag{23}$$

with $A(\theta_2) := -\int (\theta_2^d - \xi)^{-1} d\xi = \ln|\theta_2^d - \theta_2| + c_2$ and two constants c und c_2. Assuming without loss of generality $c_2 = 0$, (23) leads after some elementary transformations to

$$\lambda_3(\theta_2) = \left|\theta_2^d - \theta_2\right|^{-1} \left(c - \lambda_4 k_p \int \mathrm{sign}\left(\theta_2^d - \xi\right) \gamma(\xi) d\xi \right) , \tag{24}$$

with $sign(0) = 1$. As (24) is defined and bounded on $\mathcal{I}$ there exists always a positive and non–zero constant c such that $\lambda_3(\theta_2) > 0 \; \forall \, \theta_2 \in \mathcal{I}$ holds. This proofs the existence of a positive definite function $\lambda_3(\theta_2)$ on $\mathcal{I}$ in all three cases. ∎

32

Kinesthetic Feedback on the Human Hand Interacting with Virtual Environments

C.S. Tzafestas, A. Kheddar and Ph. Coiffet

1 Introduction

During the last decade we have witnessed a considerable progress in the development of Virtual Reality (VR) systems, especially in terms of integrating these systems in new application domains. A VR system is actually a human-computer interface implying simulation and real-time animation as well as interaction via multiple sensory channels [2]. These sensory channels are for the human being: vision, audition, touch, smell and taste. This multimodal interaction constitutes one of the key characteristics of a Virtual Environment (VE). The human operator is rarely satisfied by being a simple spectator of a virtual scene. He usually manifests an intention, from one hand, to act on the VE (for instance touch and manipulate objects or other components of the virtual world) and, on the other hand, to perceive the results of his actions. The second element, that is the perception within a VE, implies a sensory feedback addressing various human sensory modalities. This interactivity contributes to the sensation of immersion or presence of the human being within the virtual world. *Interactivity* and *immersion* constitute, therefore, the two main characteristics of a VR system.

This paper focuses on the subject of *haptic interaction* between the human operator and a virtual environment. The term haptic means sensing by touch (contact) and usually includes two distinct sensory modalities: kinesthesis (perception of forces and mouvements) and tactile sense (concerning all the cutaneous sensory information). The hand, by its dexterity and its sensory capacities, constitutes undoubtedly the most efficient 'tool' of the human being for action and perception. Integrating these functionalities within a VE still consitutes a real challenge for researchers and engineers in the field of haptic interaction systems design and implementation. In fact, this means that the system must allow the human operator: (a) to use natural hand gestures as a new means of communication with the computer, (b) to act with his own hand on the virtual world, for instance, in order to touch, grasp and manipulate virtual objects in an intuitive way, that is by using a large part of his natural manual dexterity, degrees of mobility etc., and (c) to perceive physical characteristics of the virtual objects through a haptic feedback on the human hand.

Keeping in mind the extraordinary complexity of the human haptic system, it is obvious that such an artificial haptic feedback can implicate only a small subset of the whole sensory capacities of the human hand. The human sensory system has a property of major importance, namely: an increased degree of redundancy. The connectivity between different sensory modalities and the inter- and intra-sensory interactions contribute to the fact that even some noisy or limited sensory information can often be sufficient for the creation, in the Central Nervous System (CNS), of an adequate internal representation, and permit the perception of some physical characteristics of the external world. The problem that is then raised concerns the evaluation of the performances of a haptic feedback restricted to the human hand (eg. localized on some finger joints), in terms of perception of such physical characteristics.

This paper deals more particularly with problems related to the synthesis of kinesthetic feedback for the human hand. Such a feedback has as a goal to convey haptic sensory information by the application of forces on different parts of the hand. This is here called: *hand-distributed kinesthetic feedback*. We must note here that, to make such a hand distributed kinesthetic feedback possible, an exoskeleton glove-type device has to be used, allowing: (a) good freedom of mobility for the human hand, (b) monitoring of a large part of the degrees of freedom of the hand, and (c) application of feedback forces on different areas of the human hand (forces on the phalanges or torques on the finger joints).

Such a mechanism has been developed in our laboratory (the LRP hand master, [1]) and is integrated in the virtual prehension system, presented in this paper, for the experimental evaluation of the proposed methodologies. This mechanism allows in fact the application of 14 individual joint torques resisting to the flexion of the fingers.

The first part of this paper deals with the problem of designing a kinesthetic feedback distributed on the human hand. Such a feedback involves the application of forces in different regions of the human hand, and must take into account: (a) the functionality of the hand integrated within the VE, as interpreted by an "intention of action" on the virtual objects, (b) relevant biomechanical studies concerning the relative contribution of each finger and phalanx of the human hand during various natural grasp actions, and (c) the function of the haptic feedback device, its degrees of freedom and its capacities in terms of feedback forces application. The synthesis of the kinesthetic feedback proposed is based on a quasi-static analysis of virtual grasping. The computation of the force distribution on the different contact areas between the virtual hand and objects constitutes a nonlinear optimization problem. The solution of the problem is based on the Lagrange multipliers method, and makes use of weighted pseudoinverses of the grasp matrix.

The second part of this paper concerns the implementation and the experimental evaluation of the proposed kinesthetic feedback. The particular problems related to the physical interaction between the human operator and the haptic device are first analyzed. These problems derive, from one hand, from the uncertainties and the variability of the human hand impedance and, on the other hand, from delays existing in the different control loops (dynamic simulation, graphics rendering etc.). These delays can jeopardise the stability of a direct force-feedback virtual environment system. The solutions that are usually employed to tackle this type of problems are based on a distribution of the computational load on a multi-processor (eg. parallel architectures) or multi-task computer system, and on the use of intermediate representations for the human-VE haptic interaction. In the context of the interactive virtual prehension system described in this paper, we propose the use of a virtual manipulation state feedback, instead of a direct feedback of the computed grasping forces alone.

Initial experimentations were performed to evaluate the system performance from a point of view of : (a) torque servoing for each joint of the glove mechanism worn by the human hand ; (b) haptic interaction with a VE and simulation of the rigidity of a virtual object. Experimental evaluation of the system performance in terms of human haptic perception of virtual physical characteristics is also considered. Some experimental results concerning weight perception of manipulated virtual objects are presented. These results are compared with those reported by other relevant psychophysical studies on human haptic perception (eg. weight discrimination of manipulated real objects). This comparison allows the evaluation, on a common base, of the overall performances of the system, and the investigation of the questions raised by this research work, concerning the design of a hand-distributed kinesthetic feedback and its contribution on haptic perception within a virtual world.

2 Synthesis of hand-distributed kinesthetic feedback

Kinesthesis is the human sensory modality that provides to the Central Nervous System (CNS) information concerning the mouvements and the forces applied on different parts of the human body. McCloskey has defined it as the "sense of positions and actions of the limbs of our body" [15]. Three main classes of afferent signals, emanating from the joints, the muscles and the skin, as well as signals provided by efferent mechanisms have their own contribution to kinesthetic sensibility. This is part of what is more generaly called *haptic perception* including also tactile sensing (for instance feeling the temperature or a vibration on the surface of the skin).

Providing kinesthetic feedback on the human operator therefore means constraining in some way the mouvements or, equivalently, applying forces on some part of the body and more particularly on the human hand. The goal of such a kinesthetic feedback, in the context of the applications considered here, is to provide to the human operator pertinent sensory information concerning his interaction with a VE and therefore improve the human perception of virtual physical properties. For instance, in the context of a robot telemanipulation application this could mean feeling the interaction of the robot with its environment and therefore the characteristics of the manipulated objects and in general of the remote environment. These characteristics can be related to static parameters (such as the stiffness or the weight of a manipulated object), as well as to dynamic parameters or events (such as collisions with obstacles, friction characteristics etc.).

The forces and moments related to such static or dynamic environment characteristics are in general distributed to all the contact regions between the human hand and a manipulated object. These actions are perceived by the whole kinesthetic system of the human being, concerning not only the hand but also joints and muscles of the arm and eventually the rest of the body. This property is related to the *sensory-motor redundancy* characterizing in general even the simplest living organisms. If, therefore, during haptic interaction within a VE, the applied kinesthetic feedback is restricted on the human hand, for instance in the form of torques localized on some finger joints, the subject will not benefit from the whole sensory capacities. Some of the sensations usually involved when manipulating real objects will be completely absent. In other words there will be a sensory discrepancy with respect to the real-world manipulation case. One question that can be raised is then: can such a limited kinesthetic feedback prove sufficient, and up to what point, for reconstructing in an artificial way the perception of certain virtual properties? Answering this question constitutes the subject of the last part of this paper where experimental evaluation of the system performances and limitations is performed based on haptic perception studies. Such experiments can provide usefull hints concerning the relative contribution of the kinesthetic/perceptual capacities of the human hand during manipulation tasks and eventually provide guidelines for the design of human/VR haptic interfaces.

Another question that can be raised is: how can the external wrenches, related to static or dynamic characteristics simulated within a VE, be distributed on the human operator hand in order to generate the appropriate, and maybe adequate, sensations and create the corresponding "perceptual images". This question constitutes the topic of this section. The problem is formulated as a non-linear optimization problem and is solved using the Lagrange multipliers technique. Appropriate criteria should be chosen to provide a suitable solution. This solution, that is, computing the feedback forces or torques to be applied on different regions (phalanges or joints) of the human hand, should take into account the particularities of the problem concerning a human-machine real-time interactive system. It should therefore take into account the "intention" of the human operator as interpreted by a manipulation action performed within the VE. Another important issue concerns the so-called "biomechanical resemblance", which means that the computed solution should be similar to the one usually employed by the human being when manipulating real objects. The proposed solution makes use of a quadratic optimization criterion which integrates such terms introducing human intention (squeezing coefficients for the manipulated virtual object) and biomechanical grasping data (finger-phalangeal contribution to grasping).

2.1 Load distribution on the human hand

The problem can be summarized as follows: determine a generalized method to compute the distribution of "external forces" on the human hand during a virtual prehensile task. Let's consider a virtual object being grasped by the virtual hand with n_c contact points. For each contact point we use the following notation: (fig. 2.1):

f_{ci} : i-th contact forces vector
r_{ci} : distance vector from the object's center to the i-th contact point
a_{ci} : unity vector normal on the surface of the object

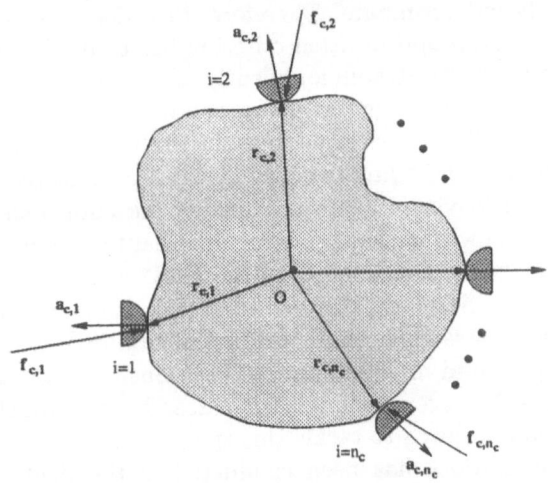

Fig. 2.1. *Grasping of an object with n_c contact points*

The static equilibrium equations for the system of contact forces applied on the grasped object can be written in the following well-known compact form:

$$G \cdot \mathbf{f_c} = \mathbf{w_e} \tag{2.1}$$

where

$$G = \begin{pmatrix} I_3 & \cdots & I_3 \\ R_1 & \cdots & R_{nc} \end{pmatrix} \text{ is the } (6 \times 3n_c) \text{ grasp matrix,}$$

I_3 is the 3x3 identity matrix,

$$R_i = \begin{pmatrix} 0 & -r_{ciz} & r_{ciy} \\ r_{ciz} & 0 & -r_{cix} \\ -r_{ciy} & r_{cix} & 0 \end{pmatrix} : \text{skew-symmetric matrix}$$

$\mathbf{f_c} = (f_{1x} f_{1y} f_{1z} \ldots f_{ncx} f_{ncy} f_{ncz})^T$: contact forces vector, and

$\mathbf{w_e} = (-\mathbf{F_{ext}}, -\mathbf{N_{ext}})$: external wrench.

Equation (2.1) is often completed by a number of constraints on the solution $\mathbf{f}_{ci}$ in order to take into account the unilateral nature of the contacts as well as limitations due to static friction (Coulomb law). These constraints can be written as follows:

$$\frac{|\mathbf{f_{ci}} \cdot \mathbf{a_{ci}}|}{|\mathbf{f_{ci}}|} \geq \frac{1}{\sqrt{1 + \mu_i^2}} \qquad (\mu_i : \text{static friction coefficient}) \tag{2.2}$$

$$\mathbf{f_{ci}} \cdot \mathbf{a_{ci}} < 0 \qquad (i = 1, \ldots, n_c) \tag{2.3}$$

In the general case ($n_c > 2$, no singularities) the system defined by the above equations is indeterminate. Therefore, in order to choose a solution for the contact forces an optimization criterion has to be defined. Definining appropriate criteria constitutes undoubdedtly one of the major difficulties of the problem. Our attention must focus on two points. Firstly, the efficiency of any kinesthetic feedback technique must be evaluated with respect to the sensations conveyed to the human operator, which constitutes the main goal of any haptic feedback system. Objective quality measures must therefore be used based on our knowledge concerning human haptic perception capacities and on systematic experimental procedures. These issues are studied in the last part of this paper. Secondly, the defined optimization criteria must reflect sensori-motor control strategies employed by the human operator during various natural grasping and manipulation actions. Such knowledge is still very limited. There exist however some relative research work which can provide usefull information and ideas to tackle this specific class of problems.

A lot of research work has been conducted in the field of multi-chain robotic mechanisms. Such mechanisms include, first of all, multifingered robot hands [10]. Force control of these mechanisms raises the problem of force distribution during the grasping and manipulation phases and of coordinated action of the motorized elements to ensure the overall stability of the system. This is often reduced to a Linear Programming (LP) problem and solved using the Simplex method (see for instance [11]). Buss et al. have proposed to formulate the problem as an optimization on a set of positive definite matrices under the application of linearized constraints [3]. They introduced a cost index which constitutes a trade-off between the total effort applied on the grasped object and a stability margin.

Another class of multi-chain mechanisms includes the multi-legged walking robots. The mathematical formulation used is practically identical with the one introduced above, and the proposed methods to solve the load distribution problem are similar. For instance, Orin and Oh have also proposed the use of the LP method to solve the problem for general locomotion systems [16]. Cheng and Orin have subsequently proposed a more efficient formulation based on the compact-dual LP method [4][5].

All the above mentionned methods, as well as other similar methods not cited above, aiming to solve the problem of force distribution for robotic mechanisms containing closed kinematic chains, use a mathematical formulation which is similar to the one introduced by equations (2.1), (2.2), (2.3). To solve the indeterminacy of the system described by these equations we can follow well-known paths and define simple optimization criteria. For instance, by minimizing a quadratic function on $\mathbf{f}_c$ we obtain a minimal-norm solution for the contact forces. To take into account the stability margin, a cost index can be introduced computing, for instance, the distance with respect to the friction constraints. However, all these criteria inspired by studies on robot grasping analysis do not use any information relative to the "intention" of the human hand during virtual grasping, which can be for instance interpreted by a local deformation of the manipulated virtual object. In an interactive virtual prehension system, with kinesthetic feedback on the human hand, such a *manipulative intention* of the human operator should be monitored on-line and taken into account for the computation of appropriate feedback forces.

This is performed by introducing what we call "squeezing coefficients" s_i, which are defined as being proportional to the intersection between the human hand and the manipulated virtual object, at each contact point i. These coefficients actually encode information concerning the desired action performed by the human hand on the virtual object. We call "squeezing forces" $\mathbf{f}_{si}$ the normal contact forces of the form: $(s_i \cdot \mathbf{a}_{ci})$, indicating how much the operator is deforming the virtual object locally. To compute these coefficients at eact time instant it is important to take into account not only the local deformation of the virtual object (interpreting the manipulative intention of the human operator) but also biomechanical data concerning the force distribution on the human hand and the contribution of each finger and phalanx during various types of natural prehensile actions [20].

A minimization function can now de written in the following quadratic form:

$$F_2 = (1/2) \sum_{i=1}^{n_c} |\mathbf{f_{ci}} - s_i \cdot \mathbf{a_{ci}}|^2 \rightarrow min \qquad (2.4)$$

The computation of the feedback forces $\mathbf{f}_{ci}$ consists therefore of solving a nonlinear optimization problem, defined by the minimization criterion F_2 (equation (2.4)), subject to the constraints described by relations (2.1), (2.2), (2.3). The solution of such a constrained optimization problem can be obtained by using various nonlinear programming methods [14] such as for instance the iterative Kuhn-Tucker method. The main drawback of applying such a technique in VR interactive applications has to do with the computation time needed to perform additional iterations, in case one or more of the constraints are not satisfied. Real-time requirements are of major importance for achieving satisfactory realism concerning such interactive simulation systems. Taking into consideration the particularities of the problem for the aplication considered in this paper, i.e. haptic interaction with a VE, some simplifications can be made to obtain an analytical solution, as discussed in the following paragraph.

2.2 Simplification of the problem and the use of pseudo-inverses

When manipulating virtual objects, the intentions of the human operator are determined by constant monitoring of the interactions between the virtual hand and the manipulated object. Control of these interactions (for instance if the object must be stably grasped or slip from the hand) is performed by the operator who acts on the haptic interface (in our case, as we will see later, an exoskeleton glove device). Therefore, it seems more reasonable to monitor the stability conditions (2.2) and (2.3), instead of imposing them as constraints to the system, and to subsequently determine the appropriate feedback forces as well as the behaviour of the virtual object for each grasping state. The problem of computing optimal feedback forces, in the case of a stable grasping (conditions (2.2) and (2.3) satisfied), can be solved by minimizing the function F_2 subject only to the constraints defined by the system of equations (2.1), which describes the static equilibrium for grasping.

To solve this problem, in the general case, we can use Lagrange theory and transform it into a system of linear equations which can be for instance numericaly solved using the Gaussian elimination algorithm. A more efficient analytical solution to the problem can be also provided by using the pseudo-inverse of the grasping matrix G. A necessary condition for the presence of a minimum for the function F_2 defined by (2.4) is the following:

$$\nabla_{\mathbf{f}_c} \{ \; \frac{1}{2}\|\mathbf{f}_c - \mathbf{f}_s\|^2 - \boldsymbol{\lambda}^T (G\mathbf{f}_c - \mathbf{w}_e) \; \} = 0 \qquad (2.5)$$

where $\mathbf{f}_s = [s_1 \cdot \mathbf{a}_1^T, \; \ldots \; , s_{n_c} \cdot \mathbf{a}_{n_c}^T]^T$ is the $(3n_c \times 1)$ vector containing the grasping forces and $\boldsymbol{\lambda}$ is the (6×1) vector of Lagrange multipliers. This equation can be developed as follows:

$$\mathbf{f}_c = \mathbf{f}_s + G^T \cdot \boldsymbol{\lambda} \qquad (2.6)$$

or equivalently:

$$G\mathbf{f}_c = G\mathbf{f}_s + (GG^T)\boldsymbol{\lambda} \overset{(2.1)}{\Longleftrightarrow} (GG^T)\boldsymbol{\lambda} = \mathbf{w}_e - G\mathbf{f}_s \qquad (2.7)$$

If the row of matrix G is equal to 6, which means that the grasping configuration is not singular, we can then write:

$$\boldsymbol{\lambda} = (GG^T)^{-1} \cdot (\mathbf{w}_e - G\mathbf{f}_s) \qquad (2.8)$$

Replacing $\boldsymbol{\lambda}$ into equation (2.6) we obtain:

$$\mathbf{f}_c = \mathbf{f}_s + G_R^+ \cdot (\mathbf{w}_e - G\mathbf{f}_s) \qquad (2.9)$$

We find therefore an analytical solution for the optimal grasping forces based on the right pseudo-inverse of G: $G_R^+ = G^T \cdot (G \cdot G^T)^{-1}$

This equation can be also written in the well-known form:

$$\mathbf{f}_c = \mathbf{f}_s \cdot (I_{(3n_c)} - G_R^+ \cdot G) + G_R^+ \cdot \mathbf{w}_e \qquad (2.10)$$

where $(G_R^+ \cdot \mathbf{w}_e)$ contains the so-called external grasping (or manipulation) forces compensating for the application of the external wrench, and $\{\mathbf{f}_s \cdot (I_{(3n_c)} - G_R^+ \cdot G)\}$ corresponds to the internal grasping forces. We can here point out that the squeezing forces $\mathbf{f}_s$, determined by the operator's action on the virtual object, control the intensity of the internal grasping forces and, therefore, the stability of the performed virtual prehensile task.

Weighted pseudo-inverses.

The solution provided above by equation (2.9) corresponds in fact to grasping forces $\mathbf{f}_{ci}$ for which the external wrench $\mathbf{w}_e$ is distributed equivalently on all the contact points. This means that the contribution of each force $\mathbf{f}_{ci}$ at compensating the external wrench $\mathbf{w}_e$ (term $\{G_R^+ \mathbf{w}_e\}$) is identical and independant of the corresponding squeezing force $\mathbf{f}_{si}$. This has an important drawback: the computed grasping forces $\mathbf{f}_{ci}$ for the contact points that correspond to small squeezing forces (small contribution d_i to grasping or small local deformation of the object) will tend to leave the friction the friction cone quite often for an increasing external wrench. In other words, the stability margin for grasping, using this method of computing contact forces, is weak. To tackle this problem, the minimization function F_2, defined by equation 2.4, is modified by introducing weight coefficients as follows:

$$F_3 = \frac{1}{2} \sum_{i=1}^{n_c} \frac{\|\mathbf{f}_{ci} - s_i \mathbf{a}_i\|^2}{s_i} \qquad (2.11)$$

where $\{s_i\}$ are the squeezing coefficients. This function can be also written as follows:

$$F_3 = \frac{1}{2}\|S \cdot (\mathbf{f}_c - \mathbf{f}_s)\|^2 \tag{2.12}$$

where S is a diagonal matrix $(3n_c \times 3n_c)$, defined as:

$$S = \mathrm{diag}[p_{si}], \quad p_{si} = (1/\sqrt{s_j}), \quad \forall j \in [1 \ldots n_c] \quad \text{and} \quad \{3j - 2 \leq i \leq 3j\}$$

Proposition 1. *The optimal forces for the minimization of function F_3 defined by equation (2.11), under the constraints imposed by (2.1), are given by the following equation:*

$$\mathbf{f}_c = \mathbf{f}_s + {}^S G_R^\# \cdot (\mathbf{w}_e - G\mathbf{f}_s) \tag{2.13}$$

or equivalently

$$\mathbf{f}_c = \underbrace{{}^S G_R^\# \cdot \mathbf{w}_e}_{(1)} + \underbrace{\left(I_{(3n_c)} - {}^S G_R^\# \cdot G\right)\mathbf{f}_s}_{(2)} \tag{2.14}$$

where ${}^S G_R^\# = S^{-1}G^T(GS^{-1}G^T)^{-1}$ is a weighted pseudo-inverse of matrix G.

(for proof see [24])

The term (1) of equation (2.14) corresponds again to the external grasping forces (also called manipulation forces). This term depends this time on the squeezing coefficients introduced by the weight matrix S. The contribution of each contact point at compensating the external wrench $\mathbf{w}_e$ increases with the squeezing coefficients s_i, that is with the contribution of each contact point at the total squeezing of the manipulated object (total grasping force).

Another criterion that can be introduced consists of minimising not the distance between the grasping forces and the squeezing forces but the sum of the tangential components of the grasping forces which constitutes a stability measure for grasping. Such an optimization function can be defined as follows:

$$F_4 = \frac{1}{2}\sum_{i=1}^{n_c} \frac{\|\mathbf{f}_{c_i} - (\mathbf{f}_{c_i}\mathbf{a}_i)\mathbf{a}_i\|^2}{\|\mathbf{f}_{s_i}\|^2} \tag{2.15}$$

or equivalently:

$$F_4 = \frac{1}{2}\|S_2 \cdot [I_{(3n_c)} - (E_a E_a^T)] \cdot \mathbf{f}_c\|^2 \quad \to \min \tag{2.16}$$

where:

$$E_a = \begin{bmatrix} \mathbf{a}_1 & 0 & \cdots & 0 \\ 0 & \mathbf{a}_2 & \cdots & 0 \\ \vdots & \vdots & \ddots & \vdots \\ 0 & 0 & \cdots & \mathbf{a}_{n_c} \end{bmatrix} \quad \text{and} \quad S_2 = \mathrm{blockdiag}\left(\begin{bmatrix} s_i & 0 & 0 \\ 0 & s_i & 0 \\ 0 & 0 & s_i \end{bmatrix}\right),$$

$i \in [1 \ldots n_c]$.

A general criterion consists therefore of minimizing the following function F_5, which constitutes a linear combination (weighted sum) of the two functions F_3 and F_4 :

$$F_5 = \alpha_1 F_3 + \alpha_2 F_4 \qquad (2.17)$$

Proposition 2. *The optimal grasping forces for the minimization of function F_5 defined by equation (2.17), under the constraints imposed by (2.1), are given by the following equation:*

$$\mathbf{f}_c = (\alpha_1 S_{12}^{-1} S_1)\mathbf{f}_s \; + \; {}^{S_{12}} G_R^{\#} \cdot \left[\mathbf{w}_e - G(\alpha_1 S_{12}^{-1} S_1)\mathbf{f}_s \right] \qquad (2.18)$$

where ${}^{S_{12}} G_R^{\#} = S_{12}^{-1} G^T (G S_{12}^{-1} G^T)^{-1}$ is a weighted pseudo-inverse of matrix G. The weight matrix S_{12} is given by:

$$S_{12} = \alpha_1 S_1 + \alpha_2 S_2 [I_{(3n_c)} - (E_a E_a^T)]$$

All the methods described above have been implemented in simulation, using C programming language, on a HP715 Apollo, 50 Mhz workstation, in order to evaluate their numerical complexity and perform a sensitivity analysis [23]. The method based on the use of the weighted pseudo-inverse ${}^S G_R^{\#}$, introduced by equation (2.13)) (proposition 1), has been found to be the most efficient, especially in terms of virtual grasp stability, and is the one that will be used in the rest of the paper for the computation of the feedback forces $\mathbf{f}_c$.

3 Haptic interaction control architecture

In the previous section, a method has been proposed for the computation of virtual manipulation feedback forces as a first step for providing kinesthetic feedback on the hand of the human operator interacting with a VE. The implementation of such a hand-distributed kinesthetic feedback has to be based on the use of some sort of dextrous hand master capable of applying forces on different areas on the human hand. Such a haptic feedback device, called the LRP hand master, has been developed in our laboratory and will be briefly described in the following paragraph.

The control of such a device during haptic interaction between the human operator and a VE presents some specific difficulties. These difficulties are related, from one hand, to the dynamic interaction between the human hand and the glove mechanism -especially in terms of the variability of the human hand impedance characteristics and the uncertainties that may result concerning the overall system dynamic parameters- and, on the other hand, to the computation time needed to perform dynamic simulation and graphics rendering, eventually resulting to significant time delays that may be therefore introduced to the closed loop system performance. The main characteristic of the control architecture presented in this section consists of a decoupling between the different control loops: graphics rendering, dynamic simulation and virtual prehension routines, active stiffness control and law level torque-servoing for the glove mechanism. Performance of the system is firstly evaluated in control terms: feedback-torque trajectory tracking, static errors, effects of nonlinearities etc. The goal of this preliminary experimental evaluation is to clear out the limitations of the system, especially due to technological constraints, and to point out the trade-offs that have to be made during the mechatronic design of such a system. The resulting dynamic characteristics, such as a hysteresis due to static friction, can certainly influence significantly the "quality" of the kinesthetic feedback provided to the operator. These issues related to the human haptic perception characteristics and to the experimental evaluation of the system based on psychophysical methods, are discussed in the last part of this paper.

3.1 The LRP Hand Master

An exoskeleton force-feedback glove has been recently developed in our laboratory [1](figure 3.1). This glove consists of five fingers and has 19 degrees of freedom (dof) (14 actuated and 5 passive dof). Each finger mechanism has a passive abduction/adduction joint that connects it to the base (fixed on the back of the palm using straps) and 9 rotational joints that constitute with the finger metacarpophalangeal and interphalangeal joints three four-bar closed kinematic chaines, allowing 3 dof (flexion/extension) to the mechanism. Total weight of the mechanism does not exceed 350 gr.

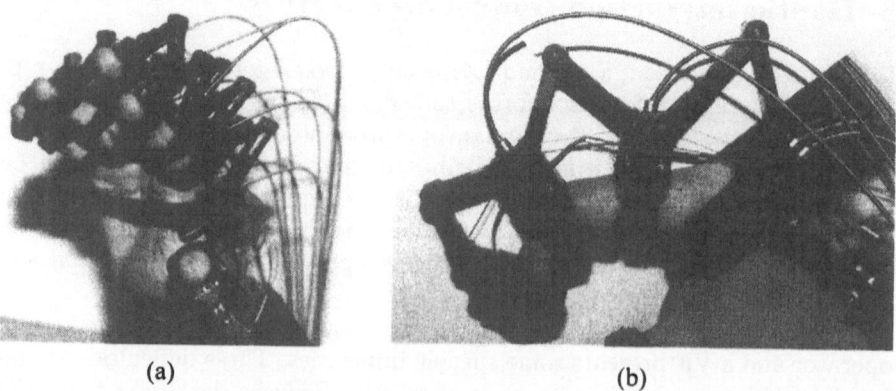

(a)   (b)

Fig. 3.1. Photo of the LRP force-feedback exoskeleton glove: (a) global vue, (b) detailed image of the index.

The mechanical structure offers an adaptability to different human hand sizes. Moreover, it allows ergonomic joint motion tracking for the fingers without imposing major constraints.

The glove is actuated by 14 DC motors that can develop a maximumm torque of 1.4 *Nt.m* and a continuous torque of 0.12 *Nt.m*. The DC motors are remotely located to minimize the weight of the mechanical structure worn by the human hand. Force transmission is performed using cables which run through flexible sheaths. The maximum length of each cable sheath is 1.8 *m* which allows free motion of the human hand in space.

Two types of sensors are integrated to the system. Optical encoders are mounted on pulleys situated close to the motor axes and measure the displacement of the cables. Based on these measures, the finger joint flexions are estimated using a simplified calibration model. Miniature force sensors, based on the use of strain gauges, are also mounted on the endpoint of each cable, at the interior of the mechanical supports which are fixed on the upper part of the phalanges using straps. These sensors allow to measure the traction force applied by each cable and to deduce the torque applied on the corresponding joint of the human hand. These sensors have a linear behaviour for measured forces up to 4.5 *Nt*.

The LRP hand master, therefore, has a double functionality: from one hand, as an input device, it monitors the joint mouvements of the human hand fingers allowing either the remote control of a mechanical hand or the animation of a virtual hand within a VE. On the other hand, as a feedback device, its main function is to apply torques on the joints of the fingers and to convey therefore haptic sensations relative to virtual or real telemanipulation tasks. This system constitutes the experimental platform for the evaluation of the hand-distributed kinesthetic feedback methods proposed above.

3.2 Overall hardware architecture

The overall hardware architecture of the experimental system is illustrated in figure 3.2. It consists of:

• two Hewlett-Packard (HP) workstations equiped with graphics accelerator. The first workstation (HP-A) performs graphic rendering of the virtual scene (virtual hand and objects), while the second one (HP-B) assures operations such as collision detection and computations concerning feedback-force distribution on the hand.

• a 3D tracking sensor, of type Isotrack™ of Polhemus. This sensor provides real-time information concerning position and orientation of the human hand in space. This information is subsequently used by the virtual engine to animate and control the motion of the virtual hand.

• the LRP hand master which is, as has been described above, a prototype force-feedback exoskeleton glove. It is controlled by a PC (Pentium 133MHz Processor) equipped with a number of DA/AD data conversion boards to perform data acquisition for the optical encoders and for the force sensors, and DC motor control. These boards are interfacable on the ISA bus of a standard PC. Communication between the control PC and the HP-B workstation is performed using a RS-232 serial communication link. The two HP workstation communicate with each other using an Ethernet network connection.

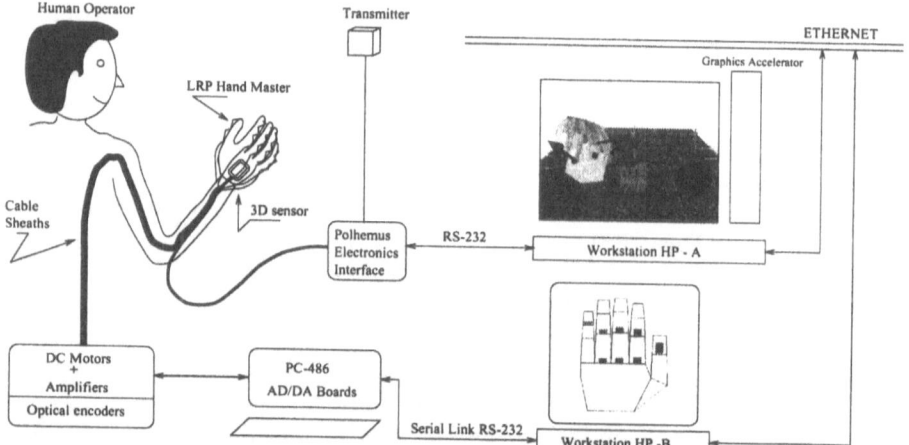

Fig. 3.2. Hardware arcitecture of the experimental system

3.3 Control architecture based on virtual manipulation state feedback

The overall control architecture of the system is illustrated in figure 3.3. The two main components are: the *virtual engine* and the *haptic interface* interacting with human operator hand.

1. The **virtual engine** is implemented on the HP workstations. The first one (HP-A), as we have already said, performs graphics rendering based on information concerning the state of the VE, information supplied via the Ethernet connection by the HP-B workstation. The state of the VE -that is, the positionning and instantaneous displacement of the virtual hand and objects and, eventually, the contacts and grasping configuration- is updated by the HP-B workstation by two main modules:

a. the **dynamic animation module**. This module computes, at each step of the simulation, the linear and angular acceleration for each one of the moving virtual objects, by solving numerically the Newton-Euler equations of motion.

b. the **virtual grasping module**, which simulates the prehensile actions of the hand on the virtual objects. *Collision detection* is based on the use of hierarchical, tree-structures (especially spherical octrees as described in [20]).

2. The **haptic interface** and its control system, which is composed of two main parts:

a. the **torque controller**, that has as an output the desired voltage u to be supplied to the motor amplifiers, which corresponds to the desired torque delivered by the DC motors. This controller is based on a feedback provided by the force sensors mounted on the glove mechanism.

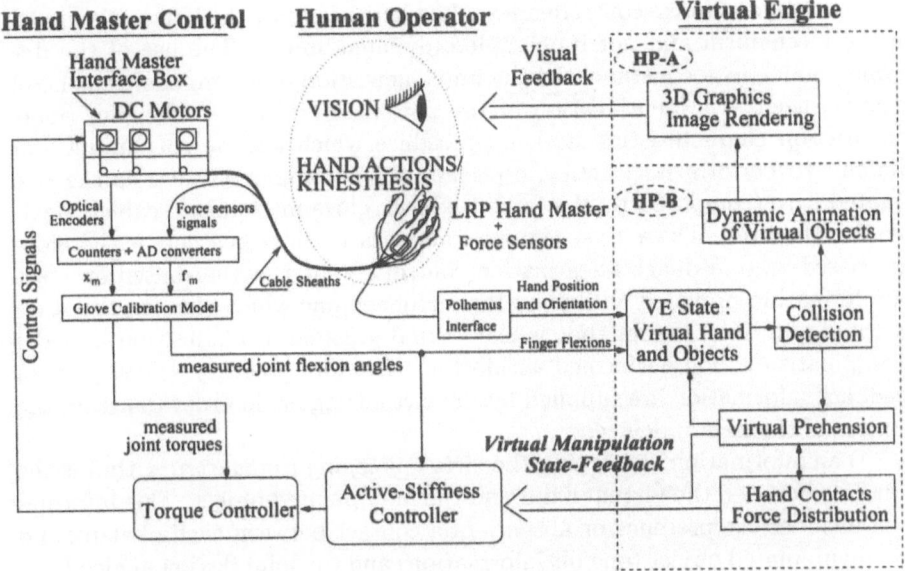

Fig. 3.3. *Control architecture for haptic interaction with a VE.*

b. an **active-stiffness controler**, which is based on a feedback provided by the virtual engine. This feedback gives information concerning the haptic interaction state, that is the grasping configuration and the manipulative action performed within the VE. As will be discussed in the following section, this can be considered as a *"virtual manipulation state feedback"*.

3.3.1 Active-stiffness control based on "virtual manipulation state feedback". The second stage of the haptic interface control system consists of an active-stiffness controller [19]. Such a controller in fact regulates the resistance of a mechanism to an external force applied to its endpoint, with respect to the deviation from a desired equilibrium position. The behaviour of the mechanism, as seen from the external environment, becomes then similar to that of a spring with controllable stiffness. The control law can here be written as:

$$\tau_d = J_c^T(q) \cdot f_c + \left[J_c^T(q) \cdot K_c \cdot J_c(q) \right] \cdot e_q \qquad (3.1)$$

where $e_q = q_c - q_m$ is the error with respect to a reference joint position q_c. The matrix $K_{qc} = \left[J_c^T(q) \cdot K_c \cdot J_c(q) \right]$ is called joint stiffness matrix, while K_c is the contact stiffness matrix expressed in a cartesian coordinate frame. $J_c(q)$ is the Jacobian matrix of the considered mechanism. In the case of a virtual object grasping, equation 3.1 gives the desired torques to be applied on the joints for each finger of the human hand. The reference forces $\{f_{ci}\}$, the reference angles $\{q_{ci}\}$ for the joint flexions, as well as the stiffnesses $\{K_{ci}\}$ for each phalanx i of the hand, are supplied by the virtual engine at each step of the simulation.

The active-stiffness controller described here uses only the kinematic model of the mechanism and not its complete dynamic model. The use of the dynamic model, as for instance for the implementation of an impedance control scheme, needs the identification of the parameters of the mechanism (inertia, friction characteristics etc.), a procedure which is time consuming and difficult to perform in practice, especially for a complex system as the one treated here (composed by the force-feedback glove mechanism, cables, flexible sheaths, etc.). The active-stiffness compensator described here is therefore not based on a feedback linearization control law. It is rather based on a set of information provided on-line by the virtual engine which concern the state of the haptic interaction, that is the virtual grasping configuration and the manipulative action performed within the VE. More precisely, three types of feedback information are supplied by the virtual engine, in order to assure the control of the haptic device:

(i) an information concerning the *virtual grasping configuration*, that is the contacts between the virtual hand and the manipulated object. This information includes the presence or absence of a contact between each phalanx and the manipulated object (one bit information) and the joint flexion angles $\{q_{ci}\}$ which correspond to the time instant when each contact is initially established and are considered as reference angles for the active-stiffness controller.

(ii) the *simulated stiffness* K_{ci} for each contact point i.

(iii) the *reference forces* $\{f_{ci}\}$ that are supplied by the load distribution module on the hand.

This information feedback, provided by the virtual engine to the haptic interface, can be considered on its whole as a *"virtual manipulation state feedback"*. This feedback constitutes in fact an intermediate representation for the coupling between the haptic feedback device and the virtual engine. It is therefore not a simple force feedback, computed and supplied directly by the dynamic simulation and virtual prehension force distribution modules, but rather a set of feedback information concerning globally the haptic interaction state between the human operator and the VE.

The motivation for designing such a haptic interaction control system based on a virtual state feedback comes mainly from considerations relative to the stability of a closed loop system in the presence of time delays, especially as far as the response of the virtual engine is concerned. This is mainly due to the computation load necessary to perform all the collision detection and dynamic animation procedures. In fact, there are four main control loops in the system, which function in parallel in different sampling frequencies:

• the graphics rendering loop, which is executed by the HP-A workstation at a typical image refresh rate of 20 Hz. This frequency is sufficient for realistic visual feedback but certainly not for real-time dynamic, force control.

• the virtual manipulation state feedback control loop, which functions at a frequency within the range between 50 and 100 Hz (sampling time: 10-20 ms), and constitutes in fact the external control loop closed around the master haptic system via the virtual prehension and force distribution modules (executed by the HP-B workstation).

• the active stiffness control loop, which simulates the rigidity of a manipulated virtual object. This loop is executed on the Pentium control PC at a fixed frequency of 500 Hz (using the interrupts mechanism), which can however increase up to 1 KHz (sampling frequency from 1 to 2 msec).

• the joint torque control loop, which assures the low level, local servoing of the force-feedback glove. Its main role is to reduce the effects of friction and other sources of disturrbance and errors, introduced to the system by the cables/sheaths transmission mechanism, and therefore assure a good tracking of the desired torques τ_d supplied by the upper control levels.

3.3.2 Control experiments. The haptic interaction control system, presented on its whole above, has been implemented on the experimental site, illustrated in figure 3.2. Many experiments have been performed to evaluate the properies and the performance, in terms of control objectives, of this haptic feedback system. Each stage of the controller has been tested separately and in interaction with the upper control levels, in order to observe the behaviour of each part and of the control system on its whole. System has performance has been more precisely tested in three cases:

(i) **Torque servoing**, which is performed by the low-level torque compensator. Figure 3.4 shows some experimental results obtained for the proximal interphalangeal (PIP) joint of the index and for the metacarpophalangeal joint (MP) of the thumb. The desired torque was set to 5 $Nt.cm$.

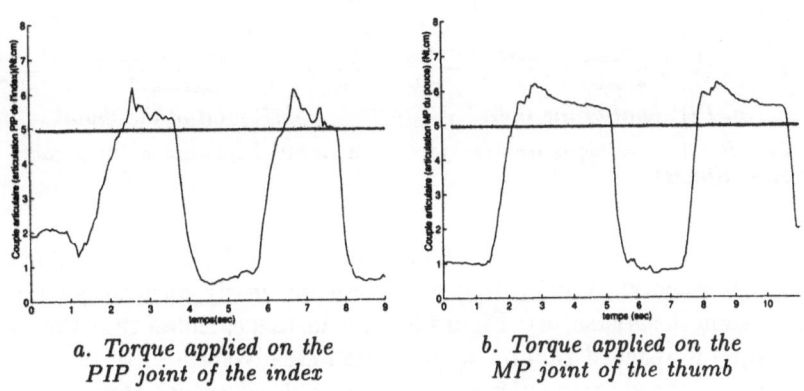

a. *Torque applied on the*   b. *Torque applied on the*
   PIP joint of the index   *MP joint of the thumb*

Fig. 3.4. *Torque servoing for the index and the thumb.*

(ii) **Stiffness simulation**, performed by the active-stiffness compensator described in section 3.3.1. This compensator has as a goal to regulate the resistance to the displacement of each cable (that is, of each joint of the mechanism) in order to emulate the behaviour of a variable stiffness spring. An example of a relation between the pulling tension force and the linear displacement, measured experimentally for two cables corresponding to the index PIP and the thumb MP joints, are shown in figure 3.5.

(iii) **Interaction with a VE:** grasping of a virtual object and virtual manipulation state feedback. The local servoing system of the glove (torque compensator and active-stiffness controller) is here connected to the virtual engine supplying information the state of the virtual manipulation task. An example of a two-finger grasping of a virtual object has been performed and the experimental results, for a virtual stiffness of: $K_c = 1$ (Nt/cm) and $K_c = 2$ (Nt/cm), are presented in figure 3.6. The subject participating in the experiments, and wearing the LRP hand master, performs active finger flexion mouvements deforming the grasped virtual object, with the index and the thumb always being maintained in opposition. The dashed lines in figure 3.6 represent the desired force (the slope of the line corresponding to the desired virtual stiffness), while the small circles correspond to samples of the force actually applied on the index (which is reconstructed by the measured applied joint torques).

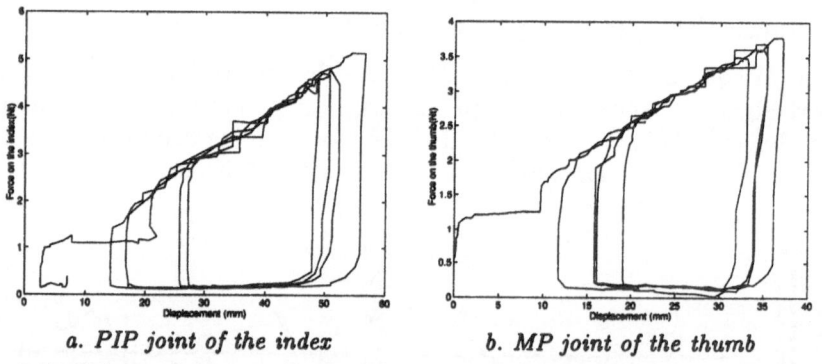

a. PIP joint of the index	*b. MP joint of the thumb*

Fig. 3.5. *Relation between traction force and linear displacement of a cable for a simulated stiffness.*

The first observation that can be pointed out analysing all the above presented experimental results is that force, or equivalently, local torque tracking of the system is satisfactory. Figure 3.4, for instance, shows that the desired joint torque is reached rapidly with a small steady-state error of about 0.5 *Nt.cm* (for a desired torque of 5 *Nt.cm*). This is also demonstrated in figure 3.6 where we can observe that the desired relation between linear displacement of a finger (deforming a virtual object) and the applied force (simulating the virtual stiffness) is established in a satisfactory way. In fact, the samples of the measured finger force approximatively reconstruct a straight line with a slope very close to the one corresponding to the desired value.

However we must point out here that this force/torque tracking performance is obtained only during active finger flexion mouvements. This is due to the static friction introduced to the system by the continuous contact between

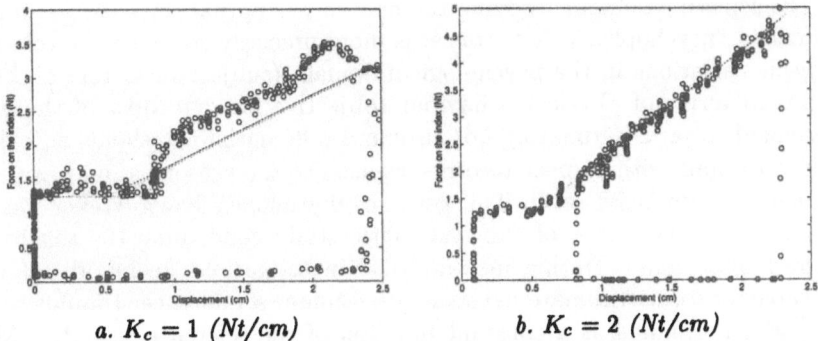

a. $K_c = 1$ *(Nt/cm)*   b. $K_c = 2$ *(Nt/cm)*

Fig. 3.6. *Force applied on the index in relation to the deformation of a grasped virtual object.*

cables and flexible sheaths. The hysteresis characteristics that result to the system are illustrated clearly in figures 3.5 and 3.6. When the finger performs a rapid extension mouvement in order to regain its initial angular position, the applied force drops to a level close to zero. The applied force/torque is servoed to the desired value only when the finger is pulling on the corresponding cables performing an active flexion motion. This problem cannot be solved with the actual hardware configuration of the system. A solution could be to change the combination of cables/sheaths using different material in order to minimize the stick-slip effects. This is a topic of future research work.

As a conclusion it can be said that the kinesthetic feedback provided by the system consists actually of applying a controllable resistance to the joint flexions of the fingers. This variable resistance force can be sensed by the haptic system of the human hand and help the operator to perceive/estimate physical characteristics of the manipulated virtual objects. Human haptic perception issues during interaction with a VE are discussed in the following section.

4 Experimental evaluation: haptic perception of virtual physical characteristics

4.1 Perception anf psychophysics

In the previous sections, the haptic interaction system (integrating the LRP hand master) has been described, together with methods to synthetize a hand-distributed kinesthetic feedback during virtual manipulation tasks. The performance of the system, in terms of control objectives, has also been evaluated. However, the goal of such a system is to reproduce in an artificial manner sensations related to haptic interaction within a virtual world. Objective criteria must therefore be established to evaluate the "quality" of the proposed kinesthetic feedback techniques. The criteria chosen are based on the estimation of the human operator's capacity to perceive physical properties of the manipulated virtual objects.

The scientific domain studying the human perceptual capacities is in general called "psychophysics". It concerns more precisely studying the relations between sensations in the psychological domain (subjective factor) and sensations in terms of physical behaviour (objective factor). Most of the work is oriented towards estimating absolute and differential thresholds as well as their variations with respect to other aspects of the sensory stimuli, such as frequency or intensity level. The german physiologist Ernst Weber can be considered as the father of the systematic study concerning the sensitivity of the human hand. During his empirical investigations he found out that the capacity to discriminate between two stimuli, a reference stimulus and a comparison stimulus, is a constant function of the reference stimulus. More precisely, the ratio between the differential threshold of perception (also called just noticeable difference or jnd) and the intensity of the reference stimulus is constant. This ratio is called *Weber fraction* and is often used as a measure of the perceptual capacities of the human haptic system.

The haptic sensations created to the human operator using the proposed hand-distributed kinesthetic feedback have been evaluated in three cases:

- application of a torque localised on an isolated human hand joint, which constitutes the basic function of the system,

- perception of the stiffness of a grasped virtual object, which demonstrates the performance of the system related to the application of internal grasping forces during interactive virtual prehensile tasks (squeezing of a virtual object), and

- perception of the weight of a manipulated virtual object, which is of particular importance, since results of such psychophysical experiments would demonstrate the capacity of the system in rendering the distribution of external virtual manipulation forces on the human hand (eg. during active lifting actions of a virtual object).

Experimental results in this last third case are briefly presented and discussed in the following section. An important issue which we tried to investigate concerns the relative contribution of the sensori-motor capacities of the human hand in perceiving physical properties related to the application of an external wrench during an active manipulation task. The sensori-motor redundancy is a fundamental property of the human haptic system which results in the fact that each external force, such as the weight of a manipulated object, is distributed on various areas of the human body, especially on muscles of the human hand and arm. Therefore, such a force is percieved as a unique set of sensory excitations which the Central Nervous System (CNS) has learned through practice to recognize. The question which can therefore be raised is: what is the role of each of these sensory elements (emanating from different regions and mechanoreceptors of the human haptic system) for the creation of a consistent "haptic perceptual image"? It seems certain that limiting ourselves to the application of a kinesthetic feedback localised on the human hand would result in the creation of a new sensation (feeling) for the human haptic system, which is more or less close to the one usually related to the manipulation of objects in the real world (as opposed to virtual manipulation tasks). Is this sufficient for the generation of a haptic image (impression, or else illusion) related to a simulated physical property (such as weight) within a virtual world? And if this results in a drop of performances, can this effect be quantified?

All these questions constitute an active research field, in terms of analysing in the microscopic scale the properties and functionnality of different human sensory organs, as well as in terms of evaluating the human perceptual capacities by using, as far as the present work is concerned, psychophysical experimental procedures and methodologies. The VR can constitute a very powerfull tool for the elaboration of such experimental studies. It offers new technological means for measuring the human performances in various simulated situations and can therefore constitute the basis to gain better knowledge of the human sensori-motor functions and strategies.

4.2 Virtual weight perception

This section presents some experimental results evaluating the performance of the proposed hand-distributed kinesthetic feedback during an interactive virtual manipulation task. The performance of the system is evaluated by estimating the difference thresholds (or else the Weber fraction defined above) related to the perception of the weight of a manipulated virtual object. These estimates constitute a measure of the resolution of the haptic feedback system and indicate with which precision the human subject can discriminate between different sensory stimuli (in this case virtual weights applied/distributed on the human hand via the described glove-based kinesthetic feedback).

The perception of the weight of a virtual object using the proposed kinesthetic feedback is of particular interest. It involves the application not only of internal grasping forces (squeezing forces during active deformation of the manipulated virtula object), but also the distribution of an external static wrench on the human hand. It is therefore particularly interesting to evaluate the performance of the haptic feedback system in this case and to estimate the capacities of the human subject in perceiving the rendered sensory signals.

The experimental procedure has been thoroughly described in [23] and is illustrated schematically in figure 4.1. The subject estimates the weight of a manipulated virtual object by performing a series of active finger flexions resulting in lifting motions of the object. The glove was used to provide kinesthetic feedback on the human hand in order to create a sensation for the virtual object's weight.

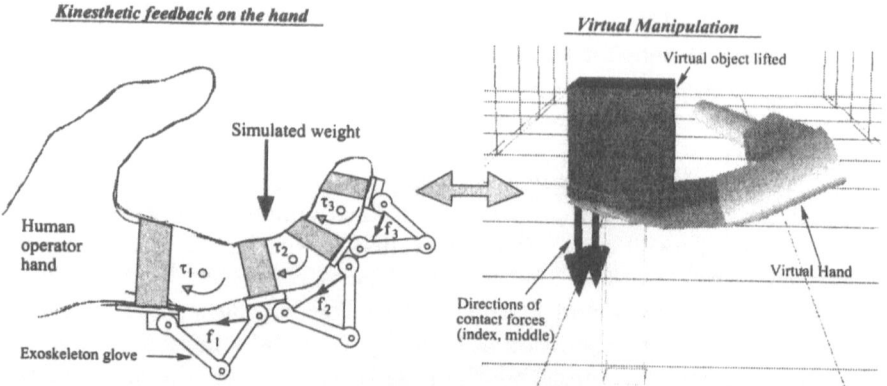

Fig. 4.1. *Manipulation of a virtual object, for the weight perception experiments, and kinesthetic feedback on the hand.*

The Weber fraction (Jnd/F_r), for each reference weight ($F_r = 1, 2, 3N$), as well as the corresponding standard deviations, have been found to be [23]:

- Weber-Fraction = 24.4%, standard-deviation = 5.2%, for $F_r = 1N$,
- Weber-Fraction = 16.3%, standard-deviation = 3.5%, for $F_r = 2N$,
- Fraction-de-Weber = 13.5%, standard-deviation = 3.6%, for $F_r = 3N$.

These results are illustrated by figure 4.2.

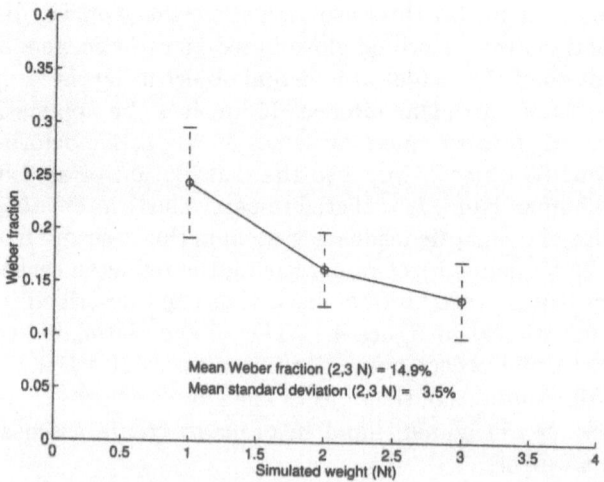

Fig. 4.2. *Weber fraction for virtual weight perception.*

The Weber fractions obtained for virtual reference weights $F_r = 2N$ et $3N$ are similar with a mean value equal to 14.9%. This result is close to the ones reported by other researchers concerning the perception of external forces applied on the human arm or fingers. For instance, [18], [9] and [17] mentioned Weber fractions in the order of 10%. In our case, an external force, as the weight of a manipulated virtual object, is distributed on the fingers of the hand and reconstructed in the form of torques applied by the glove mechanism on the human hand joints. The small increase of the perception thresholds (a mean of 15% instead of 10% approximately) can be due to the imperfections of the experimental system, and especially to disturbances caused for instance by the stick-slip effect related to the friction of the cables inside the flexible sheaths. A considerable increase of the Weber fraction is also observed for a reference weight $F_r = 1N$. This drop of the performance is significant as shown by a first-order analysis of variance and is in accordance with theory which predicts such a phenomenon when the intensity of the stimuli is approaching the absolute threshold of perception. However, this phenomenon is here observed for forces below a threshold which is greater than the one expected by other relevant studies. For instance, the results presented by [17] for forces varying between $20g$ and $200g$ showed an increase in the Weber fraction for a force magnitude greater than $50g$. The Weber fraction for forces between $100g$ and $200g$ has been found practically constant with a value approximately equal to 12 %.

This difference between the results obtained here and those reported in litterature concerning haptic perception of real objects, can be due to disturbances related to the mechanical behaviour of the physical system (glove device), which certainly has an influence on the perceptual performances of the human operator. Besides the problem of static friction (stick-slip effect) other sources of noise exist that can considerably affect the human operator's sensibility especially in the perception of small amplitude forces. For instance, the transmission time delay between the workstation performing dynamic simulation (virtual engine) and the haptic interface control system, can sometimes influence the realism of the simulation and the sensation of haptic interaction for the operator.

However, the results obtained here are globally satisfactory indicating a mean Weber fraction for virtual weight perception of approximately 15 % or less. These results are comparable to those reported in literature relative to human haptic perception. As a conclusion, it can be said that despite the limitations of the system, due mainly to technological constraints, the proposed hand-distributed kinesthetic feedback allows, through the application of torques localised on the joints of the human fingers, the perception of virtual physical characteristics (such as the weight of manipulated virtual object) in a consistent and reliable manner.

5 Conclusion

This paper has concentrated on the haptic interaction between a human operator and a virtual environment. The problem can be seen as that of integrating the functionalities of the human hand within a virtual environment. These functionalities include: (a) the degrees of mobility of the human hand, (b) its dexterity and its prehensile skills, that is its capacity to act on the virtual environment by, for instance, grasping and manipulating virtual objects, and (c) its perceptual capacities implying haptic feedack conveying a variety of sensory information.

We have proposed the synthesis of a kinesthetic feedback based on the solution of a force-distribution problem on the hand during virtual grasping. We called this type of sensory feedback: *hand-distributed kinesthetic feedback*. It consists of applying controllable, feedback joint torques on all the finger joints of the human hand. These torques are computed by a quasi-static analysis of the virtual grasping and by a distribution of external wrenches on the contact regions between the hand and the manipulated objects within the virtual environment. This allows the application of feedback forces on the hand conveying information not only about the strength of grasping and "squeezing" a virtual object (internal grasping forces) but also about the performed manipulation actions (external manipulation forces). The distribution of an externally applied wrench on feedback grasping forces (or more precisely on feedback joint torques) can presumably permit the perception of various virtual physical characteristics, such as the weight or the inertia of a manipulated virtual object.

The proposed methods have been implemented on a virtual prehension system integrating a dextrous force-feedback glove, the LRP Hand Master. The control architecture of this system is based on virtual manipulation state feedback, which constitutes in fact an intermediate representation for the human-VE haptic interaction. The main issue at this stage is that of dealing with problems related to the presence of delays in the different control loops. A set of control experiments have been performed in order to evaluate the system performance in terms of: (a) torque servoing for an individual joint of the glove mechanism worn by the human hand, and (b) haptic interaction and simulation of the stiffness of a manipulated virtual object.

Another question that is being investigated by this research work is whether such a kinesthetic feedback exciting only a part of the human hand sensibility can provide pertinent sensory information for the creation of adequate haptic sensations and for the human perception of haptic stimuli related to virtual physical characteristics. A number of psychophysical studies performed, concerning human haptic perception of virtual stiffness and weight, have demonstated that:

- the sensory resolution, as measured by the differential thresholds of perception (Weber fractions), is close to the one obtained by relevant studies on the perception of real physical characteristics (eg. weight discrimination of real objects lifted by the human hand).

- a slight decrease of the performances must be due to the imperfections and limitations of the mechanical haptic device (exosqueleton glove) used to provide kinesthetic feedback on the human hand. This fact shows clearly the constraints related to the limitations of the current technology and to the trade-offs that have to be satisfied between different ergonomical and technological factors.

The goal of the design of a hand-distributed sensory feedback is certainly not to replace existing force-feedback systems that can be very efficient in some cases (as is for instance the case of direct bilateral teleoperation of a robot manipulator). The utility of such a feedback on the human hand should be evaluated in the context of specific applications, especially those necessitating an increased degree of dexterity from the human operator, in terms of coordinated control of many degrees of freedom. Such a system could have as a goal:

- to better 'exploit' the functionalities of the human hand, and to allow the human operator to control, in a more natural and intuitive way, the execution of virtual or real telemanipulation tasks. This could be for instance the case of teleoperating a dextrous robotic hand.

- to ameliorate the realism of a virtual environment in general, by permitting a more natural haptic interaction. The use of VR techniques can have two different objectives: not only to use the skills of the human operator for the control of increased complexity tasks, but also to study these sensori-motor capacities by simulating a variety situations and monitoring/evaluating the human performances (reactions etc.). The study of the control strategies employed by the human being can help us better understand the different human sensori-motor systems, their properties and limitations. The Virtual Reality, as a human-computer interface implicating a variety of multimodal interactions, can therefore constitute a powerfull tool, from one hand, for a human-machine skill transfer and, on the other hand, for the "decoding" of our own sensori-motor functionalities.

References

1. M. Bouzit, 1996, "Conception et Mise en Oeuvre d'un Gant de Données à Retour d'Effort pour la Télémanipulation d'Objets Virtuels et Réels", Doctorat de l'Université Pierre et Marie Curie (Paris 6) (in french).
2. G. Burdea and Ph. Coiffet, 1994, "Virtual Reality Technology", *John Wiley*.
3. M. Buss, H. Hashimoto and J. B. Moore, 1996, "Dextrous Hand Grasping Force Optimization", *IEEE Transactions on Robotics and Automation*, vol. 12, no. 3, pp. 406-418, June 1996.
4. F. T. Cheng and D. E. Orin, 1991, "Optimal Force Distribution in Multiple-Chain Robotic Systems", *IEEE Transactions on Systems, Man and Cybernetics*, vol. 21, no. 1, pp. 13-24, January/February 1991.
5. F. T. Cheng and D. E. Orin, 1991, "Efficient Formulation of the Force Distribution Equations for Simple Closed-Chain Robotic Mechanisms", *IEEE Transactions on Systems, Man and Cybernetics*, vol. 21, no. 1, pp. 25-32, January/February 1991.
6. G. A. Gescheider, 1985, "Psychophysics: Method, Theory and Application", 2nd edition, Lawrence Erlbaum, New Jersey.

7. J. J. Gibson, 1962, "Observations on active touch", *Psychological Review*, Vol. 69, pp. 477-490.
8. T. Iberall, 1997, "Human Prehension and Dextrous Robot Hands", *International Journal of Robotics Research*, Vol. 16, No. 3, pp. 285-299.
9. L. A. Jones, 1986, "Perception of Force and Weight: Theory and Research" *Psychological Bulletin*, Vol.100, No.1, pp.29-42.
10. L. Jones, 1997, "Dextrous Hands: Human, Prosthetic and Robotic", *Presence*, vol. 6, no. 1, pp. 29-56.
11. J. Kerr and B. Roth, 1986, "Analysis of Multifingered Hands", *The International Journal of Robotics Research*, vol. 4, no. 4, Winter 1986.
12. A. Kheddar, 1997, "Teleoperation basée sur le concept du robot caché", Thèse de Doctorat, Université Pierre et Marie Curie (Paris 6) (in french).
13. A. Kheddar, C. Tzafestas, P. Coiffet, T. Kotoku, K. Tanié, 1998, "Multi-Robot Teleoperation Using Direct Human Hand Actions", *International Journal of Advanced Robotics*, vol. 11, no. 8, pp.779-825.
14. D. G. Luenberger, 1989, "Linear and Nonlinear Programming", *Addison-Wesley*, 1989.
15. D. I. McCloskey, 1978, "Kinesthetic Sensibility", *Physiological Reviews*, Vol. 58, No. 4, pp.763-820, Oct. 1978.
16. D. E. Orin and S. Y. Oh, 1981, " Control of Force Distribution in Robotic Mechanisms Containing Closed Kinematic Chains", *ASME Journal of Dynamic Systems, Measurement and Control*, vol. 102, pp. 134-141, June 1981.
17. D. V. Raj, K. Ingty and M. S. Devanandan, 1985, "Weight appreciation in the hand in normal subjects and in patients with leprous neuropathy", *Brain*, Vol.108, pp.95-102.
18. H. E. Ross and E. E. Brodie, 1987, "Weber Fractions for Weight and Mass as a Function of Stimulus Intensity", *The Quarterly Journal of Experimental Psychology*, Vol.39A, pp.77-88.
19. J. K. Salisbury, Jr., 1980, "Active stiffness control of a manipulator in cartesian coordinates", *IEEE Decision and Control Conference*, Albuquerque, NM, Dec. 1980.
20. C. S. Tzafestas, Ph. Coiffet, 1996, "Real-Time Collision Detection Using Spherical Octrees: Virtual Reality Applications", *IEEE Int. Workshop on Robot and Human Communication* (ROMAN'96), pp. 500-506, Nov. 11-14, Tsukuba, Japon.
21. C. S. Tzafestas, Ph. Coiffet, 1997, "Computing Optimal Forces for Generalized Kinesthetic Feedback on the Human Hand during Virtual Grasping and Manipulation", *IEEE Int. Conf. on Robotics and Automation* (ICRA'97), pp. 118-123, Albuquerque, New Mexico, 20-25 Avril.
22. C. S. Tzafestas, A. Kheddar and Ph. Coiffet, 1998, "Virtual Prehension and Hand Distributed Kinesthetic Feedback for Robot Telemanipulation: Theoretical Concepts and Experimental Results", submitted to the: *IEEE Transactions on Robotics and Automation* (special issue on Virtual Reality in Robotics and Automation).
23. C. S. Tzafestas and Ph. Coiffet, 1998, "Dextrous Haptic Interaction with Virtual Environments: Hand-Distributed Kinesthetic Feedback and Haptic Perception", in Proceedings of the *1st IARP Workshop on Humanoid and Human Friendly Robotics*, Tsukuba, Japan, October 26 and 27.
24. C. Tzafestas, 1998, "Synthèse de retour kinesthésique et perception haptique lors de tâches de manipulation virtuelle ", Thèse de Doctorat de l'Université Pierre et Marie Curie (Paris 6) (in french).

33

Modular Robotics Design: System Integration of a Robot for Disabled People

G. Bolmsjo, M. Olsson, P. Hedenborn, U. Lorentzon,
F. Charrier and H. Nasri

1   Introduction

Robots with higher degree of autonomy is a challenge in robotics research to take a further step in new applications. Today robots in industry are more or less used as programmable automatic machines with little functionality compared to the view or idea of robots as they are presented in most motion pictures. However, although most industrial robotics applications are relatively simple the trend is that each cell consisting of programmable machines including robots is operating more independent on human interaction or supervision than before. This, together with market demand on increased flexibility on products put pressure on flexibility in the production line which ultimately will advance robotics technology.

This trend as a combination of demand and need can also be seen from new robotic application areas coming up, especially those relating to the field of "service robotics". One of these areas include health care and assisting robotics for disabled and elderly people. As this application area brings together most aspects on autonomous robotics with a need from the welfare society which must be solved, the MOBINET project is a driving force with the aim to design the concept of a robot system to assist disabled and elderly people. The primarily user relating to the system presented in this paper is typically an individual with a high spinal injury within the tetrapledic group. In Sweden, which has a population of some 9 million people, about 100 individuals are injured with this type of disability every year. Typically, it is a result of diving or traffic accidents and the average age is below 30. The result of such accidents is that the individual has no controlled finger motion and possibly only limited upper arm motion enough to control a joystick to drive the wheelchair. In practice, no injury is the same and this makes it important to be able to adapt to different requirements related to the manipulator, the user interface and the different input devices.

For the user group described above it is evident that robots will contribute to increased independence and privacy as well as less dependent on 24 hour services by assistants being around all the time. This have both social and economical implications which acts as a driving force to promote robotics research in a direction to be used as an assisting tool together and in combination with people rather than apart as in the case of industrial robots.

Despite the fact that robots used in industry and robots used for caring people are used in quite different applications and with totally different work cycles and functional specifications, most robots used in rehabilitation today have high similarities with industrial robots, such as the RT-series robots and the SCORBOT which originally were developed for educational purposes.

An example of an adaptation of a robot for rehabilitation purposes is the HANDY-1 which is used to assist in eating [1] and DeVar which uses a PUMA robot for assisting disabled at home based or vocational workplaces [2]. The Handy-1 robot is further developed through improvements in the controller and attachments adapted to the user for specific purposes (shaving, tooth brushing, make up) within the RAIL project [3]. However, new designs are on the way that will include the use of compact arms, flexible arms as well as new drives/actuators. Examples of this include the wheelchair based Manus robot [4], the Tou soft (flexible) assistant arm [5], the pneumatically driven Inventaid arm [6, 7] and the compliant actuator Digit Muscle [8]. Wheelchair mounted manipulators shows an increase in interest not only through the manipulator itself, but also through the enhancements of the wheelchair control by providing it with sensors and control system like any mobile robotic base [9]. An example of mobile robotic systems with autonomous functionality within the area of health care is the MOVAID project [10] which is to develop a modular robotic system, including the mobile unit. The development of flexible arm/link systems will also have a great impact on gripper systems, which need a high degree of flexibility in terms of maneuverability and dexterity.

Despite these developments, much work is needed in the area of mechanical design, specifically the introduction of composite materials in the arm structure with inbuilt strain gauges which may be used as a flexible link with feed back of the deflection and redundant kinematics for optimal reachability. Another area closely connected to both the mechanical design and the controller design is the modularity and flexibility of such systems in terms of robot link configuration which means a true robotic system for disabled people should be able to be customized for the individual user need. This paper will specifically address issues related to such demands in flexibility and the related aspects on system integration as being a part of it; to integrate and set up a robot system with a unique configuration for an individual user.

2   Specification

Robotics for use by the disabled is an application area which integrate robots and humans in a common workspace and in the execution of the same work task. Examples of such tasks include control of different equipment through push buttons such as, computers, copy machines, turn on/turn off lights, functions in the kitchen, pick up dropped items, etc.

Therefore, the design of robots for rehabilitation must consider different specifications compared to those used in industrial applications which may affect design aspects of the mechanical structure, kinematics and the system as a whole. Example of differences are:

- Payload of the robot will be in the lower range (typically 1 kg).

- Payload/weight ratio must be far higher than today's robots giving priority to mobility and quick set-up at new workplaces.

- Lower accuracy is allowable if the resolution in the motion control is the same as today's industrial robots.

- A larger workspace and a more flexible configuration with a high degree of maneuverability will be needed compared to industrial robots.

- Life duty cycle will be lower for assisting robots than industrial robots.

- Acceleration and velocity performance can in general be much lower compared to heavy-duty robots.

- Design criteria must enable production at a low price in high volumes.

- The design must allow a modularity within the system to make customization possible.

The aspects on mobility used above refer to the possibility to mount the robot arm on an electrical wheelchair. During discussions with a wheelchair maker, it was evident that the total weight of a robot system should be as low as possible. In addition, for safety reasons 3-phase brushless DC-motors was selected and harmonic drive gear transmission was not allowed. The main arguments for these boundaries in the design process was to secure a safe system working in the same workspace as a human user where an error in the electronic system automatically will stop the 3-phase motor which, in such case, must be able to be easily pushed away by human force.

The focus on the specification and the design of a specific robot was not to create an "optimal" solution of robots for disabled as such does not exist as there are specific individual requirements that must be fulfilled. Instead, the aim was to meet the specifications in the design with the customization in mind which meets the "user need" criteria. Within the scope of the project work, only limited work has been done on the customization of the controller functionality. However, methods has been developed for rapid development of inverse kinematics modeling.

3 The Design Process

The basic need of a large working enveloped and a low weight of the robot made it important to evaluate other than the conventional design concepts where the drive system in general are designed with the motors close to the robot base and transmissions to the respective joints. This leads to a rather heavy system and a bulky base which in this case is undesirable.

To enable fast evaluation of different solutions, both with respect to robot configuration of the link combination and specific solutions of the system a concurrent engineering method was applied which included a combination of software tools ranging from CAD (Pro/ENGINEER), robotics simulation (IGRIP, envision TR), and analysis and modeling (MAPLE V, MatLab). This environment allows for fast interaction between design concepts and "realization" in a virtual world where specific problems can be attacked and analyzed instantly.

One basic characteristic feature relating to joint design and work space analysis which have been simulated extensively is the use of rotating joints having a 45 degree angle between the input and output axis. During early simulations this was found to give a large working envelope together with relatively low torque and possibilities to fold the manipulator in a small volume. Some sample studies from simulations in IGRIP are shown in Fig. 1.

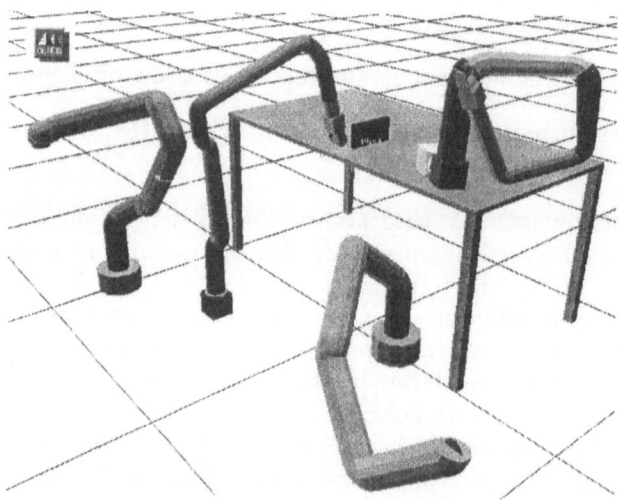

Fig. 1. *Example of early studies of a service robot from simulations in IGRIP.*

From the initial simulations a detailed design analyzes was made in an earlier pre-study project to validate the possibility to construct and assemble the robot arm with the skewed joints. The first attempt followed the conventional concept to minimize the weight in the arm and through this minimize the resulting torque needed to drive the arm. A three axis prototype was built with motors in the base and transmissions through the arm and harmonic drives in the joints. Due to the specifications relating to the selection of transmission, this concept was just to verify the joint/link configuration simulation in a physical mock-up. It was also confirmed by the first analysis of the three-axis prototype of the robot that the arrangement with motor placement in the base was not suitable. In principle, the main draw-backs of integrating transmissions to enable motor placement in the base of the robot are:

1. The gain in weight is only true at higher payloads, since we still have to have transmissions in the joints due to otherwise too high torques in the 45 degree gearing in the joints. This was solved by harmonic drives which adds to the weight in the robot arm.

2. The total weight of the robot should, according to initial specifications be less than 20 kg, and preferably less than 15 kg. This was not possible with the selected solution.

3. Transmissions through the arm adds to the friction and the selection of motors must compensate for the added weight to the robot.

4. Transmissions in the arm makes it impossible to reconfigure the robot arm and exclude a modular approach to reconfigure the robot for different application cases depending on end-user needs.

5. The robot arm must be able to be moved by hand in case of a break down for safety reasons. This exclude the use of harmonic drives as mentioned above.

6. Safety analysis confirmed our choice to use 3-phase brushless DC motors which, in case of malfunction, will stop their motion. This may not be the case for traditional DC motors, which may run in full speed in case of a short circuit in the drive electronics.

As a result of this, a new direction was taken which confirmed the user oriented specification and emphasize on the modularity principle. In conclusion, this demand for integrating the necessary components which drive each joint in the respective links. The following measures was taken to fulfill the requirements:

1. Minimizing the weight of each part. Solution: design optimization together with excessive use of carbon reinforced polymer materials and alloyed aluminum materials.

2. Integrate the drive systems in the joints. Solution: use of high performance brushless motors with integrated planetary transmissions (i = 150-100:1) which meet the reversible drive requirement.

3. Integrate the motor controllers for each drive unit in each link to minimize the cabling in the robot arm. Solution: develop a small motor control unit for servo control and commutation with power electronics for each motor. To facilitate "intelligent" distributed control every 1-3 motor controllers share a 8-bit micro controller with CAN-bus interface The CAN-bus makes it possible to reduce the cabling through the robot arm to one 24 V power line, one ground line and two digital signal lines for the motor control.

These demands could not be satisfied without further analyzes through simulations in IGRIP. Specifically, precise simulations started to evaluate effects of different designs with respect to required torques of the drive units for a "standard" work cycle defined as a simulation case in IGRIP. Typical cases was to pick up a video tape from the floor or a table and put it in a video recorder, or pick up a book and place it in front of the user.

At the same time, the new design concept led to new limitations in the final construction as well as demands from the available space on the mobile platform. Even though the manipulator itself is independent from the mobile platform it is essential to have as compact robot as possible. This increase the usability when the available space is limited or if the manipulator is to be mounted on an electrical wheelchair or elsewhere. As the design had to go in a new direction there was a need for a quick analyze of possible solutions which could work for a first working prototype to validate the design.

During this work, virtual scenarios was developed to set up working joint/link configurations of the robot. Figure 2 shows an example of such a simulation where the important aspect is to make sure the specific robot link configuration can do the specific task. Thus, the virtual scenarios was primary within the design process used to make a final decision on the robot link configuration for the construction and system integration. However, the important result of the work is not the specific solution, but the solutions obtained to create modularity and flexibility to meet user needs on an individual basis.

4   System Integration

The most important aspect within the concept of adapting robots to disabled people is the capability to customize the robot link configuration to an individual. In this section, the system integration will be described concerning both the mechanical arm and the controller. In this paper, the detailed description will be related to the wrist only.

4.1   Arm/Wrist

The basic design of the arm is based on the principle where the main orientation of the end-effector is carried out by a 3-axis wrist system and the position in Cartesian space is carried out by supporting base axes. The wrist has a traditional configuration of a roll axis, followed by a pitch axis and a final roll axes ending in the mounting plate for end-effectors, typically a gripper.

Fig. 2. *Simulation of a virtual scenario in an office environment.*

The transformation of joint motions is created by a gearing system, see figure 3. To improve the system with respect to backlash and noise, helical gears was selected instead of spur gears which was used in an early prototype. However, helical gears produce axial forces which must be accounted for in the design. Adjustments for backlash is made by adjusting the pressure on elastic washers, see figure 4.

A third axis is constructed at the other side of the wrist which is to be mounted to the basic axes of the arm which is shown in figure 5. The total length of the wrist is approximately 250 mm. The connection interface is a simple screw mechanism with pins defining the orientation and a nut with the same diameter as the arm and a left and right thread that hold two links together. All mechanical interfaces are made in a similar way providing a standard set of links which, for the base axes, can be combined in any suitable way and number of axis according to the individual user need.

4.2 Control System

The control system of the robot arm is designed as a main computer running all user interfaces and motion planner including the Cartesian space synchronized motions. All motion control in joint space is distributed and performed locally for each joint. Through the CAN-bus, up to 64 nodes can be connected within the same LAN. Within the scope of this robot arm an eight-axis system was designed with four nodes (one node control in most cases more than one motor).

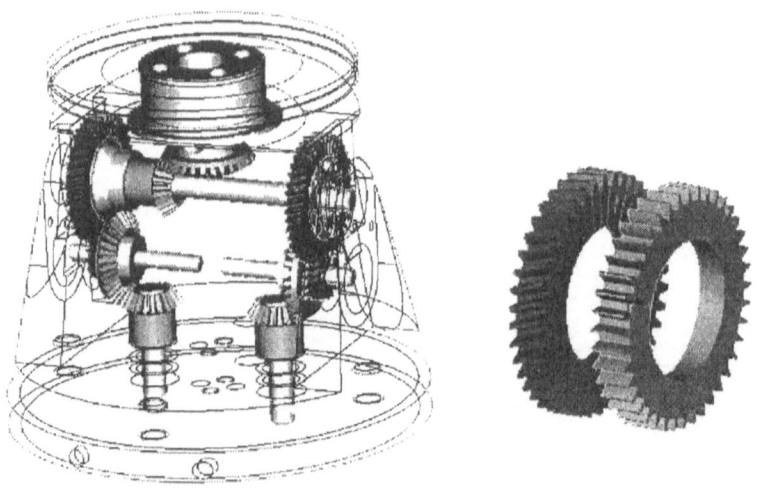

Fig. 3. *Gearing system of the wrist (left), helical and spur gear (right).*

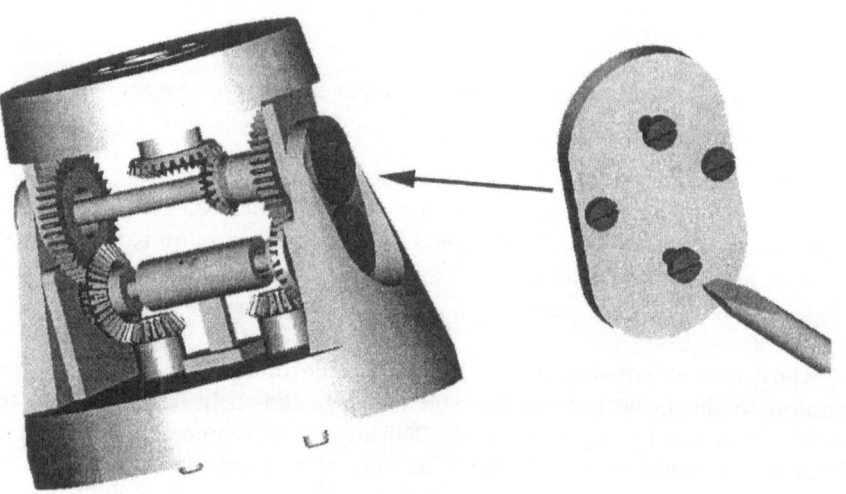

Fig. 4. *The adjustable pressure of washers due to the helical gears to avoid backlash.*

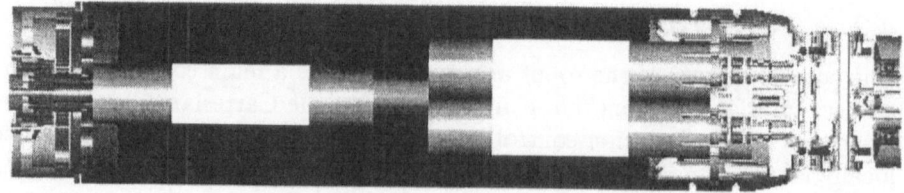

Fig. 5. *The three axis wrist.*

5 Validation of the Integrated System

5.1 Validating the mechanical robot arm

At an early stage in the validation of the mechanical arm the control system will have limited functionalities. At this stage the Low Level Interface of a simulator is used to create and send robot joint data to the real time controller which receives the data and drive the robot to the target position using the distributed joint controllers. Thus, the controller can be operating in joint level while the motions are defined or generated in Cartesian space in a simulator. The simulator used in this case was IGRIP.

Through successive software development the different modules can be tested in the simulator such as kinematics and motions controllers, before implementing them in the real time controller. A specific problem associated to this work is rapid prototyping of inverse kinematics for different manipulators including arms with redundant links with more than six degree of motions. Here, generic analytical algorithms have been used integrated to the simulator by the mechanism of shared library.

5.2 Developing and validating the control software

Software related to the controller such as kinematics, motion planner, trajectory strategies with respect to the redundancy of the manipulator is under development and is in part implemented in "C". Testing and validation of the performance of the motion related software is used through IGRIP which provides an excellent platform for such complex tasks. Typically, the kinematics models are numerical as the robot-link configuration may change from robot to robot.

To support rapid customization analysis of the dynamics must be made. The dynamics problem can be divided into two different areas;

- *Dynamics analysis*; this is important for the verification of the mechanical arm. By specifying masses and inertia of the different links of the arm and the payload on the end-effector, the required joint torques and resulting inter-link forces can be calculated for a specific planned trajectory (path and speed). This information is crucial for the qualification of the mechanical structure and the choice of the servos.

- *Dynamics simulation*; this is very important in the process of building control algorithms for the manipulator. It is possible to visualize the resulting motion of the manipulator arm and the response to the control model. Phenomena like overshooting, oscillation, and tracking error can be visualized and investigated.

In this work the dynamic module within the software tool IGRIP has been used for dynamic analyzes. For a typical working task the different joint torques has been recorded and used for servos dimensioning, see figure 6. The different interlink forces has in the same manner been used to choose type of bearings. This information can be used for further analysis of other included structural elements.

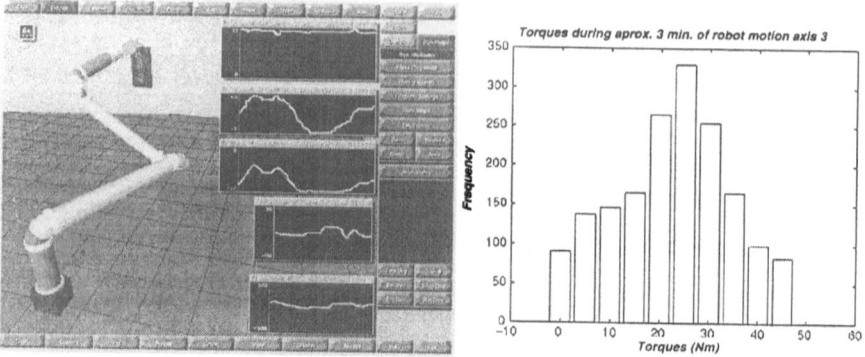

Fig. 6. *To the left the IGRIP simulation of a typical manipulator task is presented. The charts show in real-time during the simulation torques, forces, speeds etc. The information can also be recorded and further analyzed. To the right a histogram is shown created in MatLab (MathWorks Inc.) from recorded data.*

Further work will be performed in the dynamics simulation part. Fine-tuning of control parameters and refinement of the virtual manipulator model will be developed as a generic model to enable rapid customization.

5.3   Validating the control system

To validate the controller feed-back data from the robot (motor feed-back) the actual position of the joint angles are fed back to the high level controller (Cartesian space) where the robot motion can be compared and analyzed with respect to the motion commands sent to the robot. Initial tests have shown that simulations tools such as IGRIP provide a crucial tool to debug the performance of the robot which in this way can be used to simultaneously develop both the mechanics and the controller software.

The integration process of the mechanical prototype and the controller software can be a tedious and error-prone task. This process should start as early as possible so defects can be discovered in the beginning of the development process. By using the simulation system as the primary controller, and transferring information about low-level manipulator information (joint values) to the target system for the control process, the pre-defined modules in the simulation system can be used to supply the functionality that still is missing in the target system.

Examples of how *IGRIP* and *envision TR* can be used for real-time control are presented in [11] and [12]. These papers also discuss the use of sensors for continuously updating the virtual world model. The simulation model can then be used for decision making, re-planning and adaptation of the tasks to the current state of the world.

6   Conclusions

A modular and reconfigurable robot which is possible to mount on a mobile base has been designed and constructed and is now in the systems integration validation phase of the physical robot. During this work, simulations using IGRIP and other related analysis software tools have been important during the development work to analyze and evaluate the design including kinematics, dynamics, controller evaluation of conceptual design solutions and virtual scenarios.

The concurrent development of the mechanics and the controller is facilitated through a robotic simulator that speed up the project work and reduces the technical risk of failure. This is an important factor in complex projects with time constraints where a failure in any part of the project may affect the whole project. In this case, testing and validation can be done on almost every detail in the project. Furthermore, even if a failure occur in the controller which can not be resolved during the lifetime of the project the robot arm can be driven using the simulator with some degradation in real time performance only. Furthermore, virtual scenarios are modeled and simulated that will provide the project team and intended users with a realistic tool to discuss user requirements and user interfaces with the intended users before the robot exist.

Acknowledgments

The work presented in paper has been funded by the European Commission under Contract ERB FMRXCT 960070 (MOBINET). Extensive reference is also made to earlier work funded by AMS (The Swedish National Labor Market Board), RFV (National Social Insurance Board) and NUTEK (Swedish National Board for Industrial and Technical Development).

References

[1] J. Hegarty. Rehabilitation robotics: The users perspective. In *Proceedings of the 2nd Cambridge Workshop on Rehabilitation Robotics*, Cambridge, UK, April 1991.

[2] M. van der Loos, J. Hammel, and L. Leifer. DeVar transfer from R&D to vocational and educational settings. In *Proceedings of the Fourth Int. Conf. on Rehabilitation Robotics*, pages 151–155, Wilmington, Delaware, June 14-16 1994.

[3] M.J Topping, H.Heck, and G.Bolmsjö. An overview of the BIOMED 2 RAIL (Robotics Aid to Independent Living) project. In *Proceedings of the Int. Conf. on Rehabilitation Robotics, ICORR'97*, Bath University, UK, 14-15 April 1997.

[4] T. Øderud, J.E. Bastiansen, and S. Tyvand. Experiences with the MANUS wheelchair mounted manipulator. In *Proceedings of ECART 2*, page 29.1, Stockholm, May 26-28 1993.

[5] A. Casals, R. Villà, and D. Casals. A soft assistant arm for tetraplegics. In *Proceedings of the 1st TIDE Congress*, pages 103–107, Brussels, 6-7 April 1993. Studies in Health Technology and Informatics, Vol. 9, IOS Press.

[6] R.D. Jackson. Robotics and its role in helping disabled people. *Engineering Science and Educational Journal*, pages 267–272, December 1993.

[7] J. Hennequin and Y. Hennequin. Inventaid, wheelchair mounted manipulator. In *Proceedings of the 2nd Cambridge Workshop on Rehabilitation Robotics*, Cambridge, UK, April 1991.

[8] Sam Greenhill. The digit muscle. *Industrial Robot*, 20(5):29–30, 1993.

[9] S. Levine, D. Bell, and Y. Koren. An example of a shared-control system for assistive technologies. In *Proceedings of the 4th Int. Conf. on Computers for Handicapped Persons*, Vienna, Austria, September 1994.

[10] Paolo Dario et al. Movaid: a new European joint project in the field of rehabilitation robotics. *http: //www.alfea.it/movaid/Public_Domain_Area/Papers/paper1.html*.

[11] Krister Brink. *Event Based Control of Industrial Robot Systems*. Dept. of Production and Materials Engineering, Lund University, Lund, Sweden, 1996. Lic. Thesis, CODEN:LUTMDN/(TMMV-1026)/1-139/(1996).

[12] S. A. Ameduri J-Y Jo, Y. Kim and A. Podgurski. A new role of graphical simulation: Software testing. Technical Report Technical report TR-96-115, Center for Automation and Intelligent Systems Research (CAISR), Case Western Reserve University, Cleveland, Ohio, 1996.

34

Design of an Active Arm Support for Assisting Arm Movements

I. Süssemilch and W.S. Harwin

1   Introduction

By the year 2000 the number of individuals over 65 will account for 16% of our population. It is therefore widely expected that there will be a greater number of elderly individuals who require some form of medical assistance. This assistance includes long term residential care, short term hospitalisation, and provision of assistive technologies to promote greater independence for an individual. Mobile robot technologies have much to offer to promote the delivery of medical services but simply providing a robot with the ability to navigate a hospital or residential environment will not realize the full potential of robot technologies in medical and healthcare. The robot must be able to deliver some service. One category of services that are needed in hospital, home and long term residential care environments centre on the ability to assist and control a person's arm movements. This domain includes stabilising a person's arm in the presence of tremor, assisting the person's arm for intensive rehabilitation therapies, or enhancing a person's strength. This strength enhancement might allow a person with muscular dystrophy who is using a powered wheelchair, to move their arm against gravity, yet keep it stable during movements of the wheelchair, or a member of the nursing or care staff to lift a patient without risk to back injury - one of the leading causes of disability in the nursing profession. In all cases combining mobile robotics approaches with technologies to assist arm movement and manipulation has the potential to reduce costs of delivering healthcare by making the patient less dependent on healthcare resources and the care staff more efficient in the operation of their duties.

Unlike mainstream robotics technologies, the robots proposed in this paper have a very close interaction with the person guiding and controlling them. This is an under researched area and significant work must be done to ensure that any products that result are intrinsically safe, and obvious to operate. Thus to investigate issues that relate to the safety and functioning of robots in such close contact to the human user we have developed a series of prototype mechanisms culminating in a three degree of freedom Redundant Active Arm Support (AAS) test-bed. These are research machines intended to allow the evaluation of new methods of robot design and control that can assist with arm movements. The machines are designed to consider different tasks and applications by human arm control but with one focus being a device that can be used in rehabilitation for specific physical therapy methods.

2 Background

The concept of using a machine to enhance an individual strength is attractive for many applications but has still not achieved a practical reality. Possibly the most widely known and used mechanism that has a net amplification of a persons power is the hydraulic power steering mechanism used in modern cars. The concept of power amplification has been extensively elaborated by Kazerooni [1] and requires a bilateral system where power and information to and from a persons sensory-motor system relates directly to the power and information to-and from a working environment.

Early work on medical applications enhancing a persons residual work strength includes work on the Case arm. This was a 4 degree of freedom powered arm orthosis designed to meet the needs of individuals with limited arm function as a result of polio. The first version of this arm was a floor mounted exoskeleton. Control of this version was achieved using a head mounted light source to trigger light sensors in the environment. A series of preprogrammed movements were stored on a magnetic tape and were used to move the arm in response to the light sensors [2]. A second version added Cartesian movements and included light sensors on the arm to allow direct movement of the joints. This version used myoelectric signals to control the velocity of arm movements [3, 4].

A simple powered orthotic mechanism replicating much of the functionality of a upper limb prosthesis was demonstrated by the Hugh McMillan medical centre, and this idea was further developed by Anglin et al. [5] at the University of British Columbia. This upper limb powered orthosis is designed to be body worn and to assist movement in a person with limited upper arm strength. The mechanism, still in prototype form, must be custom fitted and the interface is envisaged to be a bank of chin switches that initiate programs to perform specific actions, in much the same way the Case Arm used tape loops to program procedures.

Work on a wheelchair mounted orthotic mechanisms is also ongoing at the University of Delaware which is targeting simple devices for individuals with muscle weakness resulting from conditions such as muscular dystrophy. This work has produced a test-bed mechanism with 6 degrees of freedom based on a commercial robot, and prototype mechanisms to allow passive gravity compensation of the person's forearm [6]. In the European Union a TIDE funded project 'MULOS' is seeking to develop a six degree of freedom arm orthosis for movement assistance and exercise [7, 8]. Work thus far has concentrated on the structure of of the orthosis which is closely aligned to the human arm kinematics.

Whereas the mechanisms described above are intended to provide greater independence to individuals who have a long term motor disability there is a growing interest in applications of powered orthotic mechanisms in physiotherapy and in particular in the rehabilitation process of individuals recovering from stroke. The annual incidence rate of individuals surviving a stroke but with a long term disability is 14 per 10,000. In the United Kingdom treatment of stroke represents an estimated 5% of the budget of the National Health Service. It is hypothesised that powered orthotic mechanisms used intensively in the rehabilitation process of individuals following a stroke can provide a better outcome and reduce healthcare costs. Krebs et al. [9] found a general acceptance of robot assisted therapies and found that there were no adverse effects from the therapy. The kinematics of the AAS mechanism outlined in this paper is kinematically similar to the Krebs mechanism. Similar work by Lum et al [10] looked at a variety of mechanisms and evaluated a bi-manual therapy based on a robot mechanism to assisting a person to lift a tray. Because of the substantial costs of rehabilitation there are both economic and social justifications for using AAS for stroke rehabilitation.

3   Design Approach

The design criteria for robotic mechanisms that interact with people requires substantially different design constraints than more traditional robotic devices. The safety of the mechanism in operation must be included in the mechanism design from the beginning. Further the traditional joint level position and force control servo-mechanisms do not match the expectations of the user and do not provide data to allow higher level control mechanisms to make safety decisions. A seminal paper by Hogan [11] identified a control mechanism that seeks to match the mechanical impedance of the mechanism to that of the individual. This mechanism is implemented in work by Krebs [9] and Harwin et al. [12]. This work allows the workspace of the AAS to be mapped as an impedance field and this field can be changed by the mechanism controller to reflect the clinical condition of the patient and the intent of the persons arm movement.

In this section we will also discuss first designs that have been developed, to establish prototypes for cable transmission mechanisms.

3.1   Perturbation Theory

Using robots to assist and enhance movement provides an ideal opportunity to quantify human movement. Although parameters such as speed of movement, duration of movement, number of velocity peaks, radius of curvature of movement etc. all provide interesting methods of processing this data it is difficult to correlate these measures with muscle activation and hence with level of recovery. Methods developed for system identification can provide a more intrinsic measure of recovery. Although the ideal would be to assess recovery from measurements of nerve and muscle activation these parameters are difficult to record directly.

Electromyograms (EMG) provide some indication of aggregate muscle activation but this is a crude measure and does not correlate well with measured muscle force. A method that may provide more accurate information relies on random perturbations superimposed on a patients movements. This approach was used by Bennett et al. [13] who used random torques applied by jets of air superimposed in the elbow joint torque to estimate a time varying stiffness of a person's elbow. This is the approach proposed for processing data from the elbow parameter test-bed and the AAS.

In contrast to the approach proposed in Bennett where a baseline datum must be estimated from an ensemble of results, there is a possibility to obtain the baseline measurement when arranging bi-manual therapies as prescribed for the patient. Although specific to the therapy of individuals who have hemiplegia following stroke these bi-manual therapies are envisaged to form a significant percentage of therapy applications for an AAS when used in physiotherapy. In these therapies the affected and unaffected arm attempt to perform a task that requires movements to be symmetrical around the sagittal plane. Since the brain is commanding motor patterns that are mirror image equivalents the path of the unaffected arm can be used as a datum path $d(t)$.

Again in contrast to the approach used by Bennett, the perturbing forces are now imposed by actuators and, since they are measured with two degrees of freedom, can be used to estimate planar equivalent stiffnesses for both the shoulder and elbow. These torques are resolved to estimate a perturbing torque $T(t)$. The resulting perturbed path $p(t)$ is measured and the vector difference, $x(t)=p(t)-v(t)$, used as the exogenous input in an ARX model. Using techniques in system identification [17] it is then possible to estimate time dependent model parameters for the ARX model as

$$A\left(q^{-1}\right)T = B\left(q^{-1}\right)x + e \qquad (1)$$

Where A and B are time dependent parameter matrices, and e is a noise vector.

A continuous time equivalent models with parameters having more traditional engineering meanings can then be calculated from the discrete time model. One likely model form is to assume the joints form a second order system with time varying parameters, that is

$$T = I(t)\ddot{x} + B(t)\dot{x} + K(t)x \qquad (2)$$

Where T is the perturbing torque applied resolved to the elbow or shoulder, x is the resulting perturbation in position and I, B and K are inertia, damping and stiffness matrices. Although it might be possible use more sophisticate models such as the modified Stark and Start model [18] the complexity of this model makes parameter estimation highly sensitive to error. The damping and, more particularly, the stiffness estimates are robust and physiologically meaningful estimates of joint condition and may be a more reliable monitor of a patient's progress.

3.2   Test-Bed Designs

To provide accurate information about the mechanical properties of the human arm, a single degree of freedom test-bed was required, which could be attached to the human arm to measure its response while performing simple tasks, and while being perturbed by forces applied by an electric motor.

3.2.1 Bilateral Lever

The test-bed illustrated in Figure 3.1 contains a one degree of freedom cable transmission mechanism, providing a measure of force input to keep the lever balanced. The high gain of the system ensures we get a strong output signal even when little force is applied. On the point of applied force, strain gauges provide information concerning the magnitude. Positional information is given by a potentiometer located on the axle of the cable mechanism. This information is used to provide closed loop position control. With two such devices it is possible to implement a simple one degree of freedom teleoperated system, where the movement of the second lever is controlled by the operators input to the first.

 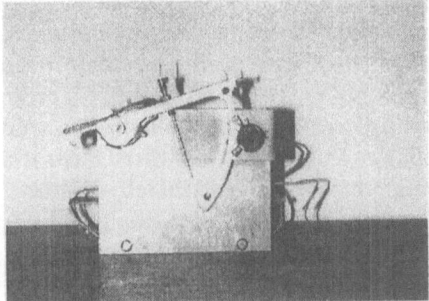

Figure 3.1a, 3.1b: Bilateral lever test-bed

Figure 3.2: Elbow parameterization test-bed

3.2.2   Elbow Parameterisation Test-Bed

With this second test-bed we are able to measure the mechanical properties of human arm   movements. In left side of the Figure 3.2. there is shown an input handle which can be adjusted to the individuals arm and give therefore an exact test situation for all participants in the trial. This database will be achieved to design a part of the control from the AAS. This test-bed design consist a high powered electric motor linked with cable transmission to the axle of the test-bed where the input handle is fixed.

3.3   Cable Approach

In the following we will discuss the merits of cable transmission. Wire cables are used for the transmission which allow a larger distance between axles, a high gear rate in less space and a good power to weight ratio. They also have an inherent compliance which limits the range of the mechanisms stiffness. Cable also allows us to keep the motors out of the mechanism which can take up the most of the weight and place them instead in a position to statically balance the robot. The transfer of power to the joint is via cable and that allows to build them even smaller and lighter. Cable technologies are well developed in areas as such as the aircraft industry and in robotics where there is a similar consideration for achieving a light weight mechanism and thereby assisting with the design of the system controller.

After the joints, the cable is the weakest link in a mechanism. A complex cable route and small pulley diameter results in a high stress on the cable and hence will shorten the work life of the cable. With a special thread design in the joint part it is possible to protect the cable from high friction while moving and thus extend their working life [13, 14, 15]. In order to transmit force via cables without slip, the tension is required to be greater than a geometrically determined minimum. However, to manage this complex problem it is necessary to design several points of adjustment to find a practical tension value.

The advantages and disadvantages of cable driven mechanisms can be summarised as follows.

Advantages:

1.   Large distances between axles are possible.
2.   There is low backlash due to negligible flexibility in the cable.
3.   Complex cable routes are possible.
4.   Wire cables have a low mass per unit weight, therefore low inertia.
5.   Cables contain several strands and minor damage to the cable is easily visible, allowing the cable to be replaced before failure.
6.   Requires very little space and allows a high gear ratio.
7.   A wide selection of cables are commercially available.
8.   Less friction in comparison to belt drives.

Disadvantages:

1. The correct tension can be hard to achieve so as to prevent slip.
2. Wire cables have a short working life.
3. Can be difficult to assemble.
4. Cable failure can be dangerous.

4 Redundant Active Arm Support Mechanism

In this section we will discuss the design of the redundant active arm support mechanism and the demands from the environment as follows.

4.1 Work Space

Human arm movement involves a complex sequence of movements by the shoulder, elbow and forearm. From this point of view you can understand the arm as a kinematic chain. When coupled to a robot the kinematic chain is closed and movement of the combined human and robot must follow a set of kinematic constraints. Furthermore it is

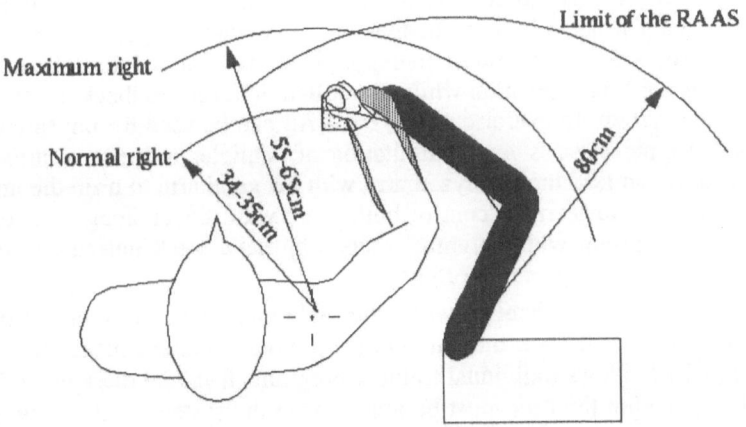

Figure 4.1: Arm reach compered to movement of the AAS

desirable to ensure that relative movement between the components supporting the person's arm and the arm itself are minimised. This must be achieved to ensure there is no damage to the skin of an already vulnerable individual. Therefore one of the most important demands is that the mechanism should allow natural movement through all of the required positions in his arm reach. Considering a workplace where the user is in a sitting position in front of a table, the wheelchair, the table and possibly the AAS are limiting arm movement. The mechanism has to use the available work space as efficiently as possible. Therefore the distances between the joints in the mechanisms are in the same proportions as for a human arm so as to get a parallel movement effect.

One result is shown in Figure 4.1. The complete mechanism, except the arm rest, is separated from the users arm. However, there is no need in this case for individual adjustment by different body mass. The first rotation point is close behind the shoulder so that the mechanisms will not limiting the work space to the right of the user. The AAS reach covers the complete normal work space at the 95th percentile, of men and women [19, 20]. For a wheelchair user this is an important point, because they often lean on a backrest instead of sitting in an active position while working. This implies that their arm reach is even shorter. In a future version we intend to improve the AAS to allow it to perform vertical movements. In this case, the table will be limiting the work space again. A 2 cm clearance between the palm of the patient's hand and the table is necessary to protect the function of the support. The design of the arm rest itself will be customized for each individual and will be considered to be the responsibility of an orthotist.

4.2   Conceptual Mechanism Design

The design of the AAS is intended to assist people with disabilities, either helping with their daily living or assisting with recovering some function. For example, people with muscle dystrophy, motor neuron disease or tremor may use the AAS for daily living tasks, where as people who have hemiplegia after a stroke may require the robot for rehabilitation therapy. In assistance mode the AAS can lead the hand to a desired fixed position while controlled by force feedback or other suitable input on the system. In exercise mode, the AAS can be used for physiotherapy. An example of typical use is for rehabilitation of hemiplegia due to a stroke. In this case, the user can lead the paralyzed arm with his good arm to train the undamaged side of the brain to learn to control both arms with mirror image movements. In contrast to the group with a tremor, users who have weak muscles in their arms need support from the powered orthosis.

They require a mechanism with adjustable compliance or damped movement to perform tasks. However, they might also provide objective measuring of patient progress, which allows individual training programs from the therapist to be set up. In each application the user must be able to stop the system at any time in case of danger or pain. Changing the fundamental arm control allows the AAS to accommodate individuals and needs without requiring mechanical modification to provide a level of therapeutic intervention for an individual with a neuromuscular disease other than adapting the forearm support cradle. Therefore the mechanism and the controller design together has a mechanical independence which can be considered as the combination of mechanism stiffness, damping and inertia. To meet the variety of needs that individuals

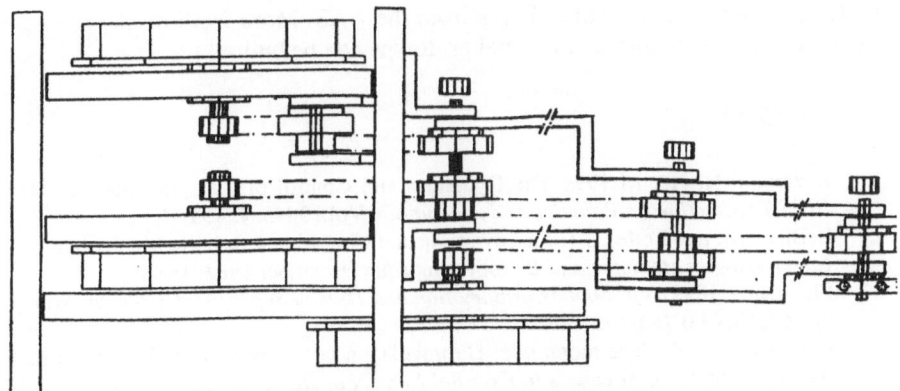

Figure 4.2: Conceptual diagram of an Active Arm Support

have when performing arm movement a range of control methods must be implemented to consider both the task and the user.

The basic idea is to build the arm mechanism in different modules, such as a base containing the motors, the mechanism including the joints and the sensors and lastly the adaptable arm rest. The design of the arm mechanisms is the most complex part. One of the most important design considerations is to keep the weight of the mechanism to a minimum. To achieve this requires either light actuators or a system to transfer power from the actuator to the joint so that the actuators do not contribute to the system inertia.

The AAS should consist of a light joint mechanism and is considered to have three degrees of freedom, which can usually provide for the movement of the arm at first in a horizontal plane, particularly when in a sitting position. The first concept is shown in Figure 4.2. Each joint is powered by an electric motor, which relays force to the operator's arm by cable transmission. The pulleys should be designed primarily from plastic. Both, the principle of cable transmission and pulleys built from plastic reduce the weight of the mechanism and guarantee low system inertia. With strain gauges on each joint we get a force feedback to control the movements and with a potentiometer we get information about the position from each joint.

The first prototype of the AAS will be built for a disabled right side but can easily be reconfigured for use on the left side. The focus has been on simple design that can easily made and assembled in a small workshop. In the second stage the AAS should support arm movements in the vertical direction against gravity.

5   Conclusion

In this paper we have explored areas of application for health care manipulators and possible user groups. We have shown the steps in the design approach to the conceptual mechanism from the AAS. The future work will be measurement from properties of the muscle with the elbow parameterization test-bed to get a database to design one part of the control area from the AAS. More work on the mechanical design is required before a functional prototype can be built.

References

1   Kazerooni H, Her M 1994 The Dynamics and Control of a Haptic Interface Device. *IEEE Transaction on Robotics and Automation* Vol.10 No. 4: 453-464
2   LeBlanc M. and Leifer L 1982 Environmental Control and Robotic Manipulation Aids. *Engineering in Medicine and Biology Magazine* December 1982: 16-22
3   Sheridan T 1992 *Telerobotics, Automation and Human Supervisory Control.* MIT press pg 102, ISBN 0-262-19316-7
4   Reswick J B 1990 The moon over Dubrovnik - a tale of worldwide impact on persons with disabilities. *In Advances in External Control of Human Extremities,* Dubrovnik
5   Romilly D P, Anglin C, Gosine R G, Hershler C and Raschke S U 1994 A Functional task analysis and motion simulation for the development of powered upper-limb orthosis. *IEEE Transactions on Rehabilitation Engineering* Vol 2(3): 119-129
6   Rahman T, Ramanathan R, Seliktar R, Harwin W S 1995 A Simple Technique to Passively Gravity-Balance Articulated Mechanisms. *Journal of Mechanical Design* Vol. 117: 655-658
7   Yardley A, Parrini G, Carus D, Thorpe J 1997 Development of an Upperlimb Orthotic Exercise System. ICORR '97, Bath: 59-62
8   Jonson G R, Buckley M A 1997 Development of a New Motorised Upper Limb Orthotic System. *RESNA '97* June 20-24: 399-401
9   Krebs H I, Hogan N, Aisen M L, Volpe B T 1998 Robot-Aided Neurorehabilitation. *IEEE Transactions on Rehabilitation Engineering* Vol. 6. NO.1: 75-86
10 Lum P S, Lehman S L, Reinkensmeyer D J 1995 The Bimanual Lifting Rehabilitator: An Adaptive Machine for the Therapy of Stroke Patients. *IEEE Transactions on Rehabilitation Engineering* Vol. 3 No.2: 166-173
11 Hogan N 1985 Impedance Control: An Approach to Manipulation. *Journal of Dynamic Systems:* Measurement and Control Vol 107: 1-24
12 Harwin W S, Leiber L O, Austwick G P G, Dislist C 1998 Clinical potential and design of programmable mechanical impedances for orthotic applications. *J Robotica* Vol. 16: 68-76
13 Bennett D J Hollerbach J M Xu Y, and Hunter I W 1992 Time-varying stiffness of human elbow joint during cyclic voluntary movement. *Experimental Brain Research* Vol 88: 433-442
14 Bätge J 1997 Auslegung von Band- und Seilantrieben für die Handhabungstechnik *Konstruktion* 49 H.9: 21-24. Springer-VDI Verlag
15 Feyrer K 1998 Lebensdauer-Vorhersage von laufenden Drahtseilen. *Konstruktion* 50 H. 4: 51-52. Springer-VDI Verlag
16 Pfeifer H 4th edition *Grundlagen der Fördertechnik.* Vieweg&Sohn, Braunschweig
17 Soderstrom T and Stoica P 1989 *System Identification.* Prentice Hall, ISBN 0138812365
18 P.A. Prokopiou, W S Harwin, and S G Tzafestas 1998 Exploiting a human arm model for fast intuitive and time-delay-robust telemanipulation. In: Tzafestas S G (ed): *Advances in Manufacturing: Decision, Control and Information Technology*, Springer, 255-265, In Press
19 Grandjean E 1980 *Fitting the task to the Man.* Taylor & Francis Ltd., London
20 Sanders M S, McCormick E J 1993 *Human Factors in Engineering and Design.* Seventh Edition. McCraw-Hill Book Company, London

35

Enhancement of a Telemanipulator Design with a Human Arm Model

P.A. Prokopiou, W.S. Harwin and S.G. Tzafestas

1   Introduction

Although current telemanipulator designs [1],[2],[3],[4],[5], succeed in performing their tasks, they are often accused of being tiring, slow, unable to cope effectively with large time delays and providing an obscure feeling of the telemanipulated objects [6]. This is an important drawback for applications such as telesurgery and rehabilitative teleoperation [7]. It in part results from the fact that the modelling of the operator arm and the brain decision tactics, due to their complexity and the ambiguity of models previously available, are either neglected or overly simplified.

Although first attempts to model the human muscles date in the 1930s, an increasing amount of models of human and animal limbs continue to appear, the latter easily adaptable to the former. These are either anthropometric [8],[9],[10], or in neural networks or other 'black-box' form [11],[12]. Inputs of the models are mostly the electromyograph (EMG) or the muscle's nervous input (NI, its measurement termed electroneurograph - ENG). Measurement devices for both, e.g. surface or implanted electrodes, are available, although (especially for the latter) in need of improvement. Implanted electrodes have been successfully used on humans for both stimulation and neural signal measurement over a large period (see e.g. [13]).

A literature survey of teleoperation-related research [6], [14], [15] has revealed that the human arm dynamics and the decision making process of the human brain are neglected or excessively simplified. Although it is acknowledged that the system is in reality more complex, most papers adopt for simulations and stability analysis a simple, time invariant, spring-damper-mass model for the human arm (exceptions include [2],[3]). Stability and passivity results are based on a (usually underlying) assumption that "the operator's input τ_{op} is independent of the state of the master-slave system" [1]. Thus current designs ensure that the mechanical part of the system is stable or passive and rely on the operator to use it successfully, like a 'low-tech' passive tool such as a hammer. As argued in [15], this is correct "most of the time" and "the best we can do", since the brain neither *can* nor *has to* be analyzed for teleoperation.

The authors have explored various ways to employ a human arm model in the telemanipulation area. A key feature of our research is the use of the neural input to the human arm as an observable variable, measured or estimated through an inverse model of the arm. We consider an "extended teleoperator system", comprising of the manipulators and the operator arm (reflex and local neural loops included), whose input is the NI rather than the "macroscopic" variables of position and force. Given the current level of research [15], this is a realistic advance. Previous research has regularly thought of the NI (often termed "intentional force") as the "starting point" of the design but only for the stability analysis. Note that the ENG has to be rectified and further processed to become suitable for use within a model (see [13] and references therein). In this chapter, we will employ the acronym NI to refer to this processed signal

Another basic feature of our designs is the use of a module that predicts the operator's arm movement as a response to the NI. This is used to direct the slave along the desired trajectory *ahead* of the master. The slave's response is fed back to the master appropriately delayed, so that the operator feels the results of his actions simultaneously with his actions, exactly as in physical "by hand" operation. This is illustrated in Fig. 1.

It is preferable to make a prediction of the NI / EMG signal, rather than the model *output*, since it has some well studied characteristics (e.g. the three pulses' sequence for fast movements, modelled by varying the amplitude *or* the period, roughly square NI, roughly sinusoidal rectified EMG, 1st order activation [6]). We can predict the output either by simulation ("running the model forward") or analytical calculations. Depending on the desired prediction accuracy and the available hardware, a lead (=T_t /2) of 100 to 500 msec is achievable [15]. Considering typical values [17] we "erase" time delay for many earth to earth applications (e.g. telesurgery through dedicated communication lines or Internet

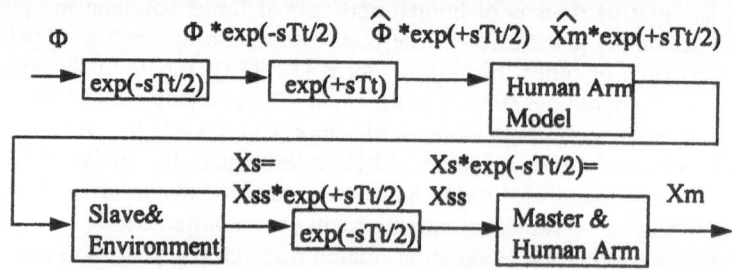

Figure 1: Teleoperation through time delay and predictor. exp(-sT_t/2) denotes delay due to transmission through the communications channel. exp(+sT_t) denotes prediction. Other blocks are free of delay. Φ is the neural input. "Hat" denotes estimate. Xss illustrates the signals' timing. Notice that by a prediction horizon of T_t time delays up to T_t/2 are "erased". In order to equalise the time indexes, artificial delay (signal buffering) might have to be introduced.

link), underwater applications for depth of 400 m, or even single-link earth-to-orbit applications. Employment of hints about the intended human movement or combination with existing techniques could increase those limits.

Three main groups of techniques robustifying against time delay have appeared up to date (refer to [17] for a thorough survey). The first is based on the use of predictive displays for the slave and the remote environment. The second entails the use of wave variables and scattering theory to passivate the communications channel (e.g. [4]), which provide stability but alter the force fed back. A final solution is to use supervisory control [17], which however leads to a non continuous form of teleoperation, i.e. it changes the basic specifications of manipulating as close to physical as possible. The scheme proposed in this chapter, effectively uses a prediction of the master state *only*, which can be much more accurate than predicting the remote environment, and incorporates this in a passive force-feedback scheme. Also, it cancels the computational burden of visually representing the slave future state, needed in predictive displays: the slave-side cameras' image is enough for optical feedback, since its timing is correct (Fig. 1). This is to the best of the authors' knowledge an unexplored approach. In [17] two early efforts predicting the control input along with the rest of the system state are reported for non-telemanipulation tasks, but are judged inadequate for telemanipulation.

One of the teleoperator designs exploiting a human arm model will be examined in this chapter. With this we elucidate how the knowledge of a model can be used to enhance existing architectures, in our case the one proposed by Yokokohji and Yoshikawa [1]. This is a classic architecture, developed through a well-defined general framework of teleoperation, and following the widely accepted design specifications of aiming at transparency and passivity of the system. In this chapter, it is robustified against time delays, by being augmented with the predictor outlined above. The other main aspects of the original design are preserved, and a theoretical analysis following their line of thought leads, assuming perfect prediction, to passivity proofs. It is also shown that the system can in principle be passive under non perfect prediction. Finally, it is argued that the proposed modification is a conservative one, suitable for use when prediction errors are too large to ignore. Through the design process following the classic thinking of [1], it will be shown that the mechanical part of the enhanced system reduces under perfect prediction to an open-loop one. This way the modified architecture of this chapter constitutes a "starting point" for, and elucidates the advantages of, a novel "from scratch" architecture developed by the authors [14],[15]. In this, termed "neuropredictive teleoperation" (NPT), we break up the control loop traditionally closed at the master robot: the master's state is still modified according to feedback from the slave, so that the operator senses the results of his actions, but it is not compared against the predictor output - rather we attempt to close the control loop in the human brain, mimicking normal human movements. The slave is also here directed by the predictor output. The obsolete, as also revealed in this chapter, bilateral constraint that the two robots are simultaneously compliant to each other's state (equal positions and forces in both sides) is lifted, resulting in a succession of only locally controlled modules and thus a much simpler control problem. Our designs were inspired by well established physiological evidence that the brain, rather than controlling the movement on-line, "programs" the arm with an action plan of a complete movement, which is then executed largely in open loop, regulated only by local reflex loops [18]. It acts as a supervisory controller of the "semiautonomous" arm, in the sense of [17].

As a model of the human arm we chose the Stark model [6],[8] (also refer to Appendix) adapted for the arm with parameters from [10], because it is well-established and, being anthropometric, can give valuable insight. The control schemes outlined in the sequence are however valid for *any* human arm model.

This chapter is organized as follows: In the next Section the enhanced Yokokohji and Yoshikawa scheme will be presented. Simulations and a comparison with NPT are reported in Section 3. After the concluding remarks of Section 4, in the Appendix the Stark model of the human arm and the modifications we enforced to it, are outlined.

2   The Enhanced Yokokohji and Yoshikawa scheme

To facilitate comparisons, the derivation and notation in this chapter follow as closely as possible that of the original paper [1]. The general system set-up is actually the same as in [1], except that the slave follows the predicted command. The goal is now modified to achieving $x_m(t) = x_s(t-T_t/2)$ (Ideal response I), or $f_m(t) = f_s(t-T_t/2)$ (Ideal response II), or both simultaneously (Ideal response III), where x denotes position, f force and the subscripts m and s master and slave respectively.

In the sequence, we will develop a scheme which "pushes" the master and slave state (position and force) to the mean values: $x'_{ms}(t) = [x_m(t) + x_s(t-T_t/2)]/2$, $f'_{ms}(t) = [f_m(t) + f_s(t-T_t/2)]/2$. These differ from [1]: the slave states are now delayed. A more important difference which will become apparent in the sequence, is that while the master is directed to $x'_{ms}(t), f'_{ms}(t)$ the slave is pushed to $x'_{ms}(t+T_t/2), f'_{ms}(t+T_t/2)$, i.e. it leads the master arm along the desired trajectory.

As in [1], the design will be carried out for the 1-DOF case. The derivation is easily extended to multiple DOFs if an underlying linearizing and decoupling controller is assumed. The dynamics of master and slave arm are given by the equations:

$$\tau_m + f_m = m_m\ddot{x}_m + b_m\dot{x}_m, \quad \tau_s - f_s = m_s\ddot{x}_s + b_s\dot{x}_s \qquad (1a,b)$$

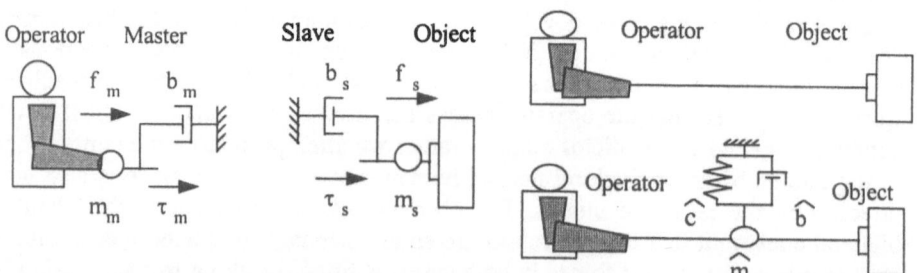

Figure 2: Teleoperator system (left). Manipulation through a rigid bar (right, upper) and an intervening impedance (right,lower). Figures from [1], modified.

where τ denotes actuator driving force and m_m, m_s, b_m, b_s are constant parameters. In contrast to [1], we do not adopt any particular form for the operator arm dynamics, i.e. f_m is not specified. The system setup is shown in Fig. 2.

2.1 Control scheme realizing the ideal responses

In the sequence we will assume that the prediction is perfect, i.e. $x_m = \hat{x}_m$, $f_m = \hat{f}_m$, which simplifies the symbolic representation of the system by decreasing the number of uncontrolled variables (system inputs) by two. Subsequent derivations will of course be "approximately correct" for "almost accurate" prediction. We consider control laws of the form:

$$\tau_m(t) = m_m u_m(t) + b_m \dot{x}_m(t) - k_{mf} \frac{f_s(t - T_t/2) - f_m(t)}{2} - \frac{f_s(t - T_t/2) + f_m(t)}{2} \quad (2a)$$

$$\tau_s(t) = m_s u_s(t) + b_s \dot{x}_s(t) - k_{sf} \frac{f_s(t) - f_m(t + T_t/2)}{2} + \frac{f_s(t) + f_m(t + T_t/2)}{2} \quad (2b)$$

where k_{mf}, $k_{sf} \geq 0$ are force gains, u_m, u_s new inputs. If the robot dynamics' parameters are exactly known, by substituting (2a,b) into (1a,b), we obtain:

$$\ddot{x}_m(t) = u_m(t) - \frac{1}{m_m}(1 + k_{mf}) \frac{f_s(t - T_t/2) - f_m(t)}{2} \quad (3a)$$

$$\ddot{x}_s(t) = u_s(t) - \frac{1}{m_s}(1 + k_{sf}) \frac{f_s(t) - f_m(t + T_t/2)}{2} \Leftrightarrow$$

$$\ddot{x}_s(t - T_t/2) = u_s(t - T_t/2) - \frac{1}{m_s}(1 + k_{sf}) \frac{f_s(t - T_t/2) - f_m(t)}{2} \quad (3b)$$

Adding both sides of (3a,b):

$$\ddot{x}_s(t - T_t/2) + \ddot{x}_m(t) = u_m(t) + u_s(t - T_t/2)$$

$$- \left(\frac{(1 + k_{mf})}{m_m} + \frac{(1 + k_{sf})}{m_s} \right) \frac{f_s(t - T_t/2) - f_m(t)}{2} \quad (4)$$

If $\ddot{x}_s(t - T_t/2) + \ddot{x}_m(t) = u_m(t) + u_s(t - T_t/2)$ can be satisfied, from (4) we obtain $f_s(t - T_t/2) = f_m(t)$, which means that ideal performance II has been achieved despite the time delay. Then, subtracting (3b) from (3a) we get:

$$\ddot{e}'(t) = u_m(t) - u_s(t - T_t/2) \quad (5)$$

where $e'(t) = x_m(t) - x_s(t - T_t/2)$ denotes the position error between the two arms. The symbol is primed to distinguish it from the same error with $T_t = 0$. Eq. (5) shows that the behavior of $\ddot{e}'$ can be specified by the control inputs' difference. If we set $u_m(t) - u_s(t - T_t/2) = -k_1 \dot{e}'(t) - k_2 e'(t)$, we get $\ddot{e}'(t) + k_1 \dot{e}'(t) + k_2 e'(t) = 0$. Thus e' converges asymptotically into zero with appropriate gains k_1 and k_2, and the ideal response III can be realized.

From (4), (5) u_m, u_s are calculated. By substituting them into (2) the resulting control (actuator) inputs for the robots become:

$$\tau_m(t) = m_m\big[\ddot{x}'_{ms}(t) + k_1(\dot{x}'_{ms}(t) - \dot{x}_m(t)) + k_2(x'_{ms}(t) - x_m(t))\big]$$
$$+ b_m\dot{x}_m(t) - k_{mf}\big[f'_{ms}(t) \cdot f_m(t)\big] - f'_{ms}(t) \tag{6a}$$

$$\tau_s(t) = b_s\dot{x}_s(t) - k_{sf}\big[f_s(t) - f'_{ms}(t + T_t/2)\big] + f'_{ms}(t + T_t/2) +$$
$$m_s\big[\ddot{x}'_{ms}(t + T_t/2) + k_1(\dot{x}'_{ms}(t + T_t/2) - \dot{x}_s(t)) + k_2(x'_{ms}(t + T_t/2) - x_s(t))\big] \tag{6b}$$

where $x'_{ms}(t) = \big[x_m(t) + x_s(t - T_t/2)\big]/2$, $f'_{ms}(t) = \big[f_m(t) + f_s(t - T_t/2)\big]/2$. As noted in [1], equations (6a) and (6b) can be interpreted as a combined scheme of a computed torque controller, which makes the manipulators follow the desired trajectory x'_{ms} and a force controller which pushes both f_m and f_s to f'_{ms}. Notice also that the gains $k_{mf}, k_{sf} \geq 0$ do not give any effects so far, since $f_s(t - T_t/2) = f_m(t)$.

The above control law can turn the system unstable if the dynamic parameters' estimation is erroneous. We can enhance it by applying the "intervening impedance" concept of Yokokohji and Yoshikawa: the force tracking is relaxed, by imposing:

$$f_m(t) - f_s(t - T_t/2) = \hat{m}\ddot{x}'_{ms}(t) + \hat{b}\dot{x}'_{ms}(t) + \hat{c}x'_{ms}(t) \tag{7a}$$

$$\ddot{e}'(t) + \dot{e}'(t) + e'(t) = \lambda\big[f_m(t) + f_s(t - T_t/2)\big]/2 \tag{7b}$$

where λ is a constant parameter. This way the operator feels as if manipulating through a "virtual rod", an "intervening impedance", whose constant parameters are denoted by "hat" symbols in (7a). With the control laws of equations (6a) and (6b) the corresponding feeling is that of manipulating through a rigid and weightless bar (Fig 2). The modified control law is:

$$\tau_m(t) = m_m\big[\ddot{x}'_{ms}(t) + k_1(\dot{x}'_{ms}(t) - \dot{x}_m(t)) + k_2(x'_{ms}(t) - x_m(t))\big]$$
$$+ b_m\dot{x}_m(t) - \frac{(1 + k_{mf})}{2}\big[\hat{m}\ddot{x}'_{ms}(t) + \hat{b}\dot{x}'_{ms}(t) + \hat{c}x'_{ms}(t)\big] + \tag{8a}$$
$$+ \frac{\lambda m_m f'_{ms}(t)}{2} - k_{mf}\big[f'_{ms}(t) \cdot f_m(t)\big] - f'_{ms}(t)$$

$$\tau_s(t) =$$
$$+ m_s\big[\ddot{x}'_{ms}(t + T_t/2) + k_1(\dot{x}'_{ms}(t + T_t/2) - \dot{x}_s(t)) + k_2(x'_{ms}(t + T_t/2) - x_s(t))\big]$$
$$b_m\dot{x}_s(t) - \frac{(1 + k_{sf})}{2}\big[\hat{m}\ddot{x}'_{ms}(t + T_t/2) + \hat{b}\dot{x}'_{ms}(t + T_t/2) + \hat{c}x'_{ms}(t + T_t/2)\big] - \tag{8b}$$
$$- \frac{\lambda m_s f'_{ms}(t + T_t/2)}{2} + k_{sf}\big[f'_{ms}(t + T_t/2) - f_s(t)\big] + f'_{ms}(t + T_t/2)$$

In the above control laws, exactly as in [1], if we set $\lambda=0$, the ideal response I is realized. If the intervening impedance is zero and $\lambda \neq 0$, the ideal response II is realized. In all other cases, neither of the two is achieved, but passivity is guaranteed.

Equations (6) and (8) above have some very interesting implications. In the slave side, the master variables that appear on τ_s are predicted ones. If on the master side for the calculation of the mean values we use local variables $x_m(t), f_m(t)$ instead of the predicted and then twice-delayed (due to the transmission) ones $\hat{x}_m(t + T_t - 2*(T_t/2)) = \hat{x}_m(t)$, $\hat{f}_m(t + T_t - 2*(T_t/2)) = \hat{f}_m(t)$, then we obtain a succession of blocks in an open-loop connection. Indeed, all predicted master variables will appear on the slave side, and all measured master variables on the master side. Under "perfect" prediction the mean values on the slave and master side, although calculated through different variables, will be equal. The open-loop feature is a direct result of using the NI and a perfect model of the human arm (refer also to Section 3). In the case of non perfect prediction, one is obliged to use a robustifying term or an online tuning of the arm model to minimize the discrepancy between measured and predicted variables. In both cases the loop is closed. However, with online tuning we can consider the adaptive mechanism part of the human arm model, so that the prediction can still be thought of as "ultimately perfect". Then the passivity analysis of the next subsection, which covers the mechanical part of the system, is valid.

2.2   Derivation of passivity

The passivity of the modified system will be now investigated. For this, we will resort to the electrical analogue and two-port network theory, specifically the impedance matrix (Z) representation. By transforming equations (1a,b) and (8a,b) in the s-domain and combining them, we finally obtain:

$$\begin{bmatrix} V_m \\ V_s \end{bmatrix} = \frac{1}{D_z}\begin{bmatrix} N_{11} & e^{-sT_t/2}N_{12} \\ e^{+sT_t/2}N_{21} & N_{22} \end{bmatrix}\begin{bmatrix} I_m \\ -I_s \end{bmatrix} = \begin{bmatrix} z_{11} & e^{-sT_t/2}z_{12} \\ e^{+sT_t/2}z_{21} & z_{22} \end{bmatrix}\begin{bmatrix} I_m \\ -I_s \end{bmatrix}$$

where voltage V and current I correspond to force and velocity respectively. D_z, N_{xx}, z_{xx} are functions of s and the system parameters, and exactly the same as those reported in [1]. Since the complete calculations are too lengthy to be included, we omit specifying them and will prove the passivity exploiting the results of [1].

The scattering matrix of the original system was shown to be:

$$S = \frac{1}{\gamma}\begin{bmatrix} \alpha & \beta \\ \beta & \alpha \end{bmatrix}$$

where, $\alpha = (s+k_1+k_2/s)(\hat{m}s+\hat{b}+\hat{c}/s) - \lambda$, $\beta = 2(s+k_1+k_2/s) - 0.5(\hat{m}s+\hat{b}+\hat{c}/s)$,

$\gamma = (s+k_1+k_2/s+\lambda/2)(\hat{m}s+\hat{b}+\hat{c}/s+2)$. The system was passive.

The passivity criterion is in fact $\sup_{\omega}\left[\lambda^{1/2}(S*(j\omega)S(j\omega))\right] \leq 1$, where $\lambda^{1/2}(.)$

stands for the square root of the maximum eigenvalue and $*$ for conjugate transpose [4]. The eigenvalues of $S*S$ above equal $\sigma_{1,2} = \left[\left(\|\alpha\|^2 + \|\beta\|^2 \pm (\alpha\bar{\beta} + \bar{\alpha}\beta)\right)\right]/\|\gamma\|^2$, where "overlined" variables are complex conjugates. For the modified architecture of this chapter, the scattering matrix S' is (α,β,γ are the same as above):

$$S' = \frac{1}{\gamma}\begin{bmatrix} \alpha & e^{-sT_t/2}\beta \\ e^{+sT_t/2}\beta & \alpha \end{bmatrix}$$

so that:

$$S'*(j\omega)S'(j\omega) = \frac{1}{\|\gamma\|^2}\begin{bmatrix} \|\alpha\|^2 + \|\beta\|^2 & e^{-j\omega T_t/2}(\alpha\bar{\beta} + \bar{\alpha}\beta) \\ e^{j\omega T_t/2}(\alpha\bar{\beta} + \bar{\alpha}\beta) & \|\alpha\|^2 + \|\beta\|^2 \end{bmatrix}$$

whose eigenvalues turn out to equal $\sigma_{1,2}$ of the original system above. So, the modified system is indeed passive, despite the time delay in the communications channel.

2.3   Imperfect prediction

In the non ideal case, $x_m \neq \hat{x}_m$, $f_m \neq \hat{f}_m$. We considered an augmentation of the general teleoperator description proposed in [1] by incorporating a term relevant to the prediction $\hat{x}_m, \hat{f}_m$ on the master side, so that a robustifying control law can be designed.

We define $\hat{x}_m = x_m + \delta x_m$, $\hat{f}_m = f_m + \delta f_m$. Then δx_m and δf_m could be considered as a new input pair (port), uncorelated to the other inputs, or as a function of x_m, f_m, or as having terms of both types. By resorting once more to the electrical analogues and Laplace-trasforming, a possible representation of the second type is (1-DOF case):

$$\begin{bmatrix} \hat{I}_m \\ \hat{V}_m \end{bmatrix} = e^{+sT_t/2}\begin{bmatrix} F_{11} & F_{12} \\ F_{21} & F_{22} \end{bmatrix}\begin{bmatrix} I_m \\ V_m \end{bmatrix} \tag{9}$$

In an ideal predictor, $F_{11}=F_{22}=1$, $F_{21}=F_{12}=0$. Generally, F_{xx} are functions of s. Since these would include random (noise-like) and nonlinear terms, one can use worst-case upper bounds. Then the overall system can be described as a 2-port.

Often in time delayed teleoperation, the singular values of the scattering matrix incorporate $\exp(j\omega)$ terms (e.g.[4]). In this case, they go to infinity as ω increases, and the system is proven active. Through tedious calculations, the singular values of the scattering matrix of the system resulting from (9) are found to be of the form:

$$\sigma_{1,2} = h_1(j\omega) + h_2(j\omega)\cos(\omega t) + h_3(j\omega)\sin(2\omega t) + h_4(j\omega)\sqrt{h_5(j\omega, \cos(\omega t), \sin(\omega t))}$$

Although a complete symbolic analysis is prohibitively complicated, it can be seen that since trigonometric functions are bounded, suitable choices of the parameters ensuring passivity can in principle be found.

If we consider δI_m and δV_m as a new input, then by employing a 3-port representation and a similar method as in the previous case, it is shown [6] that the singular values are free of exponential, trigonometric or other unbounded functions of $j\omega$, and thus passivity could result for specific design parameters.

3   Simulations and comparison with other schemes

The enhanced scheme was tested through extensive simulation. In representative results plotted in Fig. 3, the slave's lead is clearly shown. The response of the system is very smooth, despite some oscillations at the contact task. In Fig 3d, the enhanced scheme is compared with neuropredictive teleoperation (NPT). Notice that for the same NI and human arm parameters, the double distance is finally covered by NPT, i.e NPT is more efficient. Adjusting the parameters only improves the tracking between master and slave, but *cannot cover this gap*, since it is due to the fact that the two robot's position and force are pushed to their mean values. Most, if not all, other classical teleoperator systems, directly or indirectly as in [2], also follow this bilateral constraint, which is here proven obsolete.

4   Conclusions

In this chapter we described how the inclusion of a model of a human arm, combined with the measurement of its neural input and a predictor, can provide to a previously proposed teleoperator design robustness under time delay. Our trials gave clear indications of the superiority of the NPT scheme over traditional as well as the modified Yokokohji and Yoshikawa architectures. Its fundamental advantages are: the *time-lead of the slave*, the *more efficient, and providing a more natural feeling manipulation*, and the fact that incorporating an operator arm model leads to *more credible stability* results. Finally, its *simplicity* allows less likely to fail local control techniques to be employed.

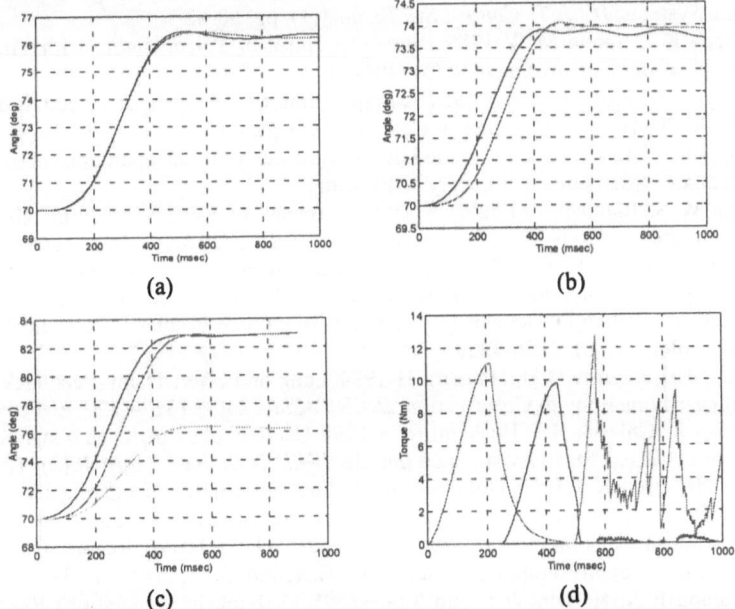

Figure 3: 1-DOF. Same input pulses (sinusoidal 1st agonist and antagonist with height 30) for all plots. <u>Plots (a)-(c):</u> slave (left line) and master (right line) position. (a) Enhanced Yokokohji and Yoshikawa scheme, free space. With the original scheme of [1], master and slave (plotted for clarity only in (d)) closely follow the black line. (b) as in (a), but contact with an object, modelled as a spring with k=40 N/rad contacted at 72 degrees. (c) Free space. Upper lines: NPT, Lower lines: master and slave as in [1]. <u>Plot (d):</u> Operator arm's extensor (left line) and flexor (central bell) force for the contact task of (b). During the movement forces are smooth (contact at around 200 msec), but due to the imperfect inverting algorithm, the clamping phase is oscillatory. In real life, the human force would be constant.

However, a significant advantage for the enhanced Yokokohji and Yoshikawa architecture results from the very fact that it's a conservative modification of current designs. Under large prediction errors, it can provide robustness through directing the master and slave states to their means and, since it relies on the passivity of the mechanical part of the system, it would not confuse the operator.

An experimental implementation of the techniques will provide further evidence for the performance of the proposed architectures. The employment of neural networks and fuzzy logic, which will provide an adaptive model of the human arm and robustifying control terms, is scheduled for the near future.

References

[1] Yokokohji Y, Yoshikawa T 1994 Bilateral Control of Master-Slave Manipulators for Ideal Kinesthetic Coupling--Formulation and Experiment. *IEEE Tr Rob&Aut* 10(5):605-619
[2] Lee S, Lee H S 1993 Modeling, Design and Evaluation of Advanced Teleoperator Control Systems with Short Time Delay. *IEEE Tr Rob&Aut* 9(5): 607-623
[3] Kazerooni H, Tsai T-I, Hollerbach K 1993 A Controller Design Framework for Telerobotic Systems. *IEEE Tr Control Sys Techn* 1(1): pp. 50-62
[4] Anderson R J, Spong M W 1989 Bilateral Control of Teleoperators with Time Delay. *IEEE Trans. on Automatic Control* 34 (5): 494-501
[5] Tzafestas S G, Prokopiou P A 1997 Compensation of Teleoperator Uncertainties with a Sliding Mode Controller. *J. Robotics & Computer-Integrated Manufacturing* 13(1): 9-20
[6] Prokopiou P A 1998 Human arm models and time delays in teleoperation Internal report, tHRIL, Dept. of Cybernetics, University of Reading
[7] Harwin W S, Rahman T, Foulds R 1995 A review of Design issues in rehabilitation robotics with reference to North American research. *IEEE Tr Rehab Engrg*, 31:3-13
[8] Ramos C F, Stark L W 1990 Postural maintenance during fast forward bending: a model simulation experiment determines a "reduced trajectory". *Exper Brain Research*, 82:651-657
[9] Prochazka A, Gillard D, Bennett D J 1997 Positive Force Feedback Control of Muscles, J. Neurophysiology 77(6): 3226-3236
[10] Gossett J H, Clymer B D, Hemami H 1994 Long and Short Delay Feedback on One-Link Nonlinear Forearm with Coactivation, *IEEE Tr SMC* 24(9):1317-1327
[11] Efranian A, Chizeck H J, Hashemi R M 1998 Using evoked EMG as a synthetic force sensor of isometric electrically stimulated muscle. *IEEE Tr Biomed Engrg* 45(2): 188-202
[12] Gollee H, Hunt K J 1997 Nonlinear modelling and control of electrically stimulated muscle: a local model network approach. *Intl. J. Control* 68(6):1259-1288
[13] Haugland M K, Sinkjaer T 1995 Cutaneous Whole Nerve Recordings Used for Correction of footdrop in Hemiplegic Man. *IEEE Tr Rehab Engrg* 3(4): 307-317
[14] Prokopiou P A, Harwin W S and Tzafestas S G (book in preparation) Exploiting A Human Arm Model For Fast, Intuitive And Time-Delays-Robust Telemanipulation, In: Tzafestas S G, Schmidt G (eds) *Progress in System and Robot Analysis and Control Design*, Springer-Verlag London, UK.
[15] Prokopiou P A, Harwin W S, Tzafestas S G 1998 Fast, Intuitive And Time-Delays-Robust Telemanipulator Designs Using A Human Arm Model In: *Proceedings Symposium Intelligent Robotic Systems*, Edinburgh, UK 1998
[16] Zangemeister W H, Lehman S, Stark L W 1981 Simulation of Head Movement Trajectories: Model and Fit to Main Sequence. *Biol Cybernetics* 41:19-32
[17] Sheridan T B 1993 Space Teleoperation through time dealy: Review and Prognosis. *IEEE Tr Rob&Aut* 9(5):592-606
[18] Houk J C, Buckingham J T, Barto A G 1996 Models of the cerebellum and motor learning. *Behavioral and Brain Sciences* 19: 368-383

[19] Hannaford B, Kim W S, Lee S H, Stark L 1986 Neurological Control of Head Movements: Inverse Modeling and Electromyographic Evidence *Mathematical Biosciencies*, 78:159-178

Table 1: Stark Model for the Human Arm - State Equations

Modified Model:

$$\dot{X}_L = \begin{cases} \dfrac{B_h(HTL-Fsl)}{0.25*HTL+Fsl}*\sigma_L+\left(v+\dfrac{H\dot{T}L}{k_2(HTL+1)}\right)*(1-\sigma_L) & \text{if } HTL \geq \text{thres\_HTL} \\ \dfrac{f_{xl}(\theta)-X_L}{\tau_L} & \text{if } HTL < \text{thres\_HTL} \end{cases} \quad (A.1a)$$

$$\dot{X}_R = \begin{cases} \dfrac{B_h(HTR-Fsr)}{0.25*HTR+Fsr}*\sigma_R+\left(v+\dfrac{H\dot{T}R}{k_2(HTR+1)}\right)*(1-\sigma_R) & \text{if } HTR \geq \text{thres\_HTR} \\ \dfrac{f_{xl}(\theta)-X_R}{\tau_R} & \text{if } HTR < \text{thres\_HTR} \end{cases} \quad (A.1b)$$

$$\dot{v} = J_p^{-1}(-B_p v - K_p\theta + F_e + F_{sl} - F_{sr}) \quad (A.2)$$

$$\dot{\theta} = v \quad (A.3)$$

with: $\sigma_L = \left(1+e^{-1.5*\left(\frac{B_h(HTL-Fsl)}{0.25*HTL+Fsl}-4.5*dXlo\right)}\right)^{-1}$, $\sigma_R = \left(1+e^{-1.5*\left(\frac{B_h(HTR-Fsr)}{0.25*HTR+Fsr}-4.5*dXro\right)}\right)^{-1}$ (A.4a, b)

$$Fsl = \max(0, k1(e^{k_2(X_L-\theta)}-1)), \quad Fsr = \max(0, k1(e^{k_2(X_R-\theta)}-1)), \quad (A.5a,b)$$

$$f_{XL}(\theta) = f_{XR}(\theta) = \theta - 0.1(\theta - X_{lo})^2 \quad (A.6)$$

$$H\dot{T}L = \frac{N_L - HTL}{T}, \quad H\dot{T}R = \frac{N_R - HTR}{T} \quad (A.7)$$

The **original Stark model** comprises of eqs. (A.2) , (A.3), (A.5), (A.7) and (A.1a,b)':

$$\frac{1.25H\dot{T}L}{B_h + |\dot{X}_L|}\dot{X}_L = H\dot{T}L - Fsl, \quad \frac{1.25H\dot{T}R}{B_h + |\dot{X}_R|}\dot{X}_R = H\dot{T}R - Fsr \quad (A.1a,b)'$$

where: the subscripts l and r denote left and right muscle, θ, v position and velocity of the arm, X_L, X_R internal model variables, K_p, J_p, B_p, passive parameters of the arm (load for the muscles), $B_h, T, k_1, k_2, X_{lo}, dX_{lo}$, X_{ro}, dX_{ro}, thres_HTL, thres_HTR are constants, N_L, N_R the neural input, HTL, HTR activation levels ("hypothetical tension"), F_{sl}/F_{sr} the left/right muscle's force, F_e an external force. The original papers also use the notation:

$$B_{VL} = B_L(V_L) = \frac{1.25HTL}{B_h + |\dot{X}_L|}, \quad B_{VR} = B_R(V_R) = \frac{1.25HTR}{B_h + |\dot{X}_R|}$$

Appendix: The Stark model of human arm

The Stark model, as outlined in [8],[16],[19], forms a good compromise between complexity and accuracy ([9] being more of both). However, it is difficult to use it for simulations, because of its nonlinear and not well defined mathematical description. Some modifications were considered necessary, which do not alter the

456

model's core, but just expand it for cases the original model was unrealistic. (Fig. 4, Table 1).

The original model collapses when the muscle activation (HTL, HTR) approaches zero. Instead of considering a minimum muscle coactivation as in [8], we produced a simplified form at small activations by ignoring the first term of (A.1)' which decreases faster. The switching between the two operating regions is smoothed by a sigmoid function. Also, in the original model X_L /X_R are not updated when the muscle receives zero input, so that subsequent activations lead to increasingly high forces. This is amended in the modified scheme, resulting in a smooth and realistic arm trajectory.

An inverse of the human arm model is of crucial importance, since it can be used to estimate the neural input when its measurement is obscure, and is needed in simulation. Stark et al. [19] have proposed an iterative inversion method, which however would be not easy to implement on line. We have developed a family of non-iterative algorithms for the modified model. The fastest requires the measurement of the relative activation of the muscles, which is easily achievable with EMG. In general, our algorithms are successful for output trajectories previously generated by the model, but often fail when just a target point is specified. This is because not all combinations of possible arm forces trajectories can be generated by the model. This would not be a problem for an experiment, since the inverse model will use as input a natural trajectory. Partially successful routines that produce a human-like output force were also devised. Because of this problem, for simulations we specifically calculated the 2nd agonist (AG2) pulse so as to stabilise the arm at its current position (when AG2 starts). This is in accord with physiological evidence that it is a 'clamping' pulse [16]. To choose suitable 1st agonist (AG1) and antagonist (ANT) pulses, we used a look-up table (Fig. 5).

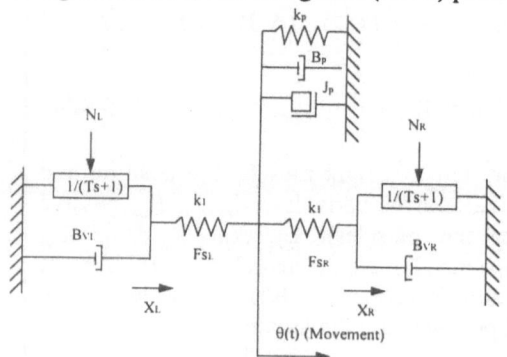

Figure 4: The (unmodified) Stark muscle model.

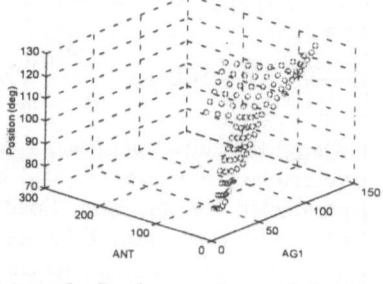

Figure 5: Stark muscle model. Final arm position for various 1st agonist (AG1) and antagonist (ANT) pulses' heights. The starting position was always 70 degrees.

Acknowledgement

We gratefully acknowledge the support of the MobiNet EU-TMR Research Network to the first author.

36

H∞ Robust Control System Design
for a 3-DOF Robot Manipulator

L.A. Gonzalez and L.T. Aguilar

---

1 Introduction

Many robot manipulators show undesirable dynamics present on their performance. A small didactic 5 DOF robot called ROBIX has this problem. ROBIX already has feedback PI control on every servo, but due to the mechanical design, nonlinearities are still present on the system. The main nonlinearities are the dry friction present on the base of the first link, and the backlash at the mechanical transmissions on every link. Effects of these nonlinearities are bad repeteability, steady state errors to step inputs and sensititvity to gravity and torque disturbances. Design of robust controllers for robots has been a subject of interest during the last fifteen years. Adaptive control as well as $\mathcal{H}_\infty$ and μ synthesis or D-K iteration methodology have been used [3], [4] for this purpose. Usually structured uncertainties are considered for open loop robot manipulators and the uncertainty bounds depend on the inertial parameters. In this robot with feedback linealization unstructured additive uncertainties were assumed and this covers the uncertainty inertial parameters. In the $\mathcal{H}_\infty$ and μ synthesis application is important to derive a family of models of the plant where the real plant resides. Here a robot with feedback linearization allows to obtain a family of non-parametric models based on frequency data. From these family of models bounds for unstructured additive uncertainties were determined and a nominal multivariable model was defined. Application of transfer function synthesis [7], [8], [9] from experimental frequency data of the nominal non-parametric model worked out a nominal matrix transfer function of the plant which was used on the design process of the robust controller. In the $\mathcal{H}_\infty$ control application to real physical systems is important to select appropriate uncertainty and performance frequency weights. Some authors suggest ways to choose these weights. For details on this consult [5] [6]. In this paper $\mathcal{H}_\infty$ methodology has been applied to a robot but the results of this experiments has not been experimentally verified

458

2 Experimental Setup

2.1 Description

ROBIX is a five link rigid robot manipulator system which combines industrial level robotics features with educational characteristics (see Figure 2.1). This system has a script language which includes specifics instructions to program robot sequences, but conditional, arithmetic and control program instructions are not available. The robot software was adapted to write algorithms for control laws, trajectory planning and/or jacobians on C language. That was done with a simple change in the configuration of the internal programming clock (8253 INTEL). For specific information on this change see [2].

Figure 2.1: Experimental robot

2.1.1 Mechanical Part

ROBIX has two vertical revolute joints (θ_1 and θ_2), two horizontal revolute joints (θ_3 and θ_4) and a rotational wrist (θ_5). The joint one has a considerable amount of friction and will have to be taken on account at the control design stage. Furthermore the load changes and actuators weight over the wrist and final effector produces gravity torques mainly on joint three.

2.1.2 Sensors

The measurement signals are obtained from potentiometers on the servo motors, which is approximately linear dependent on the arms position. This potentiometer is the main source of noise and also it will be considered in the control design. The driver includes a eight channel eight bit precision serial output ADC converter (ADC0838) which provide adequate computational speed. It includes seven switch-closure inputs that can be used for external sensors.

2.1.3 Actuators

The ROBIX power stage and driver is connected through a parallel port and it has six servomotor outputs. Each servomotor has a 42 oz/in output torque and a maximum speed of 0.22 seg/60°. There is a feedback system within the servo itself which allows it to maintain the position which the controller dictates. Within the servo as each pulse is received its width is "measured" and "compared" to the servo's position. If the servo's position does not match the position implied by the pulse's width, the servomotor is turned on for a few milliseconds to bring the servo closer to the proper position.

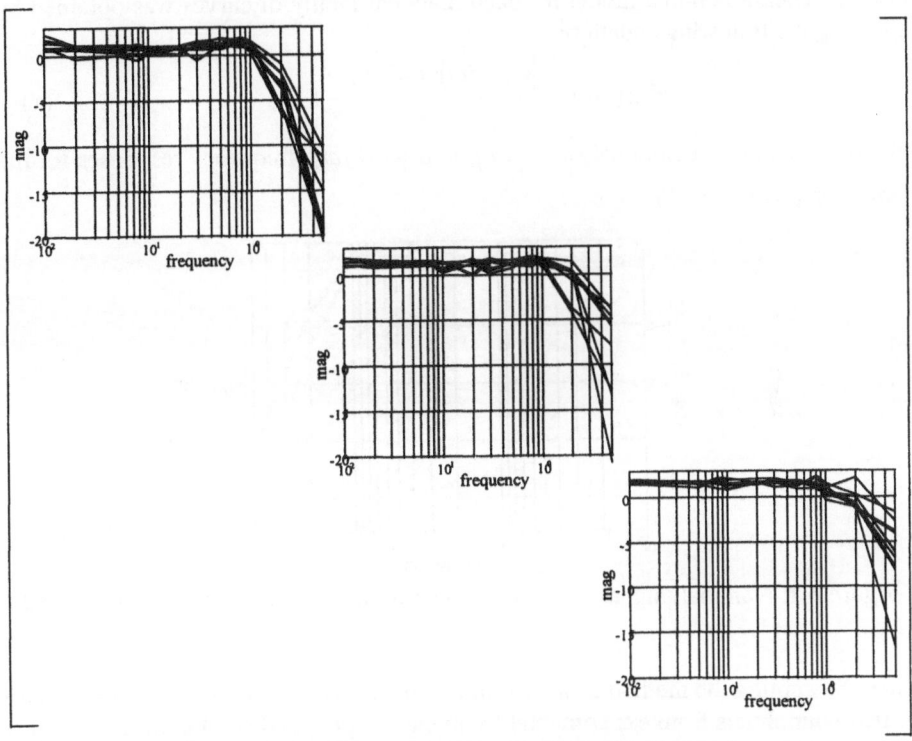

Figure 3.1: Family of models for the main diagonal

3 Non-Parametric Models

Robix consists of five rigid links connected by rotary joints. For this experiment two of the links, the wrist and the end-effector, were not taken into the design process. Therefore we worked ROBIX as a three rigid links robot. Due to the servos ROBIX has feedback linearization that made it stable and gave it some degree of robustness to the nonlinearities present on the system. Under these conditions ROBIX was taken as a linear stable multivariable system with three control inputs and three measured outputs.

Instead of derived an analytical dynamic model of the system we carried out a frequency analysis that give as result a set of non-parametric models. A set of experiments were carried out over the plant by injecting sinusoidal signal on each control input and measurements were taken on all outputs for different frequencies. These experiments were carried out for a total of nine different configurations of the robot so nine 3 x 3 matrices of models were obtained. Figure 3.1 shows the results of these experiments for the main diagonal elements of the matrix.

3.1 Nominal Model

From the above matrix of family of frequencies responses we choose those curves of highest and lowest gains. Lets name $|G_{hi}\ (j\omega)|$ and $|G_{li}(j\omega)|$ such responses for $i=1,2,3$. Then a nominal model for each diagonal family of curves was obtained by appliyng the following equation:

$$|G_{ni}(j\omega)| = \frac{|G_{hi}(j\omega)| + |G_{li}(j\omega)|}{2} \tag{1}$$

Figure 3.2 shows the nominal, $|G_{h1}(j\omega)|$ and $|G_{l1}(j\omega)|$ frequency responses for the family on position (1,1)

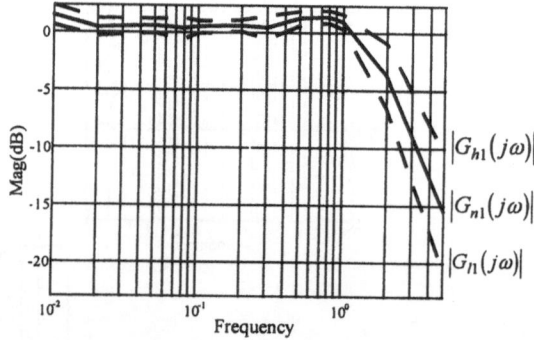

Figure 3.2: Nominal, higher gain and lower gain frequency responses for element (1,1)

Then, by appling the method proposed in [7] to derived a transfer function as a ratio of two polinomials from experimental frequency responses data of a linear system, a transfer function was obtained for each nominal non-parametric model. These transfer functions are shown below:

$$G_{11}(s) = \frac{5.64s^2 + 5.56s + 4.2}{s^4 + 5.46s^3 + 10s^2 + 8.06s + 3.9}$$

$$G_{22}(s) = \frac{5.89s^2 + 5.2s + 0.706}{s^4 + 5.13s^3 + 8.65s^2 + 5.08s + 0.606} \tag{2}$$

$$G_{33}(s) = \frac{11.96s^2 - 2.41s + 41.74}{s^4 + 6.4s^3 + 14.33s^2 + 23.86s + 34.45}$$

3.2 Additive Unstructured Uncertainty

To account for unstructured uncertainties, we consider the family of models where the real system resides. Here, the additive uncertainties were taken to be independent and stable. To estimate the additive model perturbations of the plant, an estimate of the additive perturbation for each element was carried out. We employed Nyquist frequency responses of the nominal and highest gain of each element of the matrix. The difference of gain between these two curves on every frequency resulted in the magnitude of an upper bound for the size of the corresponding additive uncertainty Figure 3.3 shows this procedure for the curves on position (1,1).

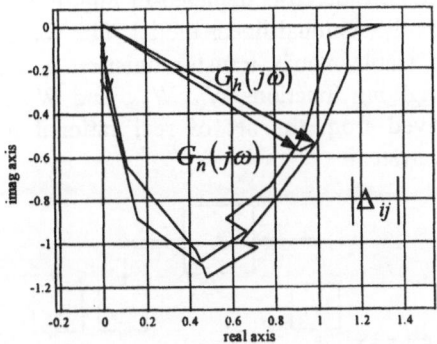

Figure 3.3: Graphics to determine an approximation of $\left| \Delta_{ij} \right|$

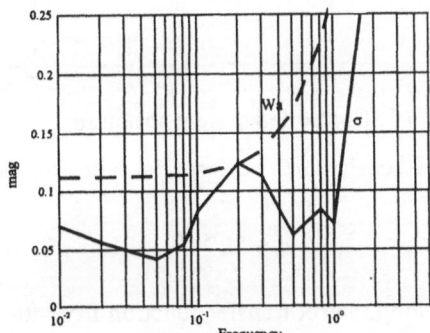

Figure 3.4: Maximum singular value of Δ

The results of the above procedure showed that the gain of the additive uncertainties for the elements outside the main diagonal can be neglected. The corresponding frequency responses of the main diagonal resulted to be the singular values frequency responses of the system. Therefore, the one with the highest gain was chosen as the corresponding upper bound for the additive uncertainties of the system. This singular value is shown in Figure 3.4.

So the set of plant models has the following form:

$$\hat{G}_{ij} := \{ G_{nij} + \Delta_a W_a : \left\| \Delta_a \right\|_\infty \leq 1 \}$$

in which the real plant resides.

4 Control Objectives

We looked for design a robust stable system to additive uncertainties present on the plant. Also we wanted robust performance of the controlled system to reject the low frequency disturbance due to gravity, high frequency noise presents in the position sensors of the servos and good tracking of the reference signals.

The feedback structure considered is depicted in Figure 4.1 which was used in the design procedure. G is the nominal linear model. The weighting functions W_1 and W_2 and the bounded stable matrix transfer function Δ are used to model plant uncertainties. The weighting functions W_d, W_{pe}, and W_n define the performance objectives. K is derived from the set of real rational matrix functions which internally stabilize the system.

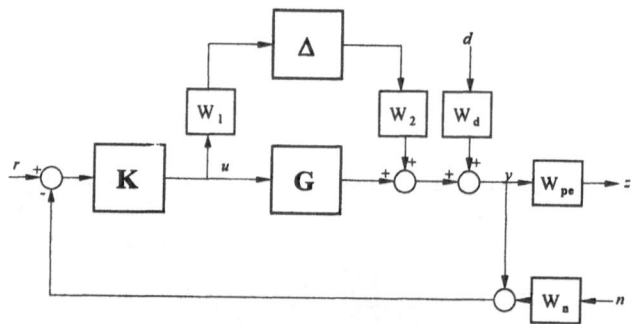

Figure 4.1: Controlled system structure

Objective of the controlled design is to perform the following optimization

$$\left\| T_{zw} \right\|_\infty < \gamma$$

where T_{zw} is the closed loop matrix transfer function from the external inputs w to objective outputs z and $\gamma > 0$ is a positive real number.

$Wa = W_1 W_2$ is the model of the additive uncertainties of the unknown dynamics of the mechanical structure. Therefore, this weighting function admits for this extra dynamics to take effect in the controller design process. W_a is chosen as 3x3 diagonal matrix transfer function which maximum singular value frequency response covers the additive uncertainty shown in figure 3.4

W_d was chosen as 3x3 diagonal transfer function matrix to model the low frequency disturbances due to gravity and other torque disturbances. The cut-off frequencies were 0.012, 0.056 and .12 rad/s for the links 1,2, 3 respectively.

W_e was chosen as well as a 3x3 diagonal transfer function matrix, such that at low frequency we have small errors in tracking the reference signals. The crossover frequencies were set to .1, .4 and 1.5 rad/s respectively.

Finally, W_n was chosen as a 3x3 diagonal matrix with high pass filters on every element because the high noise introduced into the system by the position sensors of the servos.

A robust stability test for additive uncertainties and robust performance for any plant G that belongs to the family of experimental plants is equivalent to the following inequalities be satisfied:

$$\|W_a KS\|_\infty < 1$$
$$\|W_e SW_d\|_\infty < 1$$

(5)

where K is the controller and $S = (I + GK)^{-1}$ is the corresponding sensitivity transfer function.

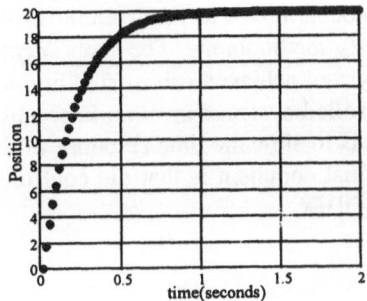

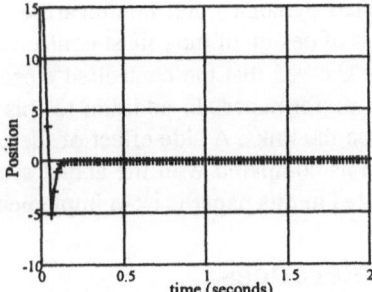

Figure 5.1: Time response to a step input   Figure 5.2: Disturbance responses

5 Experimental Results

With these weighting functions an augmented plant was formed and was used on the design procedure. A 29th-order controller K was obtained which achieved a γ of 0.12.

Tests for robust internal stability and performance using Eq.(5) were satisfied by the controlled system that guaranteed rejection of disturbances and tracking of the reference inputs within the performance specifications for every plant that belongs to the experimental models.

Figure 5.1 shows the time responses of the outputs on joints one and three for two different configurations of the robot to a 20° step on the reference input of joint one. No change was noticed on the third joint and a 0.08° and 0.4° steady-state error exists on the first joint for both configurations respectively. Similar results were derived for the other joints. Thus, the controlled system besides good tracking shows high attenuation of the interactions between the links.

A test of the system's robustness to disturbances was performed a test by adding a disturbance of 15° at the output of the third link at.where gravity's effect is present as a torque disturbance. We assumed the gravity as the main disturbance on the system. Figure 5.2 shows the effect of this disturbance on the position of the third link for the nominal and G_h configurations. A 0.06° deviation occurred with a settling time of 3.5 sec. for the nominal model. For the G_h configuration a faster time response was obtained due to the longer bandwidth of G_h and settled down to a mere 0.03° deviation. Therefore, this deviations are within the performance specifications of the W_d weighting function.

6 Conclusions

A robot with feedback linearization on each of its links was modeled from experimental frequency data and a family of non-parametric models were derived for different configurations of the robot. Based on this family of multivariable models a nominal model for the plant was determined and non-parametric model perturbations were obtained to account for additive unstructured uncertainties from neglected dynamics and nonlinearities. These models prove to be essential in the process of design of the robust controller by the $\mathcal{H}_\infty$ methodology. The experimental results showed that the controlled closed loop system achieved robust stability and robust performance. In addition to this the controlled system shows no-interaction between the links. A side effect of the controller is to slow the time response of the system as compared with the actual system. A final comment is that the controller presented in this paper is be n implemented on ROBIX.

7 References

1. Zhou K, Doyle J C 1998 *Essentials of Robust Control*. Prentice Hall, Upper Saddle River,N.J.
2. Gonzalez L A, Ivankovic B et al 1997 Desarrollo de una base experimental de control utilizando un robot de arquitectura semi-rigida. *Congreso Nacional de Robótica*, Torreón, Coah. México, pp. 43-48
3. Pannu H, Kazerooi G, Packard A et al 1996 μ-synthesis control for walking robot, *IEEE Control Systems*, 16: 20-25.
4. Spong W 1992 On the robust control of robot manipulators. *IEEE, Trans. Aut. Contr.*, 37:69-77
5. Postlethwaite I, Tsai M C, Gu D W 1990 Weighting function selection in $\mathcal{H}_\infty$ design. *IFAC 11th Triennal World Congress*, Tallin, Estonia, pp.127-132
6. Ali H, Edmunds J M 1997 C hoise of $\mathcal{H}_\infty$ weighting functions to mould signal responses. *Control System Centre Report* No 809 UMIST, U. K.
7. Sanathanan K, Koerner J 1963 Transfer function synthesis as a ratio of two complex polynomials. *IEEE, Trans. Aut. Contr.* 5:56-58
8. Bayard S 1997 Multivariable state space identification in the delta and shift operators. *NASA TECH BRIEF* 21:19-30
9. Levy C 1959 Complex curve fitting. *IRE Trans. Automat. Contr.* 4:37-43
10. Doyle J C, Glover K et al 1989 State space solution to standard $\mathcal{H}_2$ and $\mathcal{H}_\infty$ control problems. *IEEE Trans. Automat. Contr*, 34:831-847

37

Response Optimization of a Nonlinear Controlled Flexible Robot Arm Carrying a Variable Mass

A.G. Petridis, and A.E. Kanarachos

1 Introduction

The dynamic behavior of one or more links robot systems has been studied extensively for rigid link models (Refs. 1,2). However, deformations of arm links are not negligible as the speed of motion and required accuracy increase. Besides, one of the major limitations of current industrial robots is their low payload/weight ratio. Excessive arm weight, limits the maximum allowable motion speed of a robot, increases the total energy consumption and also increases the power, the size and the cost of the actuators. In recent years, programmable multi-functional manipulators, or industrial robots have become increasingly important in automation. Advanced control techniques can improve robot performance, especially motion characteristics. In the meantime, the rapid development of computing ability leads to feasible and low cost new robot control strategies. The reduction of the robot weight is the essential way to reduce the cost of industrial manipulators while improving high speed performance. In exchange for light-weight, however, one must accept an increase in system flexibility along with the associated difficulty in accurately controlling. Moreover, increased manipulator performance requires a controller which allows both nonlinear link dynamics and link flexibility.

Robot control may be categorized as collocated (actuator and sensor at the same location) and noncollocated (sensor distant to the actuator). A noncollocated system can provide more precise tracking, improved disturbance rejection and increased arm tip position bandwidth (Ref. 3) (hence providing the above performance improvements). However, the stability in noncollocated control is conditional and is not as robust as it is in collocated control (Refs. 4-6) due largely to its non-minimum phase nature (Refs. 3,7). For implementing such a control, many authors provide an accurate plant model so that confidence in analytical stability margins to be assured (Refs. 7,8) . A series of feedback control strategies for a single-link manipulator have been presented   by Cannon and Schmitz (Ref. 4). Sakawa, Matsuno and

Fukushima (Ref. 9) introduced another closed-loop control algorithm based on a detailed analytical model of the link and the sensor system. Siciliano, Calise and Jonnalagadda (Ref. 10) have also reported interesting studies in optimal control of flexible manipulators. Wang, Hsia and Wiederrich (Ref. 11) presented a general method for open-loop control of a flexible robot that is based on the closed-loop simulation.

Among the various control techniques found in the literature, none has fully accounted to the combination of the dynamic system non-linearity (hence inverse dynamic problem cannot be solved), the carried mass variation and the flexible arm initial conditions (nonzero velocity and acceleration). In this study, the response of a nonlinear and noncollocated controlled flexible robot arm is optimized for a considerable range of carried concentrated masses and under non defined initial kinetic conditions.

2 Dynamic Model

The robot system consists of a flexible aluminum arm, with ring cross-section, carrying a concentrated mass at its end and a rotating drive with hysteresis mounted at the opposite arm end. The relevant arm's data are: Outside Diameter $D_O = 0.0269$m, Inside Diameter $D_I = 0.0216$ m, Length $L_0 = 1.5$ m, Density $\rho = 7850$ Kgr/m^3, Young's Modulus $E' = 20.6 \cdot 10^{12}$ N/m^2 . The cross-section area, the cross-sectional moments of inertia and the distributed arm mass are calculated as follows.

$$A = \pi \cdot \frac{D_O^2 - D_I^2}{4} = 201.887 \cdot 10^{-6} \, m^2 \tag{1.a}$$

$$I_y = I_z = \pi \cdot \frac{D_O^4 - D_I^4}{64} = 15.0175 \cdot 10^{-9} \, m^4 \tag{1.b}$$

$$M_0 = \rho \cdot A \cdot L_0 = 2.37721 \, Kg \tag{1.c}$$

Three different concentrated mass values are taken into account $M_L = 2.0, \ 3.5, \ 5.0$ Kg. The robot arm is fragmented to four (4) finite elements (bent beams) with equivalent length. This number of elements is adequate, since its augmentation (fragmentation to five or more beams) appears to have negligible influence to the dynamic system response for the current region of operation (rotation velocities and mounted masses). The flexible arm system has nine (9) degrees of freedom, i.e. five (5) angles (A_1, A_3, A_5, A_7, A_9) between the horizontal and the tangential to the arm's center line at nodes 0 to 4 and four (4) angles (A_2, A_4, A_6, A_8) defined by the horizontal and the straight lines connecting node 0 to each one of nodes 1 to 4.

Once the (9×9) inertia and stiffness matrices (**M** and **K**) are formatted, the nine (9) eigenvectors and eigenfrequencies of the bent arm-tip system can be calculated by setting the determinant of the matrix $\mathbf{M}^{-1} \cdot \mathbf{K} - \lambda \cdot \mathbf{I}$ equal to zero. The amount of the carried mass is taken into account during the formation of the inertia matrix and thus the eigenfrequencies values depend also on the magnitude of the concentrated law. The following tables the values of all the eigenfrequencies of the modeled arm-tip system, as a function of the reduced magnitude of the mounted mass ($M_L \, / \, M_0 = 0.00,..., 2.50$).

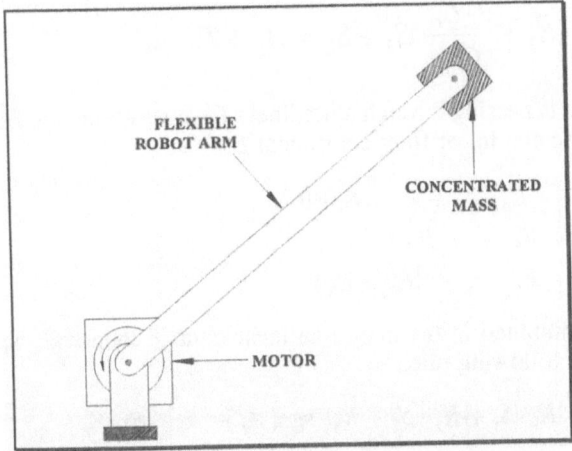

Fig. 1 Flexible Robot Arm-Tip System

Table 1. Eigenfrequencies of the Flexible Arm-Tip Dynamic System

M_L/M_0	1	2	3	4	5	6	7	8	9
0.00	108	222	986	2082	3870	6150	9592	14297	18874
0.25	91	199	836	1839	3517	5563	8762	13122	15886
0.50	86	197	812	1810	3482	5520	8709	13044	15829
0.75	81	195	802	1800	3469	5504	8690	13017	15810
1.00	78	194	796	1794	3463	5496	8661	13003	15800
1.25	77	193	793	1790	3458	5491	8675	12995	15794
1.50	77	188	791	1788	3456	5487	8671	12989	15790
1.75	76	186	788	1786	3454	5485	8668	12985	15788
2.00	75	185	787	1785	3452	5483	8666	12982	15786
2.25	74	184	786	1784	3451	5482	8664	12979	15784
2.50	73	183	785	1783	3450	5480	8663	12977	15873

The value of the lowest eigenperiod, which is taken into account for the time integration step selection, is calculated as follows.

$$T_9 = \frac{2 \cdot \pi}{\Omega_9} = 0.3390 \sim 0.3958 \text{ ms} \tag{2}$$

3 Control Law

Let A_R be the feedback signal, corresponding to the slope at the arm end, where the tip is mounted and A_C the command signal. The difference between the actual and the desired angular positions is by definition the error signal.

$$E_1 = A_C - A_R \tag{3}$$

The first derivative of the feedback signal is the input for the second order linear filter. The filter output signal S_2 is given by the following differential equation.

$$\frac{1}{\Omega_0^2} \cdot \ddot{S}_2 + \frac{2 \cdot \zeta_0}{\Omega_0} \cdot \dot{S}_2 + S_2 = \dot{A}_R + T_2 \cdot \ddot{A}_R \qquad (4)$$

The error signal is passing through a nonlinear filter resulting in the output signal S_1, depending on the non linear filter coefficient E_0.

$$S_1 = \begin{cases} -E_0 & , & -E_0 < E_1 < 0 \\ E_0 & , & 0 \le E_1 < E_0 \\ E_1 & , & |E_1| > E_0 \end{cases} \qquad (5)$$

S_1 and S_2 are combined in the main nonlinear control loop and yield the E_2 signal according to the following rule.

$$\begin{aligned} E_2 &= K_1 \cdot S_1 + K_2 \cdot S_2 + K_3 \cdot \mathrm{sgn}(A_C - A_P) \cdot S_1 \cdot S_2 \\ &+ K_4 \cdot \mathrm{sgn}(S_1) \cdot S_1^2 + K_5 \cdot \mathrm{sgn}(S_2) \cdot S_2^2 + K_6 \cdot S_1^2 \cdot S_2 \end{aligned} \qquad (6)$$

The action of the dead zone saturation filter on the above provides the motor control signal E_3 .

$$E_3 = \begin{cases} -E_{max} & , & E_2 < -E_{max} \\ 0 & , & |E_2| \le E_{max} \\ E_{max} & , & E_2 > E_{max} \end{cases} \qquad (7)$$

The motor hysteresis property is modeled as shown by the differential equation here below which defines the motor output velocity E_4.

$$T_N \cdot \dot{E}_4 + E_4 = E_3 \qquad (8)$$

The feedback signal is realized alternatively setting A_R equal either to A_8 (case A), or to A_9 (case B). Case A is easier to use in practice, due to the possibility of calculating A_8 indirectly, using common displacement sensors, which can measure the vertical D_Y, the horizontal D_X or the diagonal displacement D_{XY} of the arm end.

$$A_R = \sin^{-1}(D_Y / L_0) \qquad (10.a)$$

$$A_R = \cos^{-1}\left(\frac{L + D_X}{L_0}\right) \qquad (10.b)$$

$$A_R = \tan^{-1}(D_{XY} / L_0) \qquad (10.c)$$

4 Problem Formation

The decision variables of the response optimization problem are the six multipliers of the non linear loop, the coefficient of the error nonlinear filter and the three coefficients of the linear second order filter to which the first time derivative of the feedback is inserted. The movement scenario includes four different control commands which are activated at consecutive times.

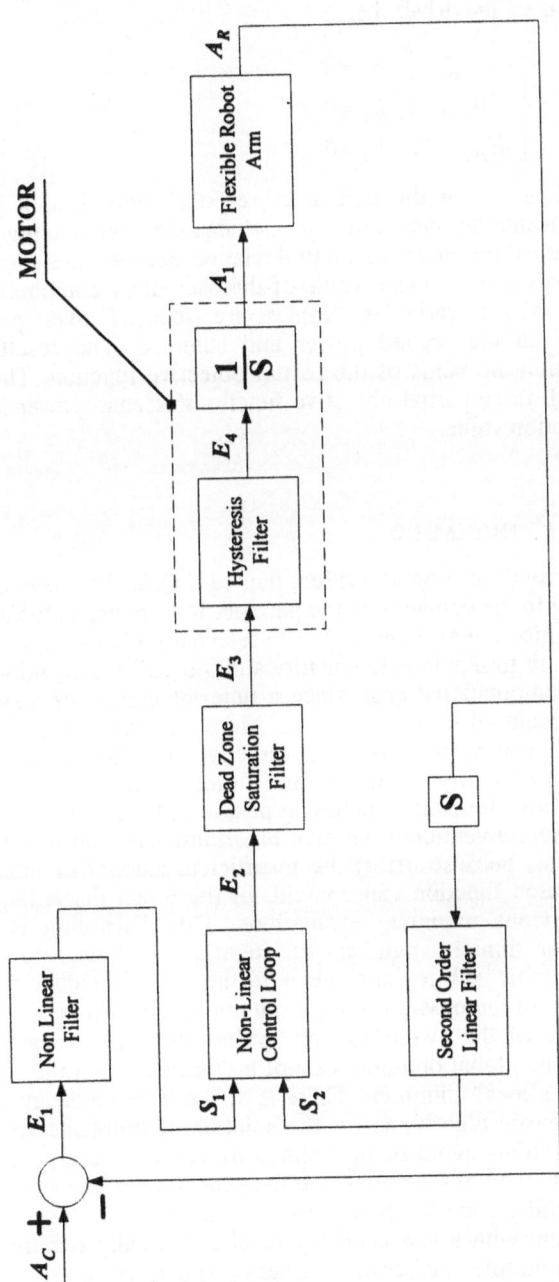

Fig. 2 Signal Flowchart

First, the nominal response curve is defined as the response of an ideally rigid system with no hysteresis on the driver, i.e.

$$\frac{d\,A_{0}}{d\,t} = \frac{d\,A_{1}}{d\,t} = \begin{cases} -E_{max} & , \quad E_{1} < 0 \\ 0 & , \quad E_{1} = 0 \\ E_{max} & , \quad E_{1} > 0 \end{cases} \tag{11}$$

The partial objective function of the parameter response optimization problem, related to a specific concentrated mass amount, includes the overshooting between two consecutive commands, the mean absolute deviation between the nominal and the actual response curves and the mean value of the oscillatory component of the system response. The above variables' values are reduced over predefined reasonable ones, raised to the second power and summed. The result of this summation is by definition the value of the partial objective function. The square root of mean value of all three partial objective functions' second powers is set as the global objective function value.

5 Optimization Procedure

The proposed parameter optimization algorithm, named S.Q.A, has been designed and developed according to the demands of the parametric response optimization of linear or nonlinear dynamic systems (Ref. 13). The relevant problems have singular objective function, difficult to manipulate equality and inequality constraints, many local optima and high computational cost, since a time integration of a system of differential equations is required for each analysis. Moreover, the derivatives of the objective function and constraints are not directly available, due to high computational cost. The peculiarity of the optimized function is intensified when a sequence of pseudo-objective functions, including penalties, is optimized instead of the original function. The conventional iterative algorithms are inefficient for the above mentioned problems because of: a) the insufficient amount of information that a typical approximation function can contain, b) the direct discarding of the available data form previous minimum estimations. This discarding is usually performed after one or limited number of iterations, leading to aimless reexaminations of acceptable sub-regions, even if the available data could be sufficient for the rejection of the possibility for a minimum existence in the specific areas, c) the local nature of the evaluated data (value, gradient vector, Hessian matrix, etc.). Therefore, the global optimum cannot be located, when the algorithm has been trapped around a local minimum. There is also a wide tendency of using semi-stochastic and stochastic algorithms for the solution of problems with many local optima. These algorithms avoid being trapped by using a cloud of analyzed points and a respectable amount of randomness at the generation of the next point or cloud of points. Meanwhile, semi-stochastic and stochastic algorithms converge slowly, locate the optimum with a low accuracy level and usually require a great number of iterations and multiple number of analyses to be performed.

For the solution of an optimization problem, a surrogate problem is usually formulated and solved. The surrogate problem has to: a) approximate the original one with a sufficient accuracy, at least at the critical sub-regions of the feasible domain b) have a simple (typical) mathematical form c) have optima that correspond with the original ones, and these only, at a sufficient accuracy level.

When the objective function is differentiable, the surrogate problem is usually modeled using first or higher order Taylor approximations. On the other hand, at some mathematical and technical problems the global minimum of non-classical functions have to be explored. Such problems appear in the approximation theory, nonlinear dynamic systems control, variable structure control systems applications, control of discrete events systems, cost optimization, etc. The results presented by numerous authors encourage the exploitation of the second order derivatives for piece-wise differentiable functions, in accordance with the evolution of normal functions numerical optimization methods, where Newton type methods tend to displace Gradient type methods.

The exploitation of information that is provided from previous iterations with simultaneous use of surrogate functions has considerable advantages, such as **a)** more accurate objective function approximation, due to higher order provided data **b)** the capability of modeling the original function through many simple local approximations, with diversified starting points **c)** the appropriate formation of the surrogate problem so as to be easily solvable by a predetermined classical optimization algorithm, through indirect approximate calculation of the necessary data.

In each S.Q.A iteration, an internal iterative optimization procedure is performed. The consequent internal optimization result is a surrogate function minimum that approximates a minimum of the original objective function. The internal procedure consists of direction and step size selection loops, that minimize the surrogate function, according to a typical Newton methodology.

The surrogate function is defined through a series of smoothed quadratic approximations of the n variables objective function through the definition of the relevant $N = (n+1) \cdot (n+2)/2$ approximation coefficients. The classical quadratic approximation evaluates a considerable amount of N information. In order to avoid the entire existence of objective function local formations and to augment the available information exploitation, the S.Q.A technique is adopted, which permits the exploitation of as many and whichever available zeroth order data. This wider capability is not succeeded thanks to the S.Q.A formation, but the procedure of data exploitation.

When the objective function has many different local optima, discontinuities and/or non defined or discontinuous derivatives, no single quadratic function can be found with adequate global approximation accuracy. Besides, the required global accuracy level can be succeeded by a set of quadratic functions with local influence. The surrogate function is defined for any feasible point via a smoothed quadratic approximation starting from the specific point. After starting point location, the stored information are partially rearranged according to their corresponding reduced Euclidean distances. This way, M of the K available points are selected ($M \geq N$) around the starting point. After the quadratic equation is asked to be fulfilled for each member of the set of selected information an $M \times N$ system of equations is provided. This over-defined system cannot be generally solved when $M > N$. Therefore, an optimization sub-problem of minimizing the second power of the norm of the corresponding error vector, is solved. The current solution defines the local S.Q.A and the specific surrogate function's data (value and first and second order derivatives). The above surrogate function is well defined. In other words, each point of the explored area corresponds to one and only one surrogate function value. This correspondence is changed between two or more sequential external iterations due to the participation of the last analyzed points. This change is imperative for the internal optimization results differentiation.

The surrogate function formation and the S.Q.A optimization methodology activation presupposes the availability of efficient number of information, that result from the analysis of a minimum number of preliminary points analyses. The corresponding feasible combinations of the decision variables should be generated, so as to facilitate the subsequent essential optimization procedure. From the examination of S.Q.A algorithm performance arise that it is expedient to augment the number of preliminary points over the minimum one (N). Two alternative methods of preliminary generations have been tested, the random and the evolutionary ones. a) *Random Generation*. Each time one point is generated with equal probability for each point located in the n-dimensional orthogonal parallelepiped feasible domain. b) *Evolutionary Generation*. A group of points is produced, corresponding to the offspring created by means of recombination and mutation of a multimembered evolution strategy (ES).

The internal optimization initialization rule affects the internal minimization results, the available data distribution and consequently influences the S.Q.A algorithm performance. Two alternative rules have been applied suitable for different classes of problems. a) *Best Point Initialization*. From the set of the available points, the one that results the best (minimum) objective function value is selected. This rule is preferable for problems without many local optima or singular local formations. b) *Random Initialization*. The starting point of the internal optimization is randomly located in the feasible domain. This way the S.Q.A algorithm is never trapped in a local optimum neighborhood. In opposite, all the optima are gradually approached by different iterations results. This rule is preferable when the location of the global optimum sub-region has not been confirmed. In order to combine the advantages of the two initialization rules, the first one can be applied in the primary stage of the optimization procedure and afterwards the second one, or both of them can be cyclically activated.

6 Results and Conclusions

The optimization results are presented in a tabular form both for case A and case B feedback signals.

	Decision Variable	Case A	Case B
Non Linear Filter	E_0	0.154442E+00	0.135221E+00
Second Order Linear Filter	T_2	0.113001E+01	0.687168E+00
	Ω_0	0.347423E+03	0.165779E+03
	ζ_0	0.142476E-01	0.878113E-01
Non Linear Control Loop	K_1	0.941692E+01	0.127641E+02
	K_2	-0.142346E+00	-0.136431E+00
	K_3	-0.799969E+00	-0.756937E+00
	K_4	0.039906E+00	0.137924E+00
	K_5	-0.258474E-03	-0.290870E-03
	K_6	0.484891E-03	-0.385420E-03
Partial Objective Function	$M_L=2.0$	0.638716E+00	0.656852E+00
	$M_L=3.5$	0.635552E+00	0.626061E+00
	$M_L=5.0$	0.730524E+00	0.740823E+00
Total Objective Function	$O_{1,2,3}$	0.669714E+00	0.676320E+00

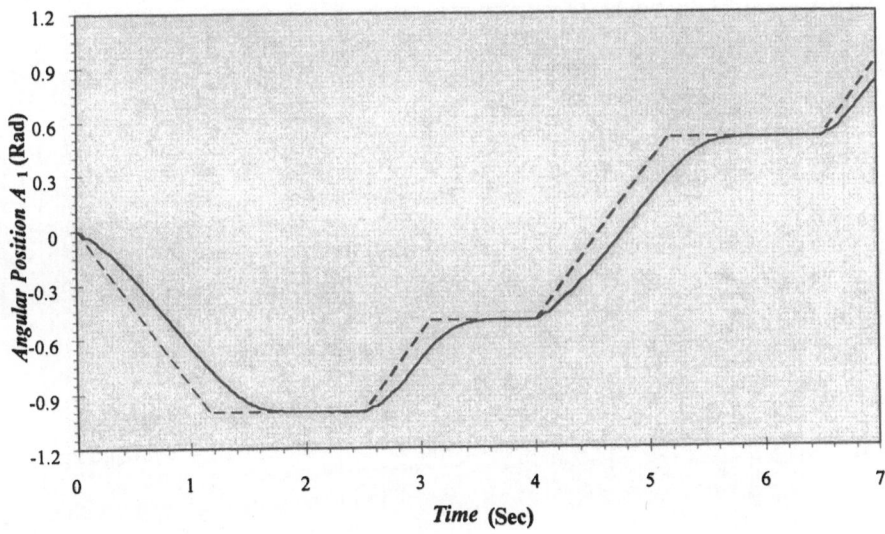

Fig. 3 Angular Displacement of Motor Rotor

474

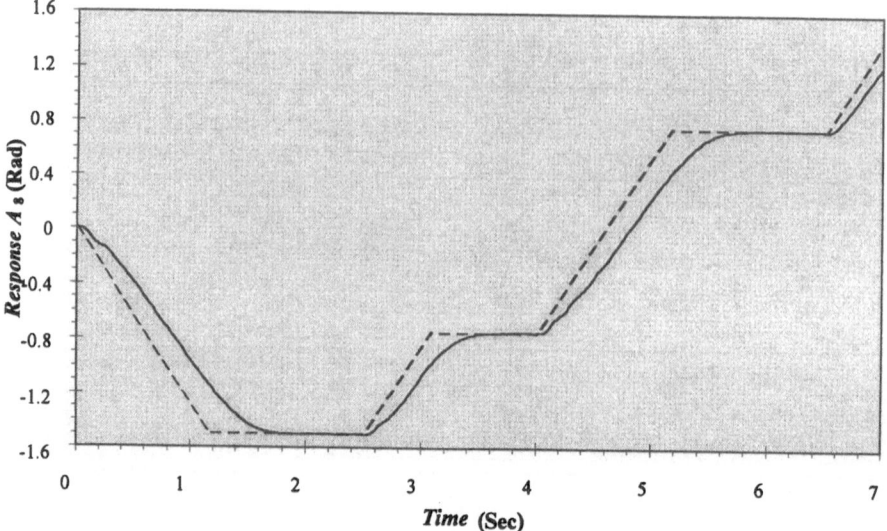

Fig 4 Virtual Rigid Angle A_8

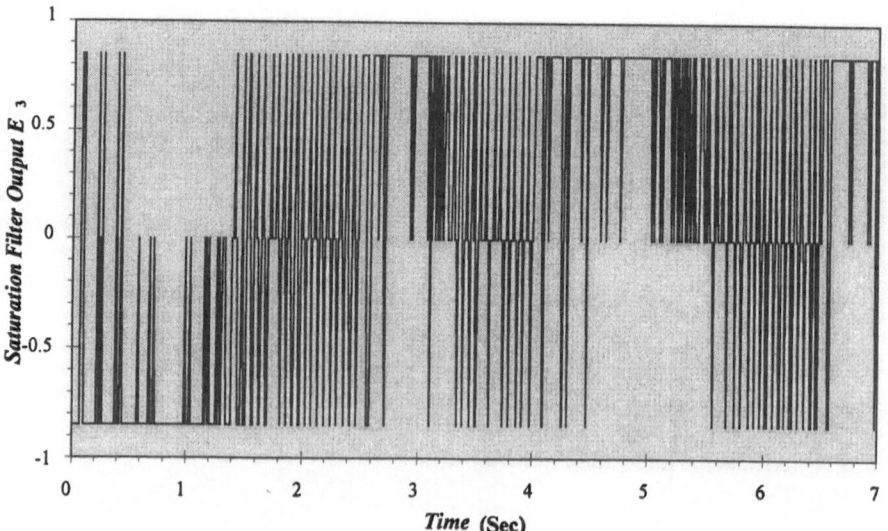

Fig 5 Saturation Filter Output

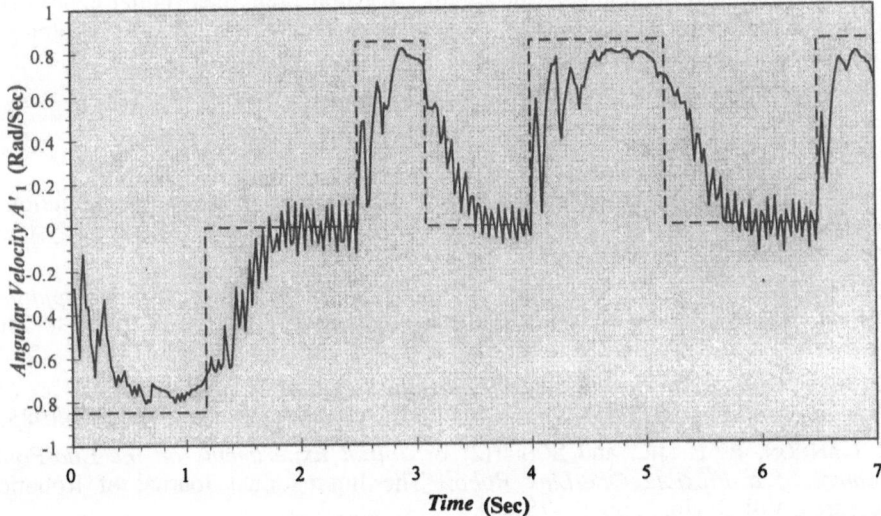

Fig. 6 Angular Velocity of Motor Rotor

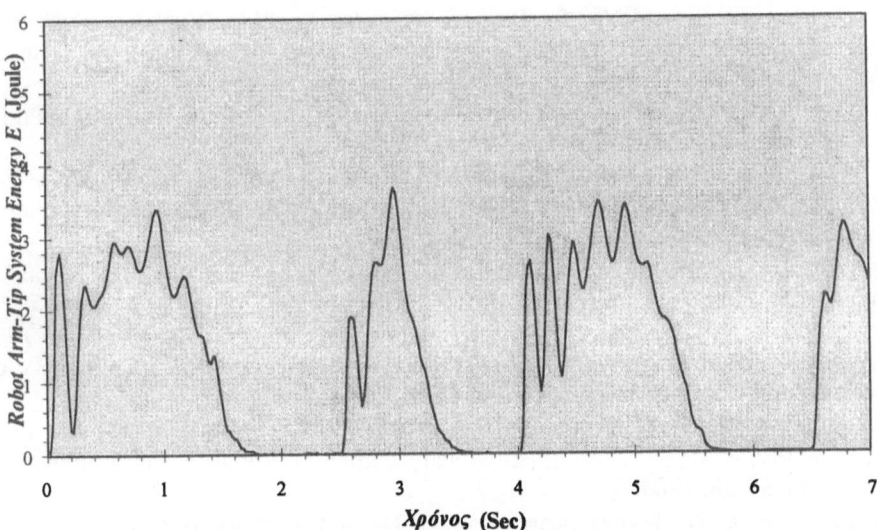

Fig. 7 Flexible Arm-Carried Mass Total Energy

It is shown that both cases lead to almost equivalent optima but for completely different values of the decision variables. This fact proves that the specific control system is highly efficient, especially when its coefficients are defined by a suitable parameter optimization algorithm, such as the S.Q.A .

It is remarkable that the system response is well accepted for the whole range of concentrated masses and all the transient periods between two consecutive control commands. It also manipulates successfully the non predefined initial conditions, since the remaining kinetic energy of a disactivated control signal constitutes a disturbance to the next command decision.

References

1. LUH, J. Y. S., WALKER, m. H., and PAUL, R. P. C., *On-Line Computational Scheme for Mechanical Manipulators*, ASME Journal of Dynamic Systems Measurement and Control, Vol. 102, pp. 69-76, 1980.

2. HOLLERBACH, J. M., *A Recursive Formulation of Lagrangian Manipulator Dynamics*, IEEE Trans. System, Man and Cybernetics, SMC-101, pp. 730-736, 1980.

3. SCHMITZ, E., *Experiments on the End-Point Position Control of a Very Flexible One-Link Manipulator*, Ph.D. Thesis, Stanford University, SUDAAR No. 548, 1985.

4. CANNON, R. H. JR., and SCHMITZ, E., *Initial Experiments on the End-Point Control of a Flexible One-Link Robot*, The International Journal of Robotics Research, Vol. 3, No. 3, pp. 62-75, 1984.

5. CENTIKUNT, S., and YU, W. I., *Closed Loop Behavior of a Feedback Controlled Flexible Arm: A Comparative Study*, International Journal of Robotics Research, 1989.

6. CENTIKUNT, S., and YU, S., *Discrete-Time Tip Position Control of a Flexible One Arm Robot*, ASME Journal of Dynamic Systems Measurement and Control, Vol. 114, pp. 428-435, 1992.

7. SPECTOR, V. A., *Modeling of Flexible Systems for Control System Design*, Ph.D. Thesis, University of Southern California, 1991.

8. SPECTOR, V. A., and FLASHNER, H., *Sensitivity of Structural Models for Noncollocated Control Systems*, ASME Journal of Dynamic Systems Measurement and Control, Vol. 111, pp. 645-655, 1989.

9. SAKAWA, I., MATSUNO, F., and FUKUSHIMA, S., Modeling and Feedback Control of a Flexible Arm, Journal of Robot Systems, Vol. 2, pp. 453-472, 1985.

10. SICILIANO, B., CALISE, A. J., and JONNALAGADDA, V. P., *Optimal Output Fast Feedback in Two-Time Scale Control of Flexible Arms*, Proc. IEEE Conference of Decision and Control, Vol. 3, pp. 1145-1150, 1986.

11. WANG, S H., HSIA, T. C., and WIEDERICH, J. L., Open-Loop Control of a Flexible Robot Manipulator, International Journal of Robotics and Automation, Vol. 2, pp. 54-58, 1986.

12. PETRIDIS, A. G., HARALABOPOULOS, G. N., and KANARACHOS, A. E., *A New Global Optimization Algorithm Combining the Natural Evolution Model and the Deterministic Newton Methodology*, EURISCON '98, 1998.

PART VI

MOBILE AND WALKING ROBOTS

38

Development of an Application Platform for Mobile Robots

O. Buckmann, M. Krömker and U. Berger

1 Introduction

It is a common understanding that mobile robots will be firstly introduced in huge numbers in buildings providing necessary technical requirements such as hospital, geriatric or rehabilitation clinics or homes for disabled and elderly people. Mobile robot systems can also be applied to prevent accidents that cause injuries, e.g. for mine removal in war zones [1].

Two fields of activity in health care purposes are of core interest: At first the execution of cleaning and home services and secondly the performance of logistical functions. These items are further specified in the following list:

- Mobile Robotic Platforms for Health Care Services can ensure adequate hygienic standards in view of dry / wet cleaning of ground floors, sanitary chambers and meal preparation areas.
- Food service from main kitchen to rooms, suites or apartments (transportation service of tablets, dishes etc.).
- Mail, parcel and magazine service or pharmaceutical products under supervision of medical staff.
- Maintenance of electronic or computer programmed devices like television, telephone, information service communication networks by downloading of storage data into individual programmable systems.

It is likely to envision that mobile robots developed for industrial applications can be used for this purpose. The industrial mobile robots, however, normally operate in a clearly defined and static environment or they require intensive man-machine interactions like mobile telerobots. They further need sophisticated sensor and computer platforms as well as skilled operators. These conditions cannot be met when a mobile robot operates within a common, altering environment which consists of stationary (e.g. furniture) and moving (e.g. people) obstacles. This makes traditional industrial mobile robots unlikely suitable for applications such as health care tasks.

The design and development of a mobile robot is a quite complex task that requires the co-operation of a multidisciplinary team of skilled developers. To qualify all members of the development team to an adequate level as well as to provide a sophisticated testing and training environment, within the European Research Project "Mobile Robotics Technology for Health Care Services Research Network – MobiNet" an Application Platform has been specified.

The following techniques have been identified to enable a general purpose for a mobile robotic platform:

- actuator components as drives and drive control units, handling and gripping devices, finally signal emissions systems (optical and acoustical)
- sensoric components like navigation and manoeuvring systems including perception, collision avoidance and obstacle bypassing technology
- human interaction devices like ergonomic user interaction devices as handles, joy-sticks, folding and unfolding systems for house to house / house to car or house to environment movement and man machine communication interfaces.
- supervisory and maintenance functions as emergency recovering (battery loading), order receiving and diagnostic communication system, error message and emergency handling.

For an efficient and cost minimising strategy, the utilisation of existing industrial robot technology was envisaged.

1.1 Basic Concept of an Application Platform

The basic concept of the Application Platform describes various elements supporting the different stages of a mobile robot development process. In addition, several generic interfaces have been defined. The Platform Processor provides a holistic integration of all components; it is depicted in figure 1.1.

The Application Platform covers two different areas: an area for adaptive functions and an area for motion functions. The area of adaptive functions comprises three elements: Task based sensoric elements, navigation based elements and communication elements.

Operation tasked based sensoric elements are sensors like e.g., tactile, optical or megnetoresistive sensors. Navigation-based elements are sensors like radar, laser, gyroscope, 3-D ultrasonic range measurement sensors optics or GPS (Global Positioning System). Even a sensor for the battery status has to be considered. The communication elements are devices such as Etherlink microwave or radiotransmission connections. To ensure the signal processing to the platform processor, each element has a dedicated operation system in front, like a task planner for the task based sensoric elements or a path planner for obstacle avoidance for the navigation elements. All necessary functions have to be implemented here. The functions itself are determined by the desired task. Above these functions is an simulation and modelling environment for

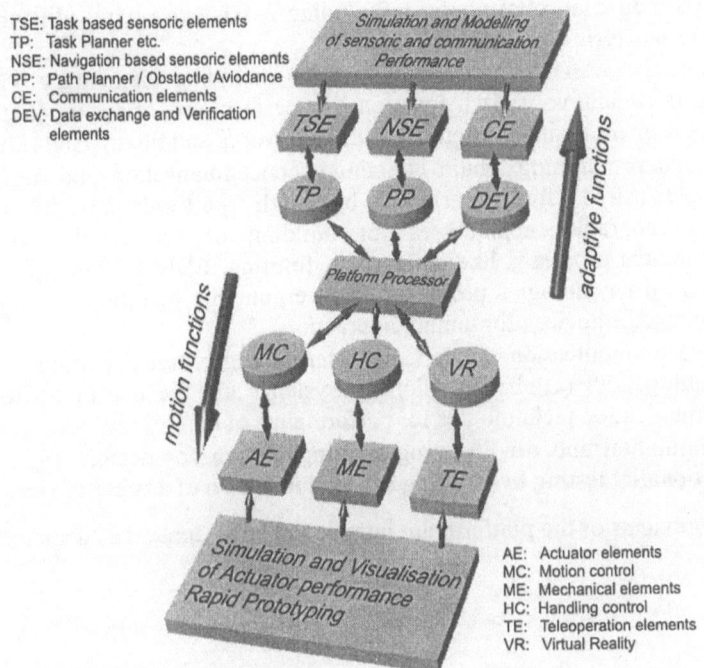

TSE: Task based sensoric elements
TP: Task Planner etc.
NSE: Navigation based sensoric elements
PP: Path Planner / Obstactle Aviodance
CE: Communication elements
DEV: Data exchange and Verification elements

AE: Actuator elements
MC: Motion control
ME: Mechanical elements
HC: Handling control
TE: Teleoperation elements
VR: Virtual Reality

Figure 1.1: Basic modular concept of an Application Platform

determination and validation of the elements to be used in the platform. To realise this concept, both existing or self-developed software can be applied.

On the other side there are the area of motion functions. These functions will also comprise several elements: actuator elements, mechanical elements and teleoperation elements. Parts of the actuator elements are devices like driving wheels, legs or caterpillar tracks. The mechanical element can comprise devices like grippers or servo tools. For the teleoperation elements devices like switchboards, joysticks or several kinds of user interfaces or even throughout these elements a big range of hardware can be used.

For the data connection to the Platform Processor, interfaces have been specified as well: a motion controller for the mechanical elements, a handling controller and for the teleoperation a virtual reality environment. Additionally, simulation systems can be added to prevent damage of real hardware or to set up virtual test beds.

1.2 Real Environment

After the determination of an basic concept, a real environment with available technology has been realised.

The technology can be encompassed as:

- 6-axis industrial robot platform including 2-axis extra payload handler to perform actoric components behaviour.
- Serial 2D or 3D optical sensor systems including elementary / binary sensors (inductive, switch functions) to perform sensor fusion strategies. Enhancement by integration of Neural Networks and Fuzzy Algorithms.
- A rapid prototyping studio containing conceptualisation and design of complex free form surfaces by high performance computing, stereolithography apparatuses for building of functional parts and postsequent processes like die casting, forming, EDM technologies. This advanced technologies provide best-fit ergonomic handling treating and maintenance devices for human interaction.
- A telecommunication network for internal and external communication including GPS (Global Positioning System) and the use of up to date communication technologies for performance of communication needs.
- A simulation and off-line programming system for design testing and functionality testing by building a digital mock-up of a robot device.

The components of the platform are intended to be integrated as depicted in figure 1.2.

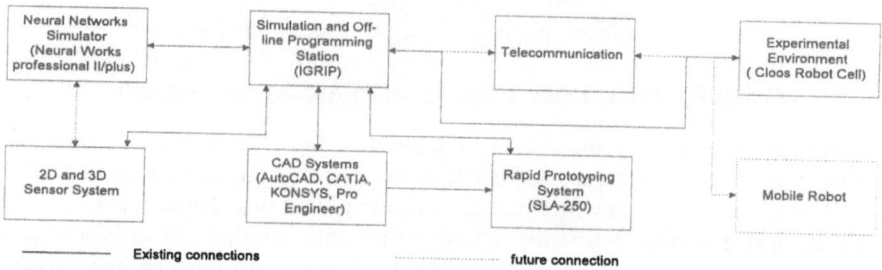

Figure 1.2: System components of the Application Platform

In comparison to the basic concept, the connection between all systems is the Platform Processor. The interfaces to the Platform Processor described in the basic concept are integrated in the subsystems itself.

The various components of the Experimental Platform are described in the following sections. Not all components are precisely specified at this stage. The inclusion of telecommunication facilities will allow the exchange of the conventional robot by a mobile unit.

2 Simulation and off-line Programming Station

In order to save expensive equipment from crashes and destruction, minimise programming of high complex moving paths and enhance understanding of function by graphical virtual display of moving sequences, simulation systems allow to test the feasibility and functionality of a robot system . Failures like design bottlenecks of robots, other kinematic components or tools of a shop floor cell, programming errors or reachability errors can be detected and changed during the design process. Through the possibility of the off-line programming functionality is it possible to save time and money as the production process does not have to interrupted because of the development of programming tasks can be performed independently from the robot system itself.

Such a station has been foreseen in the experimental platform. As a well-suited system IGRIP (Interactive Graphical Robot Instruction Program), developed by DENEB Robotics[1] has been selected.

IGRIP has a "Two-world" concept, supported by a powerful graphical user interface (as depicted in figure 2.1). The first world is an integrated CAD

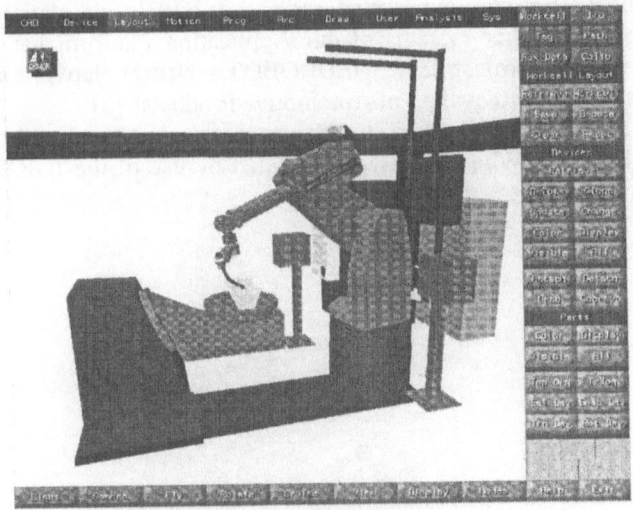

Figure 2.1: The IGRIP Simulation System

System. All parts of the workcell components can be modelled here. It is also possible to import data from external CAD Systems (AutoCAD, Pro Engineer) via standard Interfaces like IGES or VDA-FS. The second world is the Workcell-World. All parts will be assembled here to devices and arranged to a layout of the real shop floor cell or to a virtual test layout. Kinematics for appropriate devices can be assigned here. A kinematic will be described trough its degree of freedom, the motion axis of each moving part and the way of movement, i.e. linear, rotational or both. If it's necessary user defined functions and program parts can be archived trough the UNIX shared libraries function or through a socket connection to other applications. These can be applications for calculation special kinematics equations or for path planning tasks as we need by mobile units.

---

[1] Auburn Hills, MI,USA. Further information: http://www.deneb.com

Working points in the workspace of a device will be stored in so called Tag Points. The kinematic routines calculates then the movement of the device. Additional Information (e.g. Robot configuration) or User defined information can also be stored in the Tag Points. A Program for a device will be written in the IGRIP internal language GSL (Graphical Simulation Language). This language is structured like Pascal with additional commands for the process simulation. Macros can be written in GSL for e.g. mostly used tasks or for simulation of an user interface [2]. A device can be also controlled by a native language interface provided by the simulation system manufacturer of with self-written interpreters trough shared libraries of the simulation system included in external C-programs. Socket connections to other applications can also be included here. Various experience with the IGRIP simulation system in combination with other systems of our Application Platform has been gained within the ESPRIT project 8338 "NEUROBOT – Neural Network based Robots for Disassembly and Recycling of Automotive Products" [3].

For an simulation example in health care services at the Lund University in Sweden a Permobil Wheelchair was simulated by use of the IGRIP System, as depicted in figure 2.2.

Figure 2.2: Simulation of a Permobil Wheelchair

2.1 Integration between simulation and off-line Station and the experimental environment

The system connection to the Robot controller is realised with a postprocessor that translates the GSL Program and the information of the Tag Points into the native robot controller language or from the controller to IGRIP. This postprocessor for the Cloos robot is provided by Deneb.

A user can also provide a data exchange to the experimental environment with GSL -Macros or self written translators. Data between the workstation and the robot controller will be transferred by a Local Area Network (LAN). A PC connects the controller via the Cloos CarolaEdi Software to the LAN Network.

When the program is transferred into the controller, the robot can execute the program. This connection will be later replaced by the program and command transfer interface for the mobile robot unit.

3 Rapid prototyping module

Rapid Prototyping (RP) techniques are methods that allow to quickly produce physical prototypes with the important benefit to reduce the Time-to-Market. By use of these techniques, prototypes can be built needing skill of individual craftsmen for no more than just the finishing the part. Furthermore, the resulting design cost will be decreased considerably.

Within a rapid prototyping process, the object is firstly designed on a computer screen and then created based on the computer data. This eliminates inevitable errors which usually appear when a model-maker interprets a set of drawings. An essential prerequisite is the computer representation: it is usually a 3D geometrical modelling system like a CAD system, a 3D scanner, a computer tomograph, etc. Its precision is a key parameter controlling the tolerances of the future model (the different techniques allow an average accuracy of approx. 0,1 mm).

Though for various RP systems there are no restrictions concerning complexity and geometrical features, the physical objects are limited in their size. An advantage is the fact that the same data used for the prototype creation can be used to go directly from prototype to production, eliminating further sources of human errors.

Several RP techniques are available. The first commercial process, Stereolithography (SL), was brought on the market in 1987. Nowadays, more than 30 different processes (not all commercialised) with high accuracy and a large choice of materials exist. The most successfully developed techniques are, Stereolithography, Selective Laser Sintering, Laminated Object Manufacturing, Ballistic Particle Manufacturing, Fused Deposition Modelling. Information about these systems in [4], [5], [6].

In our Application Platform a Stereolithography system, was integrated. It is a 3-Dimensional printing process which uses a laser beam directed by computer onto the surface of a photocurable liquid plastic (resin) to produce copies of solid or surface models. The basic steps of the overall process are depicted in figure 3.1.

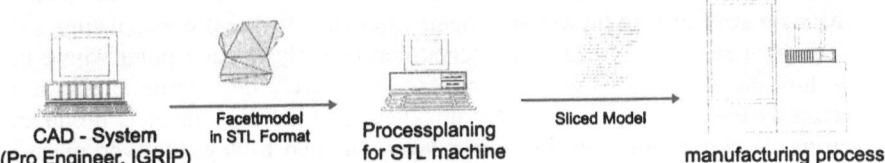

CAD - System (Pro Engineer, IGRIP)   Facettmodel in STL Format   Processplaning for STL machine   Sliced Model   manufacturing process

Figure 3.1: Steps from CAD model to real model

Within process planing, a vector scheme is calculated based on a 3D solid model in STL Format of the workpiece with a so-called "slicing program". The output describes the workpiece in a layered form. A supporting structure will be added during the process planing to the 3D Model in order to ensure that the produced workpiece has a connection in itself. In the manufacturing process, the workpieces will be build layer by layer out of a liquid photopolymer resin which is partially hardened by a monochromatic ultra violet laser, a Helium-Cadmium Laser or a Argon ion laser.

The complete system consists of a process PC witch controls the building process of the workpiece and a process chamber as depicted in figure 3.2. The process chamber mainly consists of a sink filled with a liquid photopolymer resin. In here is a carrier platform which is mounted to an elevator for vertical movement. The ultra violet laser beam is moved over the surface of the liquid by use of a x-y-scanner optical mirror device.

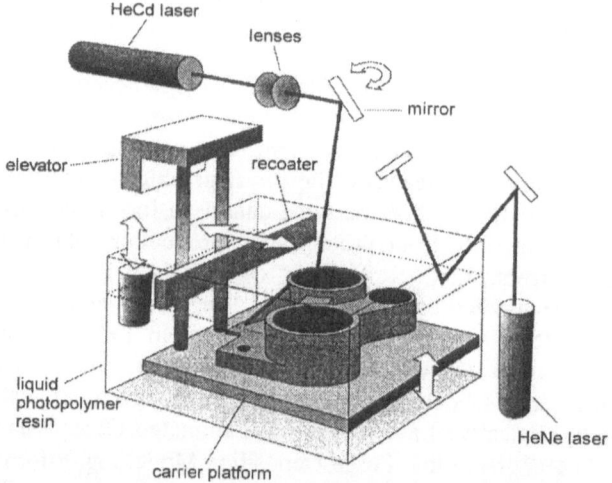

Figure 3.2: Schematic Layout of the stereolithography process chamber

An additional Helium-Neon Laser measures the level of the photopolymer resin surface. A recoater smoothes the surface before a new layer is build.

At the beginning of a building process, the carrier platform is positioned below the surface of the photopolymer resin by the thickness of the layer, e.g. 0.25 mm. The X-Y-scanning system moves the ultra violet laser over the surface of the resin according to the vector scheme calculated by the process planing PC.

The photopolymeric resin is hardened permanently at each point where the laser hits the surface. After finishing the first layer, the carrier platform is lowered by the distance of the layer thickness and the next layer is produced, ensuring its connection with the lower layer. In such a way, the workpiece is created layer by layer from its bottom to the top.

This method was developed in 1986 by the American company 3D Systems and is still distributed by them. With our System SLA-250 from 3D Systems[2], workpieces with the maximum dimension of 250 mm x 250mm x 250 mm can be built. Additionally, CAD Systems were integrated to the Rapid Prototyping Module for the creation and processing of CAD Data.

3.1 Integration between the Stereolithography system and the simulation and off-line Station

The simulation and off-line programming system IGRIP comprises a Stereolithography-interface. Data from the IGRIP CAD world can be directly exported to the Stereolithography system and can be sliced there. Using this system, a fast transfer of a model, e.g. a joystick, to the 'real' world can be realised. Functionality tests can be made or the model can be use to make a mould for this part.

3.2 Integration between the Rapid Prototyping Modules and the CAD Systems

The easiest method to provide the SLA machine with data in an adequate format is the integration of a STL-interface in the CAD System. From the systems contained in the Application Platform, only Pro Engineer has an interface which outputs satisfactory results. The other systems use neutral interfaces like IGES, VDAFS or DXF. The neutral interface data have to be pre-processed with other systems for a "clean" 3D model that can be processed by the slicing program. Such Systems are Magics RP from Materialise[3] or Rapid Work, developed by BIBA[4]. Neutral interface data usually contain errors, e.g. two surfaces are not closed at their contact points. These errors accrue by the calculation from the interior CAD format to the neutral format.

4 Neural Networks Simulator

In order to allow a mobile robot to operate and navigate autonomously, control systems based on artificial intelligence have to be integrated. The objective of artificial intelligence is the reproduction of human skills with neural networks. Neural networks are based on principles that intend to correspond to the human brain. As the functionality of a human brain depends basically on the work of neurones and the use of the incoming impulse through a neurone.
As the neural networks simulation tool, NeuralWorks Professional II/plus developed by NeuralWare, Inc.[5], has been selected as it is a well known

---

[2] Further information: http://www.3dsystems.com
[3] Ann Arbor, MI, USA. Further information: http://www.materialise.com/
[4] Further Information: http://www.biba.uni-bremen.de/projects/RapidWork/
[5] Pittsburg, PA, USA. Further information: http://neuralware.com

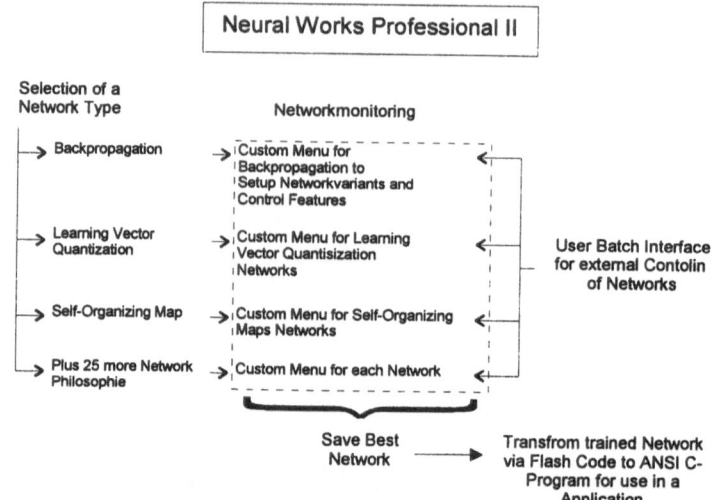

Figure 4.1: Structure of NeuralWorks professional II/plus

simulation system of neural networks. Figure 4.1 depicts the main structure of the system.

With this tool it is possible to build, train, refine, and deploy well known neural networks solutions, e.g. back-propagation networks (up to 3 hidden layers). An introduction and a description of fundamental principles of neural networks is given in [7], [8], [9] and [10].

Within the NeuralNetwork tool a program named *Flash Code* is available which converts a trained network into ANSI C code and so integration with popular commercial packages is possible. Comprehensive manuals & tutorials are included. NeuralWorks allows monitoring network performance, automatic train/test cycle, explaining of network decisions. Furthermore, a embedded IDL (InstaNet Definition Language) can be used for customising and controlling Neural Networks [8].

4.1 Integration between the Neural Networks Simulator and the off-line Programming Station

The integration of neural works into IGRIP, for e.g. to calculate inverse kinematics, is done through socket and shared library communication. A C program controls the neural network (user control program) and contains all necessary routines used for the communication with the NeuralWorks professional II/plus package. Another C program controls the simulation environment and contains all the routines used for the shared library communication of the IGRIP simulation system. Both systems are installed on the same platform and communicate via TCP/IP sockets, depicted in figure 4.2.

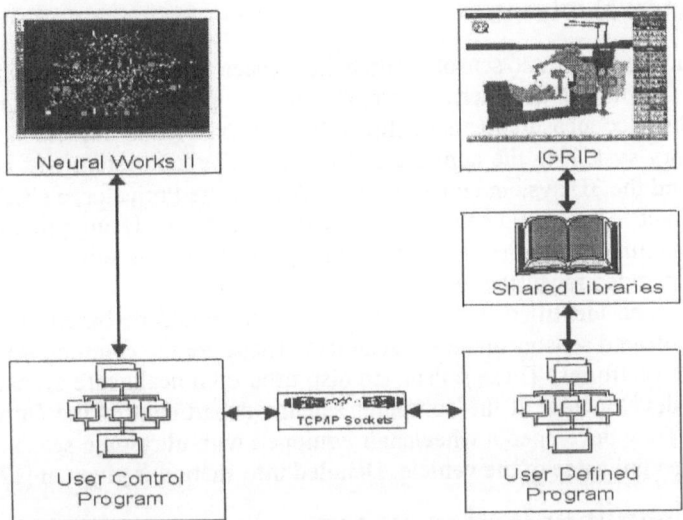

Figure 4.2: Communication between Neural Works and IGRIP

The platform is a Hewlett Packard, HP 9000 series 700 computer workstation on which the HP-UX (UNIX) operating system is running.

Within the C program for the IGRIP simulation, the user is able to define a kinematics routine which executes the movement of a robot device. The set of functions used for the definition is referred to the motion pipeline. The motion pipeline is a series of procedure calls which are executed sequentially during the simulation of a motion statement in IGRIP. With this information the user can incorporate self-developed software for performing inverse kinematics, motion planning, motion execution, dynamics analysis and dynamics simulation. The user is also able to write his own functions to work as components in the motion pipeline.

When the two systems are connected via the TCP/IP Sockets, IGRIP will send position information to NeuralWorks professional II and receive after the calculation by an appropriate Neural Network the joint values for this position. The device will be moved in the simulation to this position. This handshake procedure will continue until the desired position is reached (in the simulation) [11].

It is also intended that the NeuralWorks system will be later connected to a sensor system and can be used for learning of decision task (e.g. object recognition).

5 Sensor Systems

The most important sensors which have been selected for this application platform, are the optical sensors arrays with the possibility of topometric and range / distance measurements facilities. It has been tried to implement a 2D and 3D Sensory system to the application Platform. The 2D system was an ITRAN System and the 3D system consisted of an ABW Line Projector, a CCD Camera, a framegrabber board and an analysis software. As the analysing process was too time consuming due to the fact that too much data had to be submitted, it can not be recommended for further use.

It has been identified that integrated systems should be based on ultrasonic sensors, infrared sensors and laser scanners. These are the common sensors used on mobile platforms. These sensor are also used on a healthcare systems, which is under development at the computer science department of the University of Bremen. They developed a wheelchair equipped with ultrasonic sensors only for the assisted guidance of the vehicle. Detailed information is given in [12].

6 Telecommunication system

Within this module, the Integration of mobile communication for control of mobiles and telepresence will be addressed. The objectives of this task is to evaluate, specify and demonstrate future extension of an autonomous robot scenario using existing wireless communication technologies and computer interfaces to connect mobiles.

The intended system should allow to communicate data for robot commands, programmes and transmit sensor data as well as to realise telepresence solutions with multimedia data transmission.

The sensor systems or data system integration of mobile communication for control of mobiles and telepresence should possibly include information about robot position, orientation, speed as well as audio and video data, e. g. for applications like video conferencing. The telecommunication could also be used for navigation purpose, linkage for CAD based orientation etc.

Main aspects for communication are radio modems for wireless communication and infrared for communication with infra structure (radio transmission, infrared). This includes also the specification and assessment of methods and algorithms like digital coding or compression techniques.

7 Experimental environment (robot cell)

In the experimental platform, a six axis joint co-ordinate industrial robot, the CLOOS R 76 (figure 10), with an integrated 2 axis positioning table will be used until the development of the mobile robot is finished. This robot is normally used for welding tasks, but a E.O.A. System is mounted to change the tool for other tasks.

The CLOOS Romat 76 in upright position is specified as follows [13]:

Configuration:	Revolving joints, 6 axes
Drive:	DC-motor exited by permanent magnets
Resolution:	1024 impulses/degree
Load capacity:	10 kg
Positioning error:	± 0.2 mm
Working space:	hemispherical, diameter approx.: 3600 mm, height 3200 mm
Rotation angle:	axes 1,4 : 320° axes 2 : 240° axes 3 : 270° axes 5 : 180° axes 6 : 450°
Controller:	separate 32 bit micro controller for each axis with 8MB Ram

Figure 7.1: The Cloos Romat 76 robot

The robot can be programmed conventionally by Teach-In methods or by use of the simulation and off-line programming system IGRIP.

8 Potential use of the Application Platform in health care tasks

The main use for health care systems of this Application Platform are in the Subsystem Simulation system / CAD System in combination with the rapid prototyping system. With these subsystems, various ideas can be quickly realised and serve as a basis for further tests, like ergonomic test for joysticks or other peripherals for handicapped people.

Another average is the use of the simulation system for fast testing and design checking of kinematic parts of a robot system. The algorithms of the controller can be emulated and checked for bugs. The behaviour in the relation to the environment can be tested and a 3D model can be viewed in an early stage of development in order to improve both quality and performance of the final product, at least to perform feasibility tests.

9 Work in progress

Currently, a three finger gripper is build in co-operation with the University of Lund, by the use of the Application Platform. The gripper is depicted in figure 11. We are now starting to building the parts of the gripper by use of the RP module of our platform.

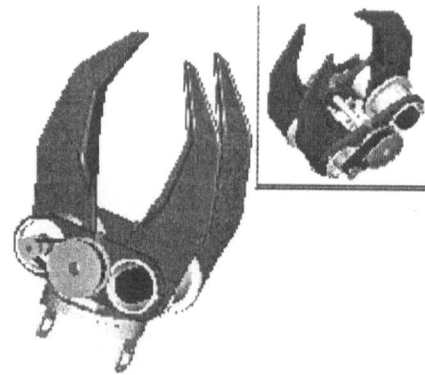

Figure 9.1: Three Finger Gripper

For evaluation and teaching purposes in the MobiNet Project, we try to integrate the new Fischertechnik Mobile Robot, depicted in figure 12. It's possible to program the control interface via C++ or Visual Basic. We just started and no results can be presented now. Fischertechnik is applied for industrial purposes to test complex systems in small scale.

Figure 9.2: Mobile Robot from Fischertechnik

10   Conclusion

The experimental platform specified in this paper provides a promising basis for training and experimenting within the development of an intelligent and autonomous mobile robot with high manoeuvre and manipulate features. The main goal of this Platform should be the formulation of a requirement specifications for a autonomous mobile robot in a clinical environment. It comprises powerful systems such as a Robot Cell, a Simulation and Off-line

Programming Station, a stereolithography system a Neural Networks Simulator, 2D and 3D Sensor Systems and telecommunication systems. During the undertaken research work, some systems have been identified as unsuited to perform the intended tasks, i.e. the 2D and 3D Sensor Systems. Other systems were integrated, such as the stereolithography system to quickly realise prototype parts for robots and the integration of the Fischertechnik mobile robot.

Current Applications within the MobiNet Project will show the reliability of the current status of the Application Platform. The main focus for our research will be on the integration of the RP Module / Simulation module in the Application Platform. The second goal of the platform is to provide project partner with systems for their work or for training courses within the MobiNet project.

11 References

1. Winning Landmine detectm, http://www.mece.ualberta.ca/landmine/winning.html
2. Deneb Robotics, Inc., 1996: "IGRIP 3.0 Manuals", 3285 Lapeer Road West, P.O. Box 21 46 87, Auburn Hills, Michigan, USA
3. Tuominen J, Berger U, Meier I.R. et al., 1995 Autonomous Robot based Disassembly of Automotive Components, *in Proceedings of the Conference on Integration in Manufacturing,*
4. Burns M *Automated Fabrication - Improving Productivity in Manufacturing*", Prentice Hall , Eaglewood Cliffs
5. Giebhardt A 1996 *Rapid Prototyping - Werkzeug für der schnelle Produktentwicklung* , Hanser, Germany
6. Jacobs P F 1992 *Rapid Prototyping and Manufacturing; Fundamentals of Streolithography,* Society of Manufactures Engineers
7. Hoffmann N 1991 *Simulation Neuronaler Netze,* Germany
8. NeuralWare Inc. 1995 *Neural Computing – A Technology Handbook for Professional II/Plus and NeuralWorks Explorer,* Pittsburgh, USA
9. Rojas R. 1993 *Theorie der neuronalen Netze: Eine systematische Einführung,* Germany
10. Schöneburg E 1990, *Neuronale Netzwerke, Einführung, Überblick und Anwendungsmöglichkeiten,* Germany
11. Schmidt T 1995 ESPRIT Project No. 8338 NEUROBOT, Workpackage 4.6: *Investigation of neural network based inverse kinematics formulations of a 6-axis articulated joint robot,* BIBA, Bremen
12. Roefer T 1998 Strategies for Using a Simulation in the Development of the Bremen Autonomous Wheelchair, In: *Proc. 12th European Simulation Multiconference* (to appear), accessable via: http://www.informatik.uni-bremen.de/~roefer/publice.htm
13. Carl Cloos Schweißtechnik, 1994, Cloos Robot manual, Haiger, Germany

39

A Framework for the Integration of Perception and Localization Systems over Mobile Platforms

F. Wawak, F. Matia, C. Peignot and E.A. Puente

The research team of the UPM-DISAM laboratory that is dedicated to the Mobile Robotics technology is implicated in the MobiNet[1] network. The purpose of the MobiNet network is to share competencies and to search for innovative solutions in the field of Mobile Robotics Technology for Health Care Services. In that way, the UPM-DISAM research team looks at the conception of software tools to ensure autonomy to a specific wheelchair.

After a lot of studies about planning and control technologies [1][2], the present preoccupation of the team is the design of a perception module, based on incoming information from heterogeneous perception systems of Mobile Platforms. The purpose of the perception module is to provide an adequate information about the environment to both modules of planning and control.

This chapter focuses more particularly on the problem of localisation in the perception module. First, the perception module is presented and the role of the localisation in such a context is defined. Then, the integration of three methods of localisation is exposed and the efficiency of their collaboration facing the perception problem is demonstrated with an experimental platform.

1   Introduction

The chapter tackles with two main topics of the Mobile Robotics Technologies that are the concept of Perception and the concept of Localisation. In order to discuss them later, these concepts are first defined.

---

[1] Mobinet — Mobile Robotics Technology for Health Care Services Research Network

1.1 Perception issue

The perception is a process that aims to interpret the information coming from sensors in a way to estimate a model of the real world. One the major area of research issues that ensure such a process is Computer Vision [3]. But, the past decade has seen a strong increase of the sensor technology allowing new generations of sensors [4]. Each of them are able to supply a kind of information about the environment. The problem is to elaborate an appropriate physical model of the sensor to ensure a good interpretation of the information. Spatial information is a fundamental requirement for Mobile Robotics Technology. Such information aims to provide a spatial model of the real world to the Mobile Platform allowing Planning and Reactive Control of the platform movements.

For the construction of a robust spatial model, researchers tend to integrate information coming from different kinds of sensors. In that case, an integration framework has to be established allowing to confront the information [5]. Basically, three frameworks are used for the integration of spatial information. First of all, the measures themselves can be used. But with that technique, the integration is made quit difficult especially due to the heterogeneity of the information coming from different sensors. To avoid that mess, an homogeneous framework can be used such as the geometric mapping [6]. The idea is to transform the sensors measures into a geometric information, for example some segments. More recently, Elfes has proposed the occupancy grid as an integration framework of the sensors information [7]. With that framework, the sensors measures expect free and occupied areas of the environment.

There are basically two steps in the Sensor Integration. The first one so called Multisensor Integration aims to integrate at a given instant and in a unique framework the piece of information brought by each sensor. The second one so called Sequential Integration updates the previous model of the world with that new information brought by the Multisensor Integration step. It comes that the most crucial information needed to ensure that second step is a good estimation of the position of the Mobile platform. This is essential to know exactly where to integrate the new information in the previous model. Here is the link between the perception issue and the localisation issue.

1.2 Localisation issue

A lot of localisation methods are available in the literature [8]. They are usually classified in two essential categories that are the absolute localisation methods and the relative localisation methods.

An absolute localisation method aims to determine the absolute position of the mobile robot. One of the standard systems is the Global Positioning System. The absolute positioning is characterised by the fact that such a system gives an estimation of the mobile robot position with a bounded error due to the imprecision of the technology used for the positioning.

A relative localisation method makes use of an incremental system to update an initial position. A representative system based on a relative positioning is the dead reckoning system. The main drawback of a relative localisation system lies in the fact that such a method accumulates the imprecision. Hence, the error of the position of the mobile platform increases continuously without bound. To overcome that problem, a step that refines the position estimation is added. The purpose is to retrain continuously the error of the position keeping it into an acceptable interval. An approach is to make use of another source of information, i.e. the information of another sensor. That new information is matched to the model of the real world and to the position given by the dead reckoning system, to adjust the estimation of the platform position. It comes that such a method is limited by the accuracy of the model of the real world.

Note that in the literature, the problems of perception and localisation are often disconnected. This chapter is going to show a way to integrate the localisation issue into the perception one.

2 Perception Module

The previous section has established that the perception module aims to estimate a spatial model of the real world allowing Planning and Reactive Control of the robot movements. The question is now to establish the framework for the spatial modelling. This chapter focuses more particularly on the methods for indoor mobile systems. That singularity makes the model of the real world easier to manipulate as we are going to see now.

The planning is based on an *a priori* knowledge of the environment, for instance maps of the plant, that is updated with the new information coming from the sensors. It is required here a stable description of the real world ignoring its dynamic aspect. The dynamic aspect is taken into account by the reactive control. In fact, the reactive control needs a current image of the environment to ensure a safety navigation of the mobile robot. These remarks lead to a structure for the perception module that is resumed in figure 1.

First, to provide the right information to the Path Planning module, a Refined Mapping is performed in the perception module. That mapping builds a geometric map as a model of the real world. The techniques used to update this kind of map ensure a strong data fusion of the information. Basically, the updating step consists in adding new obstacles in the *a priori* geometric map of the environment. A procedure allowing to deduce and to keep only the new information from the whole sensors data should be added upstream ensuring the stability of that representation of the real world.

Secondly, the information intended for the reactive control comes from a Rough Mapping. Such a mapping integrates the whole information coming from the sensor with a dynamic data fusion method. An adapted framework for that procedure is an occupancy grid. That framework is associated to fusion techniques able to perform a fast updating of such a representation of the real world.

A third procedure is integrated to the perception module: the Localisation. In the present module, the purpose of a good estimation of the position of the robot is twofold. It allows the sequential integration of the new information into the refined and rough mappings. It permits to refine the information coming from the sensors comparing it with a model of the robot. These last comments will be developed in the next section.

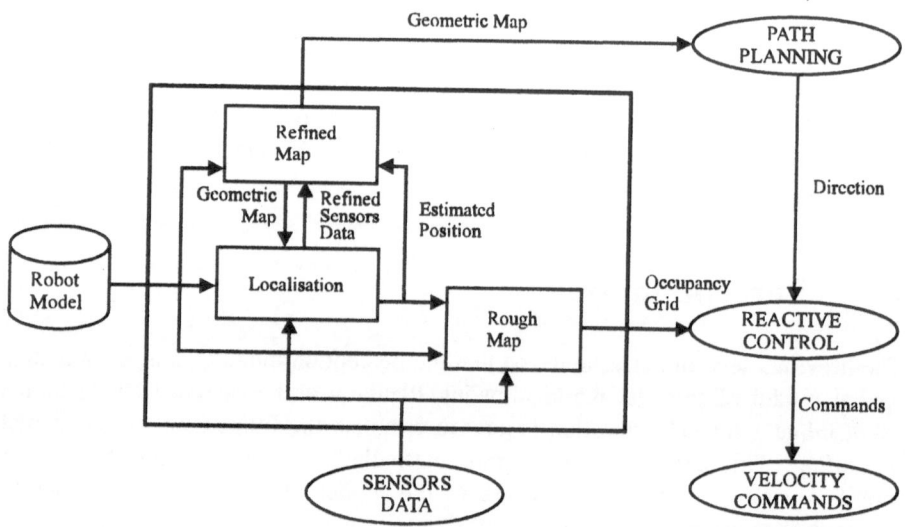

Figure 1: Perception Module

3 Localisation Problem into the Perception Module

3.1 Localisation methods

As mentioned above, in the category of the relative localisation methods, a dead reckoning method gives an initial estimation of the position. In our case, we have at our disposal an odometric system. To challenge the accumulation of the uncertainty due to that technology, we use three methods, each one based on a kind of information coming from the sensors.

A first method realises a matching between the information coming from ultrasonic sensors and a modelling of that type of sensors. An ultrasonic sensor measures the time needed for an ultrasonic pulse emitted by the sensor to come back due to a reflection on an object. That time is converted into a distance representing the distance of the closest object detected by the sensor. Because of the spherical divergence of the sound, an ultrasonic sensor is usually modelled by a beam with a conic shape. To predict the distance given by the ultrasonic sensor, the technique consists in placing the beam of the sensor into the model of the real world, and then, to determine the distance of the closest object present in the conic beam facing that model. Figure 2 illustrates such a modelling. Now, to localise the robot, the idea is to find the position of the modelled sensor, i.e. the position of the robot, that performs the best matching with the value given by the real sensor. That position gives an estimated position of the robot platform.

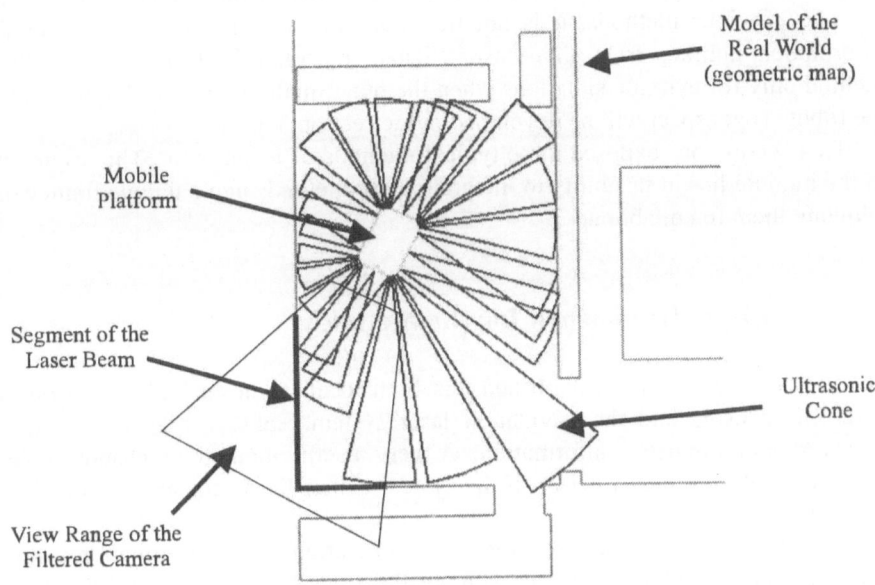

Figure 2: Sensors Modelling

With a second method, the same kind of matching is realised with a navigation laser system. The navigation laser system is composed of a source that emits a laser beam parallel to the ground in front of the mobile platform. A camera with a special filter able to see the laser infrared light looks in front of the mobile platform. Hence, the image of the camera displays some lines drawn by the laser beam on the obstacles placed in the view range of the camera. To model the information given by such a system, the idea is to determine the zone looked by the camera into the model of the real world. Then, the segments due to the shape of the obstacles into that zone are extracted (see figure 2). The localisation step here has to achieve the best matching between the segments extracted by the model and the lines seen into the image of the camera estimating a position for the sensor model, i.e. a position for the mobile platform.

A third method is available on the mobile platform. That method makes use of passive landmarks and an active vision system. From the prediction of the mobile robot position given by the odometric system and from the well known position of some passive landmarks in the environment, a mobile camera on the mobile platform searches for each of the landmark. The landmarks are red circles drawn on panels. The odometric system gives an estimation of the direction where to search for. Then, the exact angle between the mobile camera and the concerned landmark is extracted from the position of the landmark in the image of the camera. Finally, a triangulation estimation performs the estimated position of the mobile platform. The constrain of that method comes from the fact that the mobile robot has to be stopped the time for the mobile camera to find the landmarks. With the two previous methods, it is not necessary to stop the robot to capture the information coming from the sensors. That constraint leads to use that third method only for extreme situations when the other methods are not able to localise the robot. That aspect will be developed in the section 3.3.

That section has exposed three typical methods of localisation. The originality of the module lies in its ability to integrate these methods into a unique framework allowing them to collaborate.

3.2   Unique framework for the methods

The previous section has established that both localisation methods based on the ultrasonic sensors and the navigation laser system realise a matching between observed and predicted information. A very useful method to ensure such a matching is the Kalman filter or, if the system is non-linear, the extended Kalman filter [9].

The Kalman filter completes a recursive integration of new information coming from the sensors of a system to estimate the state of that system. An *a priori* knowledge about the system dynamics and the noise characteristics of the system is required. The Kalman filter is based on the calculation of a weighted average of the system state. That weighted average is manipulated through covariance matrices which are a direct indication of the error of a variable. So, there are three important covariance matrices in the Kalman filter that represent the *a priori* knowledge about the system: the matrix P that is the covariance matrix of the system state, i.e. the error of the state estimation, the matrix R that is the covariance matrix of the sensors measurements, i.e. the noise of the sensors measurements, and the matrix Q that is the covariance matrix of the system dynamics, i.e. the noise of the cinematic system.

For the application of the Kalman filter to the matching of the localisation systems, the system is defined as the mobile platform and the system state is the oriented position of the platform. The sensor information are the distances given by the ultrasonics sensors, on one hand, and the obstacle segments given by the navigation laser system, on the other hand.

Figure 3 sums up the previous comments into a general localisation module. From three sources of information that are the geometric map, the robot model, i.e. the sensors models, and the odometry, a prediction of the sensors data is evaluated.

A matching between the prediction and the real sensors data sorts out matched data for the estimation of the platform position.

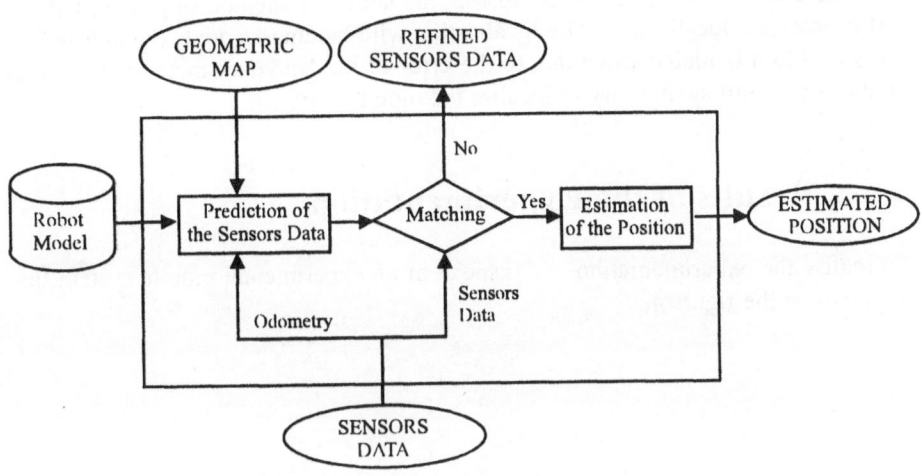

Figure 3: Localisation Module

The ultrasonic and the laser methods have in common the *a priori* knowledge characterised by the matrices P, that indicates the error of the platform position, and Q, that indicates the noise of the odometric system. A different matrix R has to be attributed respectively to the noise of each sensor system. In that way, an estimation of the position can be realised by both ultrasonic and laser methods.

The next subsection is going to stress the link between the methods realised through the matrix P and explains how intervenes the landmark method in the localisation module.

3.3   Aggregation of the methods

Now that the three methods have been specified, we have to find a way to make them to collaborate. We know that the on-line methods, i.e. ultrasonic and laser methods, make use of a Kalman filter and can share the same statistical resources that are the matrices P and Q. In fact, the matrix Q is fixed because it corresponds to the noise of the odometric system. The matrix P is updated at each iteration of the Kalman filter. That is to say that, by turns, each method updates the matrix P. It follows that the updated matrix P coming out from one of the method becomes the initial matrix P of the other one, and vice versa. Therefore, the methods collaborate in the mean that if one of the method is not able to localise the mobile platform, i.e. to reduce the matrix P, perhaps the other could to do it.

If both of the previous methods are not able to localise the mobile platform, the matrix P increases. Here intervenes the landmark method. If a threshold of the acceptable error of the position is crossed, the landmark method stops the platform and processes a localisation. The localisation with landmarks done, the matrix P is initialised to a bounded value due to the error of the landmark method. The other methods can start again to try to localise the mobile platform.

4   Results of the Experimentation

To realise the experimentation, we dispose of an experimental mobile system that is shown on the figure 4.

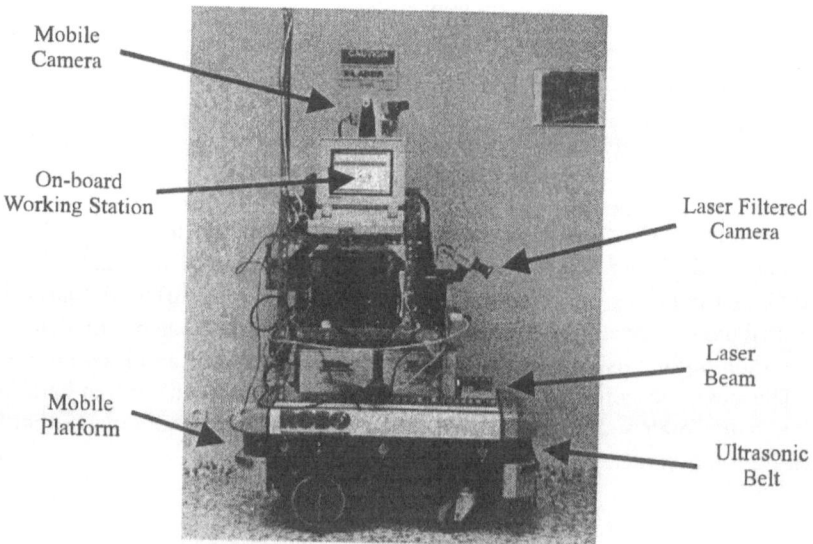

Figure 4: Experimental Mobile Platform

The mobile platform has an ultrasonic belt of twenty four ultrasonic sensors, a laser beam system, a mobile camera and an on-broad working station providing to it autonomy. With that device and after the implementation of the localisation module presented in the chapter, various experiences have be realised. Basically, that experimentation aimed to test the robustness and the accuracy of the localisation methods facing an environment free of unknown obstacles and an environment full of unknown obstacles. Here are the results of these practices.

It appears that the localisation with the Ultrasonic system is a very robust technique. Due to the geometric characteristic of the ultrasonic belt, the platform is frequently able to detected around it a known pattern of the environment to ensure the localisation. The problem of that localisation is its accuracy. Because of its technology, an ultrasonic sensor is very sensitive to the echo, especially when the obstacle is not exactly parallel to the sensor. In fact, the best results of that localisation technique are obtained when the mobile platform follows a wall.

The localisation with the Laser system is very accurate in spite of some limitations. The accuracy is limited in particular by the optical distortions and by the pixelisation of the image. But, if the detected pattern of the known environment is closed to the platform accuracy is quasi perfect, i.e. about the millimetre. The problem of that method is its low robustness. There are two cases when that sensor system is unable to provide information: when the camera looks at a free space zone, because no segment is seen on the image, and when there is an unknown obstacle in front of the platform, because no segment is matched. In fact, due to the narrow view range of the camera of the Laser system, these cases occur frequently.

The landmark method is a robust method. It is able to ensure a localisation in spite of unknown obstacle if the landmarks are not hidden. That last restriction is avoided introducing more landmarks in the environment. The accuracy of that method is quite good and depends on the proximity of the landmarks and of the angle between the camera and the landmarks.

To resume the experimentation, we must say that, in the zones of the environment relatively free of unknown obstacles, the ultrasonic and laser methods are able to localise the mobile robot. More precisely, the ultrasonic system ensures continuously a rough localisation that is reinforced by the laser system at some dedicated moments. If the platform manoeuvres in a zone of the environment full of unknown obstacles, both of the previous methods are not able to localise the mobile robot. After a period of time, the landmark method stops the robot and realises the localisation.

5   Conclusion

This chapter has presented a localisation module integrated into a perception system for mobile systems. It has been stressed the efficiency of the collaboration of three integrated methods of localisation. It appears that the main advantage of such a localisation system lies in its ability to integrate heterogeneous sensor information and so to be configurable depending on the environment encountered.

However, the purpose of that architecture is not just to estimate the position of the robot. Its matching and its integration of the information provide a sorting out of the information intended to the next two steps of the perception module that are rough mapping and refined mapping. The rough mapping issue makes use of the integrated information to provide a safety modelling of the environment. The refined mapping issue exploits the information rejected by the matching to update a static modelling of the environment.

Acknowledgements

The authors would like to acknowledge the funding support of the TMR Progamme through the MobiNet project (Contract FMRX CT96-0070), and of the spanish government through the EVS project (Contract CICYT TAP96-0600).

References

1. Matía F, Sanz R, Puente E A 1998 Increasing Intelligence in Autonomous Wheelchairs. *Journal of Intelligent and Robotic Systems* 22:211,232
2. Matía F, Moraleda E, Mena R, Puente E A 1998 Distributed Task Planner for a Set of Holonic Mobile Robots. *Proceedings of the Fourth international symposium on Distributed Autonomous Robotics Systems*, Karlsruhe, Germany, to appear
3. Horn B K P 1986 *Robot Vision*. MIT Press, Cambridge, Massachusetts
4. Everett H R 1995 *Sensors for Mobile Robots: Theory and Application*. A.K. Peters, Ltd., Wellesley, Massachusetts
5. Luo R C , Kay M G 1989 Multisensor Integration and Fusion in Intelligent Systems. *IEEE Transactions on Systems, Man, and Cybernetics* 19(5):901-931
6. Durrant-Whyte H F 1988 *Integration, Coordination and Control of Multi-Sensor Robot Systems*. Kluwer Academic Publishers, Norwell, Massachusetts
7. Elfes A 1990 Occupancy Grids: A Stochastic Spatial Representation for Active Robot Perception. *Proceedings of the Sixth conference on Uncertainty in AI*, San Mateo, CA
8. Borenstein J, Everett H R, Feng L 1996 *Where am I? Sensors and Methods for Autonomous Mobile Robot Positioning*. Edited and Compiled by J. Borenstein, University of Michigan
9. Maybeck P 1991 The Kalman Filter: An Introduction to Concepts, *IEEE International Workshop on Intelligent Robots and Systems*, pp 194-204

40
Creating Dynamic Mobile Robot Environments from an Intelligent Building

F. O' Hart and G.T. Foster

1 Introduction

Investigating the potential of robotic devices with health care environments is the primary objective of the TMR MOBINET research network. This research network aims to prototype a design for a mobile robot for health care services. So far there has been little market penetration by mobile robotic products due to the fact that fully autonomous systems are too expensive, both in initial capital cost and also in maintenance. Health care environments are by definition structured environments. Since a sizeable majority of research is involved with the design of systems for operation within a building, it therefore seems advantageous to incorporate robot design into the building process. This paper proposes the use of networked building management systems as a method of structuring the robot environment. By allowing the building to provide orientation and navigational information, the complexity and hence the cost of the robotic device can be lowered. Additionally the network also acts as a management infrastructure for multiple robots so the robotic services can be co-ordinated across the building. This is good for the marketability of mobile robots as they can be integrated with the workflow of the institution.

For a mobile robot to fit in with the building process, the robot requires information that it could not possibly gather itself. It therefore seems advantageous to incorporate robot design into the building process [1] [2]. Structuring the environment in this way should not be perceived as a step backwards in robotics research, but as the realisation that a more robust system is attainable through the utilisation of distributed computing resources present in the building.

Currently much of the research undertaken in the domain of mobile robotics is directed towards fully autonomous robots which tend to rely solely on the information gleaned about the environment from their own sensors. However, attempting to achieve a correspondence between observation and perception of physically present objects is a particularly difficult task, primarily because any delay in actuation, as a consequence of prolonged sensor interpretation, can lead to undesirable behaviour. For an autonomous robot involved in, say, a delivery task from one location of the hospital to another, navigational information is required. There are serious problems with any autonomous robot attempting to generate a route plan in isolation. For any large building, an autonomous robot needs to store a global map of its environment. The robot will attempt to derive an optimum route to the desired location from its current position based only on what it has mapped from previous experience. Although the opinion of the robot is that the route is optimal, the current state of the environment may be very different and the route far from optimal or feasible.

Researchers are beginning to see the power that may be realised from the integration of intelligent or instrumented environments into the design process for mobile robots. The notion of *intelligent space* as proposed by [3] is similar to ours. An intelligent space is defined as any area equipped with sensors, actuators, data stores, communications devices and some computational ability. The intelligent space observes its environment through its sensors and reacts via its actuators. However, Lee's approach differs from ours in that the localisation process for mobile robots in the intelligent space is achieved by tracking four targets atop the robot using a CCD camera situated in the environment. The approach also permits the environment to directly control the connection of behavioural layers in the mobile robot. Our approach is to give guidance to the robot as it traverses the environment. A model to the robot's local environment is supplied every time the robot enters the area of influence of a Reader Node.

Mobile robot systems currently installed in hospital environments are limited to delivery tasks in which the robot utilises global maps or global path generation of the building to plan a path to the target destination. The FIRST robot [4] is a prototype of a transport system for heavy loads in hospitals whereby sequences of actions associated with geometric locations are downloaded to the robot from a centralised ground station prior to actuation. This centralised ground station provides technical management for a fleet of such robots by communicating continuously via a radio link. The HelpMate™ robot [5] also utilises a global map of the building to plan a path to the target destination. The MARTHA system [6] for industrial robots allows the robots to incrementally determine the resources they require taking into account their current execution context. MARTHA comprises of a central station and a fleet of autonomous mobile robots. The robots have the ability to communicate with the central station and with one another. They receive their missions from the central station, but communicate with one another to co-ordinate their actions, refine their plans, and update their trajectories. In AMADEUS [7] intelligent AGV's communicate with job-shop cells in an agent framework to allocate an AGV to a transportation task, and to co-ordinate multiple

AGV actions so that their goals may be realised. The AGV follows a guideline. It assesses its current location based on a signpost embedded in the floor.

The following are some possible scenarios for using mobile robots within the context of health care environments.

- Much time is spent by hospital porters transferring hospital equipment and patient records throughout the hospital. A wheelchair could be used for the distribution of medical records and light equipment over short distances when not in use. The wheelchair should return to where it started from.

- An autonomous wheelchair or a robot equipped with a manipulator could be used to guide visitors or patients unfamiliar with the hospital environment to their destinations.

- Any mobile device with a manipulator attached could also be used to perform simple cleaning duties.

- In an emergency scenario where a ward or a section of the hospital has to be evacuated in the event of a fire, the robots could be used to converge on the junction of a corridor leading to a hazardous area. The robots would block access to that corridor, 'forcing' staff and patients to seek an alternative route. The robots and indeed the hospital infrastructure could provide directional information to the patients/staff.

For any of the above scenarios to be implemented the design of the robots must be such that they can interact responsibly with the building infrastructure. They must be capable of accepting instruction. The degree to which they must adopt the instruction depends largely on their task requirements. For example, an assistive robotic device such as that in [8][9] must not endanger the life of its user. If the building informs the robot that an alternative route must be taken due to a hazard then the robot must accept unequivocally. If the hazard is a spillage then a cleaning robot would ignore the danger since its function is to clean and minimise the number of hazards.

2   Intelligent Buildings

An intelligent building is any infrastructure which supports the process or function of the building. The infrastructure in this paper utilises a distributed control network populated with intelligent nodes. These nodes control various actuators based on the observation of the sensors under its control. Such networks are currently used for building management services such as fire detection, Heating Ventilation and Air-Conditioning (HVAC), and lighting control. The proposed intelligent network comprises of microwave readers installed around the building. These readers have been developed in the EU TIDE ARIADNE project [10] which aims to support people with special needs in the built environment. Many of the services provided in

ARIADNE are also applicable to the support of mobile robots. All users are identified through the use of contact-less smart cards. These contact-less smart cards are read by suitable microwave transponders distributed throughout the building. Therefore any user or device associated with a smart card can be located to a particular region of the building. Within the ARIADNE project this facility is used in the provision of services appropriate to a particular users requirements, but the same infrastructure can in fact support mobile robots.

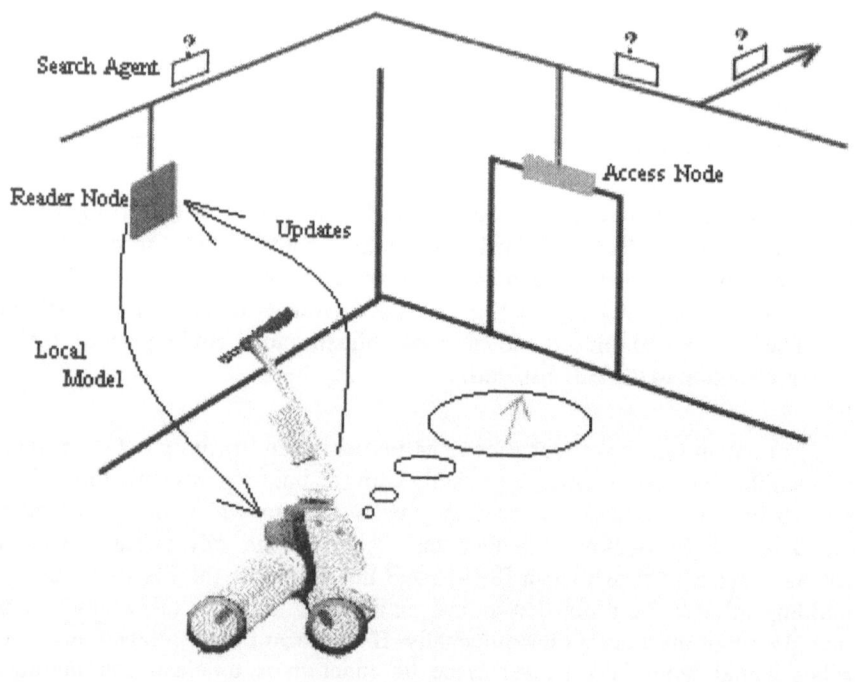

Figure 1. *The experiment installation*

The components of the experiment installation shown above are as follows:

- **Reader Node** - This identifies the robot via a contact-less smart card as it enters its area of coverage and informs the Access Node. Reader Nodes are distributed to provide maximum coverage. Maximum coverage does not imply the greatest possible physical area, but rather those areas where the robots are most likely to be required as well as those areas where activity is generally acknowledged to be the greatest.

- **Access Node** - When notified of the presence of a robot, this node transmits a local model of the environment to the robot via microwave link at the Reader Node. The Access Node initiates network based services on behalf of the robot.

The Access Node is also responsible for informing the robot of the direction it has to head with respect to the local model provided.

- *Local Model* - This model as provided by the Access Node gives the robot prior information about its immediate environment allowing the robot to correlate the data observed by its sensors.

- *Network-based services* - The network provides services in assisting the robot with its current task. For a robot involved in a transportation task, for example, the delivery of linen from the store to a nursing station, or assisting a patient to the day room, it calculates the optimum route and provides directional information as the robot travels the route.

The following sections discuss in detail issues relating to the complexity of these components, in particular, the local model, the need for filtering of this model and where it might take place, and how to maintain consistency across multiple robots and the network.

2.1 The Local Model

Rather than requiring the robot to maintain a global map, the network provides a dynamic model of the environment local to the robot as it traverses the environment. This local model contains the physical description of the environment as well as active objects that the robot can control [11]. Active objects are those features in the environment with which the robot can interact. An example of an active object is a door that opens on request. Active objects permit interaction without the need for the robot to recognise and classify them. The local model is stored by the Access Node responsible for the area described in the model. The Access Node stores a hierarchy of local sensor and actuator nodes. These nodes contain a smaller more detailed model of the area around it or the specific feature to which it relates. The Access Node accepts dynamic updates from these nodes and passes this information on to the robot via the Reader Node. The robot augments the data it observes alone with the model it has obtained from this node. This assimilated data is periodically transmitted back to the Reader Node. This simple process circumvents the need for the robot to maintain detailed maps of a large building. Every time the robot enters into the sphere of influence of a Reader Node it obtains a reasonably up-to-date map of its immediate surroundings.

There exist several standards for object description. Virtual Reality Markup Language (VRML) and AutoCad (DXF) are two of the most commonly used. There are several other proprietary modelling API's available, for example, OpenGL, 3DML, 3DS, and DirectX. These formats typically construct scene descriptions using polygons and smooth surfaces, combined with descriptions of related elements such as colour, texture, reflection coefficients, etc. This overhead is not needed by either a robot or a building user but does offer rapid and flexible inputting of the model in the first place as the model can be obtained from the architectural plans for the building. VRML is a suitable choice for use as a model descriptor. It is stored as an ASCII file. Only a subset of the language need be used, high fidelity aspects can be omitted. This results in fairly compact models being generated. These models simply contain a list of ordered vertices and a list of faces. A model of a room containing 200 3D co-ordinates organised into 134 faces occupies only 7Kb of data. For most mobile robots pre-filtering of this model to a simple 2D data set is sufficient for mapping purposes.

2.2   Filtering of Information

Due to the wide variety of service robots, different types of information and levels of detail are required to ensure safe and efficient navigation throughout the building. The level of detail in a local model of the environment given to a robot depends largely on the current task, and to a certain extent the computational load on the node. It is obviously desirable to minimise the processing at either end of the robot network link to improve speed and efficiency. There are several factors to consider when deciding whether the filtering of information should take place on the Access Node or onboard the robot.

- If a single common object model is presented to a robot, then that robot needs to be sufficiently sophisticated in order to abstract the necessary data from the common description. This places much more of a demand on the robot network link than is necessary.

- Due to the heterogeneous nature of robots, it is unreasonable to expect that all robots are capable of sophisticated reasoning. For example, many robots utilise behavioural architectures for the purposes of navigation. Their architectures do not usually lend themselves to complex reasoning. For these robots, simple instructions are probably all that is required for safe and efficient navigation.

- Filtering the information on the Access Node places much more of a computational load on that node. The node is required to profile the robot and remove any irrelevant information prior to transmitting a model to the robot. However, the demand on the robot network link is much lessened.

In any case, it is important to minimise the amount of latency between the time a robot enters an area and it being able to function there. The most pertinent information should be transmitted first, that is, the most prominent static features as well as any active objects in the region. If necessary, more detailed information may then be transmitted.

2.3   Model Consistency

To maintain consistency among multiple robots operating in the same area, dynamic changes must be propagated to all robots as quickly as possible.

- If a common model is presented to all robots in a region, then the Reader Node need only forward those changes as they occur. The onus is on the robot to decide if that information is relevant to it.

- Changes to the model stored on the robot must be sent back to the building network to allow it to modify the appropriate devices within the local environment. Conflict may arise if multiple robots attempt to update the environmental model simultaneously. Due to the asynchronous nature of the fieldbus communications mechanism, no prediction can be made as to which message will be received first. Blind conviction as to the validity of the data received at the Reader Node at any instant simply weakens the belief that the models held by devices in that area are consistent with one another. The Access Node is required to have sufficient intelligence to decide which version of the environment is most accurate.

3   Network Services

To allow the distributed control network to actively support this infrastructure, the domain of multi-agent systems is introduced as a technique for reasoning about the data necessary in the provision of services for the safe and efficient navigation of robots within the building. Within the framework of multi-agent systems, an agent engages in social discourse with one or more agents in order to achieve its goals. Two of the primary tenets in multi-agent systems research are co-operation and co-ordination. They are seen as key to an agents goal achievement. Co-operation may be necessary as an agent may not have the resources, capabilities or enough information to solve a problem and different agents may have expertise in differing areas. The Access Nodes in the control network are best viewed as agents. Co-operation and co-ordination between Access Nodes is necessary in order to provide smooth transition from one area to another. These agents co-operate through a variation of the Contract Net protocol of [12] to select the nearest idle robot for a task, and use a logical signpost mechanism to guide that robot to its destination. The Contract Net protocol provides dynamically opportunistic control where agents help each other by sharing the computation involved in the subtasks for a problem. The protocol described here provides optimal route selection through distributed search [11].

3.1   Optimal Distributed Search

When searching for the location of a feature or device the Access Node broadcasts a message to all Access Nodes on its subnet. The receiving nodes will then interrogate their local object model to see if the search feature is present. If it is then this fact will be relayed to the originator of the message in a confirm transaction. The node containing the feature then propagates a reply agent. If the reply agent has propagated back to the originator then the search is complete. Upon receiving an agent, the agent table is queried to see if this is a unique search. If so it is added to the table and propagated to adjacent nodes. The agent in the table contains a signpost to the next node. This signpost indicates the direction of travel for the most optimum route. If the search is not unique, that is, the agent has been propagated to this Access Node by some other Access Node, then a comparison is made between the two competing agents. The agent with the minimum cumulative cost is stored and the signpost adjusted accordingly. The other agent is deleted. The network will eventually converge on a solution. This route can now be followed to find the desired feature. As the user enters into the area controlled by a reader node, that node will indicate the direction of travel to complete the task.

So, for example, if a patient requests the use of a wheelchair the Access Node will query its local object model to see if a wheelchair is available locally. If not, then it will broadcast a message to all other Access Nodes requesting that an idle wheelchair should be sent to the patients location. The search strategy above results in a route being signposted to where the wheelchair is currently located and not to where it should proceed. In this situation therefore the wheelchair should request the location of the patient. This provides the wheelchair with an optimum route to the patient. The search should also be allowed to commence in the opposite direction since it would provide directions to the robot as to the optimal route back to whence it came should the request be cancelled by the patient.

This search procedure only considers one instance of an object. A class of objects can be readily handled by utilising the Contract Net protocol in conjunction with search procedures at the Access Node. When the patient requests a wheelchair, a message is broadcast by the local Access Node to all other Access Nodes on its subnet. The above algorithm assumes that the request will only be applicable to a single instance of a feature or device. In Contract Net, a task is broadcast by a manager. Agents who believe that they can service the request forward a bid to the manager. The manager then evaluates the various bids, and a contract is then awarded to the successful bidder. The Access Node local to the patient is considered to be the manager in this instance. If there are several wheelchairs available, then they will each announce themselves to the manager. In the parlance of the Contract Net protocol, the virtue of replying to the manager indicates that an interest has been expressed in the contract. It also means that any party who replied should complete the bidding process, that is, all the wheelchairs should request the location of the manager and in doing so initiate a search that will result in the optimal routes to the manager from each of the candidate wheelchairs. When the reply agents propagate back to the Access Nodes local to the wheelchairs, those Access Nodes will indicate to the manager the cumulative costs for each wheelchair' optimum route. The manager will then choose the wheelchair associated with the minimum cumulative cost.

There is a weakness in the above approach. The optimal route planned by the network for a user or device is assumed to be optimal for the lifetime of the task. However, it is quite conceivable that as the wheelchair traverses the nodes along the path decided for it, an incident occurs at some arbitrary location along that path not yet travelled which alters the optimality of travelling that route. To provide some measure of safety, Access Nodes should be allowed to dynamically adjust the cost of travelling to adjacent Access Nodes. This implies that the robot must request a new search to its target each time it comes into contact with a new reader node along the currently optimal path.

4 Conclusions

The integration of mobile robots within an intelligent infrastructure permits the services required of mobile robots to be more easily and readily realisable. The marketability of mobile robots will only improve if they can be incorporated into the building process. The infrastructure described in this paper provides services aimed at supporting the tasks of mobile robots and allows the robots to interface to existing systems. This will ultimately lead to cheaper less complex robots. Further work is necessary to provide further cues as to the nature of the robot-network communications in real healthcare environments. The effects of varying network traffic and the latencies incurred can only become evident when the robots are situated in a more active environment.

5 References

1.   Foster G T, Glover J P N 1997 Supporting Navigation Services Within Intelligent Buildings. In: Proceedings of the First Mobinet Symposium, Athens, May 1997.

2   O'Hart F, Foster G T, Lacey G, Katevas N A Structure for the Integration of Multiple Mobile Robots within Healthcare Environments. In: The 2nd Mobinet Symposium, 23rd July 1998, Edinburgh, Scotland, UK.

3   Lee J, Appenzeller G, Hashimoto H 1998 Physical Agent for Sensored, Networked and Thinking Space. In: Proceedings of the 1998 IEEE International Conference on Robotics & Automation, Leuven, Belgium, May 1998, pp838-843.

4   Smith R G, Davis A 1978 Distributed Problem Solving: The Contract Net Approach. In: 1978 Proceedings of the Second National Conference of Canada Society for Computational Studies of Intelligence.

5   Evans J 1989 HelpMate™: A Robotic Materials Transport System. In: 1989 International Journal of Robotics & Autonomous Systems, Vol. 5, pp251-256.

6   Alami R, Fleury S, Herrb M, Ingrand F, Robert F. 1998 Multi-Robot Co-operation in the MARTHA Project. IEEE Robotics & Automation Magazine, Vol. 5, No. 1, March, 36-47.

7   Kamada T, Oikawa K 1998 AMADEUS: A Mobile, Autonomous Decentralised Utility System for Indoor Transportation. In: Proceedings of the 1998 International Conference on Robotics & Automation, Leuven, Belgium, May, pp 2229-2236.

8   Lacey G, Dawson-Howe K 1996 Personal Adaptive Mobility AID (PAM-AID) for the Infirm and Elderly Blind. In: AAAI Fall Symposium, November, Boston, USA.

9   Lacey G, Dawson-Howe K 1997 Evaluation of Robot Mobility Aid for Elderly Blind. In: 1997 Symposium on Intelligent Robotic Systems, July, Stockholm, Sweden.

10   Foster G T, Solberg L A 1997 The ARIADNE Project: Supporting Access, Navigation and Information Services in Labyrinth of Large Buildings. In: 1997 Proceedings of the 4th European Conference for the advancement of Assistive Technology. AAATE'97, September, Thessaloniki.

11   Foster G T 1998 The use of local modelling techniques to facilitate information and resource management schemes in distributed control systems. PhD thesis, University of Reading, United Kingdom.

12   Sandt F, Pampagnin L 1994 Distributed Planning and Hierarchical Control for a Service Mobile Robot. In: 1994 Proceedings of the International Symposium on Intelligent Robotic Systems (SIRS'94), Grenoble, July, pp54-61.

41

Robot Path Planning Using Models
of Fluid Mechanics

C. Louste and A. Liegeois

---

1 Introduction

This chapter addresses again the problem of path planning for robotic point-vehicles using a potential field method. The idea was pioneered by Khatib [1,2] for obstacle avoidance of manipulators and vehicles in 2-dimensional (2D) planar environments, from a start location to a destination. Theoretical works on the subject [3,4,5] have further led to construction of navigation functions able to provide the planner with preferential paths, and the robot with on-line reflexive behaviours based on its sensory information.

The main difficulty of the approach using potential methods is related to avoidance of undesirable steady states which may exist between start and goal points: stable local minima and/or limit cycles. For that reason, analogues of physical behaviour of particles in continuous media have been studied. They are based on the fact that at every point of a Laplacian function there always exists a non-zero component of the gradient field [6]. Petridis and Tsiboukis [7] have used an electromagnetic analogue. Masoud *et al.* [8] have used a mechanical stress field. However, the applications have been limited in practice to 2D structured environments cluttered with polygonal obstacles. They also require long computing times, but have the advantage of being able to consider multiple goal locations within the same algorithm. Another original concept [9] has been based upon the principle of diffusion and is applicable to the problem of overtaking moving targets in a 2D universe.

Other approaches, following the works of Lozano-Perez [10], have called for a discretisation of the universe and the use of A*-like algorithms in the corresponding valued graph. For mobile robots on flat and homogeneous surfaces, the minimum-distance path from start to goal in the obstacle-free space is generally considered. Extensions have been made, which translate the problem in the Cartesian space into a similar one in the "configuration space" where the instantaneous geometrical state of the robot is defined as a point. This latter class of planning suffers from the drawback that a single optimal solution is computed, when it exists, so that entire replanning is necessary when the vehicle encounters in its course an unforeseen change of its environment. Significant improvements have then been proposed to obtain robustness of the vehicle's behaviour with respect to environmental changes and uncertainties, that induces changes in the valuations of the edges of the search graph. The corresponding algorithms have been mainly developed in the frame of projects for planetary exploration by robotic vehicles, and experimental evaluations have been performed using real vehicles and complex uneven terrain. The D* algorithm [11] provides the robot with a reflexive replanning capability, while A_ε and Genetic Algorithms principles [12,13] compute several robust sets of near-optimal paths. The latter may be used for multiple vehicles operation [14] and take into account the unevenness of the terrain, the friction between wheels and ground, given also the type of motor (thermal engine or electric motors), and the drive mechanism [15].

Following the above analysis of the abilities and limitations of the existing planning methods, the work presented in this chapter aims at using models of fluid mechanics, including viscosity and friction, in order to

i) take advantage of the Laplacian methods,

ii) include the terrain unevenness and the vehicle capabilities,

iii) obtain multiple path feasible solutions applicable to multiple vehicle missions,

iv) include multiple initial and final positions.

The rest of the chapter is organised as follows: The principles of the method and computations are given in section 2. Two-dimensional simulations illustrate the main properties in section 3. Finally, section 4 considers a real terrain and compares genetic algorithm solutions with the flows obtained by the method based on the viscous fluid equations.

2 The Equations

2.1 The Navier-Stokes Equations for an Incompressible Viscous Fluid.

Decuyper and Keymeulen [16] have used a Fluid Dynamics metaphor for generating paths between two points in a 2D factory-like environment. A "pump" between the points allows a fluid to flow from start to goal, then the "best" route is computed as the streamline following the pressure gradient when the stationary state of the fluid is attained. There cannot exist local extrema in the pressure field since the Laplace's equation is satisfied: every point of the medium is at a saddle point which is always unstable. The above work has led to a massively parallel implementation on a computer using a software developed by the specialists of fluid dynamics. It allows the rapid replanning in case of changes of goals and obstacles. However, it does not consider planning trajectories on uneven and non-homogeneous surfaces.

To handle the problem of path planning for vehicles on uneven natural terrain, some external force is to be taken into account, representing the friction between vehicle and ground and the local slope in the direction of motion. By this way, we will be able to find not only the main streamline, but also the minimum energy one, or the minimum-time one. Furthermore, considering a viscous fluid will lead naturally to pass far from obstacles where the velocity of the fluid particles is zero. The general fluid dynamics equations for viscous incompressible fluids are the Navier-Stokes equations [17] which are written as

$$\rho\frac{d\vec{v}}{dt} = \vec{f} - \nabla\vec{p} - \mu\Delta\vec{v} \tag{1}$$

$$\vec{\nabla} \bullet \vec{v} = 0 \tag{2}$$

where

ρ is the mass per unit volume
$\vec{v}$ is the velocity vector of the considered fluid particle
t is the time
$\vec{f}$ is the external force acting on the particle
p is the local pressure
μ is the viscosity coefficient
$\vec{\nabla}$ is the spatial derivation vector (the gradient operator)
Δ represents the Laplace's operator
$\bullet$ symbolises the scalar product.

2.2 The Simplified Equations

In the left-hand side of equation (1), the absolute velocity of the fluid particle may be written as:

$$\frac{d\vec{v}}{dt} = \frac{\partial\vec{v}}{\partial t} - \vec{rot}.\vec{v} \times \vec{v} \tag{3}$$

The second term in this equation allows rotation of particles to appear in the fluid behaviour, that is not admissible for a robot. Furthermore, we are interested by the stationary flow. Thus, the Stokes equations suffice to our purpose. Finally, the fluid density is of no concern since we can consider the model as an analogue and normalise the parameter values, so that $\rho=1$. Finally, one gets

$$\mu\Delta\vec{v} = \vec{\nabla}p - \vec{f} \tag{4}$$

together with equation (2) which specifies that the fluid is incompressible.

When the viscosity coefficient μ is high (low Reynolds numbers), there cannot exist any vortex having closed streamlines because of the dissipation of energy. Moreover, there cannot exist stable attractors others than the goal points since equation (2) tells us that all the fluid entering a domain must exit. The only limit case corresponds to parts where the liquid remains at rest, but no additional particle can then enter.

3 Viscous Fluid Behaviour in a 2D Environment

3.1 General Properties

The environment is modelled by a regular grid covering obstacles and free space. A difference of pressure is applied between the starting position of the vehicle (the source of fluid) and its destination point (the sink). Since the fluid is incompressible, the output flow is naturally equal to the input one. A recursive finite difference method is used for solving the Stokes equations. The boundary conditions are zero velocities on the obstacle boundaries and on the limits of the closed fluid-tight universe. In practice, setting the external friction force at a very high value at the points on the boundaries leads to the required zero velocities. If N is the number of grid points, each step of the computation requires solving a set of 3N sparse linear equations.

When compared to the other physical analogues which are based upon the pure Laplace's equation, our method incorporates the viscosity and the external forces, that gives the following advantages:

First, when the external forces are set to zero, the viscous fluid velocity increases when one goes far from obstacles, so that the streamlines following the gradient of the field are close to parts of the Voronoi diagram of a 2D binary environment. This will provide the robot with a great safety margin with respect to obstacles and impassable steeps.

Then, adding friction suppresses those of the flowlines which correspond to mechanical works greater than the potential energy due to the difference of pressure. Increasing friction restricts the admissible flowlines in a narrow domain about the near-minimum-distance path. Furthermore, it will be shown in section 4 that a fluid particle can be easily recognised as a robot navigating on an uneven terrain, taking into account the slope and the friction between the wheels and the ground.

3.2 Simulations

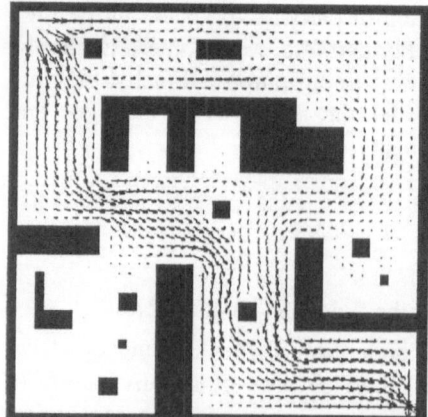

Figure 3.1: A point-to-point example without friction.

Figure 3.2: The point-to-point example with friction.

Figures 3.1 and 3.2, where the arrows are roughly proportional to the velocity, give an example of results in a 2D environment, when a single goal is given. They illustrate the robustness of the planner since several paths are generated, which allows the robot to alter its route in case of unexpected disturbances. Obtaining several paths allows also to consider multiple robot missions. The simulations confirm that the flowlines do not enter blind alleys and that no vortex is present.
Figure 3.1 shows the same environment with three goal points. Here the pressure drops are adjusted in order to get almost equal flows in the three sinks.

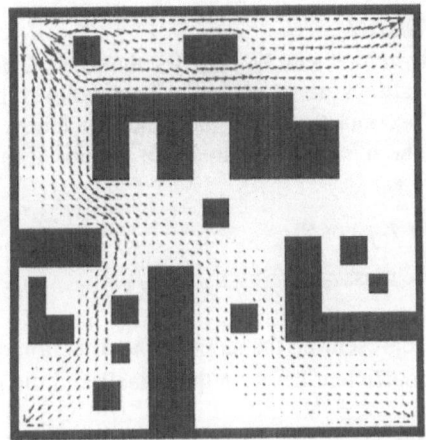

Figure 3.3: A three goal points example.

4 Path Planning on a Real Terrain

4.1 The Terrain Model

Figure 4.1: A part of the GEROMS site.

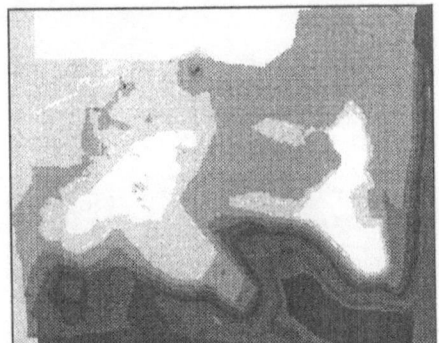

Figure 4.2: A top view of the terrain with contours.

Let us now consider a robotic vehicle navigating on an uneven natural terrain. Whatever the performance index is -distance, energy, time- one must take into account the terrain model and the corresponding interface between the wheels and the ground (elevation and friction), and the vehicle power and force. In this section, the GEROMS experimental site [18] is considered.

The navigation area dimensions of the test site are about 60mx100m. The ground represents a part of a planet surface including smooth hills, but also impassable rocks and canyons (Figure 4.1). A Digital Elevation Model is available, the best resolution of which is 0.1m. The Figure 4.2 shows some contours of the terrain. The rocks and the steeps will be considered as obstacles.

4.2 The Vehicle Model

A four-wheeled electric vehicle can be used for the tests. A detailed model has been developed [14] which allows us to compute the relationship between motor force and speed, taking into account the motor and gear efficiency. Here, we just need the resistant force, due to the terrain unevenness, to include it in the fluid equations. It can be simply modelled as:

$$f = mg(\sin \theta + K_f \cos \theta) \tag{5}$$

m is the vehicle mass
g is the gravity
θ is the angle representing the slope in the direction of motion
K_f is friction coefficient between the wheels and the ground

The slope and thus θ can be computed from the digital elevation model, given the direction of motion of the vehicle.

4.3 The Results Using The Model of Fluid Mechanics

A point-to-point planning of the robot on the GEROMS site is illustrated on Figures 4.3 and 4.4.

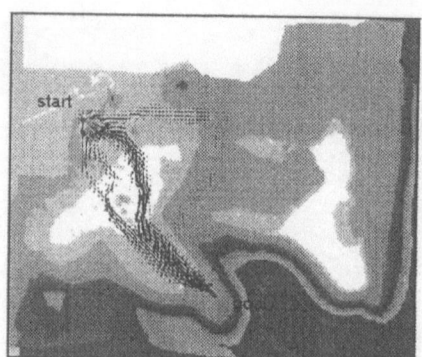

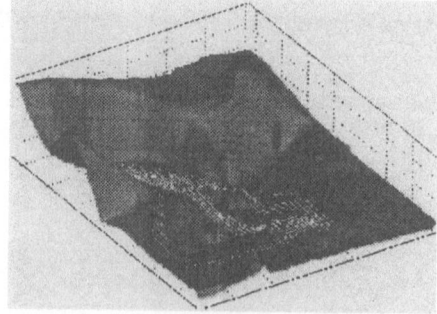

Figure 4.3: Acceptable (near-optimal) results.

Figure 4.4: A 3D representation of the paths using the fluid model.

In this example, the viscosity coefficient is constant over the area of the map, while the friction coefficient is adjusted following equation (5), where for simplicity the sine function is however replaced here by its absolute value: no energy recovery is assumed in the descents, then the cost per unit displacement is the same as when climbing. Of course, finer models are available. The pressure drop between source and sink is accordingly chosen for retaining families of paths such that the overall cost C does not exceed a predetermined threshold above the minimum cost C*:

$$C^* \leq C \leq C^*(1+\varepsilon) \tag{6}$$

The corresponding paths are labelled as acceptable with respect to the given performance criterion.

4.4 Comparison With Path Generation Based on the Principles of Genetic Algorithms

Previous research [12,13] has pointed out the advantages of using the genetic algorithms versus the computational burden and sensitivity of the classical A* and $A\varepsilon$ algorithms when searching the optimal and near-optimal paths in a valued graph associated with the terrain and vehicle models.

In relation with path planning, a path is considered as an individual in the genetic algorithm vocabulary, and a set of paths as a population. The cost of a path is taken as opposite to the performance value. An initial population is first obtained by a stochastic gradient method (or by a Metropolis algorithm) from start to goal via randomly computed intermediate points in the discrete search space. Then, the genetic algorithm operators are called for improving the performance of the population:

- Selection. The individuals are classified following the performance criterion, and the best ones are more likely to be selected in the following operations for obtaining the next generation, while the unacceptable ones are rejected. If C_{min} is the cost of the best individual of the population, the unacceptable paths have costs C_{out} which satisfy

$$C_{out} \geq C_{min}(1+\varepsilon)$$

- Cross-over. Two individuals (the parents) are selected at random in the population, the best ones having a greater chance according to the weights attributed during the selection process. The method is analogous to a roulette wheel where the area of an individual is proportional to its performance rate. Then two via point are randomly computed on each parent path, including the start and goal points, and the two crossing paths are obtained by the stochastic linking method used in the initialisation step. We thus obtain four new paths (the children) which are added to the population for future selection according to their costs.

- Mutation. Since cross-over tends to generate families ("schemata") converging to neighbour paths, we frequently use a mutation operator to explore new areas in the environment. A path is chosen at random, and two points on it, using the same methods as in the previous operators. Then a point is selected at random (with a uniform distribution) elsewhere in the environment, and is linked to the two points. Suppressing the original subpath between the two points then generates a new path between start and goal. In practice, the original path is duplicated before the mutation and is not deleted.

Figure 4.5 shows the results on the GEROMS site, using the same extremities and criterion as for experimenting the fluid method. However, the terrain has been triangulated, to reduce the cardinality of the search graph. One can observe, by comparing Figures 4.3 and 4.5, that the model of fluid mechanics provides similar topological solutions, and an additional family which has not been obtained by mutation in the genetic method within the specified number of generations (60). Furthermore, the triangulation requires a further trajectory smoothing [19] while the fluid method does it naturally.

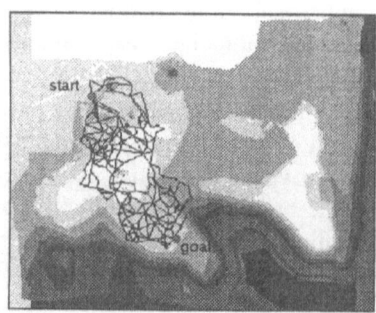

Figure 4.5: The paths generated by the genetic algorithm.

5 Conclusion

This chapter has revealed the new advantages provided by the addition of viscosity and friction to the classical implementations of well-known harmonic functions. The proposed model, similar to the Stokes equations of the fluid mechanics, allows one to get a set of admissible paths and is applicable to multi-goal problems for robotic vehicles on uneven terrain. The latter planning is obtained without additional computation.

A point-to-point application on a real terrain has been compared to the genetic algorithms principles applied to the computation of near-optimal routes and gives similar results, while directly adding continuity in the velocities and accelerations without additional trajectory smoothing.

The method is also an interesting way to get preferential routes for indoor industrial robots and motorised aids for the disabled.

Further research will include optimising the terrain meshing in order to reduce the number of equations, and experiments on the site with the robot.

The planned routes are also currently under investigation for multiple-robot behaviour optimisation, following the multiple-agent theories with communication.

Acknowledgements

The real terrain planning has been carried out thanks to the GEROMS Group and particularly the CNES Space Centre in Toulouse, France.

References

1.　　O. Khatib 1978 Dynamic control of manipulators operating in a complex environment. The 3rd CISM-IFToMM Symp. on *the Theory and Practice of Robots and Manipulators*, Udine, pp. 267-282

2.　　O. Khatib 1986 Real-time obstacle avoidance for manipulators and mobile robots. *J. of Robotics Research*, vol. 5, pp. 90-98

3.　　D. E. Koditschek 1987 Exact robot navigation by means of potential functions: some topological considerations. IEEE Int Conf. on *Robotics and Automation*, Raleigh, pp. 1-6

4.　　E. Rimon and D.E. Koditschek 1988 Exact Robot Navigation using Cost Functions: The Case of Distinct Spherical Boundaries in E^n. Proc. IEEE Int. Conf. on *Robotics and Automation*, Philadelphia, pp. 1791-1796

5.　　J. Barraquand, B. Langlois and J-C. Latombe 1992 Numerical Potential Field Techniques for Robot Path Planning. IEEE Trans. on *S.M.C.*, vol. 22, No 2, pp. 224-241

6.　　C. I. Connoly, J. B. Burns and R. Weiss 1990 Path Planning Using Laplace's Equation. IEEE Int. Conf. on *Robotics and Automation*, Cincinnati, pp. 2102-2106

7.　　V. Petridis and T.D. Tsiboukis 1992 An optimal solution to the robot navigation planning problem based on an electromagnetic analogue. S.G. Tsafestas (ed.), *Robotic Systems*, Kluwer, pp. 297-303

8.　　A.A. Masoud, S.A. Masoud and M.M. Bayoumi 1994 Robot navigation using a pressure generated mechanical stress field: the biharmonic potential field approach. IEEE Int. Conf. on *Robotics and Automation*, San Diego, pp. 124-129

9.　G. Schmidt and W. Neubauer 1992 High-speed robot path planning in time-varying environment employing a diffusion equation strategy. S.G. Tsafestas (ed.), *Robotic Systems*, Kluwer, pp. 207-215

10.　　T. Lozano-Perez and M.A. Wesley 1979 An algorithm for planning collision-free paths among polyhedral obstacles. *Communications of the ACM*, vol. 22, No 10, pp. 560-570

11.　　A. Stentz 1995 Optimal and efficient path planning for unknown and dynamic environments. *The Int. J. of Robotics and Automation*, vol. 10 (3), pp.89-100

12.   O. Pinchard, A. Liégeois, T. Emmanuel 1995 A genetic algorithm for outdoor robot path planning. Proc. of the *4th Int. Conf. on Intelligent Autonomous Systems*, Karlsruhe, pp. 413-419

13.   O. Pinchard, A. Liégeois 1996 Non deterministic methods for robot path planning in the presence of uncertainties. Proc. of *CESA '96 IMACS Multiconference (Robotics and Cybernetics)*, Lille, pp. 593-598

14.   T. Emmanuel, L. Fagegaltier, A. Liégeois Motion Planning for a patrol of outdoor mobile robots.1994. Proc. of *EURISCON '94, Malaga*, pp. 130-139

15.   A. Liégeois, L. Fagegaltier 1995 Terrain-tolerant motion planning of wheeled robotic vehicles. Proc. of the 9th World Congress on *Theory of Machines and Mechanisms*, Milano, Vol. 3, pp. 2318-2322

16.   J. Decuyper and D. Keymeulen 1990 A reactive robot navigation system based on a fluid dynamics metaphor. Springer-Verlag, (Lecture Notes in Computer Science No 496, pp.356-362)

17.   I. L. Ryhming 1985 *Dynamique des fluides*. Presses polytechniques romandes, Lavoisier diffusion

18.   M. Delail 1994 First campaigns on the GEROMS mobile robot test site. Proc. of the 2nd IARP Workshop on *Robotics in Space*, Ed. Canadian Space Agency, Montreal

19.   O. Pinchard, A. Liégeois, F. Pougnet 1996 Generalized polar polynomials for vehicle path planning. IEEE int. conf. *Robotics and Automation*, Volume 1, Minneapolis, pp. 915-920

42

Motion Planning of Mobile Robots Under a Control Constraint

P.G. Skiadas and N.T. Koussoulas

---

1 Introduction

In this paper we are going to solve the Motion Planning Problem (MPP) for nonholonomic systems without drift under a Discrete Levels Constraint (DLC), namely when their inputs take values in a given finite discrete levels set. The MPP is the problem of finding reasonable algorithms producing for every pair of points $\mathbf{x}_0$ and $\mathbf{x}_f$ an open loop control $t \rightarrow \mathbf{u}(t) = \begin{pmatrix} u_1(t) & u_2(t) & \cdots & u_m(t) \end{pmatrix}^T$ that steers $\mathbf{x}_0$ to $\mathbf{x}_f$. The DLC is a typical constraint that arises in many practical problems including the digital implementation of control policies (e.g. [1]), the hybrid control systems (e.g. [2]) and the control of physical systems (e.g. [3]).

An initial attempt to understand the effects of the DLC in the behavior of a nonlinear system was made by P. G. Skiadas and N. T. Koussoulas in [1], [4], [5] and [6]. In [1] the authors focused on the controllability properties of nilpotent systems with and without drift and in both cases they proved that the constrained system has the same controllability Lie algebra with the corresponding unconstrained one. This result is very important since it does not depend on the number of the discrete levels. Consequently, the MPP for the constrained system has at least one solution if and only if the MPP for the corresponding unconstrained system can be solved. Applications of this result appeared in [4], [5] and [6]. In [4] and [5] the authors proposed an algorithm that solves the MPP for nilpotent nonholonomic systems without drift under a DLC and they proved that it provides an exact steering. The MPP for non nilpotent systems was solved in [6], where two steering algorithms were proposed. The first algorithm [6, Algorithm 2] can be applied to nilpotentizable systems, i.e. systems that can be made nilpotent by a state or/and an input transformation, providing an exact steering. The second algorithm [6, Algorithm 3] is actually an iterated algorithm that can be applied to either nilpotentizable or general drift-free systems. A system is called general if it is neither nilpotent nor nilpotentizable. The iterated algorithm does not provide an exact steering but steers x_0 to x_f with any prescribed error.

In this paper we are going to propose another method of solving the MPP for nilpotentizable nonholonomic systems without drift under the DLC. This method is called pseudonilpotentization and uses tools from the differential-geometric control theory. The main advantage of our method is that it provides an exact steering for nilpotentizable systems.

This paper is organized as follows: In Section 2 we first establish the theoretical base of our method giving a number of results that prove the existence of the steering controls for the constrained system. Then we state the algorithm that realizes the proposed method and we prove that it provides an exact steering if the system is nilpotentizable. Finally, in Section 3 we explore the details of the pseudonilpotentization method solving the MPP for a unicycle mobile robot.

Notation

The following notation will be used throughout:

$\Re$:   the set of real numbers.

$\Re^n$:   the n-dimensional Euclidean space.

e^g: the formal exponential of the vector field g.

2   Main Results

We consider a nilpotentizable nonholonomic system without drift of the form

$$(\Sigma',Q): \quad \dot{x}(t) = f_1(x)v_1' + f_2(x)v_2' + \cdots + f_m(x)v_m',$$

where the state $x(t)$ belongs to an open set N of $\Re^n$, while the m inputs $(m \le n)$ take values in a discrete levels set $Q = \{q_1, q_2, \ldots, q_r\}$. We assume that the vector fields $f_1(x), f_2(x), \ldots, f_m(x)$, which are defined on N, are real analytic and linear independent, $\forall x \in N$. We also consider the corresponding unconstrained system

$$(S'): \quad \dot{x}(t) = f_1(x)u_1' + f_2(x)u_2' + \cdots + f_m(x)u_m',$$

where the m inputs $u_1', u_2', \ldots, u_m'$ take any value in $\Re$.

Before we state our main results we will give the following useful definitions:

Definition 1: A finite set of the form $Q = \{q_1, q_2, ..., q_r\}$ is an *admissible discrete levels set* if it satisfies the following conditions:

    (a) $q_1 \equiv 0$

    (b) $q_i \neq 0$, $i = 2, ..., r$ and

    (c)   at least one of the q_i, $i = 2, ..., r$ has sign opposite to the rest

Definition 2: A Lie Algebra $\mathcal{L}$ is *nilpotent* if there is an integer k such that all Lie brackets $[v_1, [v_2, ..., [v_k, v_{k+1}]...]]$ vanish. The smallest integer k that has this property is called *order of nilpotency of $\mathcal{L}$* and $\mathcal{L}$ is said to be nilpotent of order k.

Definition 3: A nonlinear system (S) is called nilpotent if its Controllability Lie Algebra $\mathcal{L}_S$ is a nilpotent algebra.

Definition 4: We say that the nilpotent system

$$(S')_{NE}: \quad \dot{x}(t) = g_1(x)u_1 + g_2(x)u_2 + \cdots + g_m(x)u_m$$

is a *nilpotent equivalent* of the system

$$(S'): \quad \dot{x}(t) = f_1(x)u_1' + f_2(x)u_2' + \cdots + f_m(x)u_m'$$

if there is a feedback transformation of the form $u_i' = \sum_{i=1}^{m} \beta_{ij} u_j$ that transforms (S')

to $(S')_{NE}$.

Definition 5: A control vector of the form $u = \begin{pmatrix} a_1 & a_2 & \cdots & a_m \end{pmatrix}^T$ is a *bang-bang control* if there is only one $\kappa \in \{1, 2, ..., m\}$ such that $a_\kappa \neq 0$ and $a_i \equiv 0$, $\forall i \in \{1, 2, ..., \kappa-1, \kappa+1, ..., m\}$.

<u>Remark 1:</u> We should not confuse the notion of the bang-bang control given by Definition 3 with the notion of the bang-bang control found in optimal control theory. In this paper, we use the terminology introduced by G. Lafferriere and H. J. Sussmann in [5] and we call bang-bang control every control vector of the form $v_{bb} = \begin{pmatrix} a_1 & a_2 & \cdots & a_m \end{pmatrix}^T$ where only one of the components $a_1, a_2, ..., a_m$ is different than 0.

Let us consider the vector field $G(x,v) = g_1(x)v_1 + g_2(x)v_2 + \cdots + g_m(x)v_m$. Then we denote with $v^*(a_k)$ a bang-bang control such that $G(x,v^*) = a_k G_k$, where $G_k \in \{g_1, g_2, ..., g_m\}$.

After these definitions we are ready to state our new results starting with:

Proposition 1: Let us consider a nilpotentizable system

$$(S'): \quad \dot{x}(t) = f_1(x)u_1' + f_2(x)u_2' + \cdots + f_m(x)u_m'$$

and a nilpotent equivalent

$$(S')_{NE}: \quad \dot{x}(t) = g_1(x)u_1 + g_2(x)u_2 + \cdots + g_m(x)u_m$$

of it. We assume that for each real $a \in \Re$ and for each vector field $F_i(x) \in \{f_1(x), f_2(x), \ldots, f_m(x)\}$ the solution of the differential equation $\dot{x}(t) = aF_i(x)$, $x(t=0) = x_0$ can be written in the form $x(t) = a\widetilde{F}_i(t) + x_0$, where $\widetilde{F}_i(t)$ is a time function with $\widetilde{F}_i(t=0) \equiv 0$, that does not depend on the parameter a. If the bang-bang control $u_1 = u_{bb}(a_1)$, $a_1 \in \Re$, steers $(S')_{NE}$ from x_0 to x_f in time $T > 0$ then the bang-bang control $u_1' = u_{bb}(a_1')$, where $a_1' \in \Re$, satisfies $x_f = a_1'\widetilde{F}_1(T) + x_0$ steers (S') from x_0 to x_f in time $T > 0$.

Proof: Let us consider the nilpotentizable system

$$(S'): \quad \dot{x}(t) = f_1(x)u_1' + f_2(x)u_2' + \cdots + f_m(x)u_m'$$

and a nilpotent equivalent

$$(S')_{NE}: \quad \dot{x}(t) = g_1(x)u_1 + g_2(x)u_2 + \cdots + g_m(x)u_m$$

of it. We assume that the bang-bang control $u_1 = u_{bb}(a_1)$, where $a_1 \in \Re$, steers $(S')_{NE}$ from x_0 to x_f in time $T > 0$ along the trajectory

$$x_f = x_0 e^{Ta_1 G_1(x)}.$$

We apply to the system (S') the bang-bang control $u_1' = u_{bb}(a_1')$, where $a_1' \in \Re$ satisfies

$$x_f = a_1'\widetilde{F}_1(T) + x_0 \tag{1}$$

Then at time $t = T$ the system (S') reaches

$$x_f' = x_0 + a_1'\widetilde{F}_1(T) \tag{2}$$

Combining equations (1) and (2) we get

$$x_f' = x_0 + a_1'\widetilde{F}_1(T) \equiv x_f$$

or $x_f' \equiv x_f$.

Thus the bang-bang control $u_1' = u_{bb}(a_1')$, where $a_1' \in \Re$ satisfies $x_f = a_1'\widetilde{F}_1(T) + x_0$ steers (S') from x_0 to x_f in time $T > 0$. ∎

In the case where the movement from x_0 to x_f can be performed in $M > 1$ moves we have the following corollary:

Corollary 1: Let us assume that there are M bang-bang controls of the form

$$u_1 = u_{bb}(a_1) \text{ for time } T_1 = 1$$
$$\vdots \tag{3}$$
$$u_M = u_{bb}(a_M) \text{ for time } T_M = 1$$

where $a_1, a_2, ..., a_M \in \Re$, that steer $(S')_{NE}$ from $\mathbf{x}_0$ to $\mathbf{x}_f$ through the points $\mathbf{x}_0, \mathbf{x}_1, ..., \mathbf{x}_M \equiv \mathbf{x}_f$. Then the bang-bang controls

$$\mathbf{u}_1' = \mathbf{u}_{bb}(a_1') \text{ for time } T_1' = 1$$
$$\vdots \tag{4}$$
$$\mathbf{u}_M' = \mathbf{u}_{bb}(a_M') \text{ for time } T_M' = 1$$

where the real numbers $a_1', a_2', ..., a_M'$ are such that

$$\mathbf{x}_i = a_i' \widetilde{\mathbf{F}}_i(i) + \mathbf{x}_{i-1} \tag{5}$$

steers (S') from $\mathbf{x}_0$ to $\mathbf{x}_f$ through the points $\mathbf{x}_0, \mathbf{x}_1, ..., \mathbf{x}_M \equiv \mathbf{x}_f$.

Proof: Let us assume that applying the bang-bang controls (5), we can steer $(S')_{NE}$ from $\mathbf{x}_0$ to $\mathbf{x}_f$ along the trajectory

$$\mathbf{x}_f = \mathbf{x}_0 \, e^{a_1 G_1(\mathbf{x})} e^{a_2 G_2(\mathbf{x})} ... e^{a_M G_M(\mathbf{x})} \tag{6}$$

where $\mathbf{G}_i \in \{g_1, g_2, ..., g_m\}$, $i = 1, 2, ..., M$. Trajectory (6) consists of M segments and passes through the points

$$\mathbf{x}_0, \quad \mathbf{x}_1 = \mathbf{x}_0 \, e^{a_1 G_1(\mathbf{x})}, \quad \mathbf{x}_2 = \mathbf{x}_1 \, e^{a_2 G_1(\mathbf{x})}, \quad ...., \quad \mathbf{x}_M = \mathbf{x}_{M-1} \, e^{a_M G_1(\mathbf{x})} \equiv \mathbf{x}_f$$

Then we apply to (S') the bang-bang controls (4) steering it from $\mathbf{x}_0$ to $\mathbf{x}_f$ through the intermediate points

$$\mathbf{x}_0, \quad \mathbf{x}_1' = \mathbf{x}_0 \, e^{a_1' F_1(\mathbf{x})}, \quad \mathbf{x}_2' = \mathbf{x}_1' \, e^{a_2' F_1(\mathbf{x})}, \quad ...., \quad \mathbf{x}_M' = \mathbf{x}_{M-1}' \, e^{a_M' F_1(\mathbf{x})} \equiv \mathbf{x}_f'$$

But since the real numbers $a_1', a_2', ..., a_M'$ satisfy relation (5) we can write that (see Proposition 5)

$$\mathbf{x}_1' = \mathbf{x}_0 \, e^{a_1' F_1(\mathbf{x})} = \mathbf{x}_0 \, e^{a_1 G_1(\mathbf{x})} \equiv \mathbf{x}_1$$
$$\mathbf{x}_2' = \mathbf{x}_1' \, e^{a_2' F_1(\mathbf{x})} = \mathbf{x}_1 \, e^{a_2 G_2(\mathbf{x})} \equiv \mathbf{x}_2$$
$$\vdots$$
$$\mathbf{x}_M' = \mathbf{x}_{M-1}' \, e^{a_M' F_1(\mathbf{x})} = \mathbf{x}_{M-1} \, e^{a_M G_M(\mathbf{x})} = \mathbf{x}_f \equiv \mathbf{x}_f'$$

Thus (S') follows the trajectory

$$\mathbf{x}_f = \mathbf{x}_0 \, e^{a_1' F_1(\mathbf{x})} e^{a_2' F_2(\mathbf{x})} ... e^{a_M' F_M(\mathbf{x})},$$

that steers it from $\mathbf{x}_0$ to $\mathbf{x}_f$. ∎

The above results are very useful since they can be used for the computation of the steering controls of the constrained system (Σ', Q). In particular, we have:

Proposition 2: Let us consider a nilpotentizable system of the form

$$(\Sigma', Q): \quad \dot{\mathbf{x}}(t) = \mathbf{f}_1(\mathbf{x})v_1' + \mathbf{f}_2(\mathbf{x})v_2' + \cdots + \mathbf{f}_m(\mathbf{x})v_m',$$

where Q is an admissible discrete levels set and its corresponding unconstrained system

$$(S'): \quad \dot{\mathbf{x}}(t) = \mathbf{f}_1(\mathbf{x})u_1' + \mathbf{f}_2(\mathbf{x})u_2' + \cdots + \mathbf{f}_m(\mathbf{x})u_m'.$$

530

If there exists at least one bang-bang control $\mathbf{u}_1 = \mathbf{u}_{bb}(a_1)$, $a_1 \in \mathfrak{R}$ that steers $(S')_{NE}$ from $\mathbf{x}_0$ to $\mathbf{x}_f$, then there exists at least one bang-bang control that steers (Σ',Q) $(S')_{NE}$ from $\mathbf{x}_0$ to $\mathbf{x}_f$.

Proof: Let us assume that the bang-bang control $\mathbf{u}_1 = \mathbf{u}_{bb}(a_1)$, $a_1 \in \mathfrak{R}$ steers $(S')_{NE}$ from $\mathbf{x}_0$ to $\mathbf{x}_f$ in unit time along the trajectory

$$\mathbf{x}_f = \mathbf{x}_0 \, e^{a_1 G_1(\mathbf{x})}, \tag{7}$$

where $G_1(\mathbf{x}) \in \{\mathbf{g}_1(\mathbf{x}), \mathbf{g}_2(\mathbf{x}), \dots, \mathbf{g}_m(\mathbf{x})\}$. Using Proposition 1 we can write that

$$\mathbf{x}_f = \mathbf{x}_0 \, e^{a_1 G_1(\mathbf{x})} = \mathbf{x}_0 \, e^{a_1' F_1(\mathbf{x})}, \tag{8}$$

where a_1' is a real number such that $\mathbf{x}_f = a_1' \widetilde{F}_1(1) + \mathbf{x}_0$. Since

$$a_1' F_1(\mathbf{x}) = \frac{a_1'}{Q_1} (Q_1 F_1(\mathbf{x})),$$

where the discrete level Q_1 satisfies $a_1' Q_1 > 0$, trajectory (8) can be written in the form

$$\mathbf{x}_f = \mathbf{x}_0 \, e^{\frac{a_1'}{Q_1}(Q_1 F_1(\mathbf{x}))}. \tag{9}$$

But (9) is also a trajectory of the system (Σ',Q) that steers it from $\mathbf{x}_0$ to $\mathbf{x}_f$. Thus there is at least one bang-bang control that steers the constrained system (Σ',Q) from $\mathbf{x}_0$ to $\mathbf{x}_f$. ∎

Corollary 2: The bang-bang control that steers (Σ',Q) from $\mathbf{x}_0$ to $\mathbf{x}_f$ is of the form

$$\mathbf{v}_1' = \mathbf{v}_{bb}(Q_1) \text{ for time } T_1 = \frac{a_1'}{Q_1},$$

where $\mathbf{u}_1' = \mathbf{u}_{bb}(a_1')$ is the bang-bang control that steers the corresponding unconstrained system (S') from $\mathbf{x}_0$ to $\mathbf{x}_f$ and Q_1 is a discrete level $Q_1 \in Q$ such that $a_1' Q_1 > 0$.

Proof: The proof of the corollary is included in the proof of Proposition 2 and thus is omitted.

Proposition 3: If there exist M bang-bang controls

$$\mathbf{u}_1 = \mathbf{u}_{bb}(a_1) \text{ for time } T_1 = 1$$
$$\vdots$$
$$\mathbf{u}_M = \mathbf{u}_{bb}(a_M) \text{ for time } T_M = 1 \tag{10}$$

that steer $(S')_{NE}$ from $\mathbf{x}_0$ to $\mathbf{x}_f$ then there exist M bang-bang controls of the form

$$\mathbf{v}_1' = \mathbf{v}_{bb}(Q_1) \text{ for time } T_1' = \frac{a_1'}{Q_1}$$
$$\vdots$$
$$\mathbf{v}_M' = \mathbf{v}_{bb}(Q_M) \text{ for time } T_M'' = \frac{a_M'}{Q_M} \tag{11}$$

where the real numbers a_i', $i = 1,2,...,M$ satisfy relation (5) and the discrete levels Q_i are such that $a_i'Q_i > 0$ for $i = 1,2,...,M$, that steer (Σ',Q) from $\mathbf{x}_0$ to $\mathbf{x}_f$.

Proof: Let us assume that the bang-bang controls (10) steer $(S')_{NE}$ from $\mathbf{x}_0$ to $\mathbf{x}_f$ along the trajectory

$$\mathbf{x}_f = \mathbf{x}_0 e^{a_1 G_1(\mathbf{x})} e^{a_2 G_2(\mathbf{x})} ... e^{a_M G_M(\mathbf{x})} \tag{12}$$

where $G_i(\mathbf{x}) \in \{g_1(\mathbf{x}), g_2(\mathbf{x}),...,g_m(\mathbf{x})\}$, $i = 1,2,...,M$. But according to Proposition 2 trajectory (12) can be rewritten as follows

$$\mathbf{x}_f = \mathbf{x}_0 e^{a_1' F_1(\mathbf{x})} e^{a_2' F_2(\mathbf{x})} ... e^{a_M' F_M(\mathbf{x})}, \tag{13}$$

where real numbers a_i' satisfy (5) and $F_i(\mathbf{x}) \in \{f_1(\mathbf{x}), f_2(\mathbf{x}),...,f_m(\mathbf{x})\}$ for $i = 1,2,...,M$. Trajectory (13) can also be written in the form

$$\mathbf{x}_f = \mathbf{x}_0 e^{\frac{a_1'}{Q_1}(Q_1 F_1(\mathbf{x}))} e^{\frac{a_2'}{Q_2}(Q_2 F_2(\mathbf{x}))} ... e^{\frac{a_M'}{Q_M}(Q_M F_M(\mathbf{x}))},$$

which describes a trajectory of (Σ',Q) that steers it from $\mathbf{x}_0$ to $\mathbf{x}_f$.

Thus forcing the constrained system (Σ',Q) with the bang-bang controls (11) we can steer it from $\mathbf{x}_0$ to $\mathbf{x}_f$. ∎

The above result is very important and it can be used for the construction of the algorithm that realizes the pseudonilpotentization method. This algorithm involves two steps and is given below.

Algorithm 1:
Step I: Compute the bang-bang controls that steer the corresponding unconstrained system (S') from $\mathbf{x}_0$ to $\mathbf{x}_f$.

Step II: Compute the bang-bang controls that steer the constrained system (Σ',Q) from $\mathbf{x}_0$ to $\mathbf{x}_f$ using the controls found in the previous step.

Algorithm 1 is very useful since it always converges to the bang-bang controls that steer the constrained system (Σ',Q) from $\mathbf{x}_0$ to $\mathbf{x}_f$. In particular, this is proved in the following proposition.

Proposition 4: Let us assume that there exist M bang-bang controls that steer the unconstrained system (S') from $\mathbf{x}_0$ to $\mathbf{x}_f$. Then, if Q is an admissible discrete levels set, the M bang-bang controls computed via the Algorithm 1 steer the constrained system (Σ',Q) from $\mathbf{x}_0$ to $\mathbf{x}_f$.

Proof: According to Proposition 4 if there exist M bang-bang controls that steer the unconstrained system (S') from $\mathbf{x}_0$ to $\mathbf{x}_f$ then there also exist M bang-bang controls that steer the constrained system (Σ',Q) from $\mathbf{x}_0$ to $\mathbf{x}_f$. But the steering controls of the constrained system (Σ',Q) are computed via a procedure described in the proof of Proposition 4, which coincides with Algorithm 1. Thus, if the conditions of Proposition 4 hold, then there exist M bang-bang controls that steer (Σ',Q) from $\mathbf{x}_0$ to $\mathbf{x}_f$. ∎

3 Example

In this section we are going to explore the details of the proposed method solving the MPP for the unicycle of Figure 1.

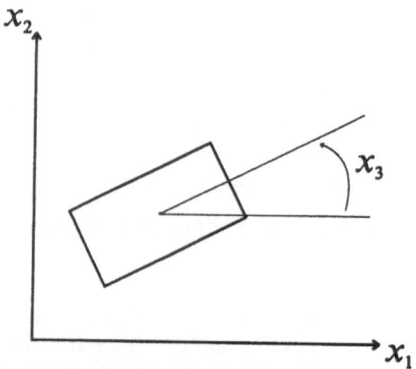

Figure 3.1: The unicycle.

This unicycle is modeled as a nilpotentizable nonholonomic system without drift and its kinematic behavior is described by the equations below

$$\dot{x}_1 = \cos(x_3)u_1'$$

$$\dot{x}_2 = \sin(x_3)u_1' \tag{14}$$

$$\dot{x}_3 = u_2'$$

where (x_1,x_2) are the Cartesian coordinates of the center of the unicycle and x_3 is the angle its main axis makes with the x_1-axis. The controls are the driving speed u_1' and the steering speed u_2'. System (14) can also be written in the form

$$(S'): \quad \dot{\mathbf{x}}(t) = \mathbf{f}_1(\mathbf{x})u_1' + \mathbf{f}_2(\mathbf{x})u_2'$$

where $\mathbf{x} = (x_1 \quad x_2 \quad x_3)^T$, $\mathbf{f}_1(\mathbf{x}) = (\cos(x_3) \quad \sin(x_3) \quad 0)^T$ and $\mathbf{f}_2(\mathbf{x}) = (0 \quad 0 \quad 1)^T$. It is easy to see that (S') is not nilpotent since

$$\mathrm{ad}_{\mathbf{f}_2}^{2n}(\mathbf{f}_1) = (-1)^n (\cos(x_3) \quad \sin(x_3) \quad 0)^T.$$

However, (S') is nilpotentizable and a nilpotent equivalent of it can be found using the feedback transformations

$$u_1' = \frac{1}{\cos(x_3)}u_1 \tag{15a}$$

$$u_2' = \cos^2(x_3)u_2 \tag{15b}$$

Denoting with $(S')_{NE}$ this nilpotent equivalent we have that

$$(S')_{NE}: \quad \dot{x}(t) = g_1(x)u_1 + g_2(x)u_2$$

where $g_1(x) = \begin{pmatrix} 1 & \tan(x_3) & 0 \end{pmatrix}^T$ and $f_2(x) = \begin{pmatrix} 0 & 0 & \cos^2(x_3) \end{pmatrix}^T$.

Let us consider the corresponding constrained system

$$(\Sigma',Q): \quad \dot{x}(t) = f_1(x)v_1' + f_2(x)v_2',$$

where the inputs v_1' and v_2' take values in the admissible discrete levels set

$$Q = \{0, \pm 0.1, \pm 1, \pm 2, \pm 3, \pm 4, \pm 5, \pm 6, \pm 7, \pm 8, \pm 9, \pm 10\}.$$

Our purpose is to steer (Σ',Q) from the initial point $x_0 = \begin{pmatrix} 0 & 0 & 0 \end{pmatrix}^T$ to the final point $x_f = \begin{pmatrix} 2 & 1 & 0 \end{pmatrix}^T$.

Note that the solutions of the differential equations

$$\dot{x}(t) = a_1 f_1(x), \; x(t = 0) = x_0$$

$$\dot{x}(t) = a_1 f_2(x), \; x(t = 0) = x_0$$

where $a_1, a_2 \in \Re$, can be written in the form

$$x(t) = a_1 \begin{pmatrix} \cos(x_{03}) \\ \sin(x_{03}) \\ 0 \end{pmatrix} + \begin{pmatrix} x_{01} \\ x_{02} \\ x_{03} \end{pmatrix} = a_1 \tilde{f}_1(t) + x_0$$

$$x(t) = a_2 \begin{pmatrix} 0 \\ 0 \\ t \end{pmatrix} + \begin{pmatrix} x_{01} \\ x_{02} \\ x_{03} \end{pmatrix} = a_2 \tilde{f}_2(t) + x_0,$$

respectively. Since Q is an admissible discrete levels set, we can use the pseudonilpotentization method in order to steer (Σ',Q) from x_0 to x_f. Using Algorithm 1 we have:

STEP I: Compute the controls that steer (S') from x_0 to x_f.

Using the method proposed by G. Lafferriere και H. J. Sussmann in [7] it is easy to compute the controls that steer $(S')_{NE}$ from x_0 to x_f. These controls are of the form

$$\mathbf{u}_1 = \begin{pmatrix} 0 & 0.5 \end{pmatrix}^T \text{ for time } T_1 = 1$$

$$\mathbf{u}_2 = \begin{pmatrix} 2 & 0 \end{pmatrix}^T \text{ for time } T_2 = 1 \tag{16}$$

$$\mathbf{u}_3 = \begin{pmatrix} 0 & -0.5 \end{pmatrix}^T \text{ for time } T_3 = 1$$

Given the controls (16) we can use Corollary 2 in order to compute the steering controls for the system (S'). These controls are of the form

$$\mathbf{u}_1' = \begin{pmatrix} 0 & a_1' \end{pmatrix}^T \text{ for time } T_1' = 1$$

$$\mathbf{u}_2' = \begin{pmatrix} a_2' & 0 \end{pmatrix}^T \text{ for time } T_2' = 1$$

$$\mathbf{u}_3' = \begin{pmatrix} 0 & a_3' \end{pmatrix}^T \text{ for time } T_3' = 1$$

where the real numbers a_1', a_2' and a_3' are given by

534

$$x_1 = x_0 \, e^{0.5g_2(x)}\big|_{t=1} = a_1' \tilde{f}_2(1) + x_0$$
$$x_2 = x_1 \, e^{2g_1(x)}\big|_{t=2} = a_2' \tilde{f}_1(2) + x_0 \qquad (17)$$
$$x_f = x_2 \, e^{-0.5g_2(x)}\big|_{t=3} = a_3' \tilde{f}_2(3) + x_0,$$

respectively. Solving (17) with respect to a_1', a_2' and a_3', we get

$$a_1' = 0.4636, \quad a_2' = 2.2789, \quad a_3' = -0.4636$$

Thus the bang-bang controls

$$u_1' = (0 \quad 0.4636)^T \text{ for time } T_1' = 1$$
$$u_2' = (2.2789 \quad 0)^T \text{ for time } T_2' = 1 \qquad (18)$$
$$u_3' = (0 \quad -0.4636)^T \text{ for time } T_3' = 1$$

steer (S') from x_0 to x_f.

STEP II: Given the controls (18) find bang-bang controls that steer (Σ', Q) from x_0 to x_f.

We choose from Q the discrete levels $Q_1 = 0.1$, $Q_2 = 2$ and $Q_3 = -0.1$. Then using Proposition 4 we can compute the steering controls of (Σ', Q)

$$v_1' = (0 \quad 0.1)^T \text{ for time } T_1' = 4.636$$
$$v_2' = (2 \quad 0)^T \text{ for time } T_2' = 1.114 \qquad (19)$$
$$v_3' = (0 \quad -0.1)^T \text{ for time } T_3' = 4.636$$

In fact, applying the bang-bang controls (19) to the system (Σ', Q) we force it to follow the trajectory

$$x_f = x_0 \, e^{4.636(0.1f_2(x))} e^{1.14(2f_1(x))} e^{4.636(-0.1f_2(x))}$$

that steers it from x_0 to x_f. The evolution of the state variables $x_1(t), x_2(t)$ and $x_3(t)$ are shown in Fig. 3.2, while controls $v_1'(t)$ and $v_2'(t)$ are shown in Figs. 3.3 and 3.4.

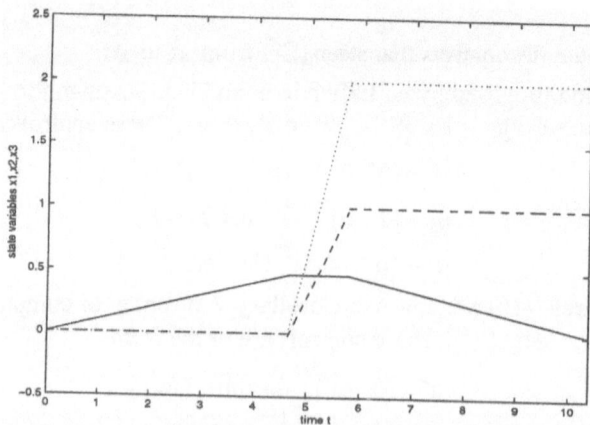

Figure 3.2: The trajectories of the state variables $x_1(t), x_2(t)$ and $x_3(t)$; $x_1(t)$: dotted line, $x_2(t)$: dashed line and $x_3(t)$: solid line.

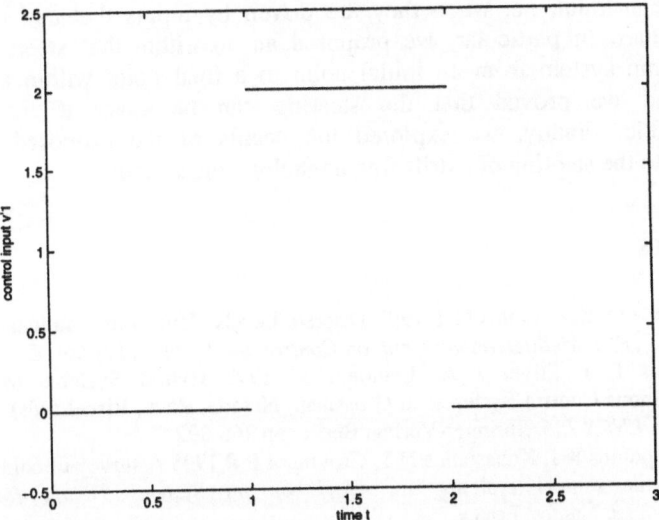

Figure 3.3: The control $v_1'(t)$ for the constrained system.

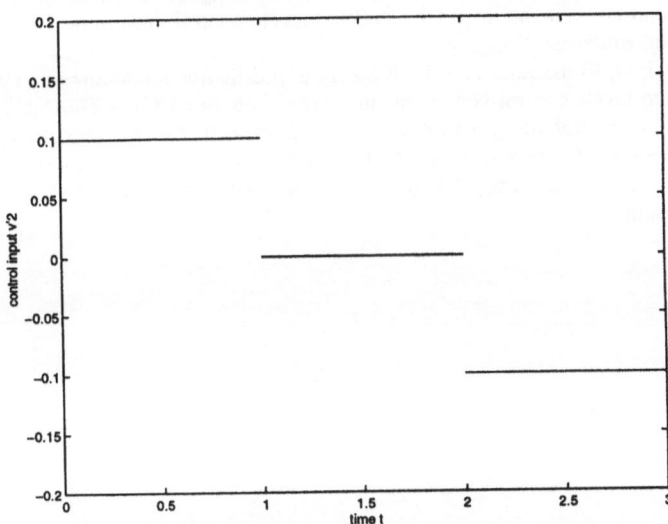

Figure 3.4: The control $v_2'(t)$ for the constrained system.

4 Conclusions

In this paper, we proposed the pseudonilpotentization method that solves the motion planning problem for nilpotentizable drift-free systems under a discrete levels set constraint, i.e. when they are driven by inputs that can assume only discrete values. In particular, we proposed an algorithm that steers a drift-free nonholonomic system from an initial point to a final point within a finite time interval and we proved that the steering can be exact if the system is nilpotentizable. Finally, we explored the details of the proposed method by applying it to the steering of a drift-free nonholonomic system.

References

[1] Skiadas P G, Koussoulas N T 1997 Discrete Levels Control of Nonlinear Systems. In *Proc. 5th IEEE Mediterranean Conf. on Control and Systems*, Paphos, Cyprus

[2] Antsaklis P J, Stiver J A, Lemmon M 1993 Hybrid Systems Modeling and Autonomous Control Systems. In Grossman, Nerode, Ravn, Riscel (eds) 1993 *Hybrid Systems. LNCS 736*. Springer-Verlag, Berlin, pp 366-392

[3] Kyriakopoulos K J, Koussoulas N T, Groumpos P P 1995 A survey of control issues on hybrid hierarchical systems. In: *Workshop on Systems Theory for Industrial Applications*, Nantes, France

[4] Skiadas P G, Koussoulas N T 1997 Motion Planning for Drift-Free Nonholonomic Systems under a Discrete Levels Control Constraint. In: *Proc. 5th IEEE Mediterranean Conference on Control and Systems*, Paphos, Cyprus

[5] Skiadas P G, Koussoulas N T 1997 Exact steering of nilpotent or nilpotentizable nonholonomic systems under the discrete levels control constraint. In: *Proc. ICIMS-ASI Conf*, Budapest, Hungary

[6] Skiadas P G, Koussoulas N T 1998 Steering of drift-free nonholonomic systems under a discrete levels control constraint. In: *Proc. Robotics 98, the Third ASCE Specialty Conference on Robotics for Challenging Environments*, Albuquerque, New Mexico

[7] Lafferriere G, Sussmann H J 1993 A differential geometric approach to motion planning. In: Li Z, Canny J F (eds) 1993 *Nonholonomic Motion Planning*, Kluwer, Amsterdam

43

Obstacle Detection and Decision Making for Intelligent Mobile Robot

A. Benmounah and H.A. Abbassi

1   Introduction

This work describes a system for autonomous navigation by an intelligent mobile robot in an unknown environment [1]. The technique described here is part of an effort to develop a low-cost intelligent mobile platform with an easy-to-maintain hardware and a relatively novel navigational concept for obstacle avoidance. Such concept is able to use its sensing, and navigation abilities to detect, measure, and make decision concerning the shortest way both to avoid the detected obstacle and contour it [5]. The general configuration of this robot is shown in Figure 1. The primary objective in the design of this robot was to experiment with a number of navigational/routing schemes for application in an environment where the locations of obstacles are not known before hand or where the position of obstacles are time-varying [2]. A detailed description of such an obstacle avoidance scheme as well as its implementation using parallel processing capabilities of a transputer to take turns on the shortest side to avoid obstacles are presented in this study [4,5].

2   Shortest Way Detection and Decision Making

The obstacle avoidance scheme described here is based on three optical (Infra-red) switches, one for forward direction (FS) and two for lateral views (LS and RS) as shown in Figure 2a. The general concept of obstacle detection and avoidance, which is based on detecting the obstacle edges and contouring around the obstacle, is illustrated in Figure 2b.
When the presence of an obstacle is detected by the forward sensor (FS), the propulsion drive comes to stop (A1) and a mechanism for detecting the relative

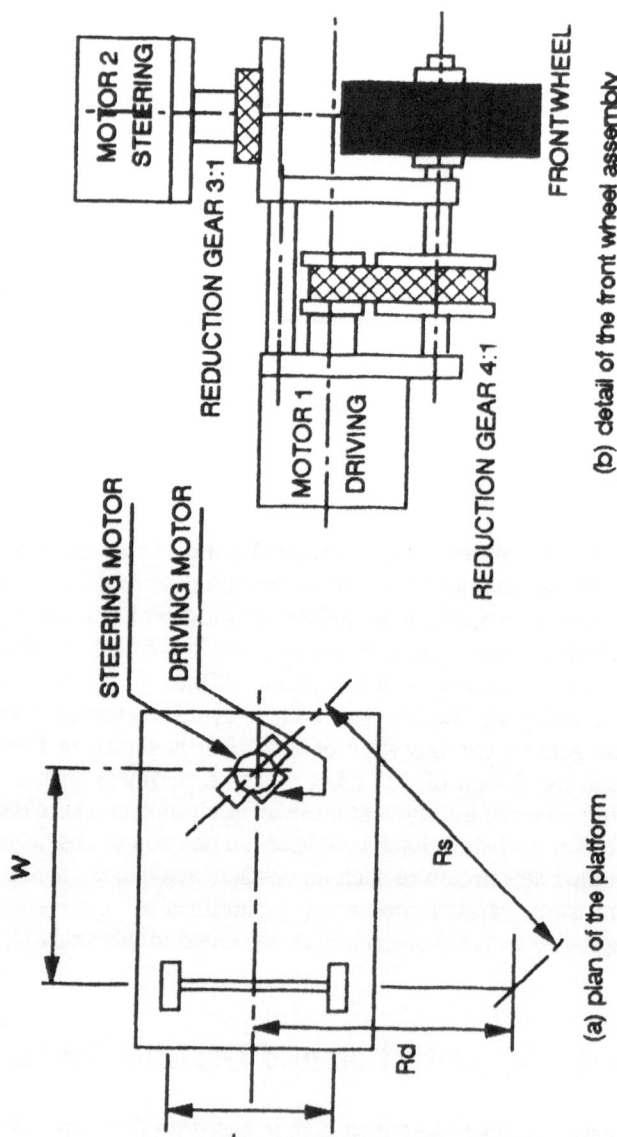

(a) plan of the platform

(b) detail of the front wheel assembly

Fig. 1 Achitecture of the mobile robot designed

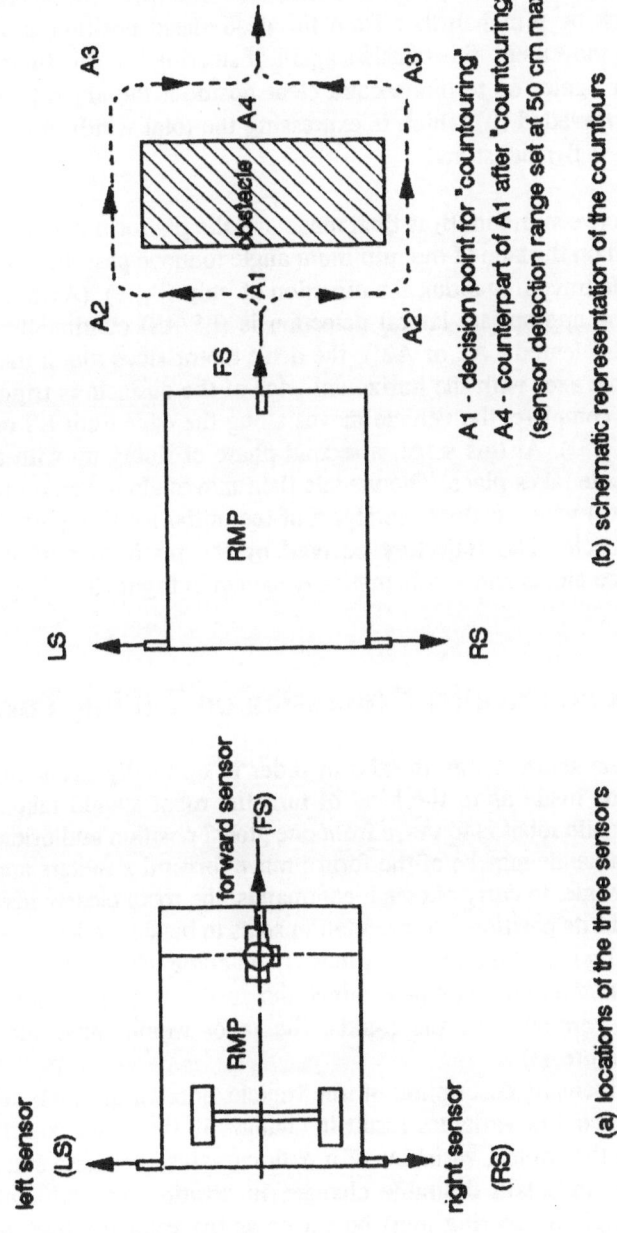

A1 - decision point for "countouring"

A4 - counterpart of A1 after "countouring"
(sensor detection range set at 50 cm max)

(b) schematic representation of the countours

(a) locations of the three sensors

Fig. 2 Outline of the obstacle avoidance concept

distance of the two edges of the obstacle is triggered. This consists of steering to the left first until FS ceases to detect the obstacle (clear). The angular movement of the steering wheel up to this clear position is then recorded (β_1). The steering wheel is then moved back by β_1 and further from the dead-ahead position to the right by another angular movement, this consists again of steering but this time to the right until FS is clear again. up to this second clear position, the angle β_2 is recorded. The total angle travelled (β) which is expressing the total width of the obstacle is calculated ($\beta=\beta_1 + \beta_2$) and stored.

A comparison between β_1 and β_2 is then made and the platform is steered along the vertical direction on the side of the minimum angle to move past the nearest (Figure 3). while travelling along the direction of min (β_1, 2), (A1 to A2 or A1 to A2') status of the appropriate lateral detection is (LS/RS) continually monitored. When the edge is cleared (A2 or A2'), the drive motor stops and a mechanism to line up the vehicle axes with the horizontal edge of the obstacle is triggered. Once this lining up is complete, the vehicle moves along the edge until LS or RS clears the edge (A3 or A3'). At this stage, a second phase of lining up with the vertical edge of the obstacle takes place. The vehicle then moves along behind the obstacle to reach point (A4) which is the counterpart of the initial starting point (A1) at the front of the obstacle. The trajectory covered by the platform during the above obstacle avoidance movement is schematically shown in Figure 3.

3   Effect on Parallel Processing on Taking Turns

After detecting the shortest way to take in order to optimally avoid the obstacle, decision should be made as to the kind of turn the robot should take. The basic behaviour of a mobile robot is to move from one global position and orientation to a another. The commands may be of the form : move forward x meters and rotate by certain constant angle. to carry out such commands, the robot clearly needs to keep track of changes in its position and orientation so as to be able to know when it has reached its spatial goal. While the robot moves, the angle which the steered wheel makes with the longitudinal axis determines the curvature of the path. When the steered wheel is parallel with the x-axis, the robot would move along a zero curvature path (Figure 4a).

Concerning the kinematic description of the vehicle, another question arises : how do changes in the control variables result in changes in the world co-ordinates. In other words, how the control variables of a path curvature limited vehicle may be altered in ordered to obtain desirable changes in position and orientation. This means how the angle of steering must be varied as the vehicle moves in order to obtain change in position and orientation (Figure 4b) [3, 4, 6]. Looking up again at Figure 1, the detail of the front wheel assembly shows that only the front wheel is motorised by two motors : one for driving the platform and the second for steering it. Because of this architectural situation one can drive the platform especially

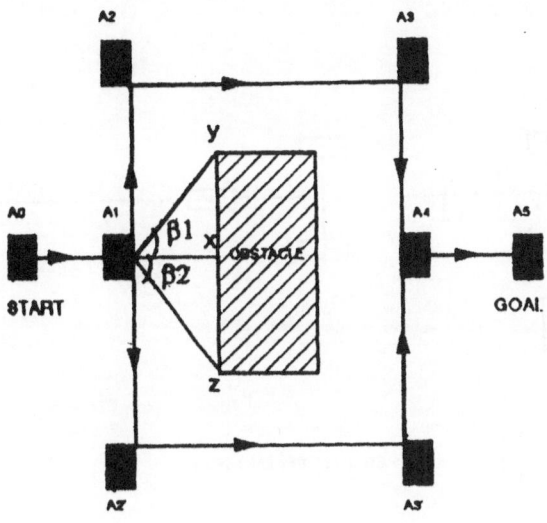

■ - Box representing the robot

A0 - Starting point of the robot

A1 - Decision point to turn left or right

A2 and A3 - Position of the robot when $\beta 1 < \beta 2$

A2' and A3' - position of the robot when $\beta 1 \geq \beta 2$

A4 - End of obstacle avoidance

A5 - Goal point

A0 to A5 - Straight path when there is no obstacle

Fig. 3 Edge detection and contouring stages

542

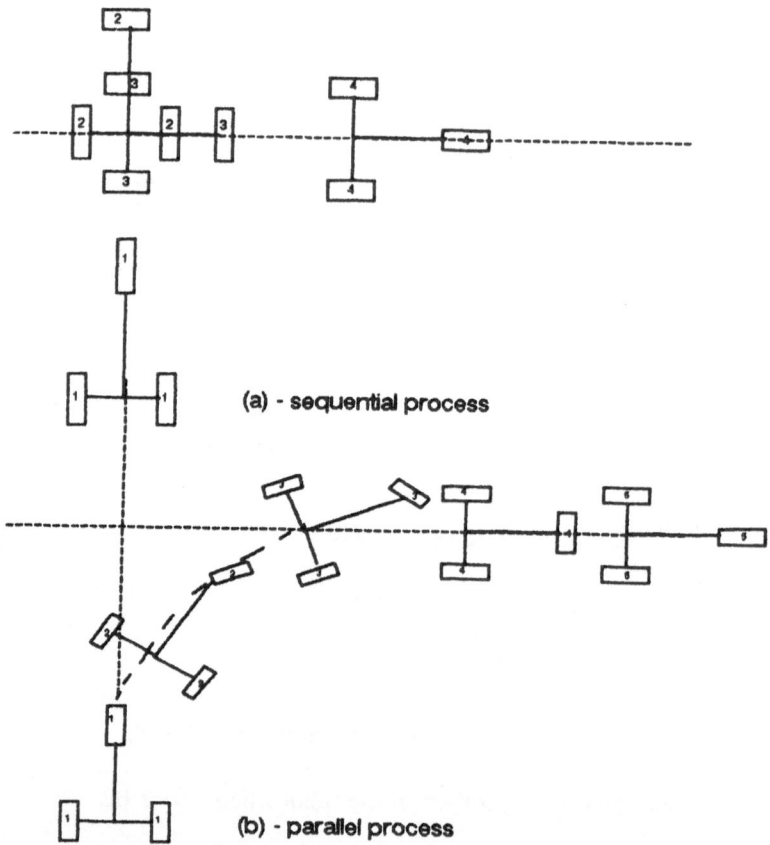

(a) - sequential process

(b) - parallel process

Fig. 4 Sequential and parallel negociation of a turn

while taking turns in two manners. One inefficient technique to ensure this would be to stop the vehicle at the appropriate position, turn the steered wheel to the desired angle Ψ and where. $R_s = W/Sin\Psi_{const}$, move the vehicle forward until the required body heading has been done, stop the body at this heading, bring the steered wheel back to its dead-ahead position, and then move again in the desired position. This is wasting time. This is practically because of processes that are executed sequentially (see Table 1).

As to the system performances and time improvement, another function technique whose initial and final value is zero and which varies smoothly is suggested. In other the steered wheel angle $R_s = W/Sin\Psi_{variable}$ must be varied while the vehicle moves along. Practically speaking, this type of function can only be obtained when processes such as in this robot control case (driving process and steering process) are executed in parallel (see Table 1). One other Strategy was to use quadratic interpolation function [7].The other function, that has attracted much attention for its use as a filtering function, is the gaussian function.

Conventional sequential programs can be expressed with variables and assignment, combined in sequential and conditional constructs. Concurrent programs are expressed with channels, inputs and outputs, which are combined in parallel and alternative constructs. Communication between two concurrent processes takes place on both inputting and the outputting processes are ready. The data to be output is then copied from the outputting process to the inputting process while both processes continue.

The schematic configuration of the hardware for the considered robot is shown in Figure 5 and the software segment for controlling the two motors are given in Table1.

4 Conclusion

From this simple tricycle vehicle architecture and its kinematic, a comparison base on the used function technique has been made between the two type of turns that can be followed by the robot while avoiding obstacles obstructing its trajectory between the start/goal point. It is concluded that for equal trajectories length, the function technique suggested where the processes are executed in parallel meaning steering while moving (R_s is a function of the steering angle $\Psi_{variable}$) is much more advantageous than the function technique where the processes are executed sequentially. This is due to movement flexibility, time consumption, and area between the start/goal point that will be occupied by the manoeuvre undertaken by the robot while taking any turn type. Depending on the control case while the robot takes turns, we can say that parallel processing permits it to take real and more natural turns, take less time, and manoeuvre in less area.

544

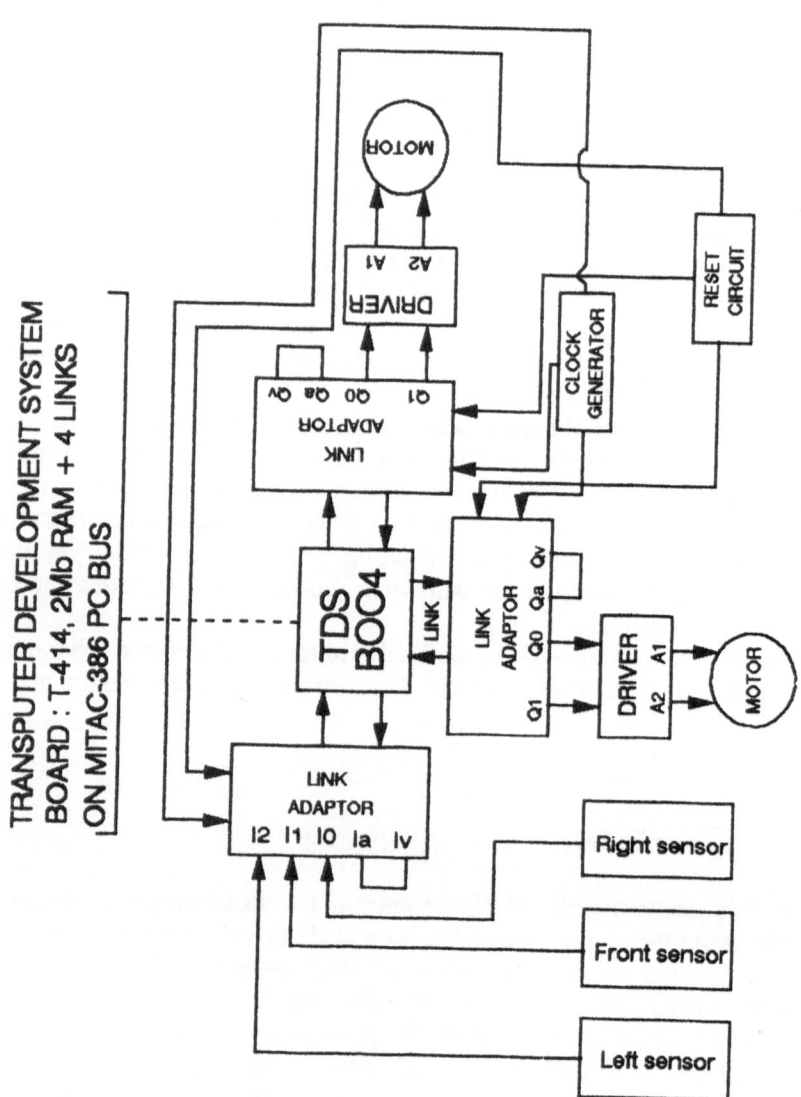

Fig. 5 Mobile robot control block diagram

```
SEQ
    IF
        obstacle is detected
        while robot is moving
        SEQ
            SEQ
                ... propulsion process
            SEQ
                ... steering process
```

a. Sequential process

```
SEQ
    IF
        obstacle is detected
        while robot is moving
        SEQ
            PAR
                SEQ
                    ... propulsion process
                SEQ
                    ... steering process
```

b. Parallel process

Table. 1 Occam shels for sequential and parallel programs

References

1.   Nitzan D 1988 Development of Intelligent Robots : Achievements and
      Issiues. IEEE J. Robotics and Automation, pp3-13

2.   Crowley J L 1988 Navigation for Intelligent Mobile Robot. IEEE J.
      Robotics and Automation, pp31-41

3.   Cox I J, Gehani N H 1989 Concurrent Programming and Robotics. Int. J.
      Rob, Res, pp3-16

4.   Inmos 1987 Transputer Architecture Reference Manual. Application note

5.   Benmounah A 1991 Transputer Control of an AGV: Design, Construction,
      and Testing of a Mobile Platform. PhD. thesis, University of Reading, UK

6.   Steer B 1985 Navigation for the Guidance of a Mobile Robot. PhD. thesis
      University of Warwick, UK

7.   Paul, L Manipulator Cartesian Path Control in Robot Motion: Planning
      and Control, Eds. M. Brady et al Cambridge, Massachusetts: The MIT
      Press

44

Dynamic Control for an Holonomic Omnidirectional Mobile Manipulator

K. Watanabe, K. Sato, K. Izumi, and Y. Kunitake

This chapter describes a control system for a holonomic omnidirectional mobile manipulator, in which the holonomic omnidirectional platform consists of three lateral orthogonal wheel assemblies and a mounted manipulator with three rotational joints is located at the center of gravity (c.g.) of the platform. We first introduce the kinematic model for the mobile manipulator and derive the dynamical model by using the Newton-Euler method, where a model which simultaneously takes account of features of both the manipulator and the mobile parts is given to analyze the effect of the movement of mounted manipulator on the platform. Then, the computed torque control and the resolved acceleration control methods are used to show that the holonomic omnidirectional mobile manipulator can be controlled so as to retain any fingertip position and orientation, irrespective of the direction of external applied force. The validity of the model and the effectiveness of the present mobile manipulator are clarified by using several numerical simulations and 3D animations.

1 Introduction

The studies on a robot with both the mobility and manipulation functions are now very attractive because they can contribute to the flexible automation in a factory, which requires the mobility of the robot at the inside and the outside of factory. Among them, in order to enhance the flexibility and degree-of-freedom of the motion for a manipulator, some studies [1]~ [6] have been proceeded on a manipulator with a mobile mechanism (hereafter, it will be called mobile manipulator). Note, however, that since almost mobile manipulators studied up to now adopt a nonholonomic system for the mobile mechanism, the movable region of the mobile part must be taken into account for the setting of desired trajectory. Additionally, such mobile manipulators have a problem that they can not retain the orientation, if a force is applied to the robot from the direction of moving constraint. Thus, the mobile manipulators mentioned above can not completely exhibit the flexibility and degree-of-freedom of the motion. From this fact, Khatib *et al.* [7] have already studied a holonomic mobile manipulator and proposed several control systems for a coordinative task using multiple mobile manipulators.

This chapter describes a mobile manipulator with a holonomic mobile mechanism, where the platform consists of three lateral orthogonal wheel assemblies [8] [9] and a manipulator with three rotational joints is located at the center of gravity (c.g.) of the platform. It is shown that the effect of the movement of mounted manipulator on the platform can be easily analyzed by introducing a model which simultaneously takes account of features of both the manipulator and mobile parts. The present mobile manipulator can be controlled so as to retain any fingertip orientation, irrespective of the direction of external force.

2 Construction of Omnidirectional Mobile Manipulator

In this study, consider the omnidirectional mobile manipulator shown in **Figure 1**. The mobile part is composed of a platform and three orthogonal-wheel assemblies [9] which are allocated at an equal distance from the center of gravity (c.g.) of the platform and with $\frac{2}{3}\pi$ [rad] between assemblies. The manipulator part is assumed to be composed of four links, but the mobile manipulator can be regarded as one with three joints mounted manipulator, because the first link is fixed on the platform.

2.1 Kinematic model of mobile part

Consider the absolute coordinates $O_w - X_w Y_w$ fixed on the plane as shown in Figure 1, as the working space for the mobile part. Additionally, in order to consider the rotational angle for the mobile part, define the moving coordinates $O_m - X_m Y_m$ having the c.g. of the platform as the origin, where let the angle between X_w and X_m axes be the rotational angle ϕ.

Defining the velocity vector for the mobile part, represented in the absolute coordinate system, as $^w\dot{p} = [^w\dot{x}_{mo} \ ^w\dot{y}_{mo} \ \dot{\phi}]^T$ and the angular velocity vector for the assemblies as $\dot{q}_{mo} = [\dot{q}_{m1} \ \dot{q}_{m2} \ \dot{q}_{m3}]^T$, it follows that the kinematic

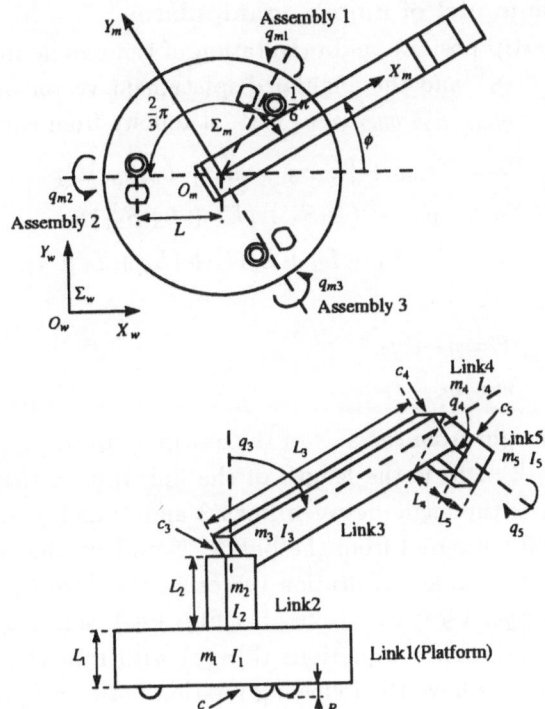

Figure 1: Mobile manipulator

model of the mobile part can be written by

$$^w\dot{p} = J_{mo}(\phi)\dot{q}_{mo} \tag{1}$$

where $J_{mo}(\phi)$ denotes the Jacobian for the mobile part given by

$$J_{mo}(\phi) = \begin{bmatrix} \frac{R}{3}(-C_\phi - \sqrt{3}S_\phi) & \frac{R}{3}(-C_\phi + \sqrt{3}S_\phi) & \frac{2R}{3}C_\phi \\ \frac{R}{3}(-S_\phi + \sqrt{3}C_\phi) & \frac{R}{3}(-S_\phi - \sqrt{3}C_\phi) & \frac{2R}{3}S_\phi \\ \frac{R}{3L} & \frac{R}{3L} & \frac{R}{3L} \end{bmatrix}$$

where R is the radius of the wheel for the assembly; L is the distance between any assembly and the c.g. of the platform; $S_\phi = \sin\phi$; and $C_\phi = \cos\phi$. From equation (1), the position and orientation vector of the platform in the absolute coordinate system, $^wp(t)=[^wx_{mo}(t) \ ^wy_{mo}(t) \ \phi(t)]^T$, is represented by

$$^wp(t) = \begin{bmatrix} ^wx_{mo}(0) + \int_0^t {}^w\dot{x}_{mo}(\tau)d\tau \\ ^wy_{mo}(0) + \int_0^t {}^w\dot{y}_{mo}(\tau)d\tau \\ \phi(0) + \frac{R}{3L}(q_{m1}(t) + q_{m2}(t) + q_{m3}(t)) \end{bmatrix} \tag{2}$$

2.2 Kinematic model of mobile manipulator

Letting the fingertip position and orientation of the mobile manipulator be $r = [{}^wx \; {}^wy \; {}^wz \; \phi \; \theta \; \psi]^T$ and the angular displacement vector of the assemblies and joints be $q = [q_{m1} \; q_{m2} \; q_{m3} \; q_3 \; q_4 \; q_5]^T$, it follows from equation (2) that

$$ {}^wx = {}^wx_{mo} + \{L_3S_3 + (L_4 + L_5)S_{34}\}C_\phi \tag{3} $$

$$ {}^wy = {}^wy_{mo} + \{L_3S_3 + (L_4 + L_5)S_{34}\}S_\phi \tag{4} $$

$$ {}^wz = R + L_1 + L_2 + L_3C_3 + (L_4 + L_5)C_{34} \tag{5} $$

$$ \phi = \phi \tag{6} $$

$$ \theta = q_3 + q_4 \tag{7} $$

$$ \psi = q_5 \tag{8} $$

where L_1 is the height from the axle of the assembly to the upper plane of the platform; $L_i (i = 2, ..., 5)$ is the length of the link i; q_3 is the angle between links 2 and 3; q_4 is the angle between links 3 and 4; and q_5 is the rotational angle of the link 5 measured from the link 4. Note here that the Euler angle $\eta_H = [\phi \; \theta \; \psi]^T$ is used as an orientation vector for the fingertip of the manipulator, and $S_3 = \sin q_3$, $C_3 = \cos q_3$, $S_{34} = \sin(q_3 + q_4)$, and $C_{34} = \cos(q_3 + q_4)$. Taking the differentiation of equations (3)~(8) with respect to the time and using equation (1), we have the following relation between $\dot{r}$ and $\dot{q}$:

$$ \dot{r} = J_r(q)\dot{q} \tag{9} $$

where $J_r(q)$ denotes the Jacobian matrix for the mobile manipulator, which is constructed from combining the element of the Jacobian matrix $J_{mo}(\phi)$ for the mobile part with the kinematic model for the manipulator part. When expressing $J_r(q)$ as

$$ J_r(q) = [\; J_{r1} \;\; J_{r2} \;\; J_{r3} \;\; J_{r4} \;\; J_{r5} \;\; J_{r6} \;] $$

each subvector can be written by

$$ J_{r1} = \begin{bmatrix} \frac{R}{3}(-C_\phi - \sqrt{3}S_\phi - \frac{AS_\phi}{L}) \\ \frac{R}{3}(-S_\phi + \sqrt{3}C_\phi + \frac{AC_\phi}{L}) \\ 0 \\ \frac{R}{3L} \\ 0 \\ 0 \end{bmatrix}, \; J_{r2} = \begin{bmatrix} \frac{R}{3}(-C_\phi + \sqrt{3}S_\phi - \frac{AS_\phi}{L}) \\ \frac{R}{3}(-S_\phi - \sqrt{3}C_\phi + \frac{AC_\phi}{L}) \\ 0 \\ \frac{R}{3L} \\ 0 \\ 0 \end{bmatrix} $$

$$ J_{r3} = \begin{bmatrix} \frac{R}{3}(2C_\phi - \frac{AS_\phi}{L}) \\ \frac{R}{3}(2S_\phi + \frac{AC_\phi}{L}) \\ 0 \\ \frac{R}{3L} \\ 0 \\ 0 \end{bmatrix}, \; J_{r4} = \begin{bmatrix} BC_\phi \\ BS_\phi \\ -A \\ 0 \\ 1 \\ 0 \end{bmatrix} $$

$$J_{r5} = \begin{bmatrix} (L_4 + L_5)C_{34}C_\phi \\ (L_4 + L_5)C_{34}S_\phi \\ -(L_4 + L_5)S_{34} \\ 0 \\ 1 \\ 0 \end{bmatrix}, \ J_{r6} = \begin{bmatrix} 0 \\ 0 \\ 0 \\ 0 \\ 0 \\ 1 \end{bmatrix}$$

where $A = L_3S_3 + (L_4 + L_5)S_{34}$ and $B = L_3C_3 + (L_4 + L_5)C_{34}$.

3 Dynamic Model for the Omnidirectional Mobile Manipulator

3.1 Assumptions

Before formulating the dynamic model for the mobile manipulator, we introduce the following assumptions on the present mobile manipulator.

(1) The wheel moves with rolling, but without slipping.

(2) The driving force induced between the wheel and the road surface is not affected by the rolling friction.

(3) Each link is rigid.

(4) The friction torque generated at the wheel and joint is due to only the viscous friction, and the effect of the Coulomb damping is not considered.

(5) The road surface is horizontal plane and there are no road disturbances.

3.2 Derivation of dynamic model

Under the assumptions, the dynamic model can be derived for the mobile manipulator by using the Newton-Euler method. By regarding the platform as one link for the manipulator, the resulting model can be calculated as a floor-fixed manipulator. Note, however, that the initial conditions must be appropriately set up for the rate and acceleration components of the coordinate systems, because the base coordinate system moves with a translation or rotation.

For the omnidirectional mobile manipulator, the coordinate systems (or frames) of the platform and each link are shown in **Figure 2**, which are defined as follows:

Σ_0: The base coordinate system, in which the absolute coordinate system Σ_w is moved translationally, having the origin on the point at which all the axles of assemblies cross.

Σ_1: The moving coordinate system having the origin on the point at which all the axles of assemblies cross.

Σ_2: The coordinate system having the origin on the root of link 2.

Σ_3: The coordinate system having the origin on the root of link 3.

Σ_4: The coordinate system having the origin on the root of link 4.

Σ_5: The coordinate system having the origin on the root of link 5.

Σ_6: The coordinate system having the origin on the fingertip of link 5.

Note here that the platform is regarded as link 1 and the fingertip of link 5 is also can be viewed as the root of a virtual link 6. The homogeneous transformation matrices $^{i-1}T_i(i = 1, ..., 6)$ from Σ_{i-1} to Σ_i can be described by

$$^{0}T_1 = \text{Rot}(z_0, \phi) \tag{10}$$

$$^{1}T_2 = \text{Trans}(0, 0, L_1) \tag{11}$$

$$^{2}T_3 = \text{Trans}(0, 0, L_2)\text{Rot}(y_2, q_3) \tag{12}$$

$$^{3}T_4 = \text{Trans}(0, 0, L_3)\text{Rot}(y_3, q_4) \tag{13}$$

$$^{4}T_5 = \text{Trans}(0, 0, L_4)\text{Rot}(z_4, q_5) \tag{14}$$

$$^{5}T_6 = \text{Trans}(0, 0, L_5) \tag{15}$$

The derivation procedure of the dynamic model is summarized below for the present mobile manipulator.

(1) Set up the initial conditions, link parameters, and dynamic parameters. When letting the gravity acceleration be $\tilde{g}$, the rotational velocity $^{0}\boldsymbol{\omega}_0$, angular acceleration $^{0}\dot{\boldsymbol{\omega}}_0$, and linear acceleration $^{0}\dot{\boldsymbol{v}}_0$ written in Σ_0 are given by

$$^{0}\boldsymbol{\omega}_0 = [0 \ 0 \ \dot{\phi}]^T \ , \ ^{0}\dot{\boldsymbol{\omega}}_0 = [0 \ 0 \ \ddot{\phi}]^T \tag{16}$$

$$^{0}\dot{\boldsymbol{v}}_0 = \begin{bmatrix} ^{w}\ddot{x}_{mo} \\ ^{w}\ddot{y}_{mo} \\ \tilde{g} \end{bmatrix} \tag{17}$$

because the base is movable for the case of mobile manipulators. It is assumed that there are no external forces and moments acting on the fingertip of manipulator, i.e.,

$$^{5}\boldsymbol{f}_5 = 0 \quad ^{5}\boldsymbol{n}_5 = 0 \tag{18}$$

(2) Iteratively compute the rotational velocity $^{i}\boldsymbol{\omega}_i$, rotational acceleration $^{i}\dot{\boldsymbol{\omega}}_i$, and linear acceleration $^{i}\dot{\boldsymbol{v}}_i$ of the root of the link written in Σ_i, from link 1 out to link 5, through

$$^{i}\boldsymbol{\omega}_i = {}^{i}\boldsymbol{R}_{i-1}{}^{i-1}\boldsymbol{\omega}_{i-1} + e_i\dot{q}_i \tag{19}$$

$$^{i}\dot{\boldsymbol{\omega}}_i = {}^{i}\boldsymbol{R}_{i-1}{}^{i-1}\dot{\boldsymbol{\omega}}_{i-1} + e_i\ddot{q}_i \tag{20}$$

$$^{i}\dot{\boldsymbol{v}}_i = {}^{i}\boldsymbol{R}_{i-1}[{}^{i-1}\dot{\boldsymbol{v}}_{i-1} + {}^{i-1}\dot{\boldsymbol{\omega}}_{i-1} \times {}^{i-1}\hat{\boldsymbol{P}}_i$$
$$+ {}^{i-1}\boldsymbol{\omega}_{i-1} \times ({}^{i-1}\boldsymbol{\omega}_{i-1} \times {}^{i-1}\hat{\boldsymbol{P}}_i)] \tag{21}$$

Here, $e_i\dot{q}_i$ and $e_i\ddot{q}_i$ denote the angular velocity vector and angular acceleration vector of rotational joint i, respectively, where e_i denotes the unit vector representing the direction of driving axle for the link i, i.e., $e_1 = 0, e_2 = 0, e_3 = e_3^y, e_4 = e_4^y, e_5 = e_5^z$, where e_i^j is the unit vector of jth direction for the link i. ${}^iR_{i-1}$ is the rotational matrix from Σ_i to Σ_{i-1} and ${}^{i-1}\hat{P}_i$ is the position vector written in Σ_{i-1} from the origin of Σ_{i-1} to the origin of Σ_i.

(3) Letting the force and moment exerted on link i by link $i-1$, written in Σ_i, be ${}^if_i$ and ${}^in_i$, compute them from link 5 back to link 1 such that

$$
{}^if_i = m_i\,{}^i\dot{v}_i + {}^i\dot{\omega}_i \times m_i\,{}^i\hat{s}_i + {}^i\omega_i \times ({}^i\omega_i \times m_i\,{}^i\hat{s}_i)
$$
$$
+ {}^iR_{i+1}\,{}^{i+1}f_{i+1} \tag{22}
$$

$$
{}^in_i = {}^iI_i\,{}^i\dot{\omega}_i + {}^i\omega_i \times ({}^iI_i\,{}^i\omega_i) + m_i\,{}^i\hat{s}_i \times {}^i\dot{v}_i + e_i^j c_i \dot{q}_i
$$
$$
+ {}^iR_{i+1}({}^{i+1}\hat{p}_{i+1} \times {}^{i+1}f_{i+1} + {}^{i+1}n_{i+1}) \tag{23}
$$

where m_i and ${}^iI_i$ denote the mass and inertia tensor at the origin of Σ_i of link i, and ${}^i\hat{s}_i$ is the position vector written in Σ_i from the origin of Σ_i to the center of mass of link i. Also, $e_i^j c_i \dot{q}_i$ in equation (23) is the viscous friction inducing at the joint located at the origin of Σ_i, where it is assumed that the actuator driving the joint is attached to the root of link i. For the force and its moment acting on the center of mass of the platform, let the X_m-, Y_m-, and Z_m-directional forces and their moments around their axes be represented by $\{f_x, f_y, f_z\}$ and $\{n_x, n_y, n_z\}$, respectively. In addition, define the driving input torques for each assembly and manipulator be $\{u_1, u_2, u_3\}$ and $\{\tau_3, \tau_4, \tau_5\}$, respectively. Since $\{f_x, f_y, f_z\}$ can be represented by x_1-, y_1-, and z_1-directional forces on Σ_1 and similarly $\{n_x, n_y, n_z\}$ can be represented by moments around $\{x_1, y_1, z_1\}$, we have

$$
f_x = {}^1f_1^T e_1^x, \quad f_y = {}^1f_1^T e_1^y, \quad f_z = {}^1f_1^T e_1^z \tag{24}
$$

$$
n_x = {}^1n_1^T e_1^x, \quad n_y = {}^1n_1^T e_1^y, \quad n_z = {}^1n_1^T e_1^z \tag{25}
$$

$\{\tau_3, \tau_4, \tau_5\}$ are balanced to the moments applied to axes $\{y_3, y_4, z_5\}$ written in Σ_3, Σ_4, and Σ_5, and hence

$$
\tau_3 = {}^3n_3^T e_3^y, \quad \tau_4 = {}^4n_4^T e_4^y, \quad \tau_5 = {}^5n_5^T e_5^z \tag{26}
$$

The driving forces $D_i (i = 1, 2, 3)$ and vertical resistances $N_i (i = 1, 2, 3)$, generated between the assemblies and the road plane, can be written by using f_x, f_y, f_z, n_x, n_y, and n_z such as
$[D_1 \; D_2 \; D_3 \; N_1 \; N_2 \; N_3]^T =$

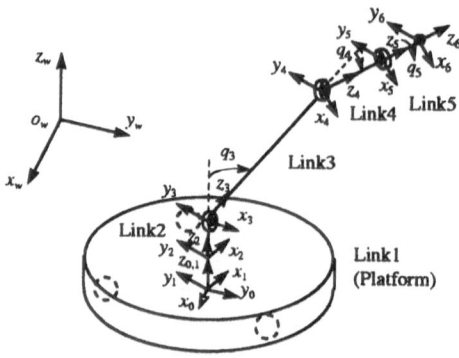

Figure 2: Coordinate frame allocation

$$
\begin{bmatrix}
-\frac{1}{3} & \frac{1}{\sqrt{3}} & 0 & 0 & 0 & \frac{1}{3L} \\
-\frac{1}{3} & -\frac{1}{\sqrt{3}} & 0 & 0 & 0 & \frac{1}{3L} \\
\frac{2}{3} & 0 & 0 & 0 & 0 & \frac{1}{3L} \\
-\frac{R}{\sqrt{3}L} & -\frac{R}{3L} & \frac{1}{3} & \frac{1}{3L} & -\frac{1}{\sqrt{3}L} & 0 \\
\frac{R}{\sqrt{3}L} & -\frac{R}{3L} & \frac{1}{3} & \frac{1}{3L} & \frac{1}{\sqrt{3}L} & 0 \\
0 & \frac{2R}{3L} & \frac{1}{3} & -\frac{2}{3L} & 0 & 0
\end{bmatrix}
\begin{bmatrix}
f_x \\ f_y \\ f_z \\ n_x \\ n_y \\ n_z
\end{bmatrix}
$$

It is assumed, from the dynamic property of driving system, that the driving input $u_i (i = 1, 2, 3)$ for each assembly can be in the following form:

$$ u_i = I_w \ddot{q}_{mi} + c \dot{q}_{mi} + R D_i \quad i = 1, 2, 3 \tag{27} $$

where I_w is the inertia moment of the assembly and c is the viscous friction factor. Introducing the total driving input vector $\tau = [u_1\ u_2\ u_3\ \tau_3\ \tau_4\ \tau_5]^T$ for the mobile manipulator and using equations (16)~(27) gives

$$ \tau = M(q)\ddot{q} + h(q, \dot{q}) + V(q)\dot{q} + g(q) \tag{28} $$

where M, h, V, and g are the inertia matrix of the dynamic model, centrifugal and Coriolis force term, viscous friction term, and gravity term, respectively.

Solving the equation of motion (equation (28)) obtained from 1. ~ 3. with respect to $\ddot{q}$, it results in

$$ \frac{d}{dt}\begin{bmatrix} q \\ \dot{q} \end{bmatrix} = \begin{bmatrix} \dot{q} \\ M(q)^{-1}(\tau - h(q, \dot{q}) - V(q)\dot{q} - g(q)) \end{bmatrix} $$

In the following, given the initial state vector $[q(0), \dot{q}(0)]$ and input torque vector $\tau(t)$, we numerically analyze the behavior of the mobile manipulator.

Table 1   Assembly parameters

Radius of each driving wheel [m]	$R=0.0245$
Moment of inertia [kgm^2]	$I_w=0.0211$
Viscous damping coefficient [Nms]	$c=0.1$

4   Model Based Control

4.1 Computed torque control

The computed torque control is a control method in which the controlled system is linearized by using the joint variables and the resultant system is compensated by a linear servo. In this approach, given the desired joint variables $q_d, \dot{q}_d, \ddot{q}_d$, the driving input torque vector τ is provided by

$$\tau = \hat{M}(q)\ddot{q}^* + \hat{h}(q, \dot{q}) + \hat{V}(q)\dot{q} + \hat{g}(q) \tag{29}$$

$$\ddot{q}^* = \ddot{q}_d + K_v(\dot{q}_d - \dot{q}) + K_p(q_d - q) \tag{30}$$

where $\hat{M}$, $\hat{h}$, $\hat{V}$, and $\hat{g}$ are the estimated inertia matrix of the dynamic model, centrifugal and Coriolis force term, viscous friction term, and gravity term, respectively, and K_v and K_p are the symmetric and positive-definite servo gain matrices.

4.2 Resolved acceleration control

In this approach, given the desired fingertip position and orientation variables $r_d, \dot{r}_d, \ddot{r}_d$, the driving input torque vector τ is computed by

$$\tau = \hat{M}(q)J_r^{-1}(q)(\ddot{r}^* - \dot{J}_r(q)\dot{q}) + \hat{h}(q, \dot{q})$$
$$+ \hat{V}(q)\dot{q} + \hat{g}(q) \tag{31}$$

$$\ddot{r}^* = \ddot{r}_d + K_v(\dot{r}_d - \dot{r}) + K_p(r_d - r) \tag{32}$$

5   Simulation Results

The physical parameters for the omnidirectional mobile manipulator are shown in **Tables 1** to **3**. Here, it was assumed that the center of mass for each link, $L_{gi}(i = 1, ..., 5)$, is located at the mid point of the link length L_i, and the mass products of inertia in the inertia tensor defined at the origin of Σ_i are all zero.

Table 2 Platform parameters

Distance between the driving wheel and the center of the platform [m]	$L=0.178$
Mass [kg]	$m_1=30.0$
Length [m]	$L_1=0.12$
Moment of inertia [kgm^2]	$I_{1xx}=0.61275$, $I_{1yy}=0.61275$, $I_{1zz}=0.93750$

Table 3 Manipulator parameters

Mass [kg]	$m_2=1.25$, $m_3=4.17$, $m_4=0.83$, $m_5=0.40$
Length [m]	$L_2=0.15$, $L_3=0.50$, $L_4=0.10$, $L_5=0.06$
Moment of inertia [kgm^2]	$I_{2xx}=0.01004$, $I_{2yy}=0.01004$, $I_{2zz}=0.00133$
	$I_{3xx}=0.34972$, $I_{3yy}=0.34972$, $I_{3zz}=0.00445$
	$I_{4xx}=0.00321$, $I_{4yy}=0.00321$, $I_{4zz}=0.00089$
	$I_{5xx}=0.00064$, $I_{5yy}=0.00064$, $I_{5zz}=0.00032$
Viscous damping coefficient [Nms]	$c_3=c_4=c_5=0.1$

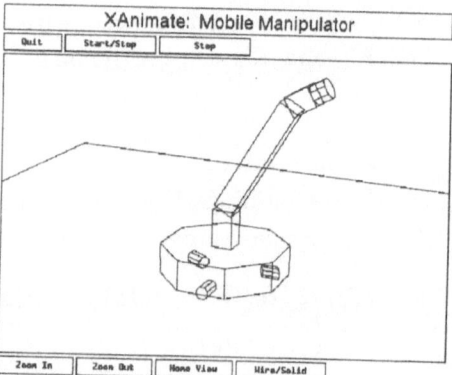

Figure 3: Initial posture of mobile manipulator

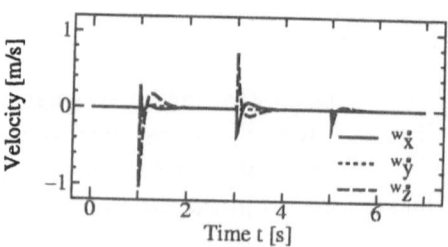

Figure 4: Tip velocities of manipulator ($^w\dot{x}$, $^w\dot{y}$ and $^w\dot{z}$)

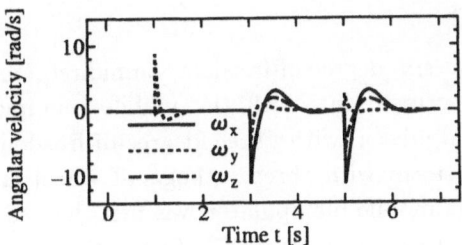

Figure 5: Tip angular velocities of manipulator $(\omega_x, \omega_y$ and $\omega_z)$

5.1 Motion in holding an orientation

As shown in **Figure 3**, it was assumed that the initial conditions are to be $r(0) = [0.389\ 0\ 0.808\ 0\ \frac{\pi}{3}\ 0]^T$, $q(0) = [0\ 0\ 0\ \frac{\pi}{6}\ \frac{\pi}{6}\ 0]^T$, and $\dot{r}(0) = \dot{q}(0) = 0$.

The torque required to retain such initial conditions is obtained by using the computed torque control method. It was assumed that the servo gains for equation (30) are given by $K_v = \text{diag}\,(16,\ 16,\ 16,\ 16,\ 16,\ 16)$ and $K_p = \text{diag}\,(64,\ 64,\ 64,\ 64,\ 64,\ 64)$. In addition, the following disturbances are assumed to be applied to the fingertip; $^0f_H = [f_x\ f_y\ f_z]$ [N] and $^0n_H = [n_x\ n_y\ n_z]$ [Nm] with

$$1.0 \le t < 1.01\ f_x = f_z = 100,\ f_y = 500,\ n_x = n_y = n_z = 0$$
$$3.0 \le t < 3.01\ f_x = f_y = f_z = 0,\ n_x = n_z = 1,\ n_y = 5$$
$$5.0 \le t < 5.01\ f_x = f_z = 100,\ f_y = 500,\ n_x = n_z = 1,\ n_y = 5$$

where ^0f_H and ^0n_H denote the force and moment acting on the fingertip written in the base coordinate, respectively. The simulation was implemented using the simulation time 7 [s] and the sampling width 0.01 [s].

This motion is aimed at showing that the robot is capable of retaining the initial orientation, irrespective of the direction of external force acting on the fingertip. Since a situation for the case of $\det J_r(q) = 0$ will appear depending on the direction and magnitude of external force acting on the fingertip, the computed torque control was used as a control method.

The simulation results of the velocity vector for the position and orientation of the fingertip, $\dot{r}_\omega(t) = [^w\dot{x}\ ^w\dot{y}\ ^w\dot{z}\ \omega_x\ \omega_y\ \omega_z]^T$, are depicted in **Figures 4** and **5**. It is seen from these figures that the fingertip can generate a velocity vector with six degree-of-freedom, against an external force in three-dimensional space, and that the robot can achieve the steady state $\dot{r}_\omega = 0$ retaining the initial orientation, after applying the external force.

6 Conclusions

In this chapter, a six degree-of-freedom omnidirectional mobile manipulator has been described, in which the mobile manipulator is composing of a mounted manipulator with three degree-of-freedom and a holonomic omnidirectional-platform with three orthogonal wheel assemblies. The dynamic model for the mobile manipulator was first derived. It was shown that the robot orientation can be controlled by using the constructed model, irrespective of the direction of external force acting on the fingertip.

References

[1] Minami, M, Fujiwara, N and Tsuge, H 1993 Position and Orientation Control of Manipulator Mounted on Autonomous Mobile Robot. *J. of the Robotics Society of Japan* 11(1): 156-164 (in Japanese)

[2] Minami, M, Asakura, T, Fujiwara, N, and Kanbara, K 1997 Effects of Inverse Dynamics Compensation for Nonholonomic Mobile Manipulators. *J. of the Robotics Society of Japan* 15(2): 216-222 (in Japanese)

[3] Minami, M, Hatano, M and Asakura, T 1997 Moving Operations of Mobile Manipulators Traveling on Unknown Irregular Terrain – Tracking Control Using Adaptive Controllers and neural Networks –. *J. of the Robotics Society of Japan* 15(7): 1004-1011 (in Japanese)

[4] Lee, J-K and Cho, H S 1997 Mobile Manipulator Motion Planning for Multiple Tasks Using Global Optimization Approach. *J. of Intelligent and Robotic Systems* 18: 169-190

[5] Huang, Q, Sugano, S and Kato, I 1995 Stability Control for a Vehicle-Mounted Manipulator – Stability Evaluation Criteria and Manipulator Compensatory Motion –. *Trans. of the Society of Instrument and Control Engineers* 31(7): 861-870 (in Japanese)

[6] Yamamoto Y and Yun, X 1994 Coordinating Locomotion and Manipulation of Manipulator. *IEEE Trans. on Automatic Control* 39(6): 1326-1332

[7] Khatib, O, Yokoi, K, Chang, K, Ruspini, D, Holmberg, R, and Casal, A 1996 Coordination and Decentralized Coorperation of Multiple Mobile Manipulator. *Journal of Robotic Systems* 13(11): 755-764

[8] Pin, F G and Killough, S M 1994 A New Family of Omnidirectional and Holonomic Wheeled Platforms for Mobile Robots. *IEEE Trans. on Robotics and Automation* 10(4): 480-489

[9] Tang, J, Watanabe, K and Shiraishi, Y 1996 Design of Traveling Experiment of an Omnidirectional Holonomic Mobile Robot. *Procs. of IROS96, Vol 1*, pp 66-73

[10] Yoshikawa, T 1990 *Foundation of Robotics: Analysis and Control.* MIT Press, Cambridge, MA

[11] Craig, J J 1989 *Introduction to Robotics: Mechanics and Control.* Addison-Wesley, Reading, MA

45

Intelligent Control of an Extension-Cableless Robotic Unicycle: A Study of Mechanical Controllabilty via Minimum Entropy

V.S. Ulyanov, T. Ohkura, K. Yamafuji and S.V. Ulyanov

1 Introduction

Emulation of intelligence or some capabilities which are inherited by human beings or animals has always been one of the important fields in the research of advanced robotics. Speaking of locomotive abilities, it's widely known that human beings and some animals have rather good on walking, running or jumping locomotion. And we believe that it is important and useful to emulate human being's walking, running or jumping ability by a robot. In fact, within the last few years several researchers have achieved certain results on emulatin human being/animals' walking, running or jumping ability by robot.

We focused on the emulation of human being unicycle riding ability by a robot. It is well-known that the unicycle system is an inherently unstable system and both longitudinal and lateral stability control are needed at the same time to maintain the unicycle's postural stability. It has an unstable problem in three dimensions (3D). However, a rider is able to achieve a postural stability on a unicycle keeping the wheel speed constant and changing the unicycle's posture in yaw or roll directions using his body flexiblity, good sensory system, skill and intelligence computational abilities. Investigating this phenomenon and emulation of the system by a robot, we aim at the construction of a biomechanical model of human motion dynamics, as at well as the evaluation of the new methods for the stability analysis and for the intelligent smart control of an inherently unstable system.

Computer simulation using dynamic equations of motion of the developed robotic unicycle has been carried out according to the proposed control scheme [1,2]. And the experiments have been conducted with the proposed fuzzy gain schedule PD-control method [1]. The usage of three rate-gyro sensors installed on the robot for the measurement of the robot's postures in 3D give us satisfactory results in posture stability and driving. Experimental results [1] show that both robot's longitudinal and lateral posture can be stabilized successfully. Thus, the proposed fuzzy gain schedule PD-control method provides one of the reasonable approaches to handle such a nonlinear problem existing in the unicycle robotic system.

In this chapter the thermodynamic approach [3] for investigation of an optimal control process and artificial life of mobile robots was used. A new physical measure, the *minimum entropy production rate* for the description of the intelligent dynamic behavior and thermodynamic stability condition [4,5,10] of a biomechanical model with an AI control system for the robotic unicycle are introduced. This physical measure is used as a fitness function in a GA for the computer simulation of the intuition mechanism as a *global* random searching measure for the decision-making process about the optimal control of a global stability on the robotic unicycle throughout the full space of possible solutions. An instinct mechanism simulation based on FNN is considered as a *local* active adaptation process with the minimum entropy production rate in the learning process of the vestibular system by teaching the control signal accordingly to the model representation results in [4]. Unlike in the papers [1,6] computer simulations in this study are carried out by the usage of *thermodynamic* equations for the motion [5,10] of the robotic unicycle. Entropy production rate and entropy measures for the robotic unicycle motion and the control system are calculated directly from the proposed thermodynamic equations of motion. From the results obtained in this study by the fuzzy simulation and soft computing (based on GA and FNN) it is obvious that the intelligent behavior controllability and postural stability of the robot is largely improved by two fuzzy gain schedule PD-controllers in comparison with those controlled only by a conventional PD and a fuzzy gain schedule PD-controller [2]. It is confirmed that the proposed fuzzy gain schedule PD-controller is very effective for the handling of the system's nonlinearity dealing with the robot's posture stability control. Furthermore an important result is that the principle of minimum entropy production rate gives a quantitative measure concerning to the controllability as well as the qualitative explanations. Thus, we provide a *new Benchmark* for the controllability of unstable essentially nonlinear nonholonomic dynamic systems by means of intelligent tools [1,4,5,9] based on a new physical concept of robust control: *the minimum entropy production rate in control systems and in control object motion in general.*

2 Biomechanical Qualitative Control Model and Design Modeling of the Extension-Cableless Robotic Unicycle

As it is observed in previous papers [1,2,4-6], human rider's controlling actions on a unicycle using his torso, shoulders and arms are quite complicated. And usually the rider's posture or altitude is not always symmetric to the wheel's principal axis. In the improved unicycle model [1] two unique and characteristic structures are contrived. One is an overhead rotor (hereafter, rotor) mounted on the torso (body) and another is the double 4 bar-closed link mechanism on both sides of the wheel. These two structures are considered to play their important roles in our biomechanical control system.

2.1 Biomechanical Qualitative Control Model

The human riding control of the unicycle as logic-dynamic hierarchical process is formed by: 1) dynamic mechanical system "human riding - unicycle"; 2) decision-making process of unicycle intelligent control with different levels of *skill* operations; 3) logic behavior for coordination of human body and foots based on *intuition, instinct, and emotion* mechanisms; and 4) distributed information system for cooperative coordination of sub-systems at biomechanical model [4]. In accordance with this representation of dynamic control process we use here a hierarchical logic structure of distributed knowledge representation for artificial life of robotic unicycle as described in [4,5]. For description of artificial life of robotic unicycle we use methods of qualitative physics for internal world representation based on mathematical model of unicycle motion.

Logic structure of biomechanical control system for description a human riding of unicycle include four levels: 1) distributed information levels with sub-levels; 2) logical system; 3) support decision-making system; and 4) dynamic mechanical system.

Distributed information levels include *four* sub-levels: 1) physical level and logic of virtual reality; 2) behavior and coordination level; 3) intelligent control levels with two sub-levels; and 4) executive biomechanical level. Intersections between horizontal lines of distributed information levels and vertical lines of *logical system, support decision-making system,* and *dynamic system* (of unicycle motion, and a *human behavior* as biomechanical control model) realize the particular models for human riding of unicycle with different skill levels of smart control tools. This structure in detail is presented and described in [4,5].

2.2 The Design Modeling of the Extension-Cableless Robotic Unicycle

In the design of a robot model (Fig. 1a) at the executive biomechanical level the rotor is composed of three bars (each length is 285mm) allocated radially from the rotor's center. On the tip of each bar a weight (0.9 *kg*) is fixed symmetrically what we call a "symmetric rotor". By the use of the 4-bar closed link mechanisms (Fig. 1b,c) the robot's posture stability in the pitch direction can be realized because the acceleration compensation in that direction is attained by the cooperative action of the link mechanisms and rotor. Thus, the stability in the pitch direction is maintained in spite of the change of the rotor and wheel's velocity or acceleration (Fig. 1b-d). In Fig. 1d the "simple" unicycle model is shown.

Remark 1. Our study of the rider's stability control on a unicycle began with the observation and analysis of the logical behavior of human riding a unicycle due to a vestibular biomechanical model [4,5], and an intelligent thermodynamic model (qualitative physical representation) including an instinct and intuition mechanism (as a logical decision-making process). From the observation and model analysis we found that the rider's body thighs and shanks construct a two closed link loop. This special mechanism playes an important role for the rider's postural stability control in a unicycle system (Fig. 1a-c). Using this idea we developed a new logic biomechanlcal model with two closed link mechanisms and one turntable (rotor) to emulate human riding a unicycle by a robot including an intuition and instinct control of the body behavior based on the soft computing. Intuition and instinct mechanisms are considered as global and local search mechanisms for the optimal solution of an intelligent behavior and realized on basis of GA and FNN accordingly. For the fitness function of the GA a new physical measure is the minimum entropy production for the description of an intelligent thermodynamic behavior in a biomechanical model. This control system in [5] is introduced.

The chapter provides a general measure to estimate the mechanical controllability qualitatively and quantitatively even if any control scheme is applied. The measure can be computed using a Lyapunov function coupled with the changed entropy rate. The interrelation between the Lyapunov function (stability condition) and the entropy production rate of the motion (controllability condition) in the internal biomechanical model is the mathematical background for the design of soft computing algorithms for the intelligent control of a robotic unicycle. Our work deals with the improvement of a fuzzy simulation of the robust intelligent control method with the minimum entropy production rate on the basis of soft computing including GA and FNN.

3 Qualitative Physics and Thermodynamic Equations of Motion of the Robotics Unicycle

First of all, for the internal world representation of an artificial life of the robotic unicycle we develop the thermodynamic equations of motion with an (a)symmetric rotor. The analysis of the robot's postural stability control is carried out and the results are compared with those of our computer simulation.

For the robotic unicycle with an (a)symmetric rotor equations of motion are given in [5] as follows,

$$
\begin{bmatrix} \ddot{q} \\ \lambda \end{bmatrix} = \begin{bmatrix} M(q) & -\dfrac{\partial c}{\partial q} \\ E(q) & 0 \end{bmatrix}^{-1} \begin{bmatrix} \tau - B(q)[\dot{q},\dot{q}] - C(q)[\dot{q}^2] - D(q)[\dot{q}] - G(q) \\ -F(q,\dot{q}) \end{bmatrix}, \tag{1}
$$

$$
\begin{bmatrix} \dfrac{dS_u}{dt} \\ \dfrac{dS_c}{dt} \end{bmatrix} = \begin{bmatrix} M(q) & 0 \\ 1 & 0 \end{bmatrix}^{-1} \begin{bmatrix} \tau_d - B(q)[\dot{q},\dot{q}] - C(q)[\dot{q}^2] - D(q)[\dot{q}] \\ -F(q,\dot{q}) \end{bmatrix} \begin{bmatrix} \dot{q} \\ 0 \end{bmatrix}. \tag{2}
$$

The parameters of Eqs (1) and (2) in detail are described in [5]. In Eq. (2) S_u is the entropy of the robot unicycle's motion and S_c is the entropy of both controllers, τ_d are dissipative parts of the control torque (for the PD-controller the dissipative part is described by $k_i(\dot{\gamma}, \dot{\beta})$). The algorithm for the entropy production calculation in the dynamic dissipative systems is described in [5]. For the stability analysis and the computer simulation of the robotic unicycle's dynamic behavior Eqs (1) are rewritten in the traditional form of ordinary differential equations as $\dot{q}_i = \varphi_i(q_i, \tau, t)$.

The analysis of results indicates that the longitudinal and lateral stability domain is a strange attractor [6]. Both are influenced by each other, so the unicycle system is essential nonlinear with it's cross braces. This analysis result helps us to understand experimental results with the galvanic vestibular stimulus for the analysis of a postural adaptation and stability for the unicycle system at all much better [9]. It is also shows that the improved model is more close to human riding a unicycle as the intelligent robotic system.

For the stability analysis of the robotic unicycle as essential nonlinear system we use the asymptotic method of a Lyapunov function and the methods of qualitative physics taking advantage of the interrelation between Lyapunov and entropy production rate functions. The new approach for the definition of the Lyapunov function is also used. The Lyapunov function for the system (1) defined as

$V = \frac{1}{2}(\sum_{i=1}^{6} q_i^2 + S^2)$, where $S = S_u - S_c$ and $q_i = (\alpha, \gamma, \beta, \dot{\alpha}, \dot{\gamma}, \dot{\beta})$. Here we use the following relation between Lyapunov function and entropy production for an *open* system like a unicycle

$$\frac{dV}{dt} = \sum_{i=1}^{6} q_i \varphi_i(q_i, \tau, t) + (S_u - S_c)(\frac{dS_u}{dt} - \frac{dS_c}{dt}).$$

(3)

From Eq. (3) the necessary and sufficient conditions for Lyapunov stability of a robotic unicycle is expressed as follows,

$$\sum_{i=1}^{6} q_i \varphi_i(q_i, \tau, t) < (S_u - S_c)(\frac{dS_c}{dt} - \frac{dS_u}{dt}), \quad \frac{dS_c}{dt} > \frac{dS_u}{dt},$$

(4)

i.e., a stable motion of a unicycle can be achieved with "negentropy" $-S_c$ (terminology from [3]) and the change of negentropy rate $-\frac{dS_c}{dt}$ in the control system

must be substracted from the change of entropy production rate $\frac{dS_u}{dt}$ in the motion

of the robotic unicycle with the second condition in Eq. (4). From Eq. (4) the stability measure for the robotic unicycle can be obtained by computing the minimum entropy production rate of the system and the controllers.

Remark 2. From qualitative physics an internal world representation of the robotic unicycle has two unstable states: 1) a local unstable kinematic equilibrium in lateral plane (angle of rolling γ); and 2) a global unstable dynamic state in longitudinal plane (angle of pitching β). The two correlation states have to be controlled with two fuzzy controllers [1]. This is necessary and sufficient conditions for the improvement of the control stability of our robotic unicycle. Approximate reasoning like the fuzzy implication A $\rightarrow$ B realized on FNN and plays the role of a local coordinator between look-up tables of two controllers with parallel-sequential data processing. A coordinated action between look-up tables of these two fuzzy controllers is accomplished with GA and FNN. Two fuzzy controllers realize the control of transfer energy with the minimum entropy production from lateral to the longitudinal planes using the dynamic of non-linear cross braces in our robotic unicycle model (compensation of transfer energy from unstable dynamic motion "in large" to longitudinal plane according Eq. (3)). The fuzzy controller in the lateral plane executes a role of human riding by organizing a special *parametric* excitation in the non-linear cross braces [8] for the generation of an energy that compensates the transfer energy from the longitudinal plane with the unstable state (thus, the unstable state "in small" compensates the unstable state "in large"). A stable motion of the robotic unicycle model results from nonlinear control on an intelligent level of correlated energy transfer between two unstable virtual states. Then two adaptive fuzzy controllers realize self-organization process of the control stability on a robotic unicycle using intuition and instinct schemes. In this case we obtain the physical model description of quantum fuzzy logic controllers. In general form the structure of quantum fuzzy logic controllers for biomechanical systems in [7] is described. The adaptive method for the feedback gains of fuzzy PD-controllers was realized in [1,2,4].

4 Fuzzy Intelligent Control of a Robotic Unicycle with Soft Computing

In our AI control system two gains schedule PD-controllers are adopted: one is for the (a)symmetric rotor, and the other is for the closed link mechanisms. The control torque to the symmetric rotor is given as,

$$\tau_\eta = kp_2 \times k_3 \times \gamma + kd_2 \times k_4 \times \dot{\gamma}, \tag{5}$$

where, τ_η is the torque to the rotor; kp_2 and kd_2 are constant feedback gains; k_3 and k_4 are fuzzy schedulers changed in [0,1] with FNN.

The control torque applied to the link 2 and link 4 are given as,

$$\tau_{\theta2} = \tau_{\theta4} = -kp_1 \times k_1 \times \beta - kd_1 \times k_2 \times \dot{\beta}, \tag{6}$$

where, $\tau_{\theta2}$ and $\tau_{\theta4}$ are the torques to the link 2 and 4 respectively, kp_1 and kd_1 are constant feedback gains; k_1 and k_2 are fuzzy values changed in [0,1] with FNN.

Remark 3. The biomechanical analysis of a posture stability shows [9] that PD-control represents the minimum complexity necessary to stable posture control. The component P (proportional) contains anti-gravitational forces and compensates the position errors. The component D (derivative) contains an anti-Coriolis compensation and provides also some kind of damping actions. The parameters (kp_1, kp_2) may be interpreted as a stiffness (spring constant) arising from passive and active muscular forces, whereas (kd_1, kd_2) might be compared with a viscous damping, as obtained with a wheel dashpot. Suffice it to mention that this is the minimum complexity anticipated for a stabilized robotic unicycle model.

The fuzzy tuning rules for k_1, k_2, k_3 and k_4 are formed by the learning system of a FNN [5]. Fuzzy controllers are hierarchical, two-level control systems which are intelligent "in small" [11]. The lower (execution) level is the same as a traditional PD-controller and the upper (coordination) level consists of a KB (with a fuzzy inference module in the form of production rules with fuzzy implication), fuzzification and defuzzification components, respectively.

5 Experimental Results

The manufactured new cableless unicycle robot is shown in Fig. 1a. It is composed of a wheel with two cranks, a main body, an overhead rotor and two closed links on both sides of the wheel. The closed link mechanism is used for a longitudinal stability and the (a)symmetric rotor is used for a lateral stability. Four motors had used for the "complicated" model, and only three motors for the new cableless unicycle model is using now. For the old -"complicated" model the wheel was driven through a ball reduction gear, a couple of spiral bevel gears and a timing belt by a DC servomotor (60W) mounted inside the robot. Now the wheel is driven through only the close links mechanism's motors.

The rotor is driven by a harmonic drive motor (for old - 34W, for new - 60W) installed on the body (Fig. 1b). The left and right closed links are driven directly by harmonic drive motors (for old - 20.3W, for new - 60W) on links 2 and 4 (Fig. 1a,b). The links motors are the same for the symmetric geometrical structure and balance of the robot.

A 32-bit personal computer is used for the system controller. The wheel, symmetric rotor and closed link mechanisms are driven by torque controlled motors with a software-servo control. All the control programs are written in C-language.

As shown in Fig. 1a, there are three rate-gyro sensors (sensor A, B and C) mounted on the three principal axes of the robot's body for measuring the angular velocities of the body inclination in the pitch, roll and yaw directions. The resolution of the angular velocity of the sensors is 0,1 degree/second. An optical rotary encoder (500 pulses/revolution) is installed on each servomotor to detect the rotation angle caused by the rotation of the servomotors. The usage of the coordinates defined in Fig. 1b enables us to calculate the robot's posture or Euler's angle (α, β, γ) related to the global reference coordinates measured from the angular velocity ω_x, ω_y, and ω_z by the three rate-gyro sensors in Eqs (10)-(12),

$$\alpha = \int((\omega_x \cos \beta - \omega_x \sin \beta) \cos^{-1} \gamma) dt \, \tag{10}$$

$$\beta = \int(\omega_y - (\omega_z \cos \beta - \omega_x \sin \beta) \tan \gamma) dt \, \tag{11}$$

$$\gamma = \int(\omega_x \cos^{-1} \beta + (\omega_z \cos \beta - \omega_x \sin \beta) \tan \beta) dt \, \tag{12}$$

where, ω_x is the angular velocity related to px_6, ω_y is the angular velocity related to py_6, ω_z is the angular velocity related to pz_6.

Using this kind of small rate-gyro sensors there is a drift on the output due to the time and the change of temperature. The drift may yield an unfavorable influence on the calculation of the postural angles in the experiment. Thus we selected the sensors with the smallest drift. The experiments are conducted within 8 seconds, because the drift of the sensor's output within this time is not so big.

The unicycle's initial posture is set by the operator and the ground's unevenness is randomly, we couldn't repeat just exactly the same result in the experiment even with the same control methods and the same feedback gains.

In Figs 2-4 shows the experimental results for the new cableless unicycle model in beginning of experiments (up line) and now presented (down line). So, the robot's lateral stability (stability in the roll direction) is obtained. In Fig.2c,f shows that the robot's posture in the yaw direction is changed quickly in the experiment. The reason is that the yaw direction control achieves the robot's lateral stability, the change of the ground's unevenness will yield the change of the robot's lateral posture and this change of the robot's roll direction requires a change in the robot's posture of the yaw direction. The wheel's average speed is about 1.2m/s, similar to that of a human rider on a unicycle.

Fig. 1a presents the photo of the robot's posture during an experiment. The initial posture is set by the operator, who removes his hand immediately after the wheel starts to go forward. Comparing the experimental results reported in [2] and in this chapter we find that it is much easier to achieve the robot's posture stability with the fuzzy gain schedule PD-controllers. If the initial posture of the robotic unicycle is near the ideal stable posture ($\beta = 0.0$ rad and $\gamma = 0.0$ rad), the postural stability can be obtained in almost all trials. However, the postural stability with two fuzzy gain

566

schedule PD-controllers can be achieved even if the posture is randomly disturbed to some extent by the ground condition. The reason seems to be, that the two fuzzy gains schedule PD-controllers are able to recover from roll angles as big as in the value of Fig. 2b,e while the PD-control of the unicycle lacks this ability. Such recovery shows the excellent characteristics of the two fuzzy gain schedule PD-controllers, so that the proposed control is very efficient in the improvement of the postural stability and controllability. In Fig.3 shown the temporal behavior of the fuzzy gains k_1, k_2, k_3 and k_4. A main result is shown in Fig.4 which described the increasing of the unicycle skill operator. The physical measure of skill operator in this case is decreasing of the control system entropy production rate.

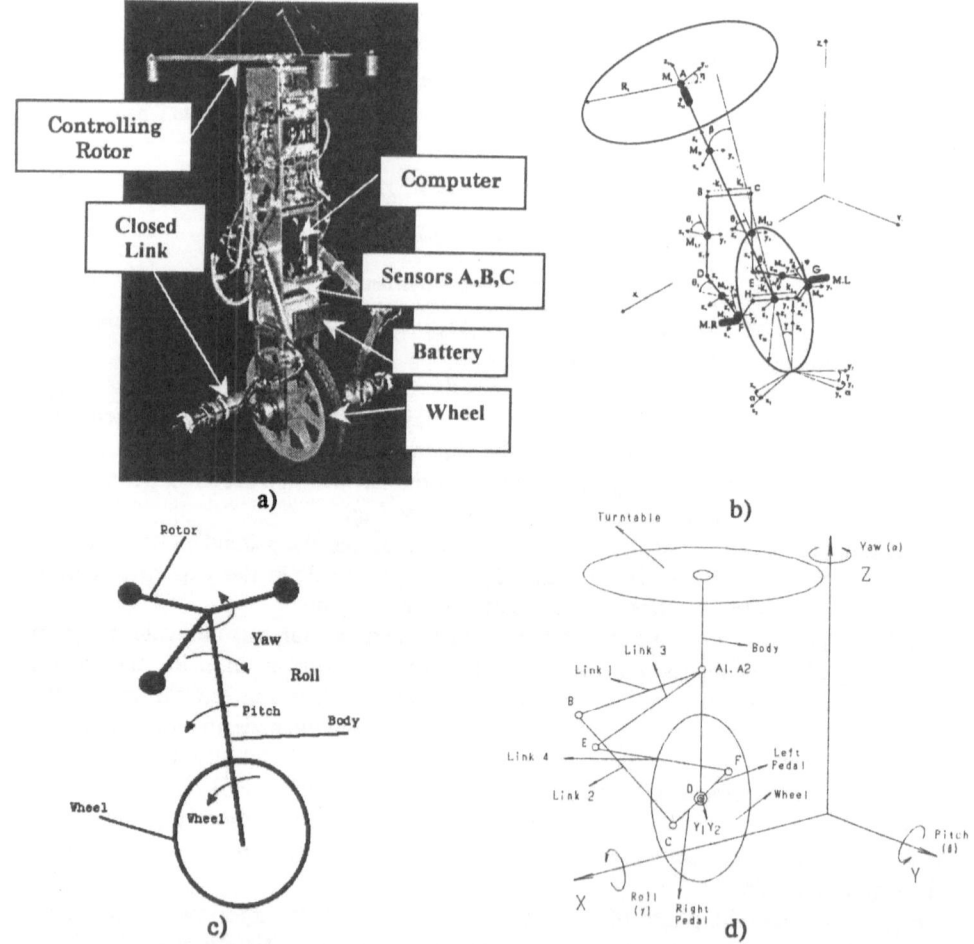

Fig. 1. a) Photo of extension-cableless unicycle; b) Coordinate description of the unicycle model; c) "Simple" model for emulating human riding of a unicycle model; d) "Complicated" model for emulating human riding of a unicycle model

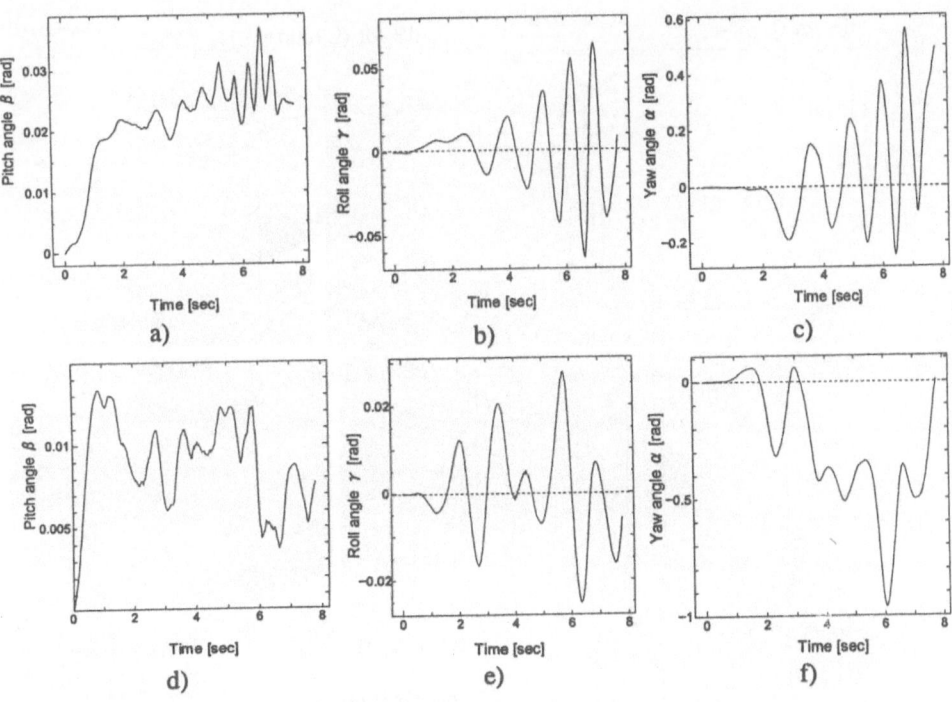

Fig.2. Experimental results of temporal behavior of a,d) Pitch angle; b,e) Roll angle and c,f) Yaw angle for the different tests

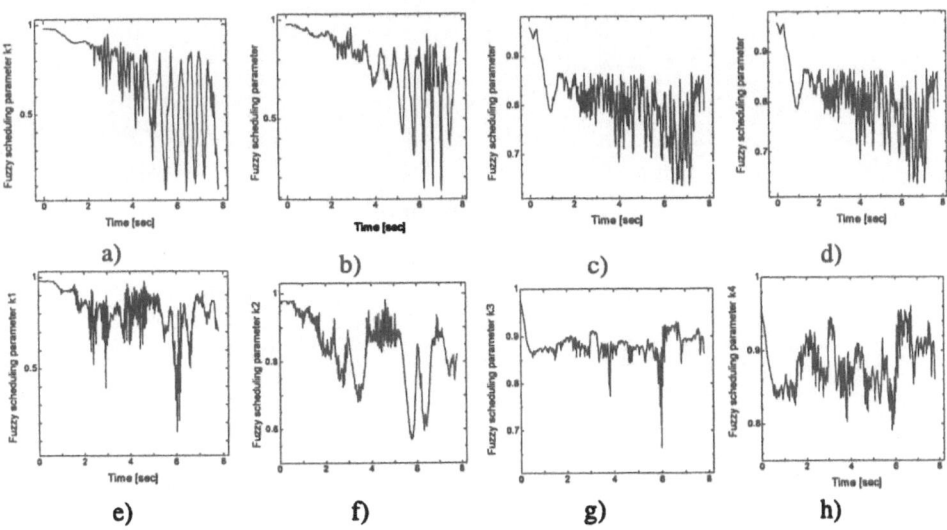

Fig.3. Experimental results of temporal behavior of the fuzzy gains a,e) k_1 b,f) k_2 c,g) k_3 d,h) k_4, for the different tests.

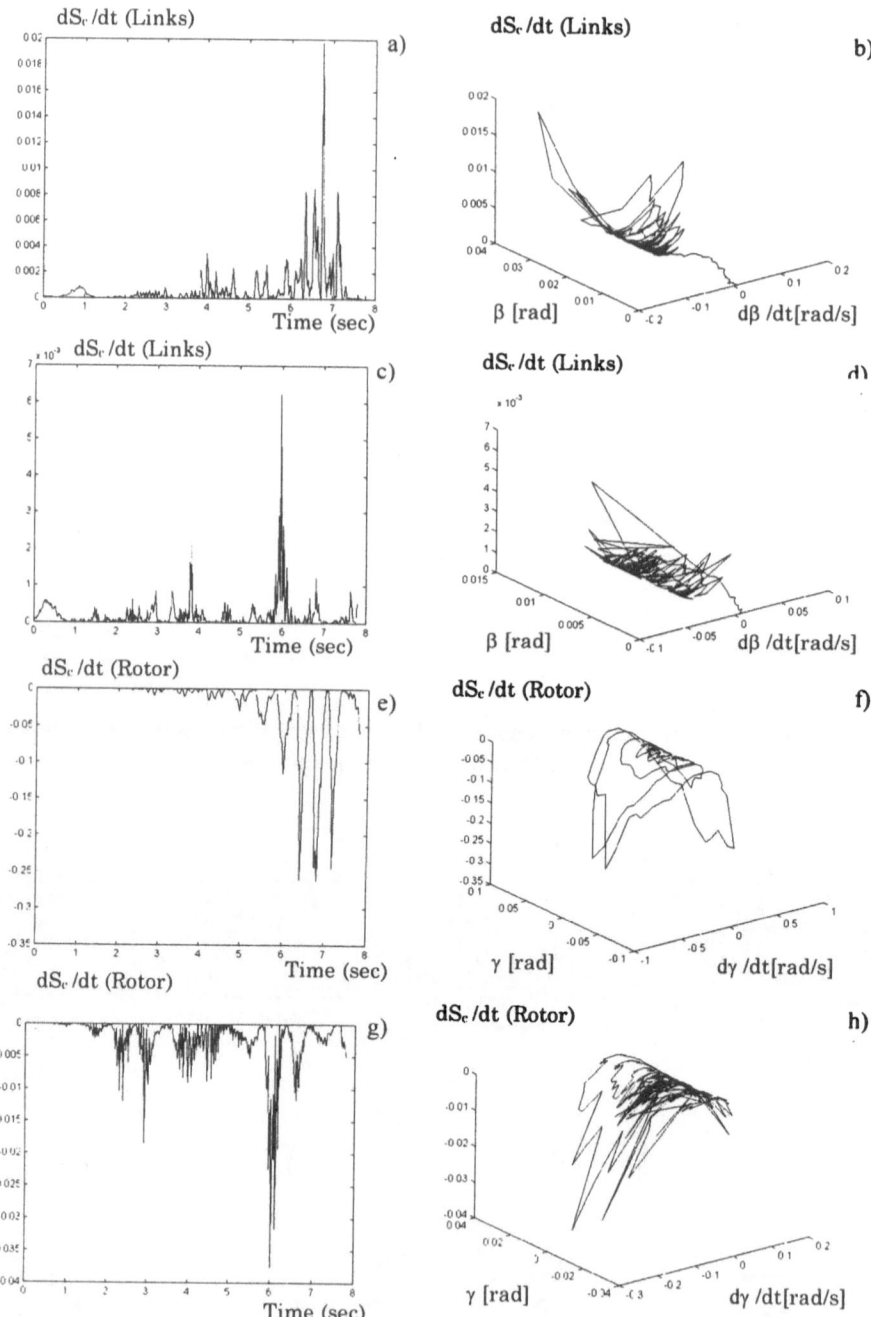

Fig.4. Temporal and 3D thermodynamic behavior (entropy production rate):
a,b,c,d) links and e,f,g,h) rotor control systems for a two different tests.

6 Conclusions

In this work we are present the main idea for a robotic unicycle intelligent robust control system. The background of this approach is the qualitative physical analysis of the unicycle's dynamic motion and the introduction of an intelligent level at the control system by realizing instinct and intuition mechanisms based on FNN and GA accordingly [4,5]. The main components of the AI control system based on soft computing and fuzzy robust control are described. Changing the parameters of two Fuzzy PD-controllers with an adaptation method to achieve a stable motion of the unicycle over a long (finite) time interval without changing the structure of the executive level of the control system using soft computing is described. The introduction of two new mechanisms to an intelligent control system is based on the principle of minimum entropy production rate in the robotic unicycle's motion and the control system itself. The simulation of thermodynamic equations of motion and the intelligent control system confirm the effectiveness to handle the system's non-linearity of the robot's postural stability control. In this case the robotic unicycle model is a *new Benchmark* for intelligent fuzzy controlled motion of a nonlinear dynamic system with two (local and global) unstable states. The usage of a fuzzy gain schedule PD-controller with look-up tables calculated by FNN, offers the ability to use instinct and intuition mechanisms in real time, so that a successful experimental result has been achieved.

References

1. Ulyanov S.V., Sheng Z.Q. and Yamafuji K. 1995 Fuzzy Intelligent control of robotic unicycle: A New benchmark in nonlinear mechanics, *Intern. Conf. on Recent Advanced Mechatronics, Istanbul, Turkey, Vol. 2*, pp. 704 – 709.
2. Sheng Z.Q., Yamafuji K. and Ulyanov S.V. 1996 "Study on the stability and motion control of a unicycle. Pts 3,4,5, *JSME International Journal, Vol. 39*, No. 3, pp. 560-568; 569-576; and 1996 *Journal of Robotics & Mechatronics, Vol. 8*, No. 6, pp. 571-579.
3. Petrov B.N., Ulanov G.M., Ulyanov S.V. and Khazen E.M. 1977 *Informational and Semantic Problems in Control Processes and Organization* (In Russian), Nauka Publ., Moscow; and Petrov B.N., Goldenblat I.I. and Ulyanov S.V. 1978 *Model Theory in Control: Thermodynamic and Information Approach*, (In Russian), Nauka Publ., Moscow.
4. Ulyanov S.V. and Yamafuji K. 1996 Fuzzy Intelligent emotion and instinct control of a robotic unicycle, *4th Intern. Workshop on Advanced Motion Control, Mie, Japan, Vol. 1*, pp. 127- 132.
5. Ulyanov S.V., Watanabe S., Yamafuji K. and Ohkura T. 1996 A new physical measure for mechanical controllability and intelligent control of a robotic unicycle on basis of intuition, instinct and emotion computing, *2nd Intern. Conf. on Application on Fuzzy Systems and Soft Computing (ICAF'96)*, Siegen, Germany, pp. 49-58.
6. Ulyanov S.V., Sheng Z.Q., Yamafuji K., Watanabe S. and Ohkura T. 1995 Self-organization fuzzy chaos intelligent controller for a robotic unicycle: A New benchmark in AI control, *Proc. of 5th Intelligent System Symposium: Fuzzy, AI*

and Neural Network Applications Technologies (FAN Symp, '95), Tokyo, pp. 41-46.

7. Vasileva O.I., Ionov I.P. and Ulyanov S.V. 1990 Dual control of artificial ventilation of lungs (AVL) process using a fuzzy quantum controller in the feedback circuit, *Biomedical Engineering, Vol. 23,* No 1, pp. 7-17; and Kantor P.S., Ulyanov S.V., Pagni A. and Rizzotto G.G. 1994 Fuzzy quantum controller on WARP for artificial ventilation of lungs, *Proc. of ICAFS'94 (Inter. Conf. on Application of Fuzzy Systems),* Tabriz, Iran, pp. 233-235.

8. Ulyanov S.V., Feng Q., Yamafuji K. and Ulyanov V.S. 1998 Stochastic analysis time-variant nonlinear dynamic systems, Pts 1,2, *Probabilistic Engineering Mechanics, Vol 13,* No. 3, pp. 183-204; pp. 205-226.

9. Ulyanov S.V., Watanabe S., Ulyanov V.S., Yamafuji K., Litvintseva L.V., and Rizzotto G.G. 1998 Soft computing for the intelligent control of a robot unicycle based on a new physical measure for mechanical controllability, *Soft Computing, Vol. 2,* No 2, (in print).

10. Ulyanov S.V., Yamafuji K., Ulyanov V.S., Fukuda T. and Kurawaki I. 1998 Interrelation between entropy production and Lyapunov stability of relaxation processes in nonlinear closed dissipative dynamic system, *Physics Letters A* (accepted).

11. Zhakharov V.N. and Ulyanov S.V. 1995 Fuzzy models of intelligent industrial controllers and control systems. Pts 2,3, *Journal of Computer and Systems Sciences International, Vol. 33,* No 2, pp. 94-108, 117-136.

46

About One Way of Increasing Stability of Climbing Robots

T. Akinfiev, M. Armada, M. Prieto and M. Uquillas

1 Introduction

In recent years, research on legged-climbing robots has become a significant subject in robotics. This is reflected in a number of publications reporting wall climbing devices. [1-2]. Electromagnetic [3-5] or vacuum grippers are used to hold on them on the wall.

There are two different unstable conditions that should be analyzed to guarantee static stability in climbing robots: turn over and sliding along the surface.

Electromagnetic grippers help the robot to hold on the wall and prevent turn over instabilities, however, metal-metal friction coefficients between electromagnet and surface use to be small, so sliding unstable condition keeps being a problem to solve.

Additional support elements, with high friction coefficient respect to metal surfaces will be used along with the electromagnets. Use of such support elements leads to redistribution of normal support reaction forces and provides additional friction forces that improve sliding unstable conditions.

Analytical calculation of robot stability under turn over and sliding instability conditions will be done in order to determine minimal reserves of robot stability and critical slopes towards both conditions.

The paper is organized as follows: First, general considerations about electromagnet grippers, geometry of the robot, forces and work space of the robot will be described. Then analytical calculations will be done and a minimal reserve and critical slopes of robot stability towards sliding and turning over conditions are determined. Finally, total reserve of stability is found equal to the minimal one of these reserves. Every calculation will be illustrated with an example of a real designed robot.

2 General considerations

Legged-climbing robots for working processes on external surface of ships under construction are considered. These surfaces can be both, vertical or inclined, including negative slopes. An electromagnet gripper to let the robot hold on these surfaces is provided in each foot of the robot.

For walking robots, gravity force makes it to be adhered to the surface during locomotion and Longitudinal Stability Margin S_L is the normal way to evaluate static stability for horizontal displacements.

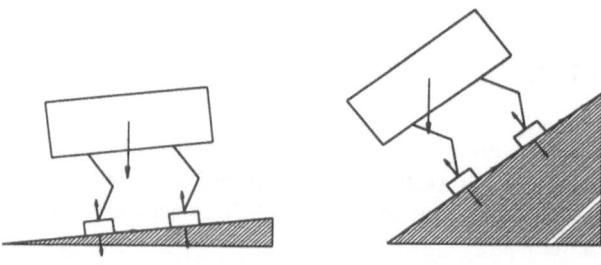

Figure 1

For climbing robots, Longitudinal Stability Margin S_L keeps changing as the slope of the surface changes and there will be a maximum slope that could be hold before the robot slides or turns over. After this slope it will be necessary to use additional gripper mechanisms to maintain the robot adhered to the surface and non-sliding.

Figure 2 shows the configuration of the gripper that will be analyzed containing an electromagnet and additional support elements.

Additional Support Elements have to bulge a little out of electromagnet surface. When the foot touches the surface they, at cost of their elasticity, will deform and let the electromagnet touches the surfaces directly

Additional elements performance will be analog to a spring, it means that forces needed to deform them will be in direct proportion with deformations. Bulge in such elements could be adjusted, so, deformation until the electromagnet touches the surface, causes a force equal to the normal force N_R needed to produce a desired additional friction force F_R that will increase sliding reserve. Normal forces N_R are directly opposed to electromagnet attraction force M and will reduce turn over reserve that should be over-dimensioned.

The electromagnet should be adhered directly to the surface without any clearance, because even minimal clearance between the surface and the electromagnet let to a fatal decrease of attraction force. Once electromagnet makes full contact with the surface, attraction forces M will be considered constant.

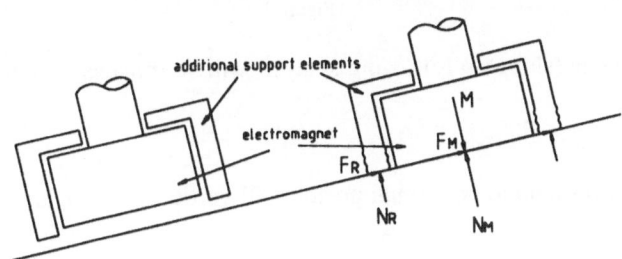

Figure 2

Figure 3 shows the basic leg arrangement and coordinate system for a legged locomotion system. Following are the basic definitions used in the later discussions.

M	Electromagnet force
P	Gravity force on the robot (payload)
F_M	Friction force action over the electromagnets
F_R	Friction force action over additional support elements
N_M	Normal force over the electromagnets
N_R	Normal force over additional support elements
μ_M	Friction coefficient surface – electromagnet
μ_R	Friction coefficient surface – additional support elements
α	Inclination angle
α^*	Critical angle for turn over condition
α^{**}	Critical angle for sliding condition
l	Height of the center of gravity of the robot
L	Distances between adjacent feet

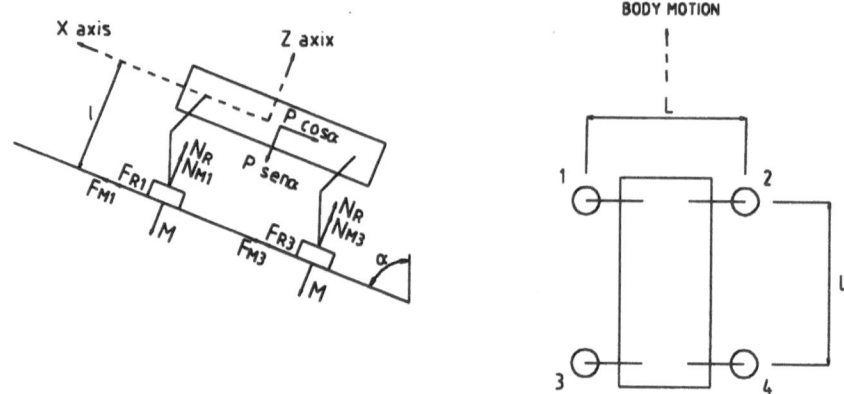

Figure 3

Inclination angle (slope) in following analysis will be considered in the range:

$$-90° < \alpha < 90°$$

where 90° correspond to horizontal position, 0° vertical position and -90° overhead position.

No oblique angles will be considered, it means that it is a symmetrical system and is possible to say that forces in legs 1 and 2, and in legs 3 and 4 will be equals.

3 Analytical calculation of robot stability

Normal forces corresponding to the additional support elements N_R shown in figure 2 could be tuned to a desired value by adjusting the bulge they surpass the electromagnets and to obtain same value in each leg:

$$N_{R1} = N_{R2} = N_{R3} = N_{R4} = N_R \qquad (1)$$

Forces summation in X and Z axis referred in figure 3 can be wrote as:

$$\sum_{i=1}^{4} \left(F_{Ri} + F_{Mi} \right) - P\cos\alpha = 0 \qquad (2)$$

$$\sum_{i=1}^{4}(N_{Mi})-4(M-N_R)-P\sin\alpha=0 \tag{3}$$

and moment summation around axis that goes through support points of legs 1 and 3 and legs 3 and 4 can be wrote as:

$$2(M-N_R)+\frac{P}{2}\sin\alpha-(N_{M2}+N_{M4})=0 \tag{4}$$

$$2(M-N_R)+\frac{P}{2}\sin\alpha-(N_{M1}+N_{M2})-P\frac{l}{L}\cos\alpha=0 \tag{5}$$

3.1 Minimal reserve towards turning over

It will determine the maximum payload for a defined slope that could be hold on before legs 1 and 2 (see figure 3) loose contact and cause that the robot turns over. Contact condition is equivalent to say that normal forces are greater than zero:

$$N_{M1}+N_{M2}>0 \tag{6}$$

and from summation force equation described in (5):

$$N_{M1}+N_{M2}=2(M-N_R)+\frac{P}{2}\sin\alpha-P\frac{l}{L}\cos\alpha$$

According with condition wrote in (6), it is possible to find an expression that relates payload P respect to the slope and normal forces in additional support elements N_R:

$$\frac{2(M-N_R)}{\frac{l}{L}\cos\alpha-\frac{1}{2}\sin\alpha}>P \tag{7}$$

From expression (7) it is possible to say that:

- For a defined slope α, positive or negative, different payloads could be hold on, depending on the tuning action over additional elements that will produce different values for their normal forces N_R.
- Normal forces in additional support elements N_R act against magnetic attraction forces M, so, as N_R increases, payload P decreases as shown in figure 4.

- Additional support elements along electromagnets reduce electromagnetic attraction forces but improve friction forces as will be demonstrated.

Minimal reserve towards turning over will be defined as the operation range below the curve for a certain slope angle. Payload values P greater than those defined by the curve will turn over the robot. For a tuned normal force N_R there will be a maximum payload P that the robot could hold on.

Figure 4 shows a numerical example obtained from expression (7), where

Electromagnetic forces: $M = 2400$ *Newtons*
Coefficient $l/L = 0.333$
Payload range: $0 < P < 2500$ *Kg*

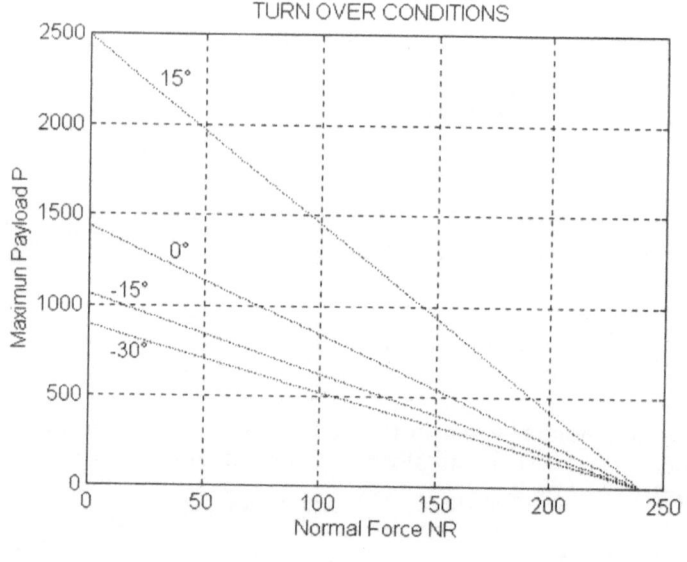

Figure 4

3.2 Turn over - critical slope $\alpha*$

It happens when legs 1 and 2 are just not touching the surface. Payload value has surpassed the maximum the robot could hold on and it is just turning over. It means that:

$$N_{M1} + N_{M2} = 0 \qquad (8)$$

In similar way, from summation forces equation described in (5):

$$2(M - N_R) + \frac{P}{2}\sin\alpha^* - P\frac{l}{L}\cos\alpha^* = 0 \tag{9}$$

According with condition wrote in (8), it is possible to obtain an expression for critical slopes α^* for defined conditions of normal forces N_R and payload P :

$$\cos\alpha^* = \frac{Qs \pm \sqrt{s^2 - Q^2 + 1}}{s^2 + 1} \tag{10}$$

where

$$Q = \frac{4(M - N_R)}{P} ; \qquad\qquad s = \frac{2l}{L}$$

Figure 5 shows the results for the same numeric example analyzed in figure 4.

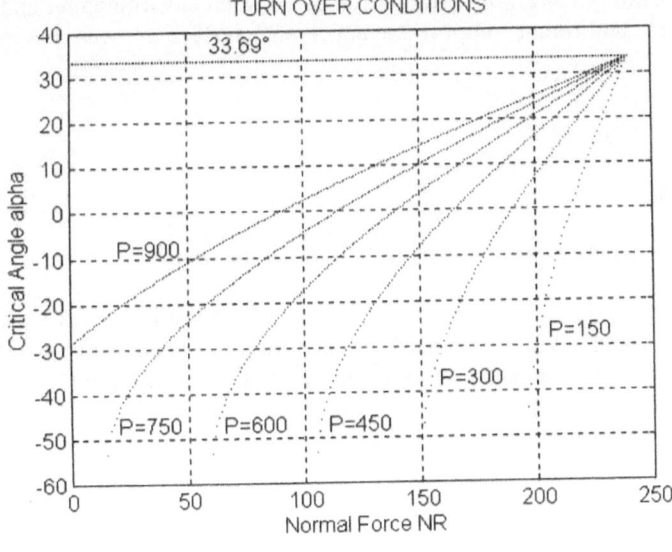

Figure 5

Critical slope α^* for turn over conditions corresponds to the angle at which denominator in expression (7)

$$\frac{l}{L}\cos\alpha - \frac{1}{2}\sin\alpha$$

gets the greatest value and corresponds to the worst case for turn over condition, it means the critical slope at which the robot could hold on the smallest payload. For the numerical example analyzed this critical slope α^* is $-56.31°$ and is the angle at which curves in figure 5 turns asymptotic.

When slope angle is $33.69°$, denominator in expression (7) is zero, so it is important to note that for slopes equal or greater than this value expression (7) is always true. It means that projection of the center of gravity will always be inside support polygon because it depends only of the geometric values of the robot l and L. In this cases robot will not turn over what ever be the values of payload P and normal forces NR. In other words, if slope is greater than $33.69°$ no grippers are necessary taking into account just turn over criteria.

3.3 Minimal reserve towards sliding

It describes stability conditions before robot slides along the surface. Additional support elements will provide additional friction forces to improve this reserve.

Under non-sliding conditions friction forces between electromagnet and surface F_M and between additional support elements and surface F_R can be expressed as follows:

$$F_{Ri} \leq \mu_R N_{Ri}$$
$$F_{Mi} \leq \mu_M N_{Mi} \tag{11}$$

for i = 1 to 4, where i corresponds to each leg of the robot

Replacing conditions described in (11) into summation forces described in equation (2), and taking into account that normal forces corresponding to the additional support elements N_R will be tuned to be the same in four legs as described in equation (1) :

$$4(\mu_R N_R) + \mu_M \sum_{i=1}^{4} (N_{Mi}) - P\cos\alpha \leq 0 \tag{12}$$

From summation forces described in equation (3), summation of normal forces in magnets can be expressed as:

$$\sum_{i=1}^{4} (N_{Mi}) = 4(M - N_R) + P\sin\alpha \tag{13}$$

Replacing in (12) and simplifying it is possible to obtain an expression to describe the *minimum reserve towards sliding*:

$$4N_R(\mu_R - \mu_M) + 4\mu_M M - P\cos\alpha + \mu_M P\sin\alpha \le 0 \qquad (14)$$

$$\frac{4N_R(\mu_R - \mu_M) + 4\mu_M M}{\cos\alpha - \mu_M \sin\alpha} \le P \qquad (15)$$

From expression (15) it is possible to state that:

- Additional support elements redistribute normal reaction forces to provide additional friction forces that help to increase *minimum reserve towards sliding*.

Minimum reserve towards sliding is defined as the operation range below the curve that defines each angle. For payload values P greater than those defined by the curve for a tuned normal force N_R robot will slide.

Figure 6 shows minimum reserve towards sliding for the same example described in figure 4, and the following friction coefficients:

Friction coefficient between additional support elements (rubber) and metal surface: $\mu_R = 0.50$

Friction coefficient between metallic surfaces: $\mu_M = 0.15$

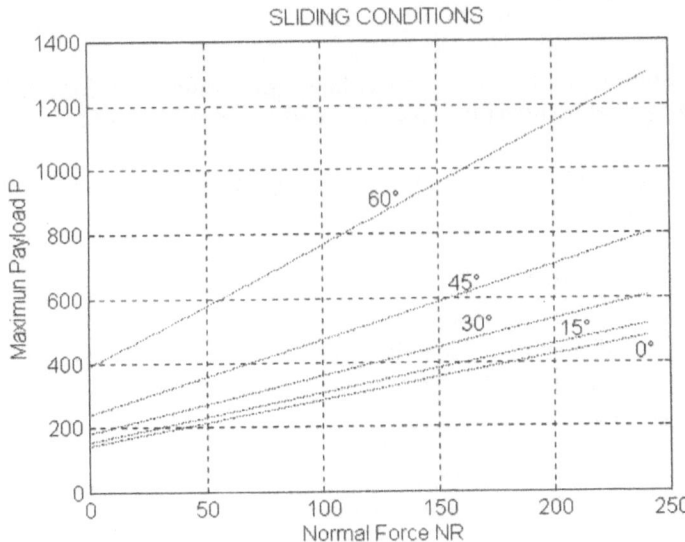

Figure 6

3.4 Critical slope for sliding conditions α**

It can be defined as the angle at which the robot is just sliding and expressions described in (11) just change to:

$$F_{Ri} = \mu_R N_{Ri}$$
$$F_{Mi} = \mu_M N_{Mi} \tag{16}$$

and payload is the smallest one.

It means that expression (14) can be wrote like an equation to determine the value of this angle:

$$4\left(N_R\left(\mu_R - \mu_M\right) + M\mu_M\right) = P\cos\alpha** - P\mu_M\sin\alpha** \tag{17}$$

and

$$\sin\alpha** = \frac{-\mu_M Z \pm \sqrt{\mu_M^2 - Z^2 + 1}}{(\mu_M^2 + 1)} \tag{18}$$

where

$$Z = \frac{4\left(N_R(\mu_R - \mu_M) + M\mu_M\right)}{P}$$

Equation (18) will be used to find critical slopes for sliding conditions, for defined conditions of payload P and a tuned normal force N_R. as shown in figure 7.

*Critical slope α** for sliding conditions* corresponds to the angle at which denominator of equation (15)

$$\cos\alpha - \mu_M\sin\alpha$$

takes the greatest value and corresponds to the worst case for sliding conditions, it means the slope when the robot could hold on the smallest payload before it begins to slide along the surface.

Figures 7 and 8 show that it happens when curves take asymptotic values and corresponds to $-8.53°$ for the numeric example analyzed previously.

When slope angle is 81.47°, denominator in expression (15) is zero, so for angles equal or greater than this value and assuming that there is not electromagnet forces

(M=0) expression (15) is always true. It means that friction forces between foot and surface provide enough reserve to avoid sliding as shown in figure 8.

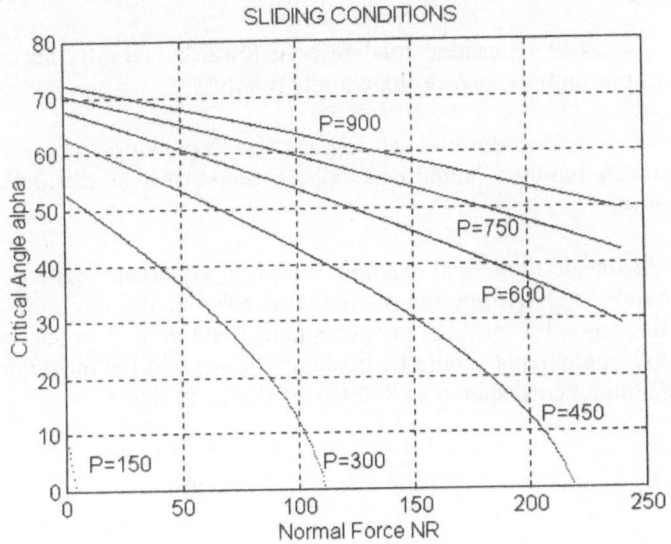

Figure 7

Additionally as normal forces *NR* keeps growing, it is easy to see in figure 8, how slope angle α keeps growing too. It means that friction forces in additional support elements contribute to increase minimum reserve towards sliding as stated before

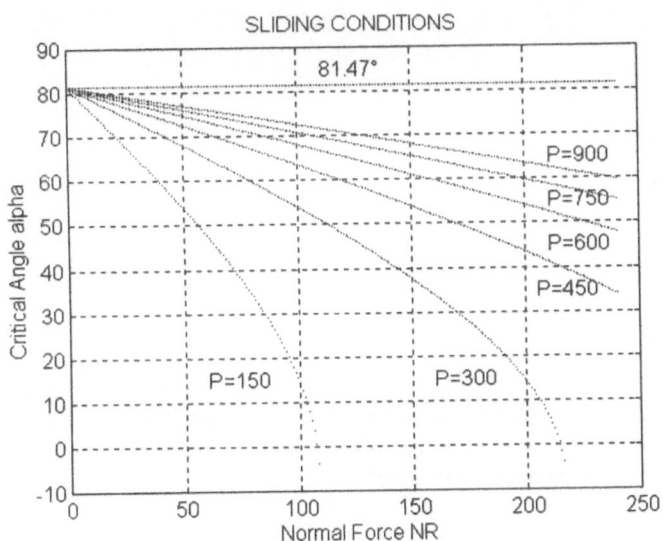

Figure 8

582

3.5 Total reserve towards stability of the robot

Total reserve corresponds to the minimal of the two reserves analyzed previously.

There is two ways to determine total reserve towards stability, according with information of the support surface slope of the robot:

1. If preliminary information about support surface slopes is available it is possible to make a tuning of additional support elements to get the highest reserve of robot stability.

If we state that for the numerical example analyzed, maximum slope is defined at 15°, it is possible to determine the maximum reserve as the intersection of lines that defines this angle for turning over and sliding conditions. Normal forces N_R in additional support elements should be tuned at 190 Kg and the maximum payload that could be hold on corresponds to 450 Kg, as shown in figure 9

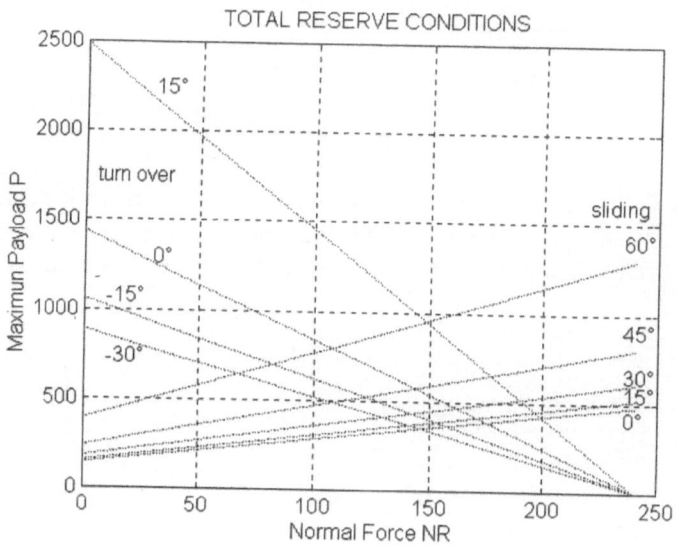

Figure 9

2. If preliminary information about support surface slope is not available total reserve for stability will be obtained by intersection between curves defined by critical angles for both turn over and sliding worst conditions. Figure 10 shows this total reserve for the numerical example analyzed.

In the numeric example analyzed, worst case for turn over condition was established for slope angles equal to -56.31°, and worst case for sliding conditions was found when slope angle equal is equal to –8.53°. According with figure 10, *total reserve toward stability* is determined by the intersection of the curves defined by these two angles. Maximum payload that could be hold on by the robot will be 340 Kg when normal force NR is tuned at 140 Kg. It means that for payload values greater than these value robot could slide if it goes through a surface with slope equal to –56.31° or could turn over if it goes though surface with slope equal to –8.53°.

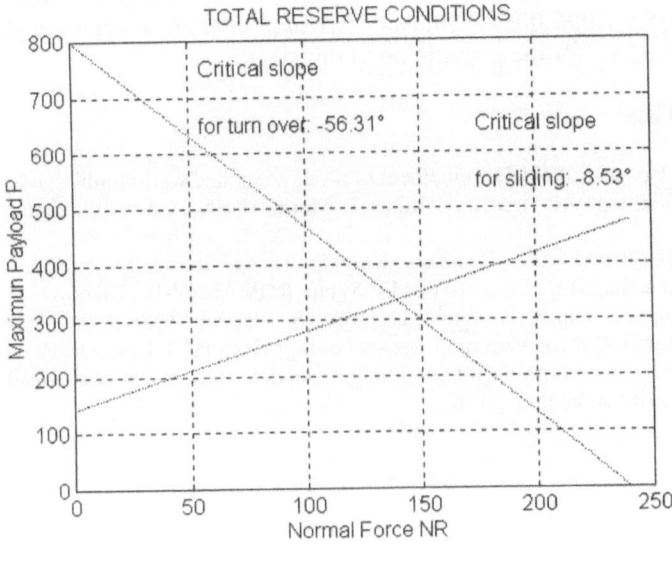

Figure 10

These results can be very important in the design phase of climbing robots allowing for a secure system to be calculated and constructed.

4 Conclusions

- The use of additional elements along with electromagnets leads to redistributing force of normal support reaction between electromagnet (which surface has low coefficient of friction) and additional support elements (which surface has high coefficient of friction). It is just what leads to increasing of total friction force, and as consequence to increasing minimal reserve towards sliding conditions.
- Additional support elements decrease minimal reserve towards turn over conditions, so it is desirable to make a fine tune of these elements to avoid turn over instabilities.

- There are two different ways for tuning additional support elements according with preliminary information availability: when support surface slope is known, it will be possible to make tuning for a higher reserve of stability. When support surface slope is unknown tuning should be done for the worst case that corresponds to the minimal total reserve for stability of the robot.

Acknowledgement

The authors would like to thank the support of INCO-COPERNICUS/ESPRIT through Project CP96-0054 Efficient start-stop intelligent drives with adaptive control - ESSIDAC for the preparation of this work.

REFERENCES

1. Ikeda K, Nozaki T 1991 Development of a self contained wall-climbing robot. ISART
2. Sato K, Honda K, Hasegawa A, Shiota T, Morita H 1991 On-wall locomotive vehicle. ISART
3. Hirose S 1986 Wall Climbing Vehicle Using Internally balanced Magnetic Units,Proceedings 6th CISM-IFTOMM Symp. ROMANSY-86, Cracow
4. S. Sugiyama, S. Naitoh, C. Satoh, N. Ozaki, S. Watahiki 1986 Wall Surface Vehicles with Magnetic Legs or Vacuum Legs,Proceedings 16th ISIR, Brussels, Belgium
5. Grieco J C 1997 Climbing robots: Design, stability and control strategies PhD thesis, University of Valladolid, Spain

Index

Lecture Notes in Control and Information Sciences

Edited by M. Thoma

1993–1998 Published Titles:

Vol. 186: Sreenath, N.
Systems Representation of Global Climate
Change Models. Foundation for a Systems
Science Approach.
288 pp. 1993 [3-540-19824-5]

Vol. 187: Morecki, A.; Bianchi, G.;
Jaworeck, K. (Eds)
RoManSy 9: Proceedings of the Ninth
CISM-IFToMM Symposium on Theory and
Practice of Robots and Manipulators.
476 pp. 1993 [3-540-19834-2]

Vol. 188: Naidu, D. Subbaram
Aeroassisted Orbital Transfer: Guidance
and Control Strategies
192 pp. 1993 [3-540-19819-9]

Vol. 189: Ilchmann, A.
Non-Identifier-Based High-Gain Adaptive
Control
220 pp. 1993 [3-540-19845-8]

Vol. 190: Chatila, R.; Hirzinger, G. (Eds)
Experimental Robotics II: The 2nd
International Symposium, Toulouse,
France, June 25-27 1991
580 pp. 1993 [3-540-19851-2]

Vol. 191: Blondel, V.
Simultaneous Stabilization of Linear
Systems
212 pp. 1993 [3-540-19862-8]

Vol. 192: Smith, R.S.; Dahleh, M. (Eds)
The Modeling of Uncertainty in Control
Systems
412 pp. 1993 [3-540-19870-9]

Vol. 193: Zinober, A.S.I. (Ed.)
Variable Structure and Lyapunov Control
428 pp. 1993 [3-540-19869-5]

Vol. 194: Cao, Xi-Ren
Realization Probabilities: The Dynamics of
Queuing Systems
336 pp. 1993 [3-540-19872-5]

Vol. 195: Liu, D.; Michel, A.N.
Dynamical Systems with Saturation
Nonlinearities: Analysis and Design
212 pp. 1994 [3-540-19888-1]

Vol. 196: Battilotti, S.
Noninteracting Control with Stability for
Nonlinear Systems
196 pp. 1994 [3-540-19891-1]

Vol. 197: Henry, J.; Yvon, J.P. (Eds)
System Modelling and Optimization
975 pp approx. 1994 [3-540-19893-8]

Vol. 198: Winter, H.; Nüßer, H.-G. (Eds)
Advanced Technologies for Air Traffic Flow
Management
225 pp approx. 1994 [3-540-19895-4]

Vol. 199: Cohen, G.; Quadrat, J.-P. (Eds)
11th International Conference on
Analysis and Optimization of Systems –
Discrete Event Systems: Sophia-Antipolis,
June 15–16–17, 1994
648 pp. 1994 [3-540-19896-2]

Vol. 200: Yoshikawa, T.; Miyazaki, F. (Eds)
Experimental Robotics III: The 3rd
International Symposium, Kyoto, Japan,
October 28-30, 1993
624 pp. 1994 [3-540-19905-5]

Vol. 201: Kogan, J.
Robust Stability and Convexity
192 pp. 1994 [3-540-19919-5]

Vol. 202: Francis, B.A.; Tannenbaum, A.R.
(Eds)
Feedback Control, Nonlinear Systems,
and Complexity
288 pp. 1995 [3-540-19943-8]